Nuclear Physics in Astrophysics - X (NPA-X 2022)

EPJ Web of Conferences Volume 279 (2023)

Geneva, Switzerland
5-9 September 2022

Editors:

Sean Freeman
Claudia Lederer-Woods
Alice Manna
Alberto Mengoni

ISBN: 978-1-7138-7248-1

Printed from e-media with permission by:

Curran Associates, Inc.
57 Morehouse Lane
Red Hook, NY 12571

proceedings.com

Some format issues inherent in the e-media version may also appear in this print version.

No changes have been made to the content of these proceedings. There may be changes to pagination and minor adjustments for aesthetics.

Printed with permission by Curran Associates, Inc. (2023)

For additional information, please contact EDP Sciences – Web of Conferences at the address below.

EDP Sciences – Web of Conferences
17, Avenue du Hoggar
Parc d'Activité de Courtabœuf
BP 112
F-91944 Les Ulis Cedex A
France

Phone: +33 (0) 1 69 18 75 75
Fax: +33 (0) 1 69 28 84 91

contact-edps@webofconferences.org

Additional copies of this publication are available from:

Curran Associates, Inc.
57 Morehouse Lane
Red Hook, NY 12571 USA
Phone: 845-758-0400
Fax: 845-758-2633
Email: curran@proceedings.com
Web: www.proceedings.com

TABLE OF CONTENTS

edp sciences

EPJ Web of Conferences

Statement of Peer review

In submitting conference proceedings to EPJ Web of Conferences, I certify to the Publisher that I adhere to the **Policy on Publishing Integrity** of the journal in order to safeguard good scientific practice in publishing.

1. All articles have been subjected to peer review administered by the proceedings editors.
2. Reviews have been conducted by expert referees, who have been requested to provide unbiased and constructive comments aimed, whenever possible, at improving the work.
3. Proceedings editors have taken all reasonable steps to ensure the quality of the materials they publish and their decision to accept or reject a paper for publication has been based only on the merits of the work and the relevance to the journal.

Title, date and place of the conference

Nuclear Physics in Astrophysics - X
CERN, Geneva
5-9 September 2022

Proceedings editor(s):

Sean Freeman
Claudia Lederer-Woods
Alice Manna
Alberto Mengoni

Date and editor's signature

6 March 2023

S Freeman
C Lederer-Woods
A Manna
A Mengoni

The long-standing connection of BBN and Indirect measurements: the ^{3}He(n,p)^{3}H reaction at Big Bang energies

R.G. Pizzone[1]*, C. Spampinato[1] for the ASFIN collaboration

[1]Laboratori Nazionali del Sud - INFN, Catania, Italy

Abstract. Nuclear reactions play a key role in the framework of the Big Bang Nucleosynthesis. A network of 12 principal reactions has been identified as the main path which drives the elemental nucleosynthesis in the first twenty minutes of the history of the Universe. Among them an important role is played by neutron-induced reactions, which, from an experimental point of view, are usually a hard task to be measured directly. Nevertheless big efforts in the last decades have led to a better understanding of their role in the primordial nucleosynthesis network. In this work we apply the Trojan Horse Method to extract the cross section at astrophysical energies for the ^{3}He(n,p)^{3}H reaction after a detailed study of the ^{2}H(^{3}He,pt)H three–body process. The experiment was performed using the ^{3}He beam, delivered at a total kinetic energy of 9 MeV by the Tandem at the Physics and Astronomy Department of the University of Notre Dame. Data extracted from the present measurement are compared with other published sets available in literature. Astrophysical applications will also be discussed in details.

1 Introduction

The ^{3}He(n,p)^{3}H process is one of the most relevant neutron induced reactions in Big Bang Nucleosynthesis (BBN) and has a strong impact on the primordial He and Li isotopes production. At the temperatures relevant for predicting Big Bang yields, the reaction rate is determined by the ^{3}He(n,p)^{3}H cross section in the energy range $0 \leq E_{cm} \leq 0.4$ MeV. The first studies of this reaction were performed by Coon et al. [1] in 1950 in the $0.1 \leq E_{cm} \leq 30$ MeV using a neutron beam. Errors turned out to be around 30%. Other measurements, more focused at lower energies, were conducted by Batchelor et al. [2] (direct one, $0.1 \leq E_{cm} \leq 1$MeV), Gibbons et al. [3] (inverse measurement) and Costello et al. [4] who measured directly in the range $0.3 \leq E_{cm} \leq 1.1$ MeV. Theoretical predictions are also available, the most recent one carried out by Drosg et al [5], which covered a wider energy range. Reaction rates were then calculated for astrophysical applications by Brune et al. [6], Adahchour et al. [7] and Smith et al. [8]. They all show a similar trend at temperatures of astrophysical interest while the reaction rate calculated by Caughlan & Fowler [9] is considerably higher. In the energy range of interest, the existing data are therefore sparse and mostly measured more than fifty years ago after facing tough experimental challenges, thus resulting several times in errors as high as 30% depending on the energy.

*e-mail: rgpizzone@lns.infn.it

 https://doi.org/10.1051/epjconf/202327901001

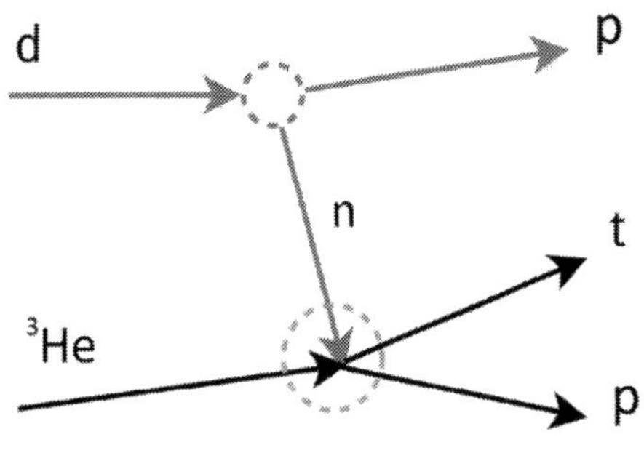

Figure 1. Diagrams describing the TH reaction ^{2}H(^{3}He,pt)H in the QF kinematics, proceeding through the direct ^{3}He(n,p)^{3}H sub-reaction.

Alternative and complementary ways to obtain the bare nucleus cross section, σ, have been provided by indirect methods such as the Coulomb dissociation method [10, 11] and the ANC (Asymptotic Normalization Coefficient) method [12]. Among those ones, the Trojan-Horse method (THM) [13, 18] is particularly suited to investigate, at astrophysical energies, binary reaction induced by neutrons or charged particles by using appropriate three-body reactions. It allows one to avoid both Coulomb and centrifugal barrier suppression and electron screening effects, thus very low energies can be reached and extrapolations are not needed. The latter has shown its great power for measuring reaction rates for the BBN in the whole energy range of interest. This is reviewed in [14] and has been extended to reactions induced by unstable nuclei of interest for BBN e.g .^{7}Be(n,α)^{4}He and ^{7}Be(n,p)^{7}Li [15–17]. The same methodology has been adopted here for the ^{3}He(n,p)^{3}H reaction. For the details on the method description please refer to [19, 20] and references therein. The THM has been successfully applied to the measurement of bare-nucleus cross sections of reactions between charged particles at sub-Coulomb energies. Many validity tests were also performed for it like the pole invariance test which was positively satisfied, for details see [21]. The method has been used in the last three decades to explore nucleosynthesis reactions other than the primordial ones in different sites, e.g., AGB stars [22–24], LiBeB depletion in stars [25, 26] as well as novae [27–29] .

In recent years, neutron induced reactions [30] have been addressed as well, using deuterons as TH nuclei to transfer neutrons, while protons act as spectators. As is sketched in Fig. 1, this work will present the investigation of the ^{2}H(^{3}He,pp)^{3}H quasi-free reaction, thus applying the THM in order to retrieve the cross section for the ^{3}He(n,p)^{3}H reaction at astrophysical energies.

2 Experiment and results

The ^{3}He beam, delivered at a total kinetic energy of 9 MeV by the FN Tandem accelerator at the Nuclear Physics Laboratory of the University of Notre Dame, was impinged on a 100 μg/cm^2 isotopically enriched (up to 98%) deuterated polyethylene target, which was manufactured at the INFN-LNS target laboratory. Detectors were placed as sketched in angular positions reported in [31]. Three silicon position sensitive detectors (PSD 1-3), 1000 μm thick were used; PSD1 was coupled with a 35 μm thin, 10 × 50 mm^2 silicon detector for particle identification, while no particle identification was required for the other two detectors,

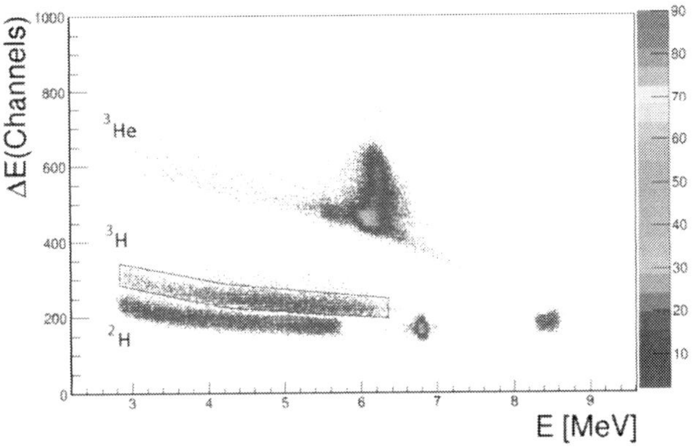

Figure 2. ΔE/E plot for PSD1 for a typical run of the ^{2}H(^{3}He,pt)H reaction. The black selection represents events identified as tritons.

thus optimizing the energy resolution of the apparatus. Two symmetrical monitor detectors were placed on both sides of the beam at 60° to check the beam symmetry. Another point-like silicon detector (PL1) was placed at 45° for on-line monitoring of the target thickness and its deuterium content during the experiment. A metal grid, with equally spaced slits, was placed in front of each PSD in order to perform an accurate angular calibration. As a first stage of the measurements, devoted calibration runs were performed for energy and angular calibration. Thanks to the spatial resolution of PSD and to the accurate positioning of the detectors which were measured by optical means, an angular resolution of about 0.15 deg. was calculated. This is required by THM for an improved spectator momentum resolution which is crucial for QF mechanism selection.

The first phase of the analysis is the identification of the three-body process of interest among those occurring in the target. For this purpose, a particle identification is performed for PSD1 via the ΔE/E technique as reported in fig. 2. Many loci are populated and in particular those related to deuterons and tritons are present, the latter being selected by means of the graphical cut. For further analysis only events corresponding to the tritium locus in ΔE/E matrices are used. The identification is then completed as reported extensively in [31] taking advantage of Q-value spectra investigations as well as kinematic loci. After it, the momentum distribution of the experimentally obtained spectator momentum distribution is taken as the main evidence of the occurrence of quasi-free process as prescribed in reference [32]. It is then verified that the proton acts as a spectator and therefore a cut on its momentum distribution is performed, using for the further analysis only events which has $p_s \leq 30$ MeV/c.

By applying the standard THM prescriptions as it is extensively reported in [31] experimental data are compared with direct data present in literature after normalization (at energies between 350 and 400 keV). The final results are then reported in the table 1. The obtained results show a good agreement with direct and indirect measurements already present in the literature, and gives, again, a strong confirmation of THM approach to neutron induced reactions.

Table 1. Cross section for the process of interest normalized to direct data as discussed in the text.

Energy (MeV)	Cross section (barn)	error on σ(E)
0.005	10.4	1.0
0.035	3.8	0.3
0.065	2.4	0.2
0.095	2.0	0.2
0.125	1.8	0.2
0.155	1.5	0.1
0.185	1.2	0.1
0.215	1.2	0.1
0.245	1.0	0.1
0.275	1.0	0.1
0.305	1.0	0.1
0.335	0.8	0.1
0.365	0.8	0.1

Acknowledgement
We thank the staff of the LNS target laboratory, for target preparation as well as the Nuclear Science Laboratory (supported by grant the National Science Foundation under grant number NSF PHY-1713857) staff at the University of Notre Dame for their invaluable efforts. J.M, G.D. and V. B. were supported by MEYS Czech Republic under the project EF16-013/0001679. The authors acknowledge "Finanziamenti di linea 2" and "Starting Grant 2020" by University of Catania. This work was also supported in part by the National Research Foundation of Korea(NRF) grant funded by the Korea government (MSIT) (Nos. 2020R1A2C1005981 and 2016R1A5A1013277).

References

[1] J.H. Coon 1950, Physical Review 80, 488.
[2] R. Batchelor, R. Aves, and T.H.R. Skyrme 1955, Review of Scientific Instruments 26, 1037.
[3] J.H. Gibbons and R.L. Macklin 1959, Physical Review 114, 571.
[4] D.G. Costello, S.J. Friesenhahn, and W.M. Lopez 1970, Nuclear Science and Engineering 39, 409.
[5] M. Drosg and N. Otuka 2015, INDC (AUS)-0019.

[6] C.R. Brune, K.I. Hahn, R.W. Kavanagh, and P.R. Wrean, 1999, Physical Review C60, 015801.
[7] A. Adahchour and P. Descouvemont 2003, Journal of Physics G: Nuclear and Particle Physics 29, 395.
[8] M.S. Smith, L. H. Kawano, and R. A. Malaney 1993, The Astrophysical Journal Supplement Series 85, 219.
[9] G.R. Caughlan and W.A. Fowler 1988, Atomic Data and Nuclear Data Tables 40, 283
[10] G. Baur, C.A. Bertulani and H. Rebel 1986, Nucl. Phys. A458, 188
[11] C.A. Bertulani and A. Gade 2010, Phys. Rep. 485, 195
[12] G.G. Kiss, M. La Cognata, C. Spitaleri et al 2020, Phys. Lett. B, 807, 135606
[13] C. Spitaleri 1991: Proceedings, (World Sci., 1991), p. 21
[14] R.G. Pizzone, R. Spartá, C. Bertulani, et al. 2014, Ap. J.,786, 112
[15] L. Lamia et al. 2017, Ap. J. 850, 175
[16] L. Lamia, M. Mazzocco, R.G. Pizzone, et al. , Astrophys. J., 879, 23 (2019).
[17] S.Hayakawa, M. La Cognata, L. Lamia et al., 2021, Astroph. J, 915, L13
[18] C. Spitaleri et al. 2003, Nucl. Phys. A 719, 99c
[19] C. Spitaleri, La Cognata, M., Lamia, L., Pizzone, R. G, Tumino A., 2019, Eur. Phys. J. A55 161
[20] Tumino A., et al., Ann. Rev. Mod. Phys. 2021
[21] R.G. Pizzone, C. Spitaleri, C. Bertulani, et al. 2013, Phys. Rev. C 87, 025805
[22] R.G. Pizzone, G.D'Agata, C. Spitaleri et al., 2017, Ap. J., 836, 57
[23] I. Indelicato, M. La Cognata, *et al.* , Astrophys. J., 845, 19 (2017).
[24] G. D'Agata, R.G. Pizzone et al., Astrophys. J., 860, 61 (2018).
[25] M. Aliotta, C. Spitaleri et al., 2000. Eur. Phys. J. A 9, 435
[26] L. Lamia, C. Spitaleri, E. Tognelli, et al. Astrophys. J., 811, 99 (2015).
[27] M. L. Sergi, C. Spitaleri, M. La Cognata, et al. 2015, Phys. Rev. C, 91, 065803
[28] R.G. Pizzone, B. Roeder, M. Mckluskey, et al. 2016, Eur. Phys. J. A52, 24
[29] M. La Cognata, R.G. Pizzone et al., 2017, Ap. J., 846, 65
[30] G. L., Guardo, C. Spitaleri, L. Lamia, et al. 2019, Eur. Phys. J. , A55, 211
[31] R.G. Pizzone, C. Spampinato et al., Eur. Phys. J. , A56, 199 (2020)
[32] R.G. Pizzone, C.A. Bertulani, L. Lamia, L. et al., *Eur. Phys. J. A*, **2020**, *56*, 283

Deuterium burning measurement at LUNA and its astrophysical and nuclear implications

Francesca Cavanna[1,*] for the LUNA collaboration

[1]Istituto Nazionale di Fisica Nucleare, sezione di Torino, Via Pietro Giuria 1, 10125 Torino

Abstract.
The deuterium burning reaction $D(p,\gamma)^3He$ at low energies is of primary importance in cosmology because it affects the primordial deuterium abundance, that in turn is very sensitive to fundamental cosmological parameters such as the baryon density and the amount of relativistic species permeating the early Universe. This reaction is also of a particular interest in theoretical nuclear physics because it offers a unique opportunity to test *ab – initio* calculations. In the following a recent study of the $D(p,\gamma)^3He$ reaction in the $30 \lesssim E_{cm}[keV] \lesssim 280$ energy range will be presented. This measurement has been performed in the underground Gran Sasso Laboratory by the LUNA (Laboratory for Underground Nuclear Astrophysics) collaboration. In addition, the need of having new and precise measurement of the $D(d,n)^3He$ and $D(d,p)^3H$ reactions will also be discussed.

1 Introduction

Big Bang Nucleosynthesis (BBN) occurs during the first minutes of cosmological time in a rapidly expanding hot and dense Universe, where a fraction of protons and nearly all free neutrons end up bound in ^4He, while D, ^3H, ^3He, ^6Li, ^7Li and ^7Be nuclei form in trace quantities. The theoretical description of BBN is based on the standard cosmological model, which assumes a homogeneous and isotropic Universe, up to small perturbations, governed by the General Relativity theory and by the Standard Model of particle physics. Under these assumptions, BBN predicts the abundances of primordial nuclides, as a function of one parameter only, the density of ordinary matter, or baryon density, $\omega_b \equiv \Omega_b h^2$, where h is the reduced Hubble constant (see [1] for a recent review). On the other hand, the primordial abundance of the light elements can be measured by observations of astronomical samples reflecting their primordial composition. Therefore, a comparison between the observed primordial abundances and those predicted by the BBN can be used to constrain the baryon density of the Universe, the only free cosmological parameter of the BBN model, under the assumption that there are three neutrino species.

Indeed, an independent estimate of ω_b can be obtained by observing the anisotropies in the Cosmic Microwave Background (CMB), the relic electromagnetic radiation left over from the Big Bang, which are measured with remarkable precision by the Planck experiment, leading to [2]: $\omega_b = 0.02242 \pm 0.00014$.

*e-mail: francesca.cavanna@to.infn.it

Assuming that there are no changes in the value of ω_b from BBN down to CMB formation epoch, the precise CMB determination of ω_b implies that nowadays the theory of BBN, in its minimal scenario where only three neutrinos species are allowed, is a parameter free model and, as such, has become a high precision tool to investigate effects of physics beyond the Standard Model and/or exotic cosmological models (see for example [3]).

With the exception of lithium, whose predictions are at odds with observations [4, 5], a good agreement between predicted and observed primordial abundances of the light elements is reached, such an agreement is continuously updated in view of new experimental results and theoretical insights. The observed values of primordial abundances are usually provided by spectroscopic observations of ancient astrophysical objects. In the case of deuterium, its primordial abundance can be obtained from astrophysical sites not affected by stellar evolution since deuterium is destroyed during stellar evolution and no production mechanism after BBN is possible. The value of deuterium abundance relative to hydrogen has now reached a percent accuracy, $D/H = (2.527 \pm 0.030) \times 10^{-5}$ at 68% of C.L. [6].

Among the light elements produced during BBN, deuterium is an excellent indicator of cosmological parameters in the early Universe because its abundance is the most sensitive to ω_b and also depends on the radiation density, usually expressed in terms of the effective number N_{eff} of neutrino species [1]. The reactions involved in the synthesis of deuterium are: production via the well known p(n,γ)D process and destruction via the D(d,n)^{3}He , D(d,p)^{3}H and D(p,γ)^{3}He reactions. Once the deuterium bottleneck is overcome and deuterium is produced by the p(n,γ)D reaction, its amount crucially depends upon destruction rates. The evaluation of the nuclear reaction rates entering the BBN network is the most important issue to solve in order to get an accurate determination of light nuclide abundances and their corresponding uncertainties.

Among the deuterium destruction reactions, the most uncertain was the D(p,γ)^{3}He . Several data sets on its cross section, or equivalently the $S(E)$-factor were available in the literature before the LUNA measurement. In the low-energy range ($E_{\mathrm{C.M.}} \sim 2-20$ keV), mostly relevant to hydrogen burning in the Sun and in protostars, cross sections were obtained with a systematic error of at most 5.3% [7] using the 50 kV accelerator (now in disuse) of the Laboratory for Underground Nuclear Astrophysics (LUNA) at the Laboratori Nazionali del Gran Sasso (LNGS) in Italy [8]. At higher energies ($E_{\mathrm{C.M.}} \simeq 30-700$ keV), available data sets were affected by systematic errors of 9% or higher [9–12]. The situation was further compounded by the fact that a recent *ab initio* calculation [13] disagrees at the 20-30% level with both the S-factor of Ma *et al.* [9] and a best fit [14] to selected data [7, 9, 15, 16], widely used in BBN calculations. These large uncertainties had significant impact on the comparison between predicted and observed primordial abundance. As a consequence of the poor experimental data for the D(p,γ)^{3}He reaction, firm conclusions on the BBN stage and cosmological model couldn't be drawn with a satisfactory precision. For all these reasons, a new experimental campaign started at LUNA in 2016.

2 Experimental setup and data analysis of the D(p,γ)^{3}He cross section measurement

The LUNA400 accelerator provided a high intensity proton beam (beam current 100-300 μA) in the 50-400 keV energy range. The beam entered in a cylindrical reaction chamber filled with high purity molecular deuterium at a 0.3 mbar pressure. The beam was stopped by a copper calorimeter that measured its intensity. Two large high-purity germanium detectors (HPGe) faced to the reaction chamber at 90° to the beam axis were used to detect the photons emitted by the D(p,γ)^{3}He reaction. The latter proceeds through direct capture to the ^{3}He ground state with the emission of a single gamma with an energy in the range 5.5 - 5.8 MeV. Since these energies are above the natural radioactivity endpoint, it is possible to fully exploit the suppression of cosmic ray induced background due to the Gran Sasso mountain [8, 17]. The incoming proton can be captured both in a *s*-wave or *p*-wave orbital angular momentum state and the emitted gamma-ray shows a not isotropic angular distribution dominated by the M1 and E1 components. To measure the cross section in the energy range of interest (30 keV < E_{cm} < 300 keV) for BBN we explored the full dynamic range of the LUNA-400 accelerator in 30-50 keV energy steps. The main source of beam induced background was due to the ^{19}F(p,$\alpha\gamma$)^{16}O reaction from the interaction of protons with fluorine contaminant usually present on collimators along the gas target and on the calorimeter. This beam-induced background was quantified in dedicated control runs in which inert ^{4}He gas was used instead of deuterium.

For an extended gas target setup, the cross section depends on: N_γ, the number of counts after background subtraction, $\rho(z)$ the gas density profile, $\epsilon(z)$ is the gamma detection efficiency as a function of position along the beam axis, $W(z)$ which accounts for the photon angular distribution and N_p the number of projectiles impinging the target.

A major effort was devoted to prevent or minimize any source of systematic uncertainty in the previous mentioned quantities to reach an accuracy of three percent on the S-factor. N_γ is essentially affected by the statistical error, that in our measurements was always lower than 1% for $E_{\mathrm{p}>100}$ keV ($E_{\mathrm{C.M.}} > 70$ keV), as a consequence of the suppression in the cosmic ray induced background with respect to surface experiments. The target density $\rho(z)$ was determined by measuring the pressure and temperature profiles along the target as well as the beam heating effect due to the power released by the incident beam in the gas target [18]. The integrated beam current N_p was measured with a consolidated technique in which the beam power was monitored during each run by measuring the power dissipated by the beam on the copper calorimeter where the beam was stopped [19]. In an extended gas target the interaction with proton beam can take place at different positions along the beam axis, resulting in different energies of the emitted photons and in different geonetrical angles subtended by the HPGe detector. Therefore, the γ-ray detection efficiency $\epsilon(z)$ must be carefully determined as a function of both position and energy. The γ-rays emitted by the D(p,γ)^{3}He reaction ($Q = 5.5$ MeV) have typical energies $E_\gamma = 5.5$-5.8 MeV, i.e. far away from the energy of the commonly used radioactive sources. The efficiency $\epsilon(z)$ of the HPGe detector was measured along the beam line, by filling the reaction chamber with N_2 and exploiting the resonance of the $^{14}N(p,\gamma)^{15}O$ reaction at $E_r = 259$ keV. This reaction produces cascade γ-rays in a large energy range (0.7-7 MeV), in such a way the efficiency were precisely determined [20]. Finally, the $W(z)$ function has been unfolded from the shape of full energy peak of photons produced by the D(p,γ)^{3}He that depends on the angular distribution of emitted photons. The detailed description of the measurements performed to determine the above mentioned quantities and uncertainties can be found in [20].

Finally, the LUNA collaboration obtained the S-factor shown in Figure 1 where the LUNA data [21] are reported together with literature data. The new data show a significantly reduced uncertainty with respect to previous experimental datasets. The polynomial best fit of S(E) for $E \sim 0-2$ MeV to previous datasets are also shown [14, 22] and the fit in which this new data set is considered (red line). The bands represent the quoted 1σ uncertainty of the best fit values.

Presently a study of the D(p,γ)^{3}He angular distribution is ongoing using a novel approach the "Peak Shape Analysis" method. The experimental differential cross section is in nice agreement with theoretical predictions.

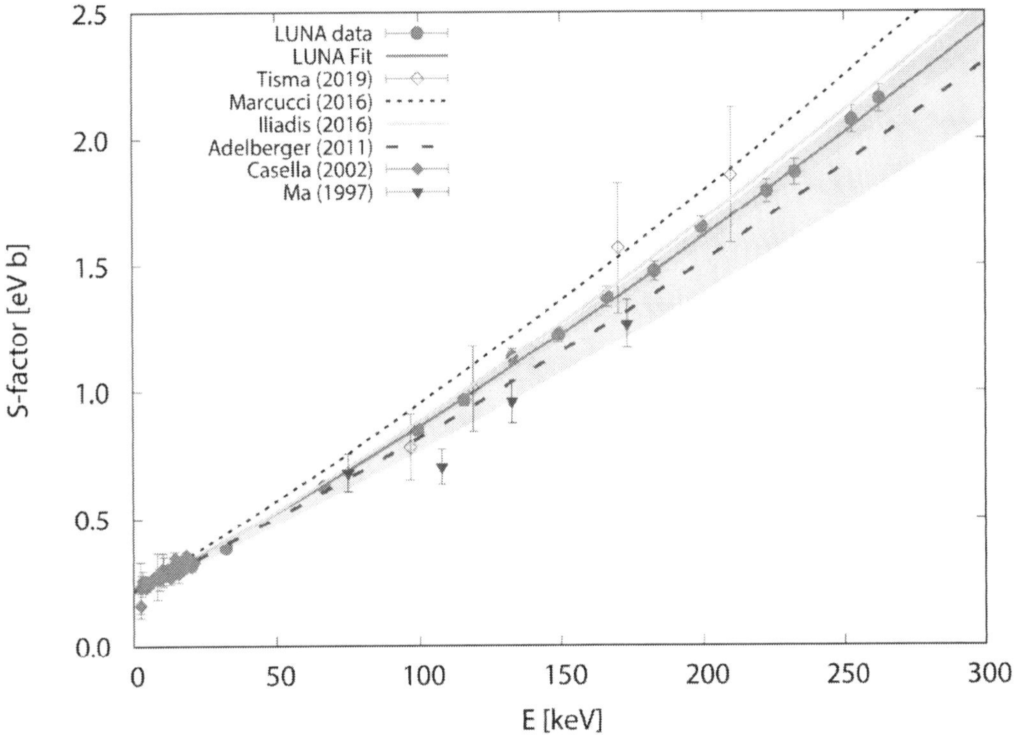

Figure 1. S-factor of the D(p,γ)^{3}He reaction; LUNA results are represented with filled red circles. Other experimental data are also shown. The best fit (red solid line) include all the reported experimental data. Band represents the 68% confidence level.

3 Conclusions and outlook

To explore the impact of our D(p,γ)^{3}He S-factor on the predicted primordial deuterium abundance, we used the second release [23] of the numerical BBN code PArthENoPE. Under the assumption of the ΛCDM model, with $N_{\rm eff}$ = 3.045, we performed a Bayesian likelihood analysis to derive $\Omega_b h^2$ using the observed deuterium abundance, $(D/H)_{\rm obs}$, and the theoretical behaviour of $(D/H)_{\rm BBN}$ (now including the new LUNA data). We obtained $\Omega_b h^2(\rm BBN) = 0.02233 \pm 0.00036$, a value which is a factor of 2 more precise than that deduced using a previous S factor [14] and now in much better agreement with the $\Omega_b h^2$ based on CMB data.

After the release of the LUNA results [21], several papers were published discussing the cosmological implications with the inclusion of the new D(p,γ)^{3}He rate. A deeper discussion on the LUNA results and their cosmological implication can be also found in [24–27]. The new challenge is now the measurement of the D(d,n)^{3}He and D(d,p)^{3}H reactions which have became the most prominent source of uncertainty in the determination of the primordial deuterium abundance.

References

[1] P.A. Zyla, R.M. Barnett, J. Beringer, O. Dahl, D.A. Dwyer, D.E. Groom, C.J. Lin, K.S. Lugovsky, E. Pianori, D.J. Robinson et al., Prog. Theor. Exp. Phys. **2020** (2020)

[2] N. Aghanim et al. (Planck), Astron. Astrophys. **641**, A6 (2020)

[3] F. Iocco, G. Mangano, G. Miele, O. Pisanti, P.D. Serpico, Phys. Rept. **472**, 1 (2009)
[4] M. Spite, F. Spite, Nature **297**, 483 (1982)
[5] B.D. Fields, Annu. Rev. Nucl. Part. Sci. **61**, 47 (2011)
[6] R.J. Cooke, M. Pettini, C.C. Steidel, Astrophys. J. **855**, 102 (2018)
[7] C. Casella, H. Costantini, A. Lemut, B. Limata, R. Bonetti, C. Broggini, L. Campajola, P. Corvisiero, J. Cruz, A. D'Onofrio et al., Nucl. Phys. A **706**, 203 (2002)
[8] F. Cavanna, P. Prati, Int. J. Mod. Phys. A **33**, 1843010 (2018)
[9] L. Ma, H. Karwowski, C. Brune, Z. Ayer, T. Black, J. Blackmon, E. Ludwig, M. Viviani, A. Kievsky, R. Schiavilla, Phys. Rev. C **55**, 588 (1997)
[10] G. Schmid, B. Rice, R. Chasteler, M. Godwin, G. Kiang, L. Kiang, C. Laymon, R. Prior, D. Tilley, H. Weller, Phys. Rev. C **56**, 2565 (1997)
[11] G. Griffiths, M. Lal, C. Scarfe, Can. J. Phys. **41**, 724 (1963)
[12] I. Tišma, M. Lipoglavšek, M. Mihovilovič, S. Markelj, M. Vencelj, J. Vesić, Eur. Phys. J. A **55**, 137 (2019)
[13] L. Marcucci, G. Mangano, A. Kievsky, M. Viviani, Phys. Rev. Lett. **116**, 102501 (2016)
[14] E. Adelberger, A. García, R. Robertson, K. Snover, A. Balantekin, K. Heeger, M. Ramsey-Musolf, D. Bemmerer, A. Junghans, C. Bertulani et al., Rev. Mod. Phys. **83**, 195 (2011)
[15] G. Griffiths, E. Larson, L. Robertson, Can. J. Phys. **40**, 402 (1962)
[16] G. Schmid, M. Viviani, B. Rice, R. Chasteler, M. Godwin, K. G.C., L. Kiang, A. Kievsky, C. Laymon, R. Prior et al., Phys. Rev. Lett. **76**, 3088 (1996)
[17] F. Ferraro, G.F. Ciani, A. Boeltzig, F. Cavanna, S. Zavatarelli, Frontiers in Astronomy and Space Sciences **7**, 119 (2021)
[18] F. Cavanna, R. Depalo, M.L. Menzel, M. Aliotta, M. Anders, D. Bemmerer, C. Broggini, C.G. Bruno, A. Caciolli, P. Corvisiero et al., Eur. Phys. J. A **50**, 179 (2014), 1411.2888
[19] F. Ferraro, M. Takács, D. Piatti, V. Mossa, M. Aliotta, D. Bemmerer, A. Best, A. Boeltzig, C. Broggini, C. Bruno et al., Eur. Phys. J. A **54**, 44 (2018)
[20] V. Mossa, K. Stöckel, F. Cavanna, F. Ferraro, M. Aliotta, F. Barile, D. Bemmerer, A. Best, A. Boeltzig, C. Broggini et al., Eur. Phys. J. A **56**, 144 (2020), 2005.00002
[21] V. Mossa, K. Stöckel, F. Cavanna, F. Ferraro, M. Aliotta, F. Barile, D. Bemmerer, A. Best, A. Boeltzig, C. Broggini et al., Nature **587**, 210 (2020)
[22] C. Iliadis, K.S. Anderson, A. Coc, F.X. Timmes, S. Starrfield, Astrophys. J. **831**, 107 (2016)
[23] R. Consiglio, P. de Salas, G. Mangano, G. Miele, S. Pastor, O. Pisanti, Comp. Phys. Comm. **233**, 237 (2018)
[24] O. Pisanti, G. Mangano, G. Miele, P. Mazzella, J. Cosmol. Astropart. Phys. **04**, 020 (2021)
[25] T.H. Yeh, K.A. Olive, B.D. Fields, J. Cosmol. Astropart. Phys. **03**, 046 (2021)
[26] C. Pitrou, A. Coc, J.P. Uzan, E. Vangioni, Mon. Not. Roy. Astron. Soc. **502**, 2474 (2021)
[27] J. Moscoso, R.S. de Souza, A. Coc, C. Iliadis, Astrophys. J. **923**, 49 (2021)

Gamma-ray spectroscopy of galactic nucleosynthesis

Roland Diehl[1,*], *Jochen* Greiner[1], *Martin G.H.* Krause[2], *Moritz M.M.* Pleintinger[1,3], and *Thomas* Siegert[4]

[1]Max-Planck-Institut für extraterrestrische Physik, Giessenbachstr. 1, D-85748 Garching, Germany
[2]Centre for Astrophysics Research, School of Physics, Astronomy and Mathematics, University of Hertfordshire, College Lane, Hatfield, Hertfordshire AL10 9AB, UK
[3]Horn & Company Financial Services GmbH, Kaistr. 20, D-40221 Düsseldorf, Germany
[4]Institut für Theoretische Physik und Astrophysik, Universität Würzburg, Emil-Fischer-Str. 31, D-97074 Würzburg, Germany

Abstract. Diffuse gamma-ray emission from the decay of radioactive ^{26}Al is a messenger from the nucleosynthesis activity in our current-day galaxy. Because this material is attributed to ejections from massive stars and their supernovae, the gamma-ray signal includes information about nucleosynthesis in massive star interiors as it varies with evolutionary stages, and about their feedback on the surrounding interstellar medium. Our method of population synthesis of massive-star groups has been refined as a diagnostic tool for this purpose. It allows to build a bottom-up prediction of the diffuse gamma-ray sky when known massive star group distributions and theoretical models of stellar evolution and core-collapse supernova explosions are employed. We find general consistency of an origin in such massive-star groups, in particular we also find support for the clumpy distribution of such source regions across the Galaxy, and characteristics of large cavities around these. A discrepancy in the integrated ^{26}Al gamma-ray flux is interpreted as an indication for excess ^{26}Al emission from nearby, distributed in cavities that extend over major regions of the sky.

1 Introduction

^{26}Al has been established as a tracer of ongoing nucleosynthesis in our Galaxy [1]. Spectroscopy at high resolution was essential for the pioneering discovery, because it provided an unambiguous line identification at 1809 keV as known from the laboratory. COMPTEL only featured scintillation detectors and hence a modest spectral resolution of ~200 keV at this energy, unlike Ge detectors, where 2-3 keV resolution can be achieved. With INTEGRAL [2] and its SPI gamma-ray spectrometer [3, 4], also featuring Ge detectors, a new step in spectroscopic detail of ^{26}Al emission could be achieved, as summarised in [5]. The INTEGRAL mission has enabled accumulation of observing time since its 2002 launch, continuing to this day, measuring the large-scale Galactic ^{26}Al emission. In this paper we summarize results from analysis of 17.5 years of data, using both single and double detector events within the SPI camera for the first time (see [6, 7] for more detail), and drawing on a significantly-enhanced version of our population synthesis code [8] as developed in [6] for interpretations of our results.

*e-mail: rod@mpe.mpg.de

2 Data and analysis

INTEGRAL operates since October 2002 in an excentric orbit located outside the radiation belts. The SPI spectrometer is one of the two main instruments on INTEGRAL, specialised for high-resolution spectroscopy over the 15 keV to 8 MeV energy range with a camera composed or 19 Ge detectors, achieving an energy resolution of 3 keV at 1809 keV. A coded mask above the camera shadows some of the detectors from emission of each region of the sky, while others obtain exposure of these regions. The instruments aboard the INTEGRAL satellite are co-aligned, and have rather large fields of view, SPI having an effective opening angle of ~30 degrees. Observations are carried out by pointing the common instrument axes towards a region of interest on the sky. Then, offsetting the instrument pointings by 2.1 degrees approximately every 30 minutes in a regular rectangular pattern (this is called *dithering*) provides additional detector shadowing variations that help to effectively achieve an imaging resolution of about 2.6 degrees at 1809 keV. For individual source regions in the plane of the Galaxy, a sensitivity (3σ) of 2.5×10^{-5} ph cm^{-2}s^{-1} has been achieved by now (from a 6 Ms exposure in the example of the Vela region) [6].

Here we use data from 2002 until 2020, i.e., 17.5 years of observations, with 2131 INTEGRAL orbits of typically 3 days duration each. After filtering out periods of abnormal conditions, 118407 spacecraft pointings from 1840 orbits remain, providing a sky exposure of about 255 Ms. For each of these pointings, the data from single Ge detector hits, as well as those where two Ge detectors triggered within their 350 ns coincidence window (called *double events, DE*) were binned into energy spectra for each of the 19 detectors named 00-18; pairs of DE detectors are assigned detector ID's above 18, i.e., 19-60. Detectors are numerated from central towards outside in left-spiral counting, so ID=0 is the central detector and detectors 07-18 are on the outside of the hexagonal dense pack forming SPI's camera. In total, during the mission four of the nineteen Ge detectors failed (ID's 01,02,05,17), at times between orbit numbers 140 and 930. The camera response differs for different numbers of operational detectors (from 19 down to 15), in particular the rates of single versus double or multiple detector triggers changes (because failed detectors do not provide a 'multiple-event' trigger any more), which needs to be accounted for in the analysis.

Event data from each detector are calibrated using known background lines, so that the distortions through gain variations are corrected for. Then these events are binned into spectra at 0.5 keV width, covering the 20–2000 keV range for response and background, and the 1790–1840 keV range for spectroscopy of the ^{26}Al line. At the ^{26}Al line energy of 1809 keV, the number of double-hit events corresponds to 56% of the amount of single-hit events, all including instrumental background and the celestial ^{26}Al signal.

These SPI detector count spectra are $d_{i,j,k}$ measured counts, with i, j, k being the indices of pointing, detector, and energy bin. The 19 physical detectors are supplemented by the respective detector combinations for multiple-detector hits, resulting in a detector variable $j \in [0, 60]$ instead of $j \in [0, 18]$. We describe these data as the result of the instrument's response to the γ-ray sky, plus an underlying instrumental background:

$$d_{i,j,k} = \sum_l R_{l;ijk} \sum_{n=1}^{N_s} \theta_n S_{nl} + \sum_{n=N_s+1}^{N_s+N_b} \theta_n B_{n;ijk} \quad (1)$$

Herein, θ_n and θ_b are model parameters scaling sky and background models, S or B, respectively. Sky models S_{nl} are formulated as photon source intensities per sky direction l, and multiple source or spectral components n can be used for sky and for background. For best sensitivity at high spectral resolution, we avoid simultaneous imaging, given our large number of free parameters of background modelling to ensure its closest match to real background. The instrument response matrix $R_{l;ijk}$ implements the shadowgram coding by mask and dithering, thus linking the emission of a pixel l in the sky to pixels in data space. The instrumental background B is modelled from detailed spectral fits to raw data. Integrating over suitable intervals results in smearing out any celestial contributions, while still retaining sufficient temporal resolution to trace temporal variations of background (see [9] and [10] for detailed descriptions of this background and its modelling). Maximum-likelihood fits then optimise model parameters through a comparison of the data as measured to predictions from these models of sky and background, thus obtaining the energy spectrum θ_n of intensities for the sky model, as a result.

Among the wide range of plausible sky models (see [1, 11, 12], for a discussion), we used the image obtained from the COMPTEL 9-year allsky survey [13, 14] as our baseline sky model. Then we aim for astrophysical implications of our data, formulating models from astrophysical considerations of the spatial distribution of ^{26}Al sources in our Galaxy. We have been using tracers that plausibly may represent these ^{26}Al sources, such as maps of molecular gas or emission from warm dust, or distributions of free electrons attributed to massive-star ionizations [12, 15]. We now advanced this approach to construct a full bottom-up model of the number and spatial distribution of the ^{26}Al sources in our Galaxy [6]. For this, we start from models of stars with their evolution and nucleosynthesis yields, building stellar groups using a mass spectrum, as we had done in our earlier population synthesis approach [8]. Now we place these groups into a spatial model for the Galaxy, using astronomical knowledge about massive star groups where accessible (i.e., catalogues derived from a variety of types of observations; these are available mainly in the vicinity of the solar system to a distance of few kpc at most), and we supplement these for the more-distant part of the Galaxy with Monte-Carlo samples of analytical models for the large-scale distribution, such as exponentially shaped disks or spiral arm models. We approximate this large-scale source distribution either through a Gaussian function based on the pulsar distribution [16], or an exponential profile with its characteristic scale radius. The vertical structure above the Galactic plane is modelled as an exponential. We superimpose spiral structure herein, as derived from observables such as the pulsar emission dispersion measure, converted into the electron distribution in the Galaxy [17]. For our description, we use a parametrisation of the four spiral arms following [18]. The main astrophysical parameters of this bottom-up modelling, which we aim to determine comparing its predictions to our data, are: (i) The absolute intensity of Galactic recent massive-star activity. This is implemented in a *star formation rate*, and scaled until the model flux matches the all-sky flux observed in the ^{26}Al line. (ii) The scale radius and width of the large-scale distribution. We decided to use five representative cases: Spiral arm models with a truncated gaussian (using widths 1.8, 2.5, and 3.5 kpc, and scale radii of 8.5, 17.5, and 5 kpc, respectively, for one four-arm and two two-arm variants), and an exponential disk (with a fixed scale radius of 5.5 kpc following the pulsar distribution but without spiral structure), plus a generic exponential centrally peaked in the Galaxy. (iii) The vertical extent of ^{26}Al emission, i.e. its scale height. Drawing random realisations of such a model by sampling the parameters included in such a bottom-up stars-groups-galaxy model, one obtains a stellar population synthesis based model of a galaxy. We call this latest approach the *PSYCO* implementation [6, 19]. We supplemented this generic and randomly-drawn model with a less-randomized treatment of massive star clusters where we have additional knowledge, either from catalogues, or from deeper studies including our own ^{26}Al measurements. For the few clusters where we were able to derive a gamma-ray brightness in ^{26}Al, we used these as brightness normalisations, rather than drawing randomly, thus without a need to draw a sample age and richness in stars. For the ones where only a position was known, we used this position. We started from these nearby clusters with some knowledge, and then continued sampling randomly from the spatial distribution model until the integrated model flux agreed with the measured flux in our dataset; this determines the normalization of our large-scale model, i.e. the corresponding star formation rate. This provides us with a predictive model for the appearance of the observed ^{26}Al sky, evaluating the Galaxy's content of ^{26}Al from theory inputs concerning source evolution and nucleosynthesis output. We then varied our model parameters, and determined a best-fit statistics based on likelihood with Poissonian statistics, to select best parameters for our measured dataset.

3 Results

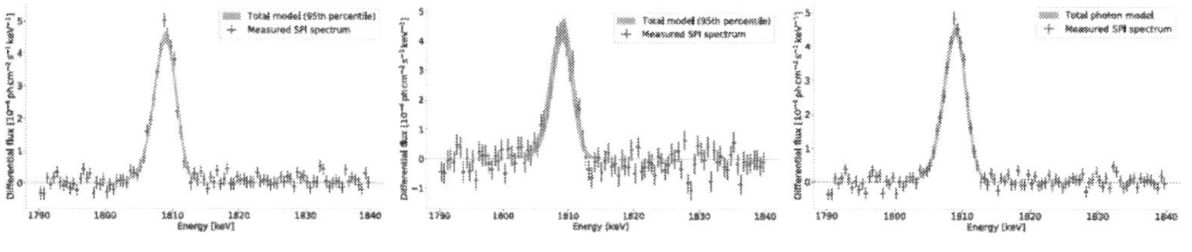

Figure 1. The energy spectrum around the ^{26}Al line, as obtained from single-hit events (left), double-hit events (middle), and combined events (right) of our observations. Data points represent the measurement, with bin width and Poissonian statistical uncertainty indicated. The uncertainties shown in shaded green have been derived from MCMC modelling of the spectral response, and show 3σ variations around the model.

From 17.5 years of data and the COMPTEL-derived sky distribution of ^{26}Al emission, we obtain spectra around the ^{26}Al line at 1808.65 keV for single and for double events, separately, and combining single and double events, as shown in Figure 1 (see [7] for more details). Intensity, line centroid, and line width are consistent within uncertainties between single-hit and double hit event results.

The spectral resolution of SPI up to now only has been determined for single events as 3.17 keV at 1.8 MeV [9]. Apparently, the spectral response appears to be less sharp for double events. This is attributed to a bias in locations of events within Ge detectors, double events favouring events detected from interactions in outer regions, rather than closer to the central anode. Charge collection properties vary across the detector with radius from the anode, and a bias in location will plausibly produce a line shape deviating from the all-detector sample in single events (which was confirmed to be nearly Gaussian after annealings). When decomposing the line width into contributions from instrumental resolution as measured with other instrumental lines versus line broadening of the celestial contribution, we obtain a consistent celestial broadening of (0.46±0.3) keV and (0.76±0.6) keV, respectively for single and double events, and a combined value of (0.62±0.03) keV. This is consistent with the Doppler broadening from regions at different bulk velocities to the observer, as incurred from large-scale Galactic rotation and the dispersal of ^{26}Al into superbubbles, as found before [20, 21]. The line centroid, found at (1809.83±0.04) keV from combined data, as compared to the laboratory value of 1808.65(7) keV, shows that the integrated ^{26}Al emission from the Galaxy appears blue-shifted by about 1.2 keV. This is understood as the net effect of ^{26}Al bulk velocities from massive star groups across the Galaxy, as viewed from our position near the Sun [see 20, for the variation of line centroids with galactic longitude].

The all-sky ^{26}Al line flux is determined as $(1.84\pm0.03)\times10^{-3}$ ph cm^{-2}s^{-1}. The flux attributed to the *inner Galaxy* is $\sim3\times10^{-4}$ ph cm^{-2}s^{-1}, or 16% of the total flux. The inner Galaxy has often been used for comparisons, and may be defined as a longitude range ±30 degrees around the Galactic centre, or, slightly differently, the *inner radian* (57 degrees). This region often had been analysed as a reference towards obtaining results representative for the Galaxy as a whole, thus focusing on the brightest region and avoiding issues from low surface-brightness contributions that often are subject to systematic uncertainties from background determinations. Different latitude ranges have been used, as well, from ±5 to 15 degrees. We note that the majority of the integrated flux is found outside the *inner Galaxy*, as reported earlier [see 22, and table 1 herein]. This is in tension with simulations and expectations, which report typically 50% of the all-sky flux coming from the inner Galaxy. Special nearby regions, such as Cygnus, Scorpius-Centaurus, and Orion, have been known before to potentially lead to distortions of such Galaxy-wide conclusions. It remains a challenge to properly account for flux contributions from nearby regions, in particular as the imaging performance of the coded mask degrades for diffuse emission that extends over large parts of the entire sky.

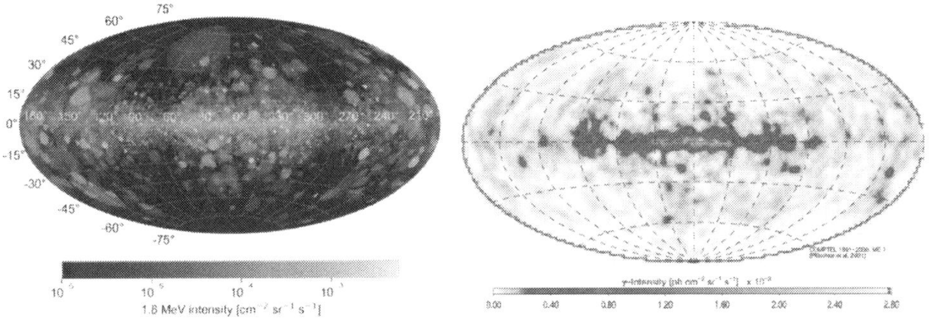

Figure 2. The spatial appearance of the ^{26}Al sky as predicted from our PSYCO bottom-up massive-cluster model (*left*) is similar to the COMPTEL ^{26}Al skymap (*right*). Note that the model image is plotted over a larger dynamic range of intensities, compared to the COMPTEL image.

The bottom-up modelling as fitted to our dataset obtains a best-matching sky distribution as shown in Figure 2. This is based on the truncated gaussian 4-armed spiral model with scale radius and width of 8.5 and 3.5 kpc, respectively, and a scale height of 700 pc. Evidently, our superbubble modelling determines many spherical features from more-nearby clusters, which appear more regular than in reality, where superbubbles are known to be elongated, distorted, and sometimes subjective to blowouts. But otherwise, the essential features and large-scale appearances of this modelling and the COMPTEL ^{26}Al image also shown in Figure 2 for comparison are rather similar. This confirms our modelling baseline of ^{26}Al emission in our Galaxy being mainly due to massive-star groups, rather than smoother spatial distributions as would be expected if more-abundant AGB stars or novae would dominate. We note that our bottom-up population synthesis model PSYCO predicts a substantially-lower flux value than observed: While a star formation rate of 2-4 $M_\odot$ y^{-1} is plausible [23, 24], our best fit is obtained with 8 $M_\odot$ y^{-1}. This is implausibly high, and we understand this as foreground emission and sources being prominent and different in appearance than our simplified spherical superbubble modelling [see 19, for details and discussion].

References

[1] N. Prantzos, R. Diehl, Phys. Rep. **267**, 1 (1996)

[2] C. Winkler, T.J.L. Courvoisier, G. Di Cocco, N. Gehrels, A. Giménez, S. Grebenev, W. Hermsen, J.M. Mas-Hesse, F. Lebrun, N. Lund et al., Astron. Astrophys. **411**, L1 (2003)

[3] G. Vedrenne, J.P. Roques, V. Schönfelder, P. Mandrou, G.G. Lichti, A. von Kienlin, B. Cordier, S. Schanne, J. Knödlseder, G. Skinner et al., Astron. Astrophys. **411**, L63 (2003)

[4] J.P. Roques, S. Schanne, A. von Kienlin, J. Knödlseder, R. Briet, L. Bouchet, P. Paul, S. Boggs, P. Caraveo, M. Cassé et al., Astron. Astrophys. **411**, L91 (2003), arXiv:astro-ph/0310793

[5] R. Diehl, M.G.H. Krause, K. Kretschmer, M. Lang, M.M.M. Pleintinger, T. Siegert, W. Wang, L. Bouchet, P. Martin, New Astron. Rev. **92**, 101608 (2021), 2011.06369

[6] M.M.M. Pleintinger, Ph.D. thesis, Technische Universität München (2020)

[7] M.M.M. Pleintinger, R. Diehl, T. Siegert, J. Greiner, M.G.H. Krause, Astron. Astrophys. **(in press)** (2023)

[8] R. Voss, R. Diehl, D.H. Hartmann, M. Cerviño, J.S. Vink, G. Meynet, M. Limongi, A. Chieffi, Astron. Astrophys. **504**, 531 (2009), 0907.5209

[9] R. Diehl, T. Siegert, J. Greiner, M. Krause, K. Kretschmer, M. Lang, M. Pleintinger, A.W. Strong, C. Weinberger, X. Zhang, Astron. Astrophys. **611**, A12 (2018)

[10] T. Siegert, R. Diehl, C. Weinberger, M.M.M. Pleintinger, J. Greiner, X. Zhang, Astron. Astrophys. **626**, A73 (2019), 1903.01096

[11] R. Diehl, Experimental Astronomy **6**, 103 (1995)

[12] J. Knödlseder, K. Bennett, H. Bloemen, R. Diehl, W. Hermsen, U. Oberlack, J. Ryan, V. Schönfelder, P. von Ballmoos, Astron. Astrophys. **344**, 68 (1999)

[13] R. Diehl, C. Dupraz, K. Bennett, H. Bloemen, W. Hermsen, J. Knoedlseder, G. Lichti, D. Morris, J. Ryan, V. Schoenfelder et al., Astron. Astrophys. **298**, 445 (1995)

[14] S. Plüschke, R. Diehl, V. Schönfelder, H. Bloemen, W. Hermsen, K. Bennett, C. Winkler, M. McConnell, J. Ryan, U. Oberlack et al., *The COMPTEL 1.809 MeV survey*, in *Exploring the Gamma-Ray Universe*, edited by A. Gimenez, V. Reglero, C. Winkler (2001), Vol. 459 of *ESA Special Publication*, pp. 55–58

[15] R. Diehl, U. Oberlack, J. Knodlseder, H. Bloemen, W. Hermsen, D. Morris, J. Ryan, V. Schonfelder, A. Strong, P. von Ballmoos et al., *Models for COMPTEL* ^{26}*Al Data*, in *The Fourth Compton Symposium*, edited by C.D. Dermer, M.S. Strickman, J.D. Kurfess (1997), Vol. 410 of *AIP Conference Series*, pp. 1114–1118

[16] I. Yusifov, I. Küçük, Astron. Astrophys. **422**, 545 (2004), astro-ph/0405559

[17] J.M. Cordes, T.J.W. Lazio, ArXiv Astrophysics e-prints **astro-ph/0207156** (2002), arXiv:astro-ph/0207156

[18] C.A. Faucher-Giguère, V.M. Kaspi, Astrophys. J. **643**, 332 (2006), astro-ph/0512585

[19] T. Siegert, M.M.M. Pleintinger, R. Diehl, M.G.H. Krause, J. Greiner, C. Weinberger, Astron. Astrophys. **(in review)** (2023)

[20] K. Kretschmer, R. Diehl, M. Krause, A. Burkert, K. Fierlinger, O. Gerhard, J. Greiner, W. Wang, Astron. Astrophys. **559**, A99 (2013), 1309.4980

[21] M.G.H. Krause, R. Diehl, Y. Bagetakos, E. Brinks, A. Burkert, O. Gerhard, J. Greiner, K. Kretschmer, T. Siegert, Astron. Astrophys. **578**, A113 (2015), 1504.03120

[22] M.M.M. Pleintinger, T. Siegert, R. Diehl, Y. Fujimoto, J. Greiner, M.G.H. Krause, M.R. Krumholz, Astron. Astrophys. **632**, A73 (2019), 1910.06112

[23] R. Diehl, H. Halloin, K. Kretschmer, G.G. Lichti, V. Schönfelder, A.W. Strong, A. von Kienlin, W. Wang, P. Jean, J. Knödlseder et al., Nature **439**, 45 (2006), arXiv:astro-ph/0601015

[24] L. Chomiuk, M.S. Povich, Astron. J. **142**, 197 (2011), 1110.4105

ANC experiments for nuclear astrophysics: the ^{26}Si(p, γ)^{27}P case

G. D'Agata[1,2,3,*], *A. I.* Kilic[3], *V.* Burjan[3], *J.* Mrázek[3], *V.* Glagolev[3], *V.* Kroha[3], *A.* Cassisa[3], *G. L.* Guardo[1], *M.* La Cognata[1], *L.* Lamia[1,2,4], *S.* Palmerini[5,6], *R. G.* Pizzone[1], *G. G.* Rapisarda[1,2], *S.* Romano[1,2,4], *M. L.* Sergi[1,2], *R.* Spartà[1], *C.* Spitaleri[1,2], and *A.* Tumino[2,7]

[1]INFN - Laboratori Nazionali del Sud, Via Santa Sofia 62, 95123, Catania, Italy
[2]Dipartimento di Fisica e Astronomia, Università degli Studi di Catania, Via S. Sofia, 64, 95123, Catania, Italy
[3]Nuclear Physics Institute of the Czech Academy of Sciences, 250 68 Řež, Czech Republic
[4]Centro Siciliano di Fisica Nucleare e Struttura della Materia (CSFNSM), Catania 77843, Italy
[5]Dipartimento di Fisica e Geologia, Università degli Studi di Perugia, via A. Pascoli s/n, 06123, Perugia, Italy
[6]INFN-sezione di Perugia, via A. Pascoli s/n, 06123, Perugia, Italy
[7]Facoltà di Ingegneria ed Architettura, Kore University, Viale delle Olimpiadi, 1, I-94100 Enna, Italy

Abstract. The Asymptotic Normalization Coefficient (ANC) method has proven to be useful in retrieving the direct part of the radiative capture cross section for a number of reactions of astrophysical interest. In this work, the study of the ^{26}Si(p, γ)^{27}P reaction, studied via the ANC in its extension for mirror nuclei will be discussed.

1 Introduction

Among the key problems in nuclear astrophysics, one the most complicated to solve has been to experimentally measure the cross section in the range of interest: Coulomb barrier in fact heavily suppresses the probability for the various reaction to happen, with the result that the cross section at astrophysical energies (between some keV and some MeV) can be very low (even at the order of magnitude of picobarn). Furtheremore, reactions involving radioactive nuclei with a short half-life in the entrance channel are frequent, and a direct measurement sometimes is hardly feasible. For these reasons, many indirect methods have been proposed, and among these the Asymptotic Normalization Coefficient (ANC)[1, 2] one has proven to be useful to evaluate the direct part of the cross section for (p, γ) [3], (n, γ) [4] and (α, γ) [5] radiative captures of astrophysical relevance. In the case of unstable nuclei with small half-lifes, an extention to the method has been developed using the so-called "mirror reactions" [6, 7]. In this work, the ANC method will be briefly presented, and recent results regarding the study of the ^{26}Si(p, γ)^{27}P (performed using the mirror nuclei procedure)[8] will be summarized.

^{26}Si production and destruction channels are deeply involved in the problem of ^{26}Al nucleosynthesis ($T_{1/2}$ = 0.72 Myrs): the 1.809 MeV γ-ray line emitted from the first excited state of ^{26}Mg – in which ^{26}Al naturally decays – has been detected along the Galactic plane, and

*e-mail: giuseppe.dagata@dfa.unict.it

has been considered as a good tracer of the recent nucleosynthesis in our Galaxy. This isotope can supposedly be formed through nucleosynthesis in massive stars and core-collapse Supernovae [see 9, and references therein], but also Wolf-Rayet objects, AGB-stars [10] and Novae [11] have been appointed as good sites for its nucleosynthesis. All of these sites share the possibility to host proton capture, and the ^{26}Al isotope can be produced through the ^{24}Mg$(p,\gamma)^{25}$Al$(\beta^+)^{25}$Mg$(p,\gamma)^{26}$Al reaction chain. Understanding such a process has been made difficult by the existence of an isomeric state ^{26}Al^m($T_{1/2}$ = 6.34 sec) which decays to the ground state of ^{26}Mg. Such an isomer can be produced via the ^{25}Al$(p,\gamma)^{26}$Si$(\beta^+)^{26}$Al^m chain, thus reducing the quantity of ^{26}Al in the interstellar medium. The unstable isotope ^{26}Si produced along the way can also be destroyed via proton capture through the ^{26}Si$(p,\gamma)^{27}$P reaction, interfering then with the production of ^{26}Al^m. The study of this reaction can therefore be useful to understand the ratio between ^{26}Al and ^{26}Al^m [12, 13].

2 The ANC method

The ANC method allows to indirectly investigate direct radiative capture reactions of astrophysical interest from the one-particle (proton, neutron or alpha) transfer. In general, a direct transfer reaction can be visualized as a $X+A \rightarrow Y+B$ process, where both the X and B nuclei can be considered as $X = Y + a$ and $B = A + a$, a being the transferred particle (Fig.1).

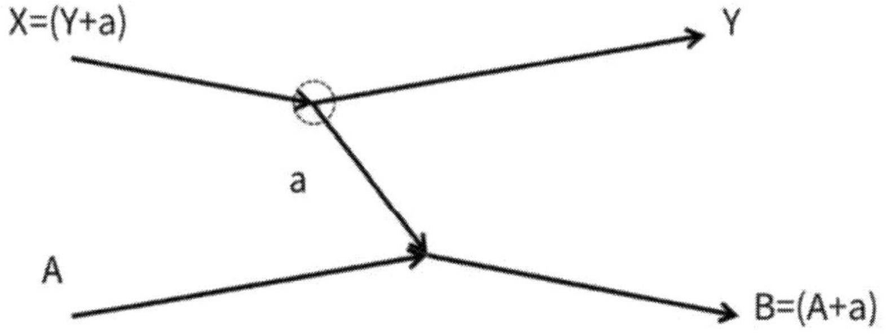

Figure 1. Sketch of a typical transfer reaction.

The experimental cross section is related to the theoretical one, using the DWBA formalism, through the spectroscopic factors (SF):

$$\frac{d\sigma^{exp}}{d\Omega} = \sum_{j_B,j_X} S_{Aa,l_B,j_B} S_{Ya,l_X j_X} \sigma^{DW}_{l_B,j_b,l_X,j_X} \quad (1)$$

Here the experimental and theoretical differential cross sections – $\frac{d\sigma^{exp}}{d\Omega}$ and $\sigma^{DW}_{l_B j_b l_X j_X}$, respectively – are related via the two SF for $A + a$ and $Y + a$ systems: while one of those is usually known, the other is left to be determined. The SF in general depends on the binding potential and on the optical model parameters (OMP)[14]: the depth of the former is in fact adjusted considering the binding energy to reproduce the bound state, while the latters are normally extracted from the elastic scattering. In general the SF is therefore very sensitive to the selection of the optical model adopted, and the experimental angular distributions can be reproduced by different families of potentials [15].
In 1990, Mukhamedzhanov and Timofeyuk [1] pointed out that suitable direct transfer reactions share the same radial overlap integral (I^c_{ab,l_c,l_j}) with the direct part of a radiative cross

section at low energies, due to the fact that such capture occurs at large distances from the nucleus, where the behaviour of the wave function can be considered asymptotic. This property can be used to get the cross section of many reactions of astrophysical interest. In fact, the radial overlap functions and bound-state wave functions can be written as:

$$I^B_{Aa,l_c,j_c}(r_{ab}) \xrightarrow{r_{ab}>R_n} C^c_{ab,l_b,j_b} \frac{W_{-\eta,l_c+\frac{1}{2}}(2k_{ab}r_{ab})}{r_{ab}} \tag{2}$$

and

$$\phi_{ab,l_c,j_c}(r_{ab}) \xrightarrow{r_{ab}>R_n} b^c_{ab,l_c,j_c} \frac{W_{-\eta,l_c+\frac{1}{2}}(2k_{ab}r_{ab})}{r_{ab}} \tag{3}$$

$W_{-\eta,l_c+\frac{1}{2}}$ being the Whittaker function, which asymptotic behaviour is equal to $W_{-\eta,l_c+\frac{1}{2}}(-2k_I r) \rightarrow e^{-k_I r+\eta_I ln(2k_I)}$. The coefficients b and C are the so-called single-particle ANC (ŠPANC) and ANC, respectively. Using equations 2 and 3 in equation 1, we can finally obtain:

$$\frac{d\sigma}{d\Omega} = \sum_{j_B,j_X} \left(C^B_{Aa,l_B,j_B}\right)^2 \left(C^X_{Ya,l_X,j_X}\right)^2 \frac{\sigma^{DWBA}_{l_B,j_B,l_X,j_X}}{b^2_{Aa,l_B,j_B} b^2_{Ya,l_X,j_X}} \tag{4}$$

where one of the two C^2 constants is the ANC for the direct part of the radiative cross section of interest. This approach is useful because the ANC's have a small dependence on the chosen nucleon binding potential [14], making the cross section nearly independent from b^2. This is true once the OMP are set: the ANC's are, in fact, dependent on the chosen OMP. Also, the relation between the ANC and the S factor sometimes is not unique: for example, in the case of strongly bound systems the final-state interaction can lead to sizable changes in the zero-energy SF [16].

In order to apply the ANC method, just one condition has to be met: the one-particle transfer from which the ANC's are deduced must be peripheral. Such a condition (called peripherality) is really important and must be ascertained.

More recently an extension of the use of mirror nuclei in transfer reactions to the ANC method has been proposed [6]: this has been useful in case of (p) or neutron (n) transfers that involve strongly unstable nuclei (such as ^{26}Si) [6–8]: if $A + p \rightarrow B$ is a certain proton capture, the proton ANC (C_p^{A+p}) can be extracted from the neutron ANC (C_n^{D+n}) of a suitable mirror partner $D + n \rightarrow E$, D and E having inverted number of protons and neutrons with respect to A and B, respectively. The two ANC are tied by the relation

$$\left(C_p^{A+p}\right)^2 = R_{mirr}\left(C^{C+n}\right)^2 \tag{5}$$

with R_{mirr} that can be written as

$$R_{mirr} = \left|\frac{F_l\left(ik_pR_N\right)}{k_pR_nj_l\left(ik_nR_N\right)}\right|^2 \tag{6}$$

In Eq.6, F_l is the regular Coulomb wave function, j_l the Bessel function of l–th order, and R_N the radius of the nuclear interior. The quantities k_p and k_n are instead related to the neutron and proton separation energy, via the relation $k = \sqrt{\frac{2\mu\epsilon}{\hbar^2}}$ [6, 7].

If in the proton channel an excited state near the threshold is present, the corresponding mirror state will exist as a resonance, and a relation exists that ties the C^2 of the direct radiative capture in the excited state and the Γ of the unbound state in the mirror counterpart exists[6]:

$$\frac{\Gamma_{p(n)}}{\left|C_{n(p)}\right|^2} = R_\Gamma \approx R_0^{res} = \frac{\hbar^2 k_{p(n)}}{\mu}\left|\frac{F_l\left(ik_{p(n)}R_N\right)}{k_{p(n)}R_Nj_l\left(ik_{n(p)}R_N\right)}\right|^2 \tag{7}$$

3 The ^{26}Si(p, γ)^{27}P reaction reaction

Experiments involving ^{26}Si are difficult, given that the half-life of such an isotope of Silicon is quite short($T_{1/2}$ = 2.24 sec). For this reason, an indirect measurement of the ^{26}Si$(p,\gamma)^{27}$P has been performed using the ^{26}Mg$(d,p)^{27}$Mg transfer reaction, applying the method described in Section 2 that involves mirror nuclei [8]. The experiment has been carried out at the Nuclear Physics Institute of the Czech Academy of Sciences, using the deuteron beam (E_{beam} = 19.2 MeV, I≈ 14 enA) available at the U-120M isochronus cyclotron operated by the CANAM infrastructure (Řež, Czech Republic). Such a beam has been sent on a ^{26}MgO (with ^{12}C as backing) target, produced by the target laboratory of the Laboratori Nazionali del Sud - Istituto Nazionale di Fisica Nucleare (LNS–INFN). The experimental apparatus was composed by five ΔE–E telescopes composed of a thin (250 μm) and thick (5000 μm) anular silicon detectors, covering the 7° – 60° range thanks to a rotating plate.
In order to extract the ANC related to the ^{26}Mg + n → ^{27}Mg capture, we need to retrieve the OMP's for the ^{26}Mg$(d,p)^{27}$Mg reaction. Therefore, we need to extract the OMP's of both the entrance(^{26}Mg + d) and the exit (^{27}Mg + p) channels. To do so, the ΔE–E telescopes have been used to select the deuteron and proton loci. Those two have been studied separately: using the FRESCO code [17], the angular distribution for the scattered deuterons has been fitted to extract the OMP's for the entrance channel. Those have been used – again by fitting procedures using the OMP's from [20] as seed values, using FRESCO – to finally retrieve the OMP's from the proton angular distribution of the exit channel, in this case ^{27}Mg + p.
The selected optical model is

$$U = V_c(r_c) - V_0 f(x_0) - \left[W f(x_w) - 4W_D \frac{d}{dr} f(x_D) \right] + \\ + \frac{\hbar^2}{m_\pi c} V_{LS}(L\sigma) \frac{1}{r}\frac{d}{dr} f(x_{LS}) . \tag{8}$$

$V_c(r_c)$ being the Coulomb potential, V_0 and W the depth of the real and imaginary parts of the volume potential, V_{LS} the depth of the real spin-orbit one, and W_D the depth of the surface term for the imaginary part of the potential, respectively. The term $f(x_i)$ represents the radial form of the Woods-Saxon potential, and can be written as

$$f(x_i) = (1 - e^{x_i})^{-1} \tag{9}$$

where $x_i = \left(r - r_i A^{1/3}\right)/a_i$, r_i being the radius parameter and a_i the diffuseness one, while A the atomic mass number. Once the OMP's are set, the ANC values for the ^{26}Mg + n → ^{27}Mg capture in the ground and first excited states have been found using Eq.4, in which the value of C^2_{pn} = 0.77 fm^{-1} [18] has been used (he results of the calculation are reported in [8]).
In order to apply the ANC method, the peripherality for the process – in this case the (d,p) reaction – in exam must be ascertained. The main way to do so is to check the dependence of the C^2 to the geometry – r_0 and a parameters – of the Wood-Saxon potential: if its values variate weakly in a wide range of b^2, then the reaction can be considered peripheral. In the same range of b^2 the SF must show a bigger variation.
In this case (Figure 2), the C^2 for the ^{27}Mg→ ^{26}Mg+n shows a variation of ≈ 13%, while the SF varies by ≈ 55%, and the process can be therefore considered predominantly peripheral.
Once the neutron ANC's are extracted, the procedure for mirror nuclei has been used with

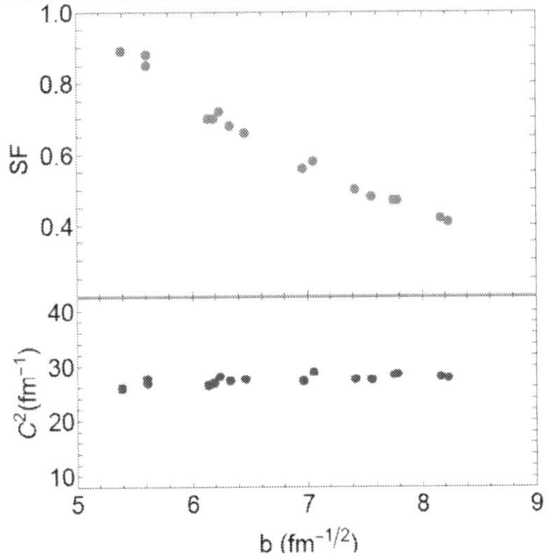

Figure 2. Spectroscopic factors (upper panel, red circles) and ANC values (lower panel, blue circles)extracted for the ground state of ^{27}Mg versus the value of the SPANC b. The calculations have been done using the optical potential P1 of [8]. In this calculations the r_0 parameter has been varied between 1.1 and 1.4 fm and a_0 between 0.5 and 0.7 fm.

Table 1. Values found in [8] and compared with [19] and [20]. The ANC values for the capture in the ground (first column) and the 1st excited state (second column) for the neutron capture ^{26}Mg$(n,\gamma)^{27}$Mg from the three publications are reported, along with the proton ANC (third column) for the capture in the ground state and the proton width Γ_p for the ^{26}Si$(p,\gamma)^{27}$P (fourth column).

ref.	$(C^{Mg}_{g.s.})^2$ [fm^{-1}]	$(C^{Mg}_{1^{st}})^2$ [fm^{-1}]	$(C^{p}_{g.s.})^2$ [fm^{-1}]	$\Gamma^p_{1^{st}}$ [MeV]
[8]	28.26 ± 5.30	3.40 ± 0.32	1420 ± 255	$(5.23 \pm 1.05) \times 10^{-9}$
[19]	24.50 ± 4.90	1.1 ± 0.15	1058 ± 273	$(4.04 \pm 0.77) \times 10^{-9}$
[20]	44.00 ± 5.30	3.40 ± 0.32	1840 ± 240	$(5.4 \pm 0.1) \times 10^{-9}$

the aim to determine the proton ANC and the proton width Γ_p for the ^{26}Si$(p,\gamma)^{27}$P.
The results (Table 1) show a substantial agreement between [8] and [19] , if the different binding energies are properly taken into account: in this calculation, the value of 0.807 MeV for the proton separation energy of ^{27}P experimentally extracted by [21] has been used (instead of the 0.859 MeV used in [19]). This updated value of the separation energy comes from the first experimental evidence for the observation of a 1.125 MeV β-delayed γ-ray from the decay of ^{27}S: the excitation energy for the first excited state of ^{27}P and the proton decay energy from the first excited state of ^{27}P to the ground state of ^{26}Si retrived in [21] brought to a more precise evaluation of the proton-separation energy and the mass excess. Now that the C^2 and Γ_p have been correcly extracted, the reaction rate for the direct capture in the ground state and the resonant one in the first excited state have been calculated using the formulas from [22], finding an enhancement by a factor 1.4 for the ground state and 2.2 for the first excited state contribution with respect to [21]. To calculate the latter, the value of Γ_γ has been calculated using the ratio I_γ/I_p taken from [21, 23]

References

[1] A. M. Mukhamedzhanov and N. K. Timofeyuk, JETP Lett. **51**, 282 (1990).
[2] H. M. Xu et al., Phys. Rev. Lett. **73**, 2027 (1994).
[3] V Burjan et al., EPJ A **55**, 114 (2019).
[4] A. M. Mukhamedzhanov et. al., Phys. Rev. C. **84**, 024616 (2011).
[5] Y. P. Shen et al, Phys. Rev. C. **99**, 025805 (2019).
[6] N. K. Timofeyuk et al., Phys. Rev. Lett. **91**, 232501 (2003) .
[7] L. Trache et al., Phys. Rev. C. **67**, 062801 (2003).
[8] G. D'Agata et al., Phys. Rev. C. **103**, 015806 (2021).
[9] L. Bouchet et al., Astrophys. J. **801**, 142 (2015).
[10] N. Prantzos and R. Diehl, Phys. Rep. **267**, 1 (1996).
[11] J. José et al., Astrophys. J. **560**, 897 (2001).
[12] M. Wiescher et al, Astron. Astroph **160**, 56-72 (1986).
[13] A. Coc et al, Phys. Rev. C **61**, 015801 (1999).
[14] A.M. Mukhamedzhanov et al., Phys. Rev. C **56**, 1302 (1997).
[15] X.D. Liu et al., Phys. Rev. C **69**, 064313 (2001).
[16] S. Typel & G. Baur, Nucl. Phys. A **759**, 245 (2005).
[17] I. J. Thompson, Computer Physics Reports **7**, 4 (1988).
[18] R. V. Reid, Ann. Phys. **50**, 411 (1968).
[19] N. K. Timofeyuk et al., Phys. Rev. C **78**, 044323 (2008).
[20] B. Guo et al., Phys. Rev. C **73**, 048801 (2006).
[21] L. J. Sun et al, Phys. Rev. C **99**, 064312 (2019).
[22] C. Iliadis, *Nuclear Physics of Stars* (Wiley-VCH Verlag, New York, 2007).
[23] L. J. Sun et al., Phys. Lett.B **802**, 135213 (2020).

EPJ Web of Conferences **279**, 02003 (2023) https://doi.org/10.1051/epjconf/202327902003
NPA-X 2022

Short-lived Radionuclides in the Milky Way Galaxy

Tejpreet Kaur[1,*]

[1]Department of Physics, Punjab University, India

Abstract. The short-lived radionuclides (SLRs) have a half-life ≤ 100 Myr. The γ-ray observations and excess abundance of their daughter nuclides in various meteoritic phases confirm the existence of SLRs in the Galaxy and early solar system (ESS), respectively. In this work, we have developed Galactic Chemical Evolution (GCE) models for SLRs, ^{26}Al, and ^{60}Fe along with ^{36}Cl, ^{41}Ca, and ^{53}Mn. These models predict the temporal and spatial evolution of SLR abundance trends in the Galaxy from 2-18 kpc. The abundance of two SLRs, ^{26}Al, and ^{60}Fe, are investigated further, as their γ-ray observations are available for comparison with the model predictions. The predictions for the abundance per unit area for each ring show the decreases from the inner to outer regions of the Galaxy.The GCE predictions for the total mass of alive ^{26}Al, and ^{60}Fe in 2-18 kpc of the Galaxy at present time are 0.2 $M_{\odot}$ and 0.08 $M_{\odot}$, respectively.

1 Introduction

Short-lived radionuclides (SLRs) are elements with a half-life of the order of a few million years. These nuclides can be formed in various stellar environments in the Galaxy. The presence of the SLRs is vital evidence that star formation is an ongoing process in the Galaxy. The main production sites of ^{26}Al are Asymptotic Giant Branch stars (AGB), Core collapse Supernovae (CCSNe), nova and Wolf-Rayet (WR) stars. For ^{60}Fe, the primary production happens during the neutron capture reaction inside the CCSNe. These are the only two short-lived radionuclides for which the γ-ray observations can provide evidence of their presence in the star–forming regions. For the other SLRs, ^{36}Cl, ^{41}Ca and, ^{53}Mn, CCSNe are major production sites except for ^{53}Mn, which can also be produced in Supernova Ia (SNIa). The details of the characteristics of these nuclides are given in Table 1. The γ-ray COMPTEL and INTEGRAL observations detected the presence of ^{26}Al in the Galaxy. The results show the map of ^{26}Al which is concentrated in the galactic plane [1] [2]. The ^{26}Al– rich regions in the γ-ray map coincide with the star-forming regions of the disc region. The ^{26}Al emits γ-rays when it decays to ^{26}Mg at 1809 keV. The ^{26}Al emission regions also contain the observations of the presence of ^{60}Fe, which emits γ-rays while decaying into ^{60}Co and further ^{60}Ni at 1173 keV and 1332 keV, respectively. With the ability of INTEGRAL to resolve the star-forming region, the ^{26}Al can be observed in the stellar ejecta, and nucleosynthesis studies can be performed inside the star-forming regions. The observations of ^{26}Al and ^{60}Fe from the same star-forming regions are being used to constrain the stellar nucleosynthesis models, identifying more such regions will also help identify the source of these SLRs in the ESS. Also, the observations from the other galaxies will provide more stringent constraints for the

*e-mail: tejpreet@pu.ac.in;tejpreetkaur95@gmail.com

stellar and galactic chemical evolution modelling community. We have studied the abundance of SLRs using the Galactic Chemical Evolution (GCE) models discussed in [3].

Short-lived radionuclide (SLR)	^{26}Al	^{36}Cl	^{41}Ca	^{53}Mn	^{60}Fe
Half-life ($t_{1/2}$ (Myr))	0.717	0.301	0.0994	3.74	2.62
Mean-life(τ (Myr))	1.035	0.434	0.1434	5.40	3.78
Decay Process	β_+	β_-	β_+	β_+	β_-
Daughter Product	^{26}Mg	^{36}S, ^{36}Ar	^{41}K	^{53}Cr	^{60}Ni
Stable Isotope	^{27}Al	^{35}Cl	^{40}Ca	^{55}Mn	^{56}Fe

Table 1. Details of Short-Lived Radio nuclides considered in this work. The data of half-life and decay modes is taken from [4].

2 Galactic Chemical Evolution Models

The study of the abundance distribution of SLRs plays a vital role in understanding the Galaxy's active star-forming regions. The SLR prediction for the Solar System also gives insights into its origin and the pre-solar molecular cloud. Despite precise observations of SLR abundances in the Solar System from meteorites, their stellar sources are still debated. To explain the Galactic scale steady-state distribution of SLRs, we divided the Galaxy into rings of 2 kpc width from 2–18 kpc, based on Monte Carlo simulations described in [5]. The homogeneous GCE model presented here is based on three-infall accretion for the formation of the Galaxy. In the case of the three infall model, the halo, thick and the thin disc all form in separate accretion episodes. The first two episodes occur on a shorter time scale, followed by the third episode in which thin disc forms over the extended time scale in inside–out scenario [6] [7]. In inside-out scenario the inner regions of the thin disc forms first and then the outer regions. The normalisation constants involved in the accretion are determined by reproducing the observed value of total surface mass density in the solar neighbourhood and other parts of the Galaxy. Various generations of stars are formed according to the prevailing star formation rate (SFR), a function of the gas and prevalent total surface mass densities. Also, the temporal and radial star formation efficiencies, $\nu(t)$ and $\eta(r)$, respectively, account for the temporal and spatial variation in the SFR. The stellar mass in each ring at every time step is distributed in the stars according to the initial mass function (IMF) in the mass range 0.8–100 $M_\odot$ in various stellar generations. These stars evolve according to their mass and metallicity. Then, at every time step T, an assessment is made for all stars formed before T ($t<T$), to estimate their nucleosynthetic yields and remnants. A binary stellar population is synthesised at the time of star formation. With the binary fraction f, some stars out of the total stars formed at each time step evolve into a definite progenitor of SNe Ia. The progenitor stars have mass in the range of 3–8 $M_\odot$ and 11–16 $M_\odot$. The radioactive nuclides produced are subject to decay at each time step. The stellar yields of AGB and massive stars contribute the radioactive nuclides along with the stable nuclides to the interstellar medium (ISM). The contribution to the ISM for CCSNe is considered from [8], and AGBs from [9] and [10].

3 Results and Discussion

Results shown in figures 1-3 are based on the models I of homogeneous GCE models presented in [3]. Figure 1 shows the absolute abundance and abundance per unit area for ^{26}Al in

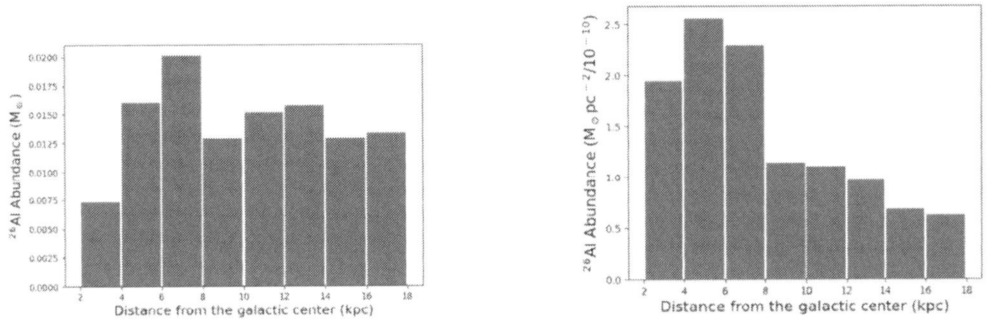

Figure 1. (Left) Absolute abundance of ^{26}Al abundance, (Right) ^{26}Al per unit area in the Galaxy for eight annular rings from 2-18 kpc.

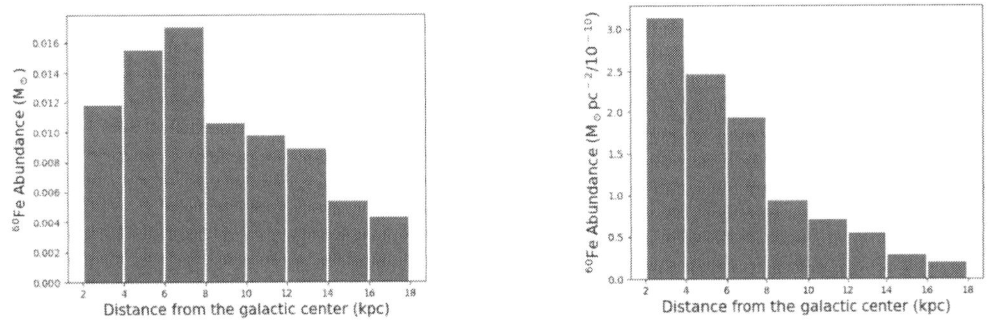

Figure 2. (Left) Absolute abundance of ^{60}Fe abundance, (Right) ^{60}Fe per unit area in the Galaxy for eight annular rings from 2-18 kpc.

the Galaxy. Similar results for ^{60}Fe are presented in Figure 2. Then the ratio of the absolute abundance of ^{60}Fe and ^{26}Al in the Galaxy is presented in Figure 3. The GCE model shown here evolves each ring of the Galaxy of width 2 kpc from 2-18 kpc over the galactic time scale. The model also predicts the metallicity and [Fe/H] for the Galaxy over the galactic time scale. Each ring experiences the accretion of gas and forms various generations of stars as explained in section 2 above and in [3]. The SLRs trends mainly depend upon the star formation rate in the region,which is higher in the inner regions and decreases as we move towards the outer regions of the Galaxy. This trend is mainly because of the higher accretion in the inner regions and high the star formation efficiency. However, ^{26}Al does not follow the decreasing trend as well as ^{60}Fe because of the contribution of ^{26}Al from AGB stars. On the contrary, ^{60}Fe is mainly coming from massive stars. The sum of the mass of ^{26}Al and ^{60}Fe over the 2-18 kpc of the Galaxy is 0.2 $M_\odot$ and 0.08 $M_\odot$, respectively.

The abundance of other SLRs, ^{36}Cl, ^{41}Ca and ^{53}Mn, also follow a similar trend of decrease in the abundance per unit area of the ring as we move towards the Galaxy's outer regions as shown in figure 1 of [3]. The areas of the rings from 2-18 kpc, which have 2 kpc width, from the inner first to outer eighth ring, are given as 37.68, 62.8, 87.92, 113.04, 138.16, 163.28, 188.4, and 213.52 kpc^2 respectively. In this GCE model, the inventories of every generation of stars are all mixed with the ISM gas and homogenised with the entire ring area. Hence

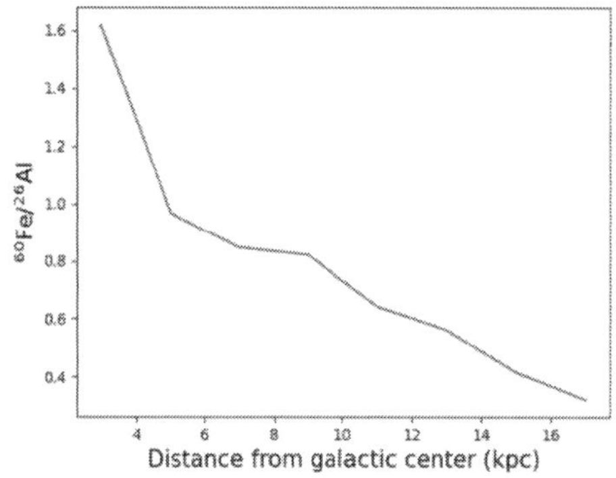

Figure 3. The ratio of absolute abundance of $^{60}Fe/^{26}Al$ in the Galaxy for eight annular rings from 2-18 kpc.

abundance per unit area gives better insight into the abundance distribution of the SLRs in the Galaxy.

The ^{26}Al and ^{60}Fe can be synthesised inside two different evolutionary phases of the massive stars [11]. So there are suggestions that the ratio of these two SLRs ^{26}Al and ^{60}Fe can be an important indicator of the stellar population in the star-forming region. The γ ray observations show that the flux ratio of $^{60}Fe/^{26}Al$ is 0.4 , which limits the steady–state mass ratio for $^{60}Fe/^{26}Al$ to be < 0.9 [11]. Figure 3 shows the trend of this ratio with the distance from the galactic center. There is significant decrease in the ratio with an increase in distance from the galactic center. The ratio is < 0.9 in the Galaxy except for the first two galactic rings from 2-6 kpc. In the solar neighborhood the ratio is ≈ 0.8 and decreases afterwards in the outer regions which could arise from the AGB contribution to ^{26}Al abundance in the Galaxy.

4 Conclusion

The SLRs abundance distribution presented here are based on the homogeneous GCE models. The GCE models represent a good approximation to understand the overall trend in the Galaxy. The homogeneous models show the abundance per unit area of ^{26}Al and ^{60}Fe in the Galaxy is higher in the the inner regions and lower in the outer. This trend is in agreement with the γ-ray observation. The estimates for the mass of ^{26}Al and ^{60}Fe in the entire Galaxy from the γ-ray observations are 1.8–2.4 $M_{\odot}$, and 1–6 $M_{\odot}$, respectively, which are higher than the predicted values from the GCE model explained above [12]. The low mass of SLRs could be outcome of the approach used to mix the stellar ejecta with the ISM. The stellar ejecta of every star is mixed with the entire annulus ring taken into account. This approximation is valid for the stable nuclides and have a little effect on their abundance predictions. However, in case of the SLRs, because of the short half-life the mixing of the ejecta with ISM can significantly reduce the abundance predictions. The second reason for low predictions from the GCE model is that the central 2 kpc is excluded from the model calculation. Finally, the ratio of $^{60}Fe/^{26}Al$ from the GCE can provide clues to the contribution of different stellar sources to these two SLRs.

5 Acknowledgement

TK is thankful to Dr Thomas Siegert from the University of Würzburg, Germany, for valueable discussions and suggestions related to the γ-ray observations for SLRs in the Milky Way Galaxy. This manuscript was written during a visit to the Konkoly Observatory in Budapest, Hungary, hosted by Dr Maria Lugaro, for which TK is also grateful.

References

[1] S. Plüschke, R. Diehl, V. Schönfelder, H. Bloemen, W. Hermsen, K. Bennett, C. Winkler, M. McConnell, J. Ryan, U. Oberlack et al., *The COMPTEL 1.809 MeV survey*, Vol. 459 of *ESA Special Publication* (2001)
[2] R. Diehl, Reports on Progress in Physics **76**, 026301 (2013)
[3] T. Kaur, S. Sahijpal, Monthly Notices of the Royal Astronomical Society **490**, 1620 (2019)
[4] M. Lugaro, U. Ott, Kereszturi, Progress in Particle and Nuclear Physics **102**, 1 (2018)
[5] S. Sahijpal, T. Kaur, Monthly Notices of the Royal Astronomical Society **481**, 5350 (2018)
[6] A. Micali, F. Matteucci, D. Romano, Monthly Notices of the Royal Astronomical Society **436**, 1648 (2013)
[7] C. Chiappini, Gratton, R, The Astrophysical Journal **477**, 765 (1997)
[8] S.E. Woosley, T.A. Weaver, Astrophysical Journal Supplement **101**, 181 (1995)
[9] A.I. Karakas, Monthly Notices of the Royal Astronomical Society **403**, 1413 (2010)
[10] S. Cristallo, O. Straniero, L. Piersanti, D. Gobrecht, The Astrophysical Journal Supplement Series **219**, 40 (2015)
[11] R. Diehl, A.J. Korn, B. Leibundgut, M. Lugaro, A. Wallner, Progress in Particle and Nuclear Physics **127**, 103983 (2022)
[12] M.M.M. Pleintinger, Dissertation, Technische Universität München, München (2020)

Potential experimental evidence of an Efimov state in ^{12}C and its influence on astrophysical carbon creation

Giuseppe Cardella[1,*], *Aldo* Bonasera[2,3], *Nunzia, Simona* Martorana[4,2], *Luis* Acosta[6,7], *Enrico* De Filippo[1], *Elena* Geraci[4,1], *Brunilde* Gnoffo[4,1], *Chiara* Guazzoni[6], *Concettina* Maiolino[2], *Angelo* Pagano[1], *Emanuele, Vincenzo* Pagano[2], *Massimo* Papa[1], *Sara* Pirrone[1], *Giuseppe* Politi[4,1], *Fabio* Risitano[5,1], *Francesca* Rizzo[4,2], *Paolo* Russotto[2], and *Marina* Trimarchi[5,1]

[1]INFN Sezione di Catania, Italy
[2]INFN- Laboratori Nazionali del Sud, Catania, Italy
[3]Cyclotron Institute Texas AM University college station, USA
[4]Dip. di Fisica e Astr. "Ettore Majorana", Università degli Studi di Catania, Italy
[5]DSMISFT Università di Messina, Italy
[6]DEIB Politecnico Milano and INFN Sez. Milano , Italy
[7]Instituto de Física, Universidad Nacional Autónoma de México

Abstract. We describe the search of the Efimov state in ^{12}C performed by analysing a large set of 3-α particles coincidence data measured at the LNS laboratories with the CHIMERA detector. The data are compatible with the presence of an Efimov state at the excitation energy of 7.485 MeV, with an excitation probability smaller than 0.3% of the population probability of the Hoyle state. The Efimov state gamma decay probability should be around 30%, much larger than the one of the Hoyle state. The population of this level is compatible with the Suda limit for the 3-α reaction rate in stars because its population should be possible only through direct 3-α reaction. More investigations improving the detector angular resolution are needed to confirm these preliminary observations.

1 Introduction

In the 70th of last century Efimov [1], following an idea of Thomas [2], demonstrated that if there is a two body system experiencing a resonance, it will generate a bound level in the 3 body system and, in principle, infinite excited levels over it. Efimov aimed at the demonstration that the Hoyle state [3] belongs to this class of levels but it was unable to reach its objective. However, later on, the presence of Thomas-Efimov (TE) levels was demonstrated in atomic physics experiments [4]. The search of this effect in nuclear physics was also very active [5–7] but up to now no convincing evidences were found both for Carbon and lighter systems. An attempt to describe the decay characteristics of such a state in ^{12}C was performed by Zheng and collaborators [8]. More recently a paper presenting some evidences against the existence of this level in ^{12}C was published by Bishop et al [9]. In particular, Bishop et al presented an astrophysical objection based on the so called Suda limit for the 3-α reaction rate in

*e-mail: cardella@ct.infn.it

stars [10]. In practice the presence of the TE level would enhance the production yield of ^{12}C in stars at much lower temperatures than what was expected up to now. The TE state could be in fact a very efficient way to build carbon with respect to the Hoyle state having a gamma decay probability nearly 100% [9], while in the case of the Hoyle state this probability is only around 1 over 2500 [11]. In this short report, we present some evidences observed in a recent experiment performed with the CHIMERA detector [12] at the INFN - Laboratori Nazionali del Sud (LNS) in Catania (Italy). In the α+^{12}C reaction at 64 MeV, we have detected a possible signature of the direct decay of the TE state in 3 α particles [13, 14]. This observation combined with an extra-yield of the gamma decay in the region of the Hoyle state [15, 16], could be explained by the population and decay of the TE state. However, we do not find any evidence of a sequential decay of such level in the α+^{8}Be channel. This means that in stars the TE state - if it exists - cannot be populated through this channel, but only by direct 3-α reactions. This fact decreases of many order of magnitude its population probability in stars, overcoming the problem underlined by Bishop et al of the Suda limit. Also the gamma decay probability seems smaller than that deduced by Bishop et al. We underline that these results are based only on accurate simulations of the experimental findings, therefore a more precise measurement will be performed soon to definitely prove or exclude the existence of this level. In the next section we shortly present the experimental results and data analysis, while in the last section we show possible improvements coming with a new experimental setup.

2 Experimental results and simulations

The experiment was performed at INFN-LNS, by using an α beam of 64 MeV delivered by the Cyclotron accelerator of the Laboratories. The beam impinged on a carbon target placed in the target holder of the CHIMERA detector. CHIMERA is a 4π detector for charged particles, mainly devoted to isospin studies (see for instance [17–20]), that in the last years was used also as a nearly 4π gamma ray detector. This was possible thanks to the sensitivity to gamma rays of the second stage (CsI(Tl) scintillators with photodiode readout) of the telescopes of the spherical part of CHIMERA covering the angles from 30° to 176° [21–23]. The high efficiency of the detector allowed the collection of a large number of coincidences between 3 α particles from the decay of the Hoyle state (about 85000 events with an average detection efficiency of the order of 50%), and also a reasonable statistics (30% error bar) in the coincidence measurement of the scattered beam, the recoiling ^{12}C, and the two decay gamma rays from the Hoyle state. The measured relative gamma decay width of the Hoyle state was of the order of $1.8{\cdot}10^{-3}$ [15]. This number is more than 3 times larger than the most recent measurement on this level [16], that is already 50% larger than the accepted value of this decay probability [11]. Due to the low energy resolution of the measurement, the possibility of some contamination by other levels was investigated also by looking at the 3-α decay channel. We emphasize that the scattered beam can be discriminate by the α particles emitted from carbon decay because of the higher energy. A detailed simulation of the detected events was performed obtaining a rather accurate modelling of the detected excitation energy spectrum in the region of the Hoyle state [13]. The ratio between the simulated and the experimental spectra shows only small differences at high excitation energy, due to random coincidences not included in the simulations. Some discrepancy is present also in the low excitation energy region. This could be a signature of the presence of the TE state. The same simulation tool was used to understand the expected behaviour of decay events from the TE state both for sequential α+^{8}Be and for direct 3-α decays. Guided by the results of the simulations we could search for such events in our data. More in detail we analyse the center of mass (CM) energy of the 3 couples of α particles obtained combining the 3 detected α particles selecting only the events in which one of these energies is equal to 92

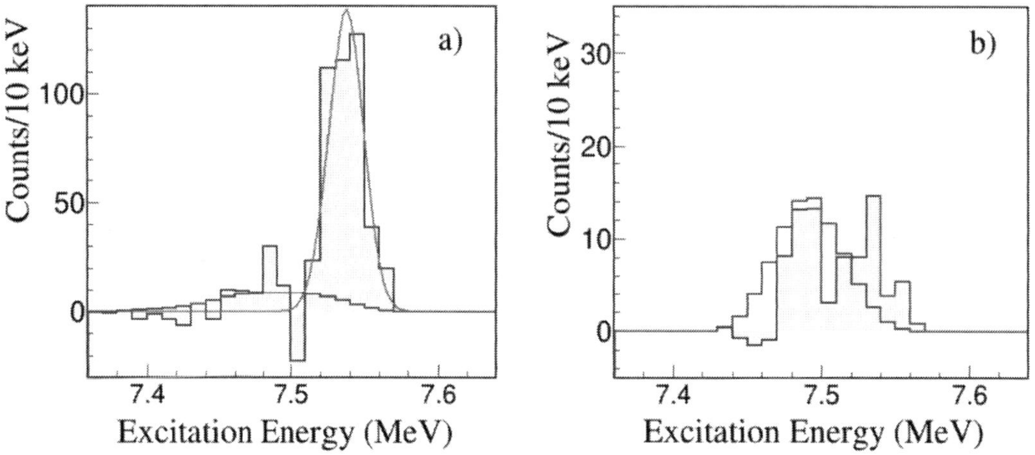

Figure 1. The filled yellow spectra are the experimental events compatible with the decay of the TE state; a) assuming sequential α+^{8}Be decay; b) assuming direct 3-α decay. The blue histograms are simulated TE decay spectra in the two decay modes. The red Gaussian line is a fit of the extra-peak observed in the sequential decay mode.

keV as the average of the other two. More than 4500 events were observed compatible with the sequential TE decay while only about 400 were found for the direct decay. However, as shown by simulations, most of the events are genuine Hoyle level decays badly measured because of the uncertainty in the detection angle of the α particles. Fig. 1a and fig. 1b show, as filled yellow histograms (adapted from ref. [14]), the spectra obtained from the experimental data subtracted of the aforementioned badly assigned events. They are compared to the blue spectra that are simulated decays of the TE state arbitrarily normalized. We note that the sequential decay spectrum of fig.1a is dominated by a peak (fitted in the figure with a red Gaussian) that it is not comparable to the decay of a TE state. This peak shows that our simulation is somewhat incomplete. Apart this peak, we note that the lower energy tail of the spectrum presents a small contribution, similar to the behaviour expected for a sequential TE decay. However this tail is much smaller than the error bar of the background evaluation of badly reconstructed events (approximately 40 events with an error bar of ±90). We conclude that our data are fully compatible with zero contribution from the sequential TE decay within error bars. The plot of fig.1b is obtained assuming the direct decay. This decay mode requires that the 3 α particles are emitted with similar energy - 60 keV in their CM - and with a relative angle around 120°. These conditions are very difficult to be fulfilled by events coming from the decay of the Hoyle state. Only few events are observed compatible with this behaviour in our data set (around 400). The background subtracted spectrum looks quite similar to the expected one from a TE state direct decay (blue histogram). A residual contribution at the high energy side could be due to simulation incompleteness, but statistics is low. The integral of the events compatible with the TE sequential decay is around 47±27 events. The statistical error comes from the subtraction of the simulated events. The enhancement of the gamma decay given by the presence of the TE state was evaluated in ref. [14] to be of the order of 270 events each 200000 Hoyle state population events. Therefore about 1 event every 1000 Hoyle events could be due to a TE level decaying by gamma emission. Taking into account selection efficiencies our data are compatible with the population and 3-α decay of 1 TE state each approximately 400 Hoyle events. Therefore summing both gamma and α decays it is deduced that the probability of the TE state population is of the order of 3 events each 1000 Hoyle state population events. Moreover we can also say that the gamma decay probability of this state is at least two times smaller than the 3-α decay probability. The large error bars and the strong dependence of these results from the details of the simulations do not allow a more precise and final answer on the existence of this level.

3 Conclusions and outlook

In this paper we present the results of the analysis carried out on a large data set of 3-α particles coincidence data measured at the INFN-LNS laboratories with the CHIMERA detector and related simulations, showing the possible evidence of the direct 3-α decay of the TE state. No evidence was observed for its sequential decay, inside our large error bars. The presence of some Gaussian prominent peak on the difference between simulated and analysed events tells us that there are still some experimental details not well included in our simulation method. This fact and the incoherent results that we obtain with other analysis methods reported in ref. [13], prevent us to extract a firm conclusion. However we underline that if the sequential α+^{8}Be channel is suppressed and only the direct 3-α decays is possible there are consequences in the 3-α process in stars. This could be the solution to the problem of Suda limit [10] reported by Bishop's work [9]. The direct 3-α reaction needs in fact a rather large density of α particles not available in stars.

Since from our data the α decay channel is not negligible with respect to the gamma decay one, a more precise experiment looking at the 3-α decay channel, could better highlight the presence of the TE state. An attempt was already performed by Bishop et al in their work with an active target [9], however the level of badly reconstructed events was too large to clearly observe the TE region in its spectrum notwithstanding the 55 keV energy resolution obtained and the detection of about 23000 Hoyle decays.

Recently we have coupled to the CHIMERA detector an array of high energy resolution telescopes named FARCOS, (see one of the adopted configuration in fig. 2-right) each one made of two double sided silicon strip detectors (DSSSD) followed by a final CsI(Tl) stage [24, 25]. With FARCOS we can determine the detection angle of an α particle with a precision of about 0.3°. In a preliminary test experiment we have already detected 3-α events in the FARCOS telescopes with an energy resolution of the order of 28 keV. With this energy resolution the contribution of the tail of the Hoyle spectrum to the TE region is reduced by more than 6 orders of magnitude. Fig. 2-left shows the obtained CM energy spectrum (filled yellow) compared to the previous one measured with CHIMERA (filled green). This enhanced resolution should be enough to discriminate the few expected events of the TE decay from the ones of the dominant Hoyle decay channel (we need to detect 100000 Hoyle events in order to observe at least 100 TE events). An experiment is planned in the next future hoping to definitively determine the existence of the TE state in Carbon.

4 Acknowledgments

This work was supported, in part, by CHIRONE experiment of INFN, the U.S. Department of Energy under Grant No. DE-FG03-93ER40773 and NNSA Grant No. DENA0003841 (CENTAUR). 061602- and, in part, by the DGAPA-UNAM IN107820, the CONACyT 315839 grant and the “Programma ricerca di ateneo UNICT 2020-22 linea 2”.

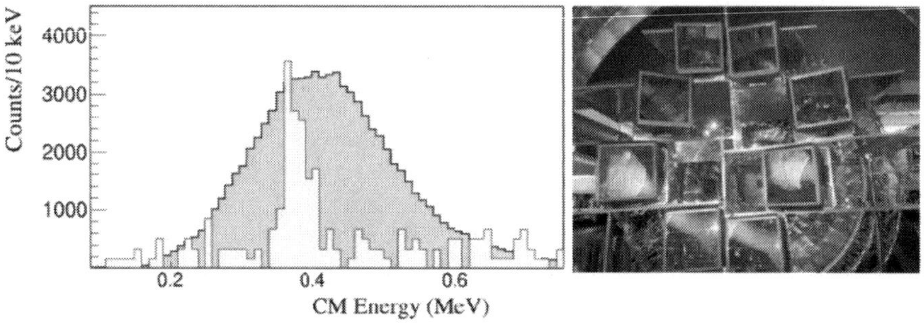

Figure 2. Left - The green spectrum is the Hoyle CM energy spectrum reconstructed with CHIMERA data, the yellow spectrum is the one reconstructed by using FARCOS telescopes in a preliminary test experiment, (random coincidences are not subtracted). Right - photo of one of the adopted configuration for the FARCOS array inside CHIMERA.

References

[1] V. Efimov, Physics Letters B **33**, 563 (1970)
[2] L. Thomas, Phys. Rev. **47**, 903 (1935)
[3] F. Hoyle, Astrophys. J. Suppl. Ser. **1**, 121 (1954)
[4] T. Kraemer, M. Mark, P. Waldburger et al., Nature **440**, 315 (2006), cited By 765
[5] P. Naidon, S. Endo, Reports on Progress in Physics **80**, 056001 (2017)
[6] A. Tumino, A. Bonasera et al., Phys. Lett., B **750**, 59 (2015)
[7] S. Zhang et al., Phys. Rev. C **99**, 044605 (2019)
[8] H. Zheng et al., Physics Letters B **779**, 460 (2018)
[9] J. Bishop, G. Rogachev et al., Phys. Rev. C **103**, L051303 (2021)
[10] T. Suda, R. Hirschi, M.Y. Fujimoto, Astrophys. J. **741**, 61 (2011)
[11] M. Freer, H. Fynbo, Progress in Particle and Nuclear Physics **78**, 1 (2014)
[12] A. Pagano et al., Nuclear Physics A **734**, 504 (2004)
[13] G. Cardella, A. Bonasera, N. Martorana et al., Nuclear Physics A **1020**, 122395 (2022)
[14] G.Cardella et al., Il Nuovo Cimento **45 C**, 56 (2022)
[15] G. Cardella, F. Favela, N. Martorana et al., Phys. Rev. C **104**, 064315 (2021)
[16] T. Kibédi et al., Phys. Rev. Lett. **125**, 182701 (2020)
[17] M. Papa, I. Berceanu et al., Phys. Rev. C **91**, 041601 (2015)
[18] S. Pirrone, G. Politi, B. Gnoffo et al., Eur. Phys. J. A **55**, 22 (2019)
[19] R. Najman, R. Płaneta, A. Sochocka et al., Phys. Rev. C **92**, 064614 (2015)
[20] P. Russotto, E. De Filippo, E. Pagano et al., Eur. Phys. J. A **56**, 12 (2020)
[21] G. Cardella et al., Nucl. Instr. and Meth. in Phys. Res. Sect. A **799**, 64 (2015)
[22] N.S. Martorana et al., Physics Letters B **782**, 112 (2018)
[23] A. Badalá, M.L. Cognata, R. Nania et al., La Rivista del Nuovo Cimento (2022)
[24] Pagano, E.V. et al., EPJ Web of Conferences **117**, 10008 (2016)
[25] L. Acosta et al., Jour. of Phys. Conf. Series **730**, 012001 (2016)

Low-energy Cross Section Measurements of ^{12}C(p, γ) Deep Underground at LUNA

Jakub Skowronski[1] and *Axel* Boeltzig[2]* for the LUNA collaboration

[1]Università degli Studi di Padova and INFN, Sezione di Padova, Via Francesco Marzolo 8, 35131 Padova, Italy
[2]Helmholtz-Zentrum Dresden-Rossendorf, Bautzner Landstraße 400, 01328 Dresden, Germany

Abstract. The ^{12}C(p, γ)^{13}N reaction cross section is currently under investigation in the low-background environment of the Laboratory for Underground Nuclear Astrophysics (LUNA). It is being studied using different types of solid targets, and employing two complementary detection techniques: HPGe spectroscopy and activation counting. To reduce systematic uncertainties, targets have been accurately characterized and their degradation under the intense beam of the LUNA-400 accelerator monitored. We present the experimental techniques and the corresponding analyses used to extract the reaction cross section.

1 Introduction

The ^{12}C(p, γ)^{13}N reaction is part of the CNO cycle, which is active in the hydrogen burning regions of main sequence, Red Giant Branch (RGB), and Asymptotic Giant Branch stars (AGB). It does not only contribute to the energy production of the stars, but also governs the abundances of several nuclei in their interiors, e. g. the ^{12}C is produced by the ^{15}N(p, α)^{12}C reaction and then depleted by proton capture.

The ^{12}C/^{13}C ratio is a useful tracer of nucleosynthesis inside stars [1], since it can be readily derived by studying the stellar spectra. AGB stars, in particular, are remarkably prolific centers of nucleosynthesis. Unfortunately the AGB phase of stellar evolution is marked by intense mixing phenomena, which make the theoretical models particularly difficult to establish [2]. The mixing heavily impacts the internal and external abundances of both the ^{12}C and ^{13}C, and can arise from several different sources: convective motion inside the star, the angular momentum of the star [3], magnetic buoyancy [4] or gravitational waves [5]. Hence, by obtaining a precise cross section for the ^{12}C(p, γ)^{13}N and the ^{13}C(p, γ)^{14}N reactions the theoretical models can be improved and will give further insight into the evolution of AGB stars.

Additionally, the proton capture on ^{12}C produces ^{13}N, which is β^+ unstable and one of the main neutrino sources from the CNO cycle. The CNO neutrinos from our Sun have been recently measured for the first time by the Borexino collaboration [6]. Since the neutrino rates are directly proportional to the reaction rates, it could be possible to extract the information about the solar metallicity directly from the Borexino result. Up to now, two different estimates of the solar metallicity exists that appear to be in disagreement [7] creating the

*e-mail: a.boeltzig@hzdr.de

so-called Solar Metallicity Problem. Thus a precise value of the $^{12}C(p,\gamma)^{13}N$ cross section could help in providing a further independent estimate [8].

The measurement of the $^{12}C(p,\gamma)^{13}N$ cross-section at astrophysical energies was the goal of a recently conducted experiment at LUNA. Different experimental methods (targets and detection setups) were employed, in order to characterize and limit systematic experimental uncertainties.

2 Current Status

Several previous studies of the $^{12}C(p,\gamma)^{13}N$ reaction cross section are available, with their results summarized in Fig. 1. There are two complete studies that report the cross-sections down to 200 keV, performed by Vogl [9], and Rolfs *et al.* [10]. Both data sets covered a broad energy region and are generally in good agreement, but show a discrepancy in the precise value of the energy of the observed broad resonance. Detection of prompt γ-rays was used in both these experiments, using thin evaporated targets. A recent measurement by Burtebaev *et al.* [11] used a thick graphite disk, and obtained data points at relatively high energies, found to be in agreement with the previous data sets. In the region of astrophysical interest, however, only older measurements are available: Bailey and Stratton [12] detected charged particles from the decay of ^{13}N, whereas Lamb and Hester [13] measured prompt γ-rays from the reaction. Both of these measurements used thick graphite targets. Their results for the cross section at low energies have large uncertainties, and show significant scatter.

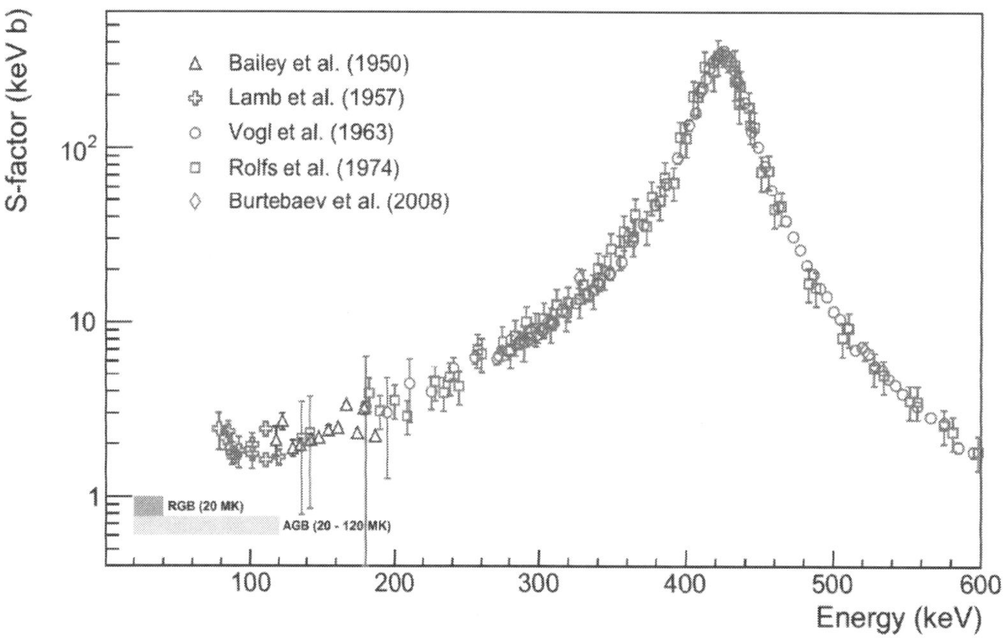

Figure 1: State of the art for the $^{12}C(p,\gamma)^{13}N$ reaction cross section. Two different energy ranges of interest are highlighted for RGB and AGB stars, respectively. Direct data available in these energy ranges comes from early measurements with significant scatter and uncertainties.

3 Experimental Setup

The LUNA-400 kV accelerator was chosen for this measurement, as it provides proton beams with high intensity (up to $\simeq 400\,\mu A$ in this experiment) at excellent beam energy resolution and stability [14]. Its deep-underground location at the Gran Sasso National Laboratory (LNGS) results in a drastic reduction of cosmic backgrounds, thus improving experimental sensitivity.

Two types of targets were used for the study of $^{12}C(p,\gamma)^{13}N$: thin targets produced by evaporation of carbon powder onto Tantalum backings, and thick targets of solid graphite. The graphite targets were sufficiently thick to stop the proton beam, providing the conditions for the measurement of the infinitely-thick-target yield. The carbon in both targets had the natural isotopic composition, i. e., containing about 1 % of ^{13}C. Targets were characterized on-site by analyzing the width and shape of the primary γ-ray peaks measured with the HPGe detector [15]. Target stability was further monitored in-situ by repeated measurements at a given reference proton energy. After the measurements at LUNA, targets were characterized by NRRA at the Institute for Nuclear Research (ATOMKI).

For the detection of the $^{12}C(p,\gamma)^{13}N$ reaction, two detectors were used in separate experimental campaigns (Fig. 2): a High-Purity Germanium detector (HPGe), and a Bismuth Germanium Oxide (BGO) detector. The HPGe detector was placed at 0° or 55° towards the target. With its excellent energy resolution it allowed for γ-ray spectroscopy of the single primary γ-ray of the reaction. With the reaction Q-value of 1943.5(3) keV, the primary γ-ray energy region is affected by environmental γ-rays in the measured spectrum. A 15 cm lead shielding [16] around target and detector setup was used to reduced these backgrounds. The large segmented BGO detector [17], providing almost 4π coverage around the target, was used to detect the reaction in two ways: i) by detecting the primary γ-ray, and ii) by detecting the signature of annihilation γ-rays from the β^+-decay of ^{13}N. The former method provides a large detection efficiency and is rather insensitive to angular distribution effects, thanks to the large solid angle covered by the detector. For this reaction, however, it is limited in sensitivity, owing to the non-negligible background of the BGO detector in the energy region of interest (largely given by intrinsic radioactivity [16]), and could only be used at the higher energies of this experiment. For the second method, the signature of two γ-rays with an energy of 511 keV in opposite segments of the detector was exploited to detect these low-energy events in the detector background. Activation (beam-on) and counting (beam-off) phases were run in the same experimental configuration, i. e., counting was performed in-situ.

Figure 2: Left: BGO detector in lead shielding, partially retracted (shielding open) to access target. Right: HPGe detector, here shown retracted (fully outside of the shielding) at 55° to the target.

4 Data Analysis

The data taken were analyzed with techniques corresponding to the characteristics of the two detection setups. Examples for spectra taken with both detectors are shown in Fig. 3. In case of the HPGe campaign, the primary γ-peak was parameterized as described in [15]. By fitting to the experimental data, the target degradation could be studied and the S-factor could be directly extracted from the γ-peak fitting procedure. Additionally, the target degradation obtained from the fits was compared with the results of the NRRA analysis. This approach was only followed for the case of the thin target. In case of the thick graphite disk, the yield was extracted from the primary γ-peak and compared to the results for the thin targets. The efficiency was obtained with a multiparametric fit to the data obtained with ^{137}Cs, ^{60}Co and ^{133}Ba calibration sources and resonances in ^{14}N(p, γ)^{15}O [18, 19] and ^{27}Al(p, γ)^{28}Si [20] with well-known strengths and branching ratios. The majority of the HPGe data was taken at 0°, whilst measurements at 55°, allowed to check and confirm the expected angular distribution of the primary γ-ray.

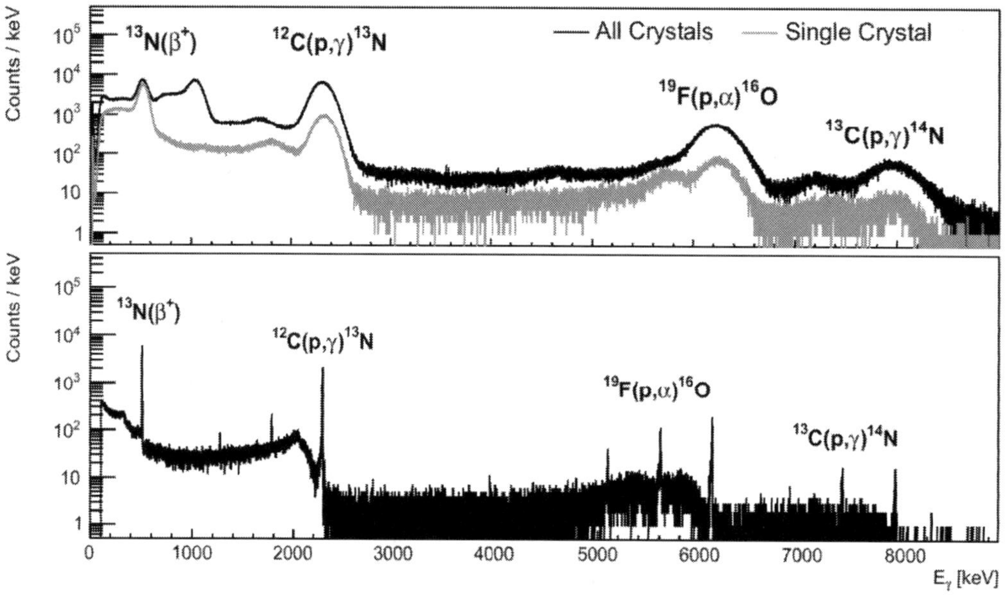

Figure 3: Examples of the spectra acquired at E_p = 380 keV. The upper panel shows the spectrum from the BGO. The lower panel shows the HPGe spectrum.

For the BGO data, the major part of the analysis focused on the β^+ decay of ^{13}N from the reaction. Since the detector is segmented in 6 different crystals, it was possible to exploit the distinct signature of the 511 keV γ-rays from positron annihilation by searching for two γ-rays in opposite crystals, thus reducing the background significantly. Since the target was inside the detector both during the irradiation and the counting, the full activation curve was analyzed by fitting the data to the solution of the differential equation that governs the amount of ^{13}N inside the target. This approach not only permitted the measurement of the cross-section down to 70 keV (center-of-mass) but also to check for any potential contaminant in the target. In this case, only thick graphite disks were employed in order to maximize the ^{13}N production. Additionally, the prompt-γ peak was analyzed for runs with E_p > 180 keV, where the signal was still above the background. At these energies, the yields from prompt-γ and activation measurements were compared and found to be compatible. The efficiency for the activation measurements was determined by using the ^{14}N(p, γ)^{15}O reaction, measuring the β^+ decay of ^{15}O. Additionally, the activation efficiency was verified by ex-situ counting of an activated graphite sample with an HPGe detector. The obtained efficiency value was further validated with the use of a simulation based on Geant4. In the simulation, the positron energy spectrum of the both decays studied here, ^{13}N and ^{15}O, was included; the differences in efficiency between the positrons of both nuclei were found to be small compared to other factors such as a variation of the position of the beam spot on target.

The data analysis for the measurement of ^{12}C(p, γ)^{13}N at LUNA-400 has been completed. The results of the different parts of this experiment are found to be in good agreement between the different targets and detection setups. The obtained cross section data will be presented in a forthcoming publication.

Acknowledgements

D. Ciccotti, the technical staff and mechanical workshops of the LNGS and INFN Divisions of Padua and Naples are gratefully acknowledged for their support for this experiment. This work was supported by INFN. J. S. acknowledges support by the Italian Ministry of Education, University and Research (MIUR) through the "Dipartimenti di eccellenza" project "Physics of the Universe". A. B. was supported by the European Union (ERC-StG *SHADES*, no. 852016; and *ChETEC-INFRA*, no. 101008324).

References

[1] F.L. Schöier, H. Olofsson, Astron. Astrophys. **359**, 586 (2000)
[2] F. Herwig, Annu. Rev. Astron. Astrophys. **43**, 435 (2005)
[3] N. Langer et al., Astron. Astrophys. **346**, L37 (1999)
[4] D. Vescovi et al., Astrophys. J. Lett. **897** (2020)
[5] P.A. Denissenkov, C.A. Tout, Mon. Notices Royal Astron. Soc. **340**, 722 (2003)
[6] M. Agostini et al. (Borexino Collaboration), Nature **587**, 577 (2020)
[7] S. Basu, H.M. Antia, Phys. Rep. **457**, 217 (2008)
[8] G.D. Orebi Gann, K. Zuber, D. Bemmerer, A. Serenelli, Annu. Rev. Nucl. Part. Sci. **71**, 491 (2021)
[9] J.L. Vogl, Ph.D. thesis, California Institute of Technology (1963)
[10] C. Rolfs, R. Azuma, Nucl. Phys. A **227**, 291 (1974)
[11] N. Burtebaev et al., Phys. Rev. C **78**, 035802 (2008)
[12] C.L. Bailey, W.R. Stratton, Phys. Rev. **77**, 194 (1950)
[13] W.A.S. Lamb, R.E. Hester, Phys. Rev. **107**, 550 (1957)
[14] A. Formicola et al., Nucl. Instrum. Methods Phys. Res. A **507**, 609 (2003)
[15] G.F. Ciani, L. Csedreki et al., Eur. Phys. J. A **56** (2020)
[16] A. Boeltzig et al., J. Phys. G **45**, 025203 (2018)
[17] C. Casella et al. (LUNA Collaboration), Nucl. Instrum. Methods Phys. Res. A **489**, 160 (2002)
[18] G. Imbriani et al. (LUNA Collaboration), Eur. Phys. J. A **25**, 455 (2005)
[19] S. Daigle et al., Phys. Rev. C **94**, 025803 (2016)
[20] C. Iliadis et al., Nucl. Phys. A **512**, 509 (1990)

Precision measurements of the $^{12}C(\alpha,\gamma)$ reaction with gamma beams and TPC detector(s)

Moshe Gai [1*]

[1] LNS at Avery Point, University of Connecticut, 1084 Shennecossett Rd, Groton, CT 06340, USA

Abstract. The cross section of the $^{12}C(\alpha,\gamma)^{16}O$ reaction, well-recognized as the most important nuclear input of stellar evolution theory, has eluded precise measurements over the last few decades, as evident in the lack of precise confirmation of an elementary prediction of quantum mechanics of the *E1*-*E2* mixing phase angle (ϕ_{12}). More troubling are modern data, measured with 4π array of HPGe detectors, that are in conflict with this prediction. A new method to measure this cross section with TPC detectors and gamma-beams, by measuring the time reverse $^{16}O(\gamma,\alpha)^{12}C$ reaction discussed here, shows great promise, for example the measured data do not disagree with the predicted ϕ_{12} and agree with the predicted trend.

1 Introduction

In stellar helium burning carbon and oxygen are formed and the C/O ratio is determined by the single process of the fusion of ^{12}C with an alpha-particle to form ^{16}O; the $^{12}C(\alpha,\gamma)$ reaction. This reaction needs to be extrapolated to stellar conditions of a stellar plasma with temperature of kT = 20 keV, and the most efficient burning energy of 300 keV (the Gamow peak). The reaction is dominated by the two partial waves of l = 1 and 2 that are denoted by the spectroscopic *E1* and *E2* amplitudes, respectively. The challenge in this field are precise measurements of angular distributions of the $^{12}C(\alpha,\gamma)$ reaction from which the *E1* and *E2* cross sections and amplitudes are extracted, for accurate extrapolation to stellar conditions.

The *E1* and *E2* amplitudes interfere with a mixing phase angle (ϕ_{12}) that require precisely measured angular distributions. The phase angle was calculated at first in the frame of R-Matrix theory [1], $\phi_{12} = \delta_2 - \delta_1 + \text{atan}(\eta/2)$, where δ_2 and δ_1 are the d- and p- waves phase shifts and η is the Sommerfeld parameter. However, more recently, Brune [2] and Gai [3] noted that the theoretical prediction of ϕ_{12} is a more general consequence of the Watson theorem, which is derived assuming the unitarity of the scattering matrix. Hence the predicted ϕ_{12} is an elementary consequence of quantum mechanics, that must be observed in all measured angular distributions.

* Corresponding author: moshe.gai@uconn.edu
Supported by USDOE grant Number DE-FG02-94ER40870

Fig. 1, prepared by Wolfgang Hammer for his retirement presentation of a review of the field (2006), demonstrates the state of affairs then and now. So far, no accurate measurement of ϕ_{12} was achieved over the resonance energy region, Ecm = 2.0 – 2.6 MeV, where the cross section is large and ϕ_{12} exhibit large variations. More troubling is the disagreement of the EUROGAM data [4], where all the 9 low energy data points that are measured with high precision, disagree with the ϕ_{12} value labeled as R-matrix fit, which in fact, is also predicted by unitarity [2,3]. We note that as shown in Table I of [4] the extracted *E1/E2* values vary by up to a factor of 5, depending whether they perform a three parameter free fit (*E1, E2* and ϕ_{12}), or fix ϕ_{12} for a two parameter fit. Indeed, such large variations of the *E1/E2* values were also noted by Gai [3] in his reanalysis of the EUROGAM data [4].

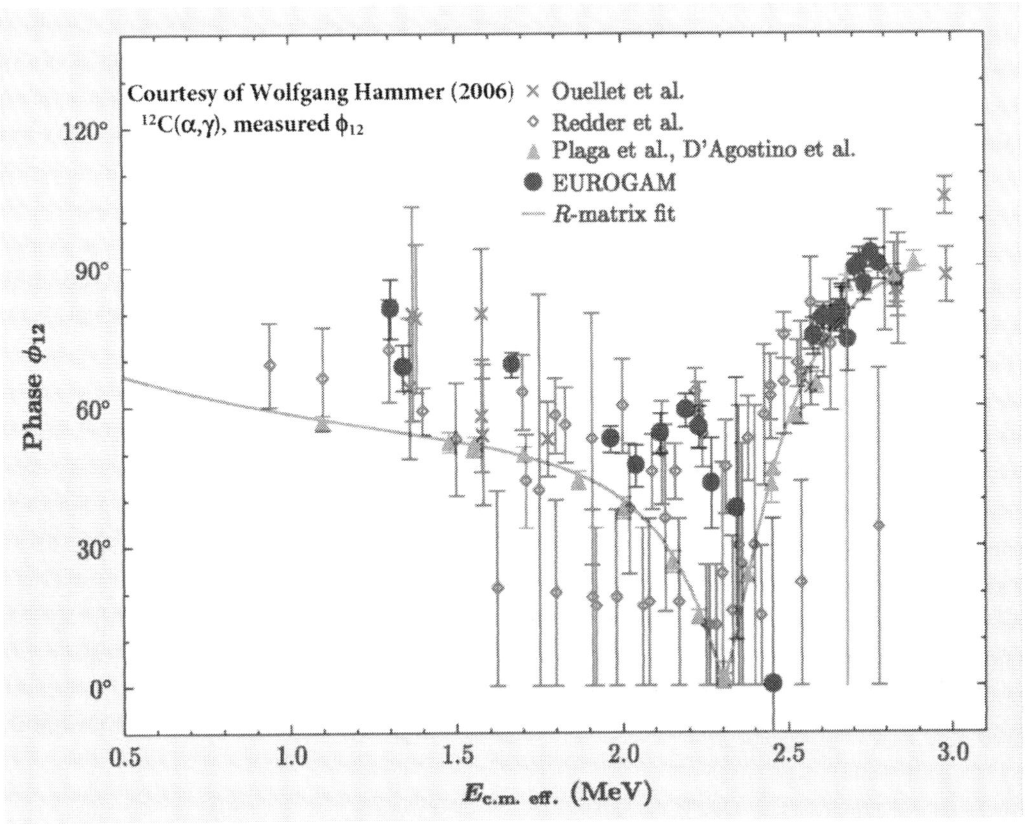

Fig 1: World data of measured ϕ_{12} compiled by Wolfgang Hammer (2006) [4]

2 The UConn-TUNL O-TPC Detector

The UConn-TUNL O-TPC was described in [5] and the analyses of the 2012 O-TPC data "on tape", that was performed during the COVID19 pandemic, are discussed in [6]. Briefly, we measure the ^{12}C(α,γ) reaction, by measuring the time reverse reaction of the photo-dissociation of ^{16}O in the ^{16}O(γ,α) reaction. As shown in Fig. 2 taken from [6], gamma-beams produced by backward scattering of free electron laser (FEL) photons from the GeV electrons in the ring, produce the HIγS quasi-monochromatic gamma-beam. The gamma-beam enters the O-TPC after a collimator 11 mm in diameter and 15 cm long. The O-TPC was operated with pure (grade 5) CO_2 (80%) + N_2 (20%) gas mixture at 100 torr. Tracks of alpha-particles and light nuclei emanating from the photo-dissociations of $^{16,17,18}O$ and ^{12}C were recorded in an array of PMT and a CCD camera, as shown in Fig.2.

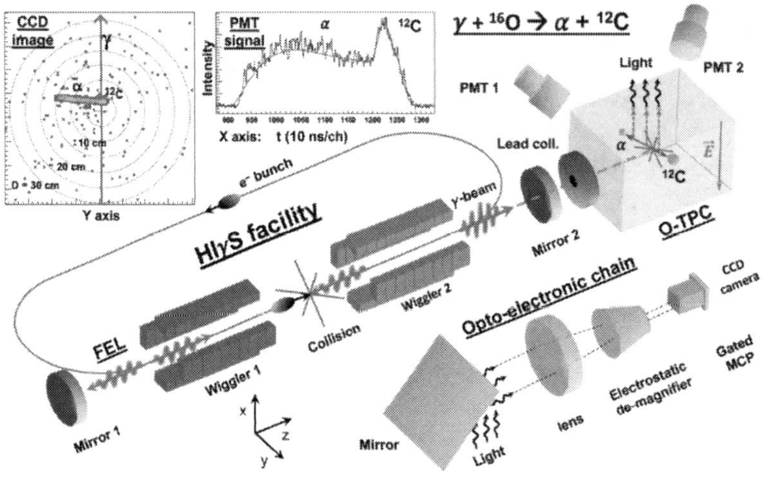

Fig. 2: A schematic diagram of the HIγS gamma-beam system, the O-TPC detector and the opto-electronic readout chain, taken from [6]. A Typical track recorded by the CCD camera and the PMT signals, are shown together with the line shape analysis of the PMT time projection signal.

3 Results

One of the many achievements of the analyses of the O-TPC data, is the automated line shape analyses that allowed separating the ^{12}C and ^{16}O dissociation events. In our automated analysis that is described in great length in our Nature paper [6], the measured time projection PMT signals are compared to a table of calculated line shapes, 180 line shapes for the ^{16}O dissociation events and 1,000 line shapes for the ^{12}C dissociation events. The resulting so-defined "reduced χ^2/ν values" are shown in Fig. 3. The ^{12}C and ^{16}O dissociation events are very well separated from each other and are measured with a large signal to background ratio, as high as 20:1, as shown in Fig. 3.

A major advantage of using the O-TPC is that all data points are measured in one detector, with very low background, if any. Hence the relative efficiencies of data points are very simple to simulate using standard Monte Carlo techniques. The accurate angular distributions measured with the O-TPC, allowed us to measure the *E1* and *E2* amplitudes and perform "phase shift like" analyses of the *E1-E2* interference angle (ϕ_{12}). We measure ϕ_{12} with an accuracy of $\pm 6°$, or better.

In contrast to the data shown in Fig. 1 our recently extracted ϕ_{12}, measured with low statistics [6], *is not in disagreement with the predicted* ϕ_{12}, *and it follows the predicted trend*, as shown in Fig. **4**. The low statistics measurement of the O-TPC "data on tape" [6] was repeated in our measurements at HIγS in August 2022 with considerably higher statistics [7], using the Warsaw eTPC detector. The analyses of these data are in progress and initial results were shown in this conference [7].

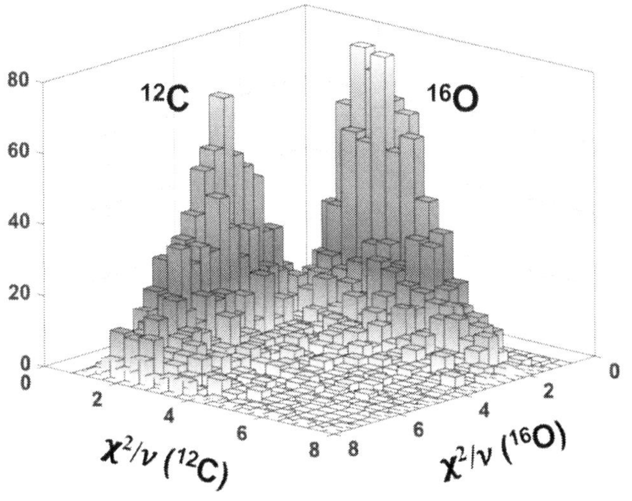

Fig. 3: Two-dimensional Lego plot of the lowest value of the so-defined "reduced χ^2/ν values" when fitting the measured PMT line-shape with the predicted line shapes of ^{16}O and ^{12}C dissociation events, taken from [6].

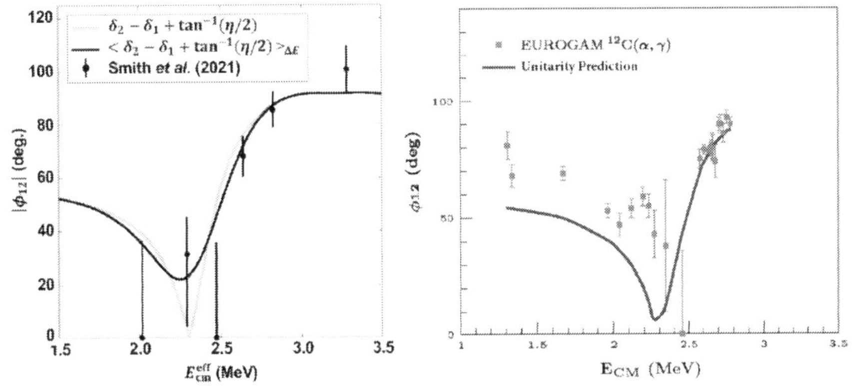

Fig. 4: The *E1-E2* mixing phase angle (ϕ_{12}) measured by the EUROGAM collaboration [4], and our low statistics O-TPC results [6]. New higher statistics data were measured in August 2022, with the Warsaw eTPC [7]. Analysis of the new data is in progress.

4 Conclusions

Our newly developed method for measuring the cross section of the $^{12}C(\alpha,\gamma)$ reaction with TPC(s) and gamma-beams shows great promise for progress in the field. New data measured with Warsaw eTPC have been extended down to $E\gamma = 8.51$ MeV [7]. We plan to continue to measure further down with lower gamma-beam energies.

Reference

[1] F.C. Barker and T. Kajino, Aust. J. Phys. **44**, 369 (1991).
[2] C.R. Brune, Phys. Rev. C 64, 055803 (2001).
[3] Moshe Gai. Phys. Rev. C **88**, 062801(R) (2013).
[4] M. Assunção, *et al.*, Phys. Rev. C 73, 055801 (2006).
[5] M. Gai *et al.*, Jour. Instr. **5**(2010)12004.
[6] R. Smith *et al.*, Nature Communications **12**, 5920 (2021).
[7] Mikołaj Cwiok *et al.*, contribution to this conference, 2022.

Studies of photo-nuclear reactions at astrophysical energies with an active-target TPC

Mikołaj Ćwiok[1], *Wojciech* Dominik[1], *Aleksandra* Fijałkowska[1], *Mateusz* Fila[1], *Zenon* Janas[1], *Artur* Kalinowski[1], *Krzysztof* Kierzkowski[1], *Magdalena* Kuich[1], *Chiara* Mazzocchi[1,*], *Wojciech* Okliński[1], *Marcin* Zaremba[1], *Moshe* Gai[2], *Deran K.* Schweitzer[2], *Sarah R.* Stern[2], *Sean* Finch[3,4], *Udo* Friman-Gayer[3,4,**], *Samantha R.* Johnson[5,4], *Tyler M.* Kowalewski[5,4], *Dimiter L.* Balabanski[6], *Catalin* Matei[6], *Adrian* Rotaru[6], *Kristian C.Z.* Haverson[7], *Robin* Smith[7], *Ross A.M.* Allen[8], *Mark R.* Griffiths[8], *Stuart* Pirrie[8], and *Pedro* Santa Rita Alcibia[8]

[1]Faculty of Physics, University of Warsaw, Warsaw, Poland
[2]University of Connecticut, CT, USA
[3]Physics Department, Duke University, Durham, NC, USA
[4]Triangle Universities Nuclear Laboratory, Durham, NC, USA
[5]Department of Physics & Astronomy, University of North Carolina, Chapel Hill, NC, USA
[6]IFIN-HH / ELI-NP, Bucharest-Magurele, Romania
[7]Department of Engineering and Mathematics, Sheffield Hallam University, Sheffield, UK
[8]School of Physics and Astronomy, University of Birmingham, Birmingham, UK

Abstract. An experiment was conducted at the High Intensity γ-ray Source (HIγS) facility at the Triangle Universities Nuclear Laboratory (TUNL) in Durham, NC, USA to measure the cross-section of the key astrophysical thermonuclear reaction $^{12}C(\alpha,\gamma)^{16}O$ by means of its inverse photo-disintegration process. A high-intensity monochromatic γ-ray beam interacted with the CO_2 gas in the active volume of the Warsaw active target TPC detector. The reaction products were detected and their momenta reconstructed, so to also determine angular correlations. Data were collected at 15 beam energies ranging from 8.51 to 13.9 MeV. Preliminary results are presented.

1 Introduction

One of the fundamental questions in nuclear astrophysics is the creation of oxygen and carbon in the Universe. These two elements are not only the third and fourth most abundant elements in the Universe, respectively, but also the two most abundant elements in the human body. Our understanding of their formation mechanisms, and consequently of the stars' life cycle, is therefore important for understanding the origins of life on Earth.
Synthesis of carbon and oxygen happens in helium-burning thermonuclear reactions in the star's core, among which are the triple-alpha reaction and $^{12}C(\alpha,\gamma)^{16}O$. Star evolution models require, as input, details of the reaction rates as a function of the temperature/environmental conditions. The fundamental observable to determine the reaction rates is the cross-section,

*e-mail: chiara.mazzocchi@fuw.edu.pl
**Present address: European Spallation Source ERIC, Partikelgatan 2, 22484 Lund, Sweden

which has to be determined at the relevant energies (Gamow peak) [1]. Measuring the reaction probability at the Gamow peak is not possible, therefore extrapolations based on theoretical models, typically R-matrix, have to be employed. In order both to validate the models and constrain the extrapolations/R-matrix fits, hence reducing the uncertainties, accurate measurements at energies as-close-to the Gamow peak as-experimentally-achievable have to be carried on.
In this work, we focus on the measurement of the key helium-burning reaction $^{12}C(\alpha,\gamma)^{16}O$ at low energies, i.e. below 2 MeV in the centre-of-mass (c.o.m.), which controls the carbon-to-oxygen ratio in the Universe. From the nuclear physics point of view, two reaction mechanisms are available in the Gamow window region (300 keV): non-resonant direct capture and capture into the tail of nearby 1^- or 2^+ resonances in ^{16}O (-0.045, -0.245, 2.42 and 2.68 MeV). In stellar environments, capture via ℓ=1 or ℓ=2 will dominate, hence not only the total E1+E2 cross section needs to be determined, but also its E1 and E2 components.

2 Methodology

Measuring the reaction cross-section for $^{12}C(\alpha,\gamma)^{16}O$ at or close to the Gamow peak for helium-burning may still be too big a challenge from the experimental point of view; nevertheless, extrapolations with the help of R-matrix fits can help overcome this difficulty. The faster-than-exponential decrease of the cross section with decreasing c.o.m. energy (E_{cm}) makes such measurements particularly challenging, especially below 2 MeV.
The limitations in beam-intensity and target-thickness combinations, coupled with the large beam-induced γ background from contaminant reactions, such as $^{13}C(\alpha,n)^{16}O$, can be partially overcome thanks to the invariance of the strong and electro-magnetic interactions with respect to time-reversal. If, instead of the direct capture reaction $^{12}C(\alpha,\gamma)^{16}O$, its time-reversal photo-disintegration $^{16}O(\gamma,\alpha)^{12}C$ is measured, a gain of a large factor in cross section can be achieved, thanks to the principle of detailed balance in nuclear reactions,

$$\sigma_{\alpha\gamma} \cdot \omega_{\alpha\gamma} \cdot p^2_{\alpha\gamma} = \sigma_{\gamma\alpha} \cdot \omega_{\gamma\alpha} \cdot p^2_{\gamma\alpha}, \quad (1)$$

where $\sigma_{\alpha\gamma}, \omega_{\alpha\gamma}, p_{\alpha\gamma}, \sigma_{\gamma\alpha}, \omega_{\gamma\alpha}, p_{\gamma\alpha}$ are the cross section, spin factors, and momenta for the (α,γ) and (γ,α) reactions, respectively. At $E_{cm}\sim1$ MeV, the gain in cross section is of the order of 40. The advantages of such an approach, in addition to the gain in cross-section, are the inherent low-background, the negligible target deterioration, the different and negligible beam-induced background, the different definition of effective beam energy and, consequently, the different systematics.
In order to study the $^{16}O(\gamma,\alpha)$ reaction, intense, monochromatic and well collimated γ-ray beams with energies in the range 8–15 MeV are needed in tandem with a detection set-up that allows reconstruction of the momenta of all the reaction products with close to 100% efficiency. It was shown [2] that a very good tool is an active-target time-projection chamber (active-target TPC), operated below atmospheric pressures (densities). The Warsaw active-target TPC was developed for studying nuclear photo-disintegration reactions where the products are charged particles, at astrophysically-relevant energies. It allows for full and unambiguous reconstruction of the momentum for multiple-particle events [3, 4], hence not only to determine the type of particles, and their energy, but also angular distributions to disentangle the contributions of the different multipolarities.
The Warsaw TPC has an active volume of 33 × 20 × 20 cm^3 and is immersed in a vacuum vessel equipped with a control system to maintain a constant gas pressure inside. The Warsaw TPC detector is composed of a cathode, a drift cage consisting of 12 aluminium electrodes, 2 mm-thick and 16 mm apart, for a total drift-length of 196 mm, an amplification structure

consisting of three 50 μm-thick Gas Electron Multiplier (GEM) foils, and a planar electrode (anode) segmented along 3 axes to obtain 3 independent linear sets of strips (U, V, W) with 1.5 mm pitch and oriented 60° with respect to each other for redundant readout. Such readout requires only about 1000 channels of electronics. The arrays of (U, V, W) strips enable dis-entangling the hit in 2 dimensions by generating virtual pixels. The time distribution of the charge collected at the electrode, combined with the drift velocity of the electrons in the given gas mixture and field in the drift cage, allows the determination of the third coordinate. The gas composition and density (pressure) are optimised for the reaction of interest. The data acquisition is based on the General Electronics for TPCs (GET) [5] complemented by a customised FPGA readout adaptation, which has been developed at the University of Warsaw. It allows to use sampling frequencies from 1 to 100 MHz (typically 12.5 or 25 MHz are used) and has adjustable gain and filtering for each channel. An external trigger can be used.

3 First measurements: ^{16}O photo-disintegration experiment

The first measurements were conducted in 2022 at the High Intensity γ-Ray Source (HIγS) facility at the Triangle University Nuclear Laboratory (TUNL) in Durham, NC, USA [6, 7]. Monochromatic γ-ray beams with energies ranging from 8.51 to 13.9 MeV were produced in Compton back-scattering of photons from a free-electron laser (FEL) with a relativistic electron beam. The beam energy was modulated by tuning the FEL wavelength and the electron beam. A collimator with diameter 10.5 mm was used to shape the beam spot on the target. The chamber was aligned with respect to the beam by means of a laser collinear with the beam trajectory as well as using attenuated beam and a γ camera for fine tuning of the alignment. Vertical alignment was such that the beam fit fully in between two consecutive electrodes of the TPC, hence avoiding generating background through the interaction of the γ beam with the electrodes.
The beam intensity and its variation in time was monitored throughout the experiment and determined through scintillation counters. Absolute calibration of the event rate determined by the scintillation detectors was obtained by means of (γ,n) activation measurements on ^{197}Au targets conducted synchronously with data taking with the chamber and with the auxiliary beam monitoring detectors. Preliminary analysis of the data yielded intensities well above $1.3 \cdot 10^8$ γ/s throughout the experiment, with a maximum of a few 10^8 γ/s at $E_\gamma < 10$ MeV.
The beam energy, E_γ, was determined by impinging the attenuated γ beam into a calibrated HPGe detector, the spectrum was then deconvolved to determine the beam energy and energy spread, which was around 350 keV at $E_\gamma = 8.5$ MeV. Such measurements were conducted every 8-12 hours to monitor the stability with time.
The data acquisition was triggered externally by using the integrated charge from the anode of the last GEM foil in the amplification stage.
The active gas in the chamber was pure CO_2, which allowed for the simultaneous measurement of ^{12}C photo-disintegration, with the pressure value chosen to optimise the length of the trajectories for the α and ^{12}C reaction products in the chamber. The CO_2 gas had natural isotopic composition, hence reactions on ^{12,13}C and ^{16,17,18}O isotopes are possible. ^{12}C and ^{16}O photo-disintegration reactions were the subject of the experiment, while ^{13}C and ^{17,18}O photo-disintegration would constitute beam-induced background. Photo-disintegration of ^{12}C and ^{16}O can be distinguished on the basis of the event topology, i.e. 3-particles vs. 2-particles disintegration, respectively. In the case of ^{13}C and ^{17,18}O, the (γ,α) reaction channels present very different Q-values with respect to ^{12}C and ^{16}O, respectively, and can be distinguished after analysis on the basis of the reconstructed momenta, see Section 4. The (γ,n) channel in ^{13}C and ^{17,18}O can be identified on the basis of the event topology, given the non-sensitivity of the detector to neutrons, while the (γ,p) channel in ^{16}O, which opens at E_γ = 12.1 MeV,

can be distinguished thanks to the different Q-values combined with the different stopping power of protons and α particles.

4 Preliminary results

A preliminary analysis of the data was conducted on a portion of the statistics collected. All data were collected and analysed with dedicated software developed at the University of Warsaw. Events were visually inspected in order to assign them to categories according to their respective topology (1-, 2-, 3-particle events, background and noise). Background consisted of alpha particles from radon in the environment and of recoils following (γ,n) reactions on the CO_2 atoms. Such events are easily separated from the rest.
Those events classified as two- or three-particles were analysed further to determine the momenta of the particles. Two example events are presented in Fig. 1. The 2-particle case shown in Fig. 1 (left) is identified as a ^{16}O(γ,α)^{12}C reaction, while the 3-particle case in Fig. 1 (right) corresponds to a ^{12}C(γ,3α) event, both at E_γ = 8.66 MeV. Assignment of the two-particle event to ^{16}O(γ,α)^{12}C or to ^{17,18}O(γ,α)^{13,14}C is done by comparing the track length for two-particle events. In Fig. 2 the histogram of the sum of the α and ^{12}C track length is shown for E_γ = 8.66 and 11.5 MeV.
Momentum reconstruction of the particles emitted in the reaction also allows a reconstruction of the angular distribution of the events. In Fig. 3 (left), the polar angular distribution is shown for events identified as ^{16}O(γ,α) at 11.5 MeV beam energy. A distribution characteristic for E2 multipolarity is observed, as expected for the 11.5 MeV 2^+ resonance in ^{16}O.
The γ beam delivered to the measuring station was circularly polarised to a degree well above 90%, as determined by measurements of the polarisation of the laser beam [8]. The reconstruction of the momenta of the reaction products for 2-particle events allowed for confirmation of the degree of circular polarisation of the γ beam, see Fig. 3 (right). Preliminary evaluation gives a degree of circular polarisation in good agreement with expectations [8].

5 Summary and outlook

An active-target time-projection chamber operating at pressures below atmospheric was developed at the University of Warsaw for studying photo-disintegration reactions of astrophysical interest at or close-to the relevant energies. The methodology exploits the principle of

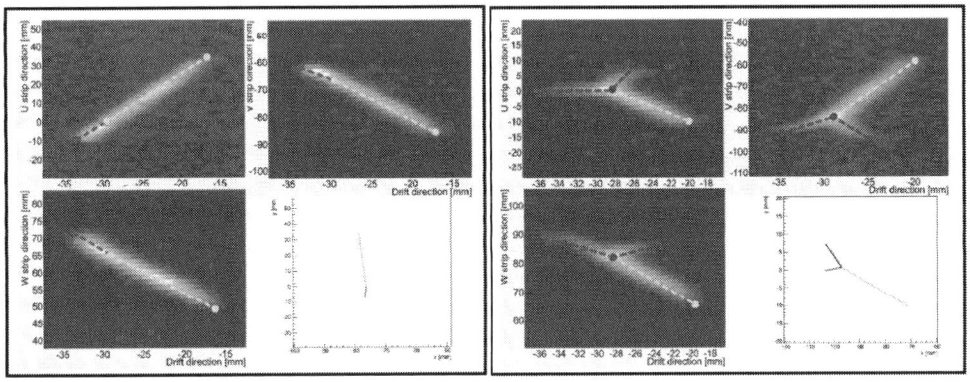

Figure 1. Event reconstruction example for an ^{16}O (left) and a ^{12}C (right) photo-disintegration event at E_γ=8.66 MeV. The raw data for the charge collected by each strip (U, V, W) strip is plotted as a function of strip position and time (both rescaled to mm), while the bottom-right plot shows the projection of the reconstructed tracks on the anode plane.

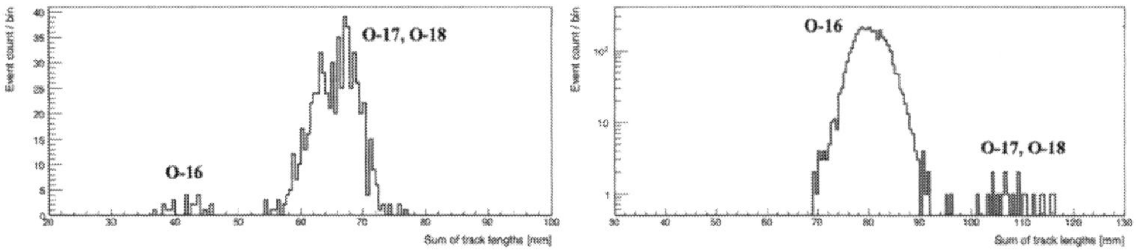

Figure 2. Identification plot, based on the sum of the α and ^{12,13,14}C track lengths for two-particle events at E_γ = 8.66 MeV (left) and 11.5 MeV (right). Partial statistics and laboratory reference frame.

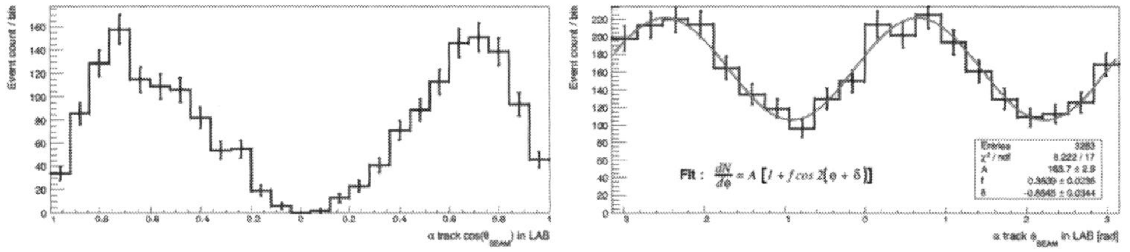

Figure 3. Polar (θ) and azimuthal (ϕ) angle distributions (left and right panels, respectively) in the laboratory reference frame for ^{16}O(γ,α) events at E_γ=11.5 MeV. Partial statistics. See text.

detailed balance in nuclear reactions to determine the cross section of ^{12}C(α,γ)^{16}O by measuring that of the inverse reaction. The first experiments were conducted in 2022 at the HIγS facility at TUNL, where intense, monochromatic γ-ray beams were made to impinge on the active gas target inside the TPC and the reaction products were detected. Data were collected for beam energies ranging from 8.51 to 13.9 MeV. Preliminary analysis shows that the complete evaluation will provide the cross sections as well as angular distributions, to determine the E1 and E2 contributions. Future experiments are being planned to increase the statistics and reach lower energies.

We would like to thank the HIγS staff for delivering high-quality beam throughout the experiments and for the good working conditions during the preparation phases. We also would like to thank H. Czyrkowski and R. Dąbrowski for their support in the preparation of the equipment. Scientific work supported by the National Science Centre, Poland, contract no. 2019/33/B/ST2/02176, by the US Department of Energy, Office of Science, Office of Nuclear Physics, grant no. DE-FG02-94ER40870 and DE-FG02-97ER41033, by the UK STFC, grant no. ST/V001086/1, by the Romanian Ministry of Research and Innovation under contract PN 19 06 01 05, by the the University of Warsaw, Poland, through the Interdisciplinary Centre for Mathematical and Computational Modelling, comp. alloc. no. G89-1286 and through the Excellence Initiative Research University (IDUB) program.

References

[1] W.A. Fowler, Rev. Mod. Phys. **56**, 149 (1984)
[2] R. Smith et al. Nature Communications 12, 5920 (2021)
[3] M. Ćwiok, Acta Phys. Pol. B **47**, 707 (2016)
[4] M. Ćwiok et al., Acta Phys. Pol. B **49**, 509 (2018)
[5] E. Pollacco et al., Nucl. Instr. and Methods **A 887**, 81 (2018)
[6] H.R. Weller et al., Prog. Part. Nucl. Phys. 62, 257 (2009)
[7] https://tunl.duke.edu/research/our-facilities (12.X.2022)
[8] Y.K. Wu, priv. comm. (2022)
[9] M. Wang et al., Chin. Phys. C **45**, 030003 (2020)

Coulomb dissociation of ^{16}O into ^{4}He and ^{12}C

Lukas Thomas Bott[1,*], *Kathrin* Göbel[2,1], *Michael* Heil[3], *Aleksandra* Kelić-Heil[3], *René* Reifarth[1], *Marialuisa* Aliotta[1,4], *Tahani* Almusidi[5,6], *Hector* Alvarez-Pol[7], *Leyla* Atar[8], *Liam* Atkins[5], *Thomas* Aumann[8,3], *Daniel* Bemmerer[9], *José* Benlliure[7], *Carlos* Bertulani[10], *Konstanze* Boretzky[3], *Benjamin* Brückner[1], *Leonhard* Brandenburg[1], *Giovanni* Bruni[23], *Pablo* Cabanelas Eiras[7], *Christoph* Caesar[3], *Enrique* Casarejos[11], *Joakim* Cederkall[12], *Leonid* Chulkov[13], *Dolores* Cortina-Gil[7], *Andrey* Danilov[13], *Enrico* De Filippo[14], *Sophia Florence* Dellmann[1], *Isabell* Deuter[1], *José Antonio* Dueñas Díaz[15], *Meytal* Duer[8], *Zoltan* Elekes[16], *Philipp* Erbacher[1], *Sonia* Escribano Rodriguez[5], *Zsolt* Fülöp[16], *Ashton* Falduto[8], *Manuel* Feijoo[7], *Stefan* Fiebiger[1], *Igor* Gašparić[17], *Daniel* Galaviz[18], *María José* García Borge[19], *Gabriel* García-Jiménez[7], *Elena* Geraci[20], *Roman* Gernhäuser[21], *Jan* Glorius[3], *Brunilde* Gnoffo[20], *David* González Caamaño[7], *Antia* Graña González[7], *Alexander* Grein[1], *Anna-Lena* Hartig[8], *Tanja* Heftrich[1], *Henning* Heggen[3], *Marcel* Heine[22], *Andreas* Heinz[23], *Corinna* Henrich[8], *Thomas* Hensel[9,24], *Matthias* Holl[8,23], *Ilja* Homm[8], *Ákos* Horváth[25], *Andrea* Horvat[8], *Andrea* Jedele[8], *Desa* Jelavic Malenica[8], *Tobias* Jenegger[21], *Håkan T.* Johansson[23], *Björn* Jonson[23], *Julian* Kahlbow[8], *Nasser* Kalantar-Nayestanaki[26], *Armel* Kamenyero[27], *Kafa* Khasawneh[1], *Oleg* Kiselev[3], *Philipp* Klenze[21], *Marco* Knösel[8], *Karsten* Koch[3], *Marvin* Kohls[1], *Daniel* Körper[3], *Thorsten* Kröll[8], *Sabina* Krasilovskaja[1], *Dmytro* Kresan[3], *Deniz* Kurtulgil[1], *Nikolaus* Kurz[3], *Bastian* Löher[3], *Christoph* Langer[1,28], *Claudia* Lederer-Woods[4], *Christopher* Lehr[8], *Yuri A* Litvinov[3], *Enis* Lorenz[8], *Nunzia Simona* Martorana[29], *Tohru* Motobayashi[30], *Silvia* Murillo Morales[5], *Enrique* Nacher[31], *Thomas* Nilsson[23], *Emanuele Vincenzo* Pagano[29], *Valerii* Panin[3], *Joochun* Park[12,32], *Stefanos* Paschalis[5], *Angel* Perea[19], *Marina* Petri[5], *Sara* Pirrone[14], *Ralf* Plag[3], *Lukas* Ponnath[21], *Romana* Popočovski[17], *Markus* Reich[1], *Han-Bum* Rhee[8], *Jose Luis* Rodriguez Sanchez[7], *Dominic* Rossi[3,8], *Paolo* Russotto[29], *Ángel-Miguel* Sánchez-Benítez[15], *Christian* Sürder[8], *Deniz* Savran[3], *Heiko* Scheit[8], *Konrad* Schmidt[9], *Hendrik* Schulte[1], *Haik* Simon[3], *Johannes* Simon[8], *Viktor* Starostin[13], *Sonja* Storck-Dutine[8], *Hans Toshihide* Törnqvist[3,8], *Junki* Tanaka[8], *Olof* Tengblad[19], *Benedikt* Thomas[1], *Marina* Trimarchi[14,33], *Stefan* Typel[8,3], *László* Varga[3], *Klaus* Volk[1], *Meiko* Volknandt[1], *Vadim* Wagner[8], *Felix* Wamers[3], *Mario* Weigand[1], and *Lorenzo* Zanetti[8]

[1]Goethe Universität Frankfurt, Germany
[2]Facility for Antiproton and Ion Research in Europe GmbH, Germany
[3]GSI Helmholtzzentrum für Schwerionenforschung Darmstadt, Germany
[4]University of Edinburgh, United Kingdom
[5]University of York, United Kingdom
[6]King Saud University, Saudi Arabia
[7]University of Santiago de Compostela, Spain
[8]Technische Universität Darmstadt, Germany
[9]Helmholtz-Zentrum Dresden-Rossendorf, Germany
[10]Texas A&M University-Commerce, USA
[11]CINTECX, Universidade de Vigo, Spain
[12]Lund University, Sweden

[*]e-mail: bott@iap.uni-frankfurt.de

[13]NRC Kurchatov Institute Moscow, Russia
[14]INFN-Sezione di Catania, Italy
[15]Universidad de Huelva, Spain
[16]ATOMKI Debrecen, Hungary
[17]Ruđer Bošković Institute, Zagreb, Croatia
[18]University of Lisboa, Portugal
[19]CSIC Madrid, Spain
[20]Università di Catania, Italy
[21]Technische Universität München, Germany
[22]Institut Pluridisciplinaire Hubert CURIEN, France
[23]Chalmers University of Technology, Sweden
[24]Technische Universität Dresden, Germany
[25]Eötvös Loránd University, Hungary
[26]ESRIG, University of Groningen, The Netherlands
[27]GANIL, France
[28]University of Applied Science Aachen, Germany
[29]INFN Laboratori Nazionali del Sud, Italy
[30]RIKEN, Nishina Center for Accelerator-Based Science, Japan
[31]IFIC, CSIC-Universidad de Valencia, Spain
[32]Center for Exotic Nuclear Studies, Institute for Basic Science, Republic of Korea
[33]Università degli Studi di Messina, Italy

Abstract. We measured the Coulomb dissociation of ^{16}O into ^{4}He and ^{12}C within the FAIR Phase-0 program at GSI Helmholtzzentrum für Schwerionenforschung Darmstadt, Germany. From this we will extract the photon dissociation cross section ^{16}O(α,γ)^{12}C, which is the time reversed reaction to ^{12}C(α,γ)^{16}O. With this indirect method, we aim to improve on the accuracy of the experimental data at lower energies than measured so far.
The expected low cross section for the Coulomb dissociation reaction and close magnetic rigidity of beam and fragments demand a high precision measurement. Hence, new detector systems were built and radical changes to the R^{3}B setup were necessary to cope with the high-intensity ^{16}O beam. All tracking detectors were designed to let the unreacted ^{16}O ions pass, while detecting the ^{12}C and ^{4}He.

1 The ^{12}C(α,γ)^{16}O fusion reaction

During their evolution stars will undergo several burning phases and reach eventually the so called triple-alpha process where three ^{4}He nuclei are fused into carbon. At the same time several other processes may take place, whereby the fusion of carbon and helium into oxygen is of particular interest. DeBoer [1] and Aliotta [2] give an excellent overview of the recent developments. The ^{12}C(α,γ)^{16}O fusion reaction determines the carbon-to-oxygen ratio at the end of the helium burning phase and, hence, the further evolution and fate of the star. Furthermore, this reaction is also crucial for the carbon-to-oxygen ratio in the universe and thus for the creation of the essential elements for life as we know it on earth.

The fate of stars in the mass range between about 14 $M_\odot$ and 60 $M_\odot$ is still unknown because of the uncertainty of the ^{12}C(α,γ)^{16}O reaction [3, 4]. To generate reliable nucleosynthesis simulations for these astrophysical purposes, the reaction rate of ^{12}C(α,γ)^{16}O in the astrophysical relevant energy region around 300 keV has to be known with uncertainties of less than 10%.

Theoretical calculations and extrapolation from higher energies result in an extremely low value of about 10^{-17} b, which poses a significant challenge for direct measurements. Available experimental data down to 1 MeV show sometimes large uncertainties of up to 100% [1].

With their larger cross sections indirect measurements promise to improve the experimental data quality in the low center-of-mass energy region.

2 Coulomb dissociation of ^{16}O at R^{3}B

In order to determine the $^{12}C(\alpha,\gamma)^{16}O$ fusion reaction, there are different experimental approaches available, see e.g. [5, 6]. The indirect method of Coulomb dissociation of ^{16}O is particularly promising and can bridge the gap to the stellar energy regime. Baur, Bertulani and Rebel proposed this measurement in 1986 [7] for the first time and performed detailed calculations later [8, 9]. The Coulomb dissociation cross section is far larger than the direct measurement and profits from the large number of virtual photons if it is performed at relativistic energies.

We measured the Coulomb dissociation of ^{16}O into ^{4}He and ^{12}C in inverse kinematics at the R^{3}B setup, within the FAIR Phase-0 campaign at GSI Helmholtzzentrum für Schwerionenforschung in Darmstadt Germany. A high-intensity ^{16}O beam impinged on a selection of different targets. In the Coulomb field of the target nuclei the ion can be excited and eventually break up into ^{4}He and ^{12}C. At beam energies of 300 MeV/nucleon, using a lead target of 80 mg/cm^2, and an ^{16}O beam of $2{\cdot}10^9$ ions per second we estimated a count rate of 30 counts per day at a center-of-mass energy of $E_{\rm cm}$ = 1 MeV. With the recorded data in the energy range of 800 keV to 5 MeV we expect a significant reduction of statistic uncertainties compared to the available data. Typical systematic uncertainties of Coulomb dissociation measurements in the past were about 10% to 15%. However the systematic uncertainties in our case will be small since we can use the well measured resonances at higher energies for an absolute normalization.

A low Z target like carbon contributes mainly to the nuclear breakup reaction. By subtracting the nuclear contribution from breakup reactions occurring in the lead target we are able to extract the desired Coulomb dissociation reaction cross section. Possible interference effects can be studied by using an intermediate mass target like Sn [10].

2.1 The setup

The magnetic rigidity of the unreacted ^{16}O beam and the fragments ^{12}C and ^{4}He are close since the mass-to-charge-number ratio $A/Z = 2$. The scintillation detectors can measure intensities of up to 10^6 ions per second. To allow the high-intensity unreacted beam to pass the tracking detectors and measure both fragments, radical changes to the R^{3}B setup [11, 12] were necessary. All detectors where designed and positioned so as to let the unreacted ions pass.

Figure 1 shows a sketch of the modified R^{3}B setup. All detectors relevant for tracking, charge identification and intensity measurement are positioned inside a vacuum chamber connected to each other. This allows a direct connection to the accelerator without the use of windows which would drastically increase unwanted reactions with the beam. Two active collimators, ROLU (Rechts-Oben-Links-Unten), in front of the target cut the beam dimension by creating a veto. During the beam setup phase they helped to center the ^{16}O beam on the target. Around the target the CALIFA (CALorimeter for In-Flight detection of gamme-rays and high energy charged pArticles) calorimeter measures γ-rays from excited fragments.

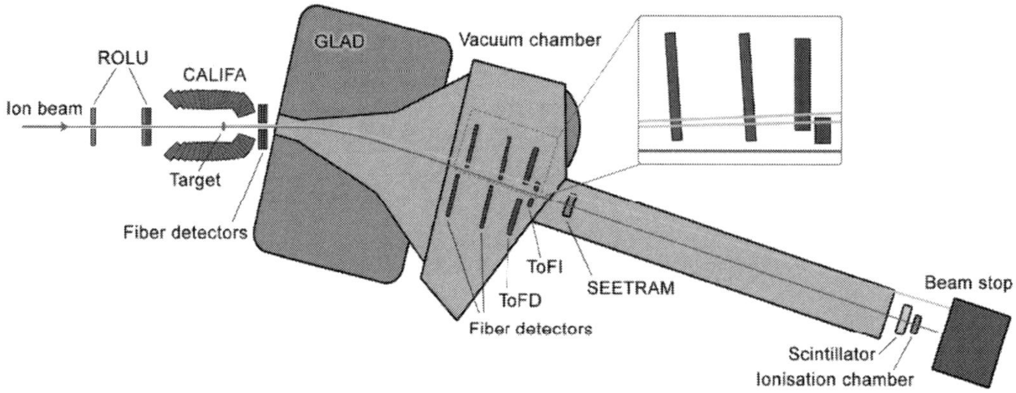

Figure 1. Sketch of the experimental setup at R^3B as it was used for the Coulomb dissociation of ^{16}O. Two active collimators, ROLU, in front of the target cut the beam dimension by creating a veto focus and help to center the ion beam on the target. γ-rays from excited ^{12}C are measured in the CALIFA calorimeter. For tracking of the fragments two fiber detectors in front of the magnet GLAD measure the x- and y-position. Two pairs of fiber detectors downstream from GLAD measure the x-position. The time-of-flight walls ToFD and ToFI allow to extract the charge of the fragments, based on their energy loss. All tracking detectors had slits or a hole and were mounted on drives to allow the unreacted beam to pass without damaging the detectors or causing high dead time by its high intensity. A secondary electron transmission monitor (SEETRAM) detector as well as a scintillator and an ionization chamber measured the beam intensity.

In front of the superconducting magnet GLAD (GSI Large Acceptance Dipole) a pair of 250 μm thick 10x10 cm^2 organic fiber scintillators measures the x- and y-positions of the fragments. Mounted on x/y drives a 3 mm diameter hole can be adjusted to let the unreacted ^{16}O beam pass. The x-position of the deflected fragments are measured with 1000 μm thick 50x50 cm^2 fiber scintillators positioned behind GLAD in two pairs. These detectors are mounted on x-drives to adjust the gap. A small rotation about the y-axis increases their detection efficiency and each fiber is read out by two multianode photomultipliers on each end. The fiber detector setup can be seen in figure 2.

The time-of-flight walls ToFD (Time of Flight Detector) and ToFI (Time of Flight Inner detector) generate the trigger and measure the flight time and charge Z of the fragments. ToFD consists of two layers plastic scintillator bars each 27 mm wide, 5 mm thick and 1000 mm long read out by two single anode photomultipliers from each end. Several bars in the center are dismounted to let the unreacted beam pass. The gap in layer one is three bars wide while in the second layer, which is shifted by the width of half a bar, a gap-width of two bars is sufficient. Mounted in a light-tight housing the active area is roughly 1200x80 cm^2. ToFI consists of one layer with twelve 5 mm square scintillation bars covering the innermost bars of ToFD for higher granularity at high intensities. Six bars on each side of the unreacted beam path are mounted on x-drives to adjust the gap.

A SEETRAM (SEcondary Electron TRAnsmission Monitor) detector, a scintillator and an ionization chamber positioned behind the time-of-flight walls are used to quantify the beam at different intensities. This allows a careful calibration in the overlapping range whereby SEETRAM can measure the highest intensities.

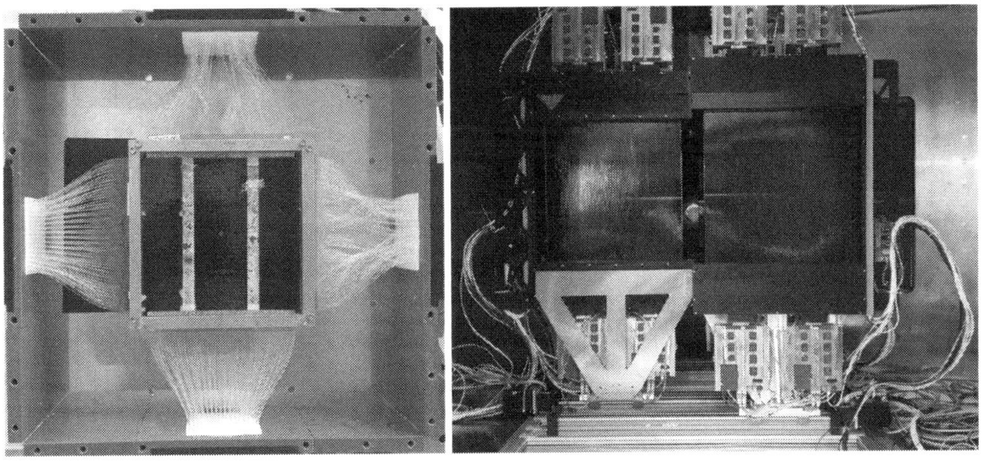

Figure 2. Fiber scintillator detectors for fragment tracking. Left: Two fiber detectors with 250 μm fibers, 10x10 cm^2 active area, and a 3 mm diameter hole, mounted in front of the magnet. Right: Four fiber detectors with 1000 μm fibers 50x50 cm^2 active area mounted inside the vacuum chamber behind GLAD. All detectors are mounted on drives to adjust the hole or gap.

2.2 The experimental campaign

The detection of all relevant fragments is of utmost importance. Therefore, prior to the experiment with ^{16}O beam, ^{4}He beam was taken in May 2021. Sweeping all detectors through the beam ensured careful threshold finding and calibration for particles with low Z. During the experiment in June/July 2021 with ^{16}O beam, the magnetic field was varied for a later reconstruction of the detector positions while tracking the well known momenta of unreacted ions. In total, we recorded 67 TB of data, where 89.6% of the runs were physics runs and the rest tuning and sweep runs. In table 1, the targets used and the corresponding measured hours are listed.

Unfortunately, in the tuning phase, it became clear that the accelerator could not provide the ^{16}O beam at 300 MeV/nucleon in a condition appropriate for the experiment. A switch to 500 MeV/nucleon improved the conditions, but limited the recorded statistics due to a lower number of virtual photons in the region of interest. In addition, beam focus and position were unstable. Therefore, we had to run with thinner targets at lower beam intensities and higher dead times of the data acquisition system than expected, which further reduced the recorded statistics.

Table 1. Targets, relative file size compared to recorded data of 67 TB and measured time. In total, we measured for 247 hours.

Target	Relative size in %	Measured time in h
Pb 38 μm	74.03	167.5
Pb 77 μm	6.39	17.3
C 252 μm	6.85	15.5
C 423 μm	2.85	5.4
Sn 104 μm	5.77	12.1
empty	2.36	22.1
other targets	1.75	7.2

3 Data analysis

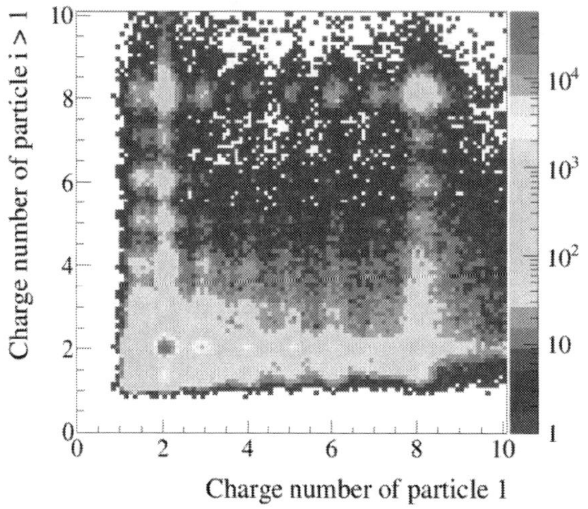

Figure 3. Charge number of first particle measured by ToFD in one event vs. charge number of every other particle in this event.

Figure 3 shows the calibrated charge number of the first hit in ToFD in an event versus all other hits of this event. The Coulomb dissociation reaction products can be identified as (6,2) or (2,6). Other charges that are visible are a result of other reactions or pile-up of events. To identify matching pairs of the desired fragments, further cuts are necessary.

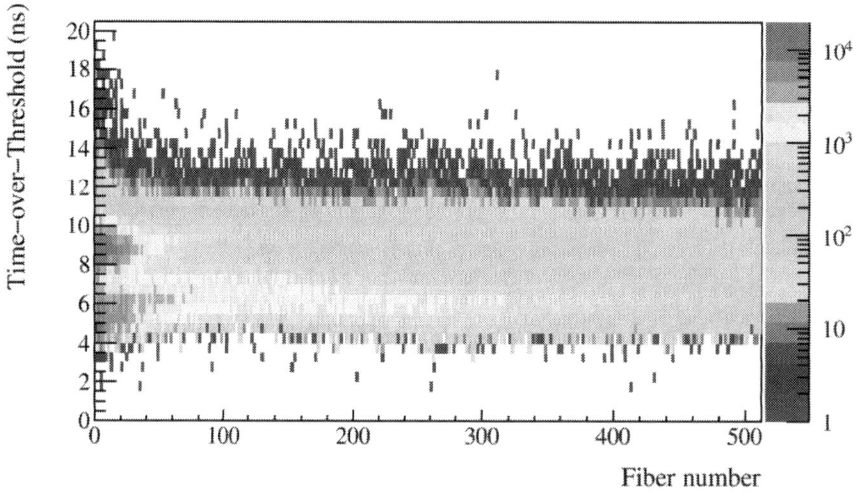

Figure 4. Fiber number of one detector behind GLAD vs. deposited energy as Time-over-Threshold. ^{12}C deposits more energy than ^4He, and they can be easily identified.

With time-of-flight cuts on the fiber detectors and selecting only matching pairs, we can separate ^{12}C and ^4He hits in the fiber detectors. In figure 4, the Time-over-Threshold for each fiber of one large fiber detector is shown. ^{12}C deposits more energy than ^{4}He and the two can easily be separated. ^{4}He shows a larger angular distribution than ^{12}C.

We can use the well known momentum of the unreacted ^{16}O beam to calibrate our detector positions for tracking of the Coulomb dissociation reaction products. Using the information from all detectors we can then reconstruct the momentum of the reaction products at a common target point and hence the center-of-mass energy in figure 5. We normalize the tracked pairs to the number of incoming ^{16}O particles measured by SEETRAM, the number of atoms for the different targets used, the dead time, and a nuclear scaling factor. This scaling factor is necessary since the nuclear contribution measured with carbon target is dependent on the size of the target nuclei and thus needs to be rescaled for a lead target. By subtracting the nuclear breakup reaction we can then extract the Coulomb dissociation cross section.

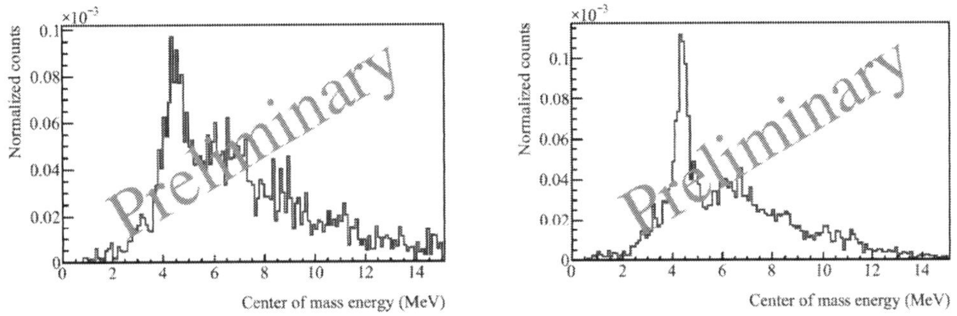

Figure 5. Counts of tracked ^{12}C and ^{4}He pairs as a function of the center-of-mass energy normalized to the number of incoming ^{16}O, number of target atoms, dead time, and a nuclear scaling factor.
Left: Lead target, Right: Carbon target

The result is a very preliminary attempt in analyzing a small percentage of the recorded data and needs more careful calibration and tracking in the near future.

Acknowledgements

This project was carried out within FAIR Phase-0 at GSI Helmholtzzentrum für Schwerionenforschung Darmstadt, Germany. We thank the beam provider and operators very much for their support.
This project was supported by Bundeministerium für Bildung und Forschung (BMBF) (05P19RFFN1, 05P19WOFN1, 05P15RFFN1, 05P15RDFN1), HGS-HIRe, HIC for FAIR and the GSI-TU Darmstadt cooperation agreement.
Marialuisa Aliotta acknowledges support by the EMMI Visiting Professor Scheme.
Jan Glorius and Yuri A Litvinov acknowledge support by the State of Hesse within the Research Cluster ELEMENTS (Project ID 500/10.006).

References

[1] R. deBoer, J. Görres, M. Wiescher, R. Azuma, A. Best, C. Brune, C. Fields, S. Jones, M. Pignatari, D. Sayre et al., Reviews of Modern Physics **89** (2017)
[2] M. Aliotta, R. Buompane, M. Couder, A. Couture, R.J. deBoer, A. Formicola, L. Gialanella, J. Glorius, G. Imbriani, M. Junker et al., Journal of Physics G: Nuclear and Particle Physics **49**, 010501 (2021)

[3] T. Sukhbold, S. Adams, Monthly Notices of the Royal Astronomical Society **492**, 2578 (2020)
[4] R. Farmer, M. Renzo, S.E. de Mink, P. Marchant, S. Justham, The Astrophysical Journal **887**, 53 (2019)
[5] R. Plag, R. Reifarth, M. Heil, F. Käppeler, G. Rupp, F. Voss, K. Wisshak, Phys. Rev. C **86**, 015805 (2012)
[6] R. Farmer, M. Renzo, S.E. de Mink, M. Fishbach, S. Justham, The Astrophysical Journal Letters **902**, L36 (2020)
[7] G. Baur, C. Bertulani, H. Rebel, Nuclear Physics A **458**, 188 (1986)
[8] C.A. Bertulani, Phys. Rev. C **49**, 2688 (1994)
[9] G. Baur, H. Rebel, Annual Review of Nuclear and Particle Science **46**, 321 (1996)
[10] C. Bertulani, A. Gade, Physics Reports **485**, 195 (2010)
[11] R. Reifarth, S. Altstadt, K. Göbel, T. Heftrich, M. Heil, A. Koloczek, C. Langer, R. Plag, M. Pohl, K. Sonnabend et al., Journal of Physics: Conference Series **665**, 012044 (2016)
[12] M. Heil, A. Kelić-Heil, L. Bott, T. Almusidi, H. Alvarez-Pol, L. Atar, L. Atkins, T. Aumann, J. Benlliure, K. Boretzky et al., The European physical journal / A **58**, 248 (2022)

Turbulence and nuclear reactions in 3D hydrodynamics simulations of massive stars

Federico Rizzuti[1,*], *Raphael* Hirschi[1,2], *Cyril* Georgy[3], *David* Arnett[4], *Casey* Meakin[5], and *Alexander* Murphy[6]

[1]Astrophysics Group, Lennard-Jones Laboratories, Keele University, Keele ST5 5BG, UK
[2]Kavli IPMU (WPI), University of Tokyo, 5-1-5 Kashiwanoha, Kashiwa 277-8583, Japan
[3]Geneva Observatory, Geneva University, CH-1290 Sauverny, Switzerland
[4]Steward Observatory, University of Arizona, 933 N. Cherry Avenue, Tucson AZ 85721, USA
[5]Pasadena Consulting Group, 1075 N Mar Vista Ave, Pasadena, CA 91104, USA
[6]School of Physics and Astronomy, University of Edinburgh, Edinburgh EH9 3FD, UK

Abstract. Our knowledge of massive star evolution and nucleosynthesis is limited by uncertainties related to multi-dimensional processes taking place in stellar interiors. Recently, theoretical works have started to improve 1D stellar evolution codes through the implementation of results from 3D hydrodynamics models, which are used to study multi-D processes on a short time range (minutes or hours) and improve 1D prescriptions. In these proceedings, we present results coming from a new set of high-resolution hydrodynamics simulations of the neon-burning shell in a massive star, employing the `PROMPI` code. We focus in particular on the interplay between turbulence and nuclear reactions, discussing the impact that different boosting factors of the nuclear rates have on the results. This has important implications for supernova studies, nucleosynthesis, the physics of neutron stars and black holes.

1 Introduction

The lifetime of a star can be modelled employing one-dimensional (1D) stellar evolution models. These simulations are run efficiently thanks to the simplified prescriptions that describe the physics of the processes inside the star. Such prescriptions need to be carefully calibrated to observations. One example is the mixing length theory [1], used in stellar models to reproduce the convective regions of the star. To better understand and reproduce these phenomena, multi-dimensional hydrodynamics models can be employed. They can simulate the fluid motions in great detail, and combine it with physical processes like nuclear burning, rotation, and magnetic fields. However, the large computing power required to run these simulations sets a strict limit on the time scales and spatial extents that can be reproduced. It is the combined strength of 1D and multi-D models that can really push forward progress in understanding stellar evolution.
One of the most interesting aspects of this field is the interplay between turbulent convection and nuclear reactions. We present in these proceedings results from a new set of hydrodynamics models run with the `PROMPI` code [2], simulating the convective neon-burning shell

*e-mail: f.rizzuti@keele.ac.uk

of a 15 $M_\odot$ star. Similar studies have been published with the same code for the oxygen- and carbon-burning shells [2, 3], but not yet using a realistic nuclear burning network combined with initial conditions extracted from state-of-the-art 1D stellar evolution models, as we do for this new set [4]. We study here the effects of the nuclear burning on the convective shell and its evolution with time.

2 Methodology

The new set of simulations has been run with the stellar hydrodynamical code `PROMPI` [2]. All simulations are started from the same initial conditions, taken from a 1D stellar evolution simulation of a 15 $M_\odot$ star calculated with the `GENEC` code [5], a state-of-the-art code widely used to produce large grids of 1D models [6]. The most important assumptions are solar metallicity, Schwarzschild criterion for convective boundaries, and penetrative overshoot for the hydrogen- and helium-burning cores (see [4] for more details). The fundamental variables (e.g., density, temperature, composition) that characterize the second neon-burning shell that forms in 1D, chosen for its fast evolution and later phase, have been mapped onto a cubic box of side 0.64×10^8 cm, from which the 3D simulations are started with a numerical resolution of 512^3 cells.

We include in our simulations an explicit nuclear network that fuels convection through energy generation, with the key isotopes for neon burning: ^{4}He, ^{16}O, ^{20}Ne, ^{24}Mg, ^{28}Si (the ^{4}He abundance is assumed to be at nuclear equilibrium, since this is a late burning stage). This simplified network includes only the main reactions for neon burning: $^{20}\mathrm{Ne}(\gamma,\alpha)^{16}\mathrm{O}$ and $^{20}\mathrm{Ne}(\alpha,\gamma)^{24}\mathrm{Mg}(\alpha,\gamma)^{28}\mathrm{Si}$. The individual simulations have been run with modifications to the nuclear reaction rates. Specifically, we included a constant "boosting factor" that multiplies the rates for the primary reactions involved in the energy generation, i.e. $^{20}\mathrm{Ne}(\gamma,\alpha)^{16}\mathrm{O}$ and $^{20}\mathrm{Ne}(\alpha,\gamma)^{24}\mathrm{Mg}$, with respect to the nominal case. Such boosting factors were originally introduced to accelerate the simulations by increasing their convective speed, and therefore reducing the computing cost. In addition, they allow us to study the different evolution of the burning shell with time, tracking the amount of material that is entrained from the stable regions into the convective zone ("turbulent entrainment", see [7, 8]).

We studied the effect of the boosting factors by running three different simulations with boosting factor equal 1, 10, and 100 (see [4] for more details). The introduction of a boosting factor is common practice in stellar hydrodynamics simulations, so it is important here to consider also the case where it has not been used (boosting factor = 1), and the simulation has been run exactly as taken from the 1D model, without changing its original state.

3 Results

As a visual representation of our simulations, we present in Figure 1 the vertical cross-section of the nominal-luminosity simulation (boosting factor = 1) at time $t = 2730$ seconds, showing the velocity magnitude in colour scale. From this image, it is easy to recognize the convective region, i.e. the central zone where the fluid is dominated by plumes and eddies with high speed. The rest of the domain is a stable, radiative region characterized by low-speed gravity waves.

To study the time evolution of the convective zone, we plot in Figure 2 the horizontally-averaged mean atomic mass (in colour scale) as a function of time, for the 10-times boosted simulation. We can see the convective zone (in yellow) growing due to entrainment, easily distinguished from the upper and lower stable regions using the chemical composition. The fact that the chemical composition distinguishes so easily the convective zone from the stable

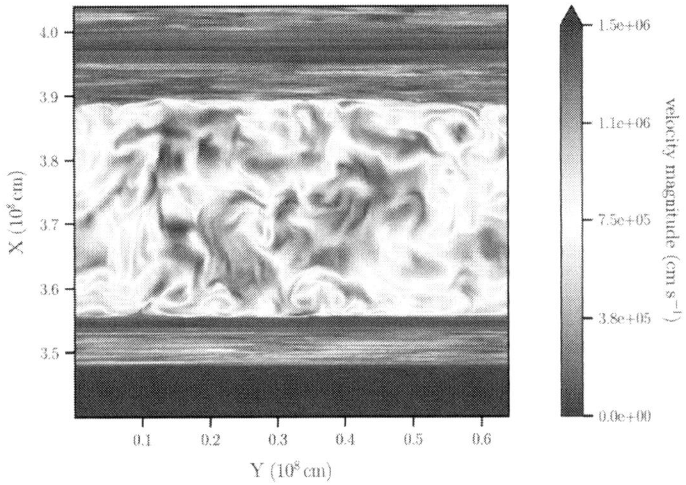

Figure 1. Vertical cross section of the velocity magnitude (in colour scale, cm s^{-1}) taken from the nominal-luminosity simulation at 2730 seconds. The convective zone is identified by high-velocity eddies. Figure taken from [4].

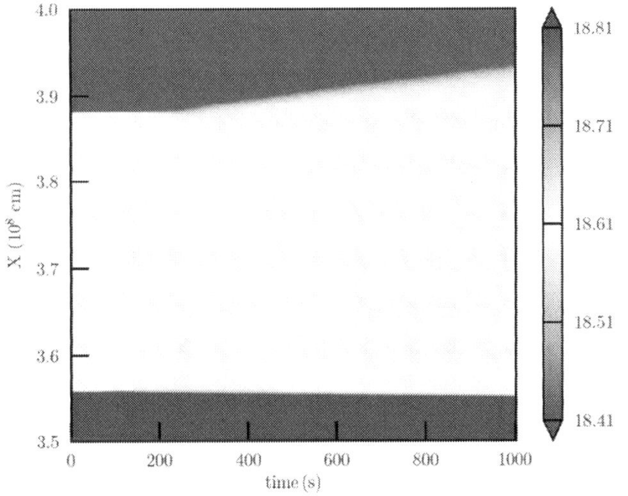

Figure 2. Time evolution of the horizontally-averaged mean atomic mass, for the simulation with energy boosting factor of 10. In yellow, the convective zone grows with time due to entrainment, after an initial transient of ~250 s. Figure taken from [4].

layers shows the importance of implementing nuclear abundances in hydrodynamics simulations of stellar interiors.

We are interested in analysing the effects of the nuclear energy on the convective shell growth. For this reason, we parametrize turbulent entrainment with a simple law expressing the entrainment rate E (i.e., the entrainment over the convective velocity = $v_e/v_{\rm rms}$) as a function of the bulk Richardson number $Ri_{\rm B}$ (a measure of the convective boundary stiffness): $E = A\, Ri_{\rm B}^{-n}$ (as defined by [2, 9]). Then, we use the data from our hydrodynamics simulations to estimate the free parameters A, n, by measuring E from the time evolution of the bottom and top boundary locations visible in Figure 2, and computing $Ri_{\rm B}$ by integrating the Brunt-Väisälä frequency across the convective boundary (see [4] for the detailed implementation).

Figure 3 shows our measurements of entrainment rate versus bulk Richardson number for the three neon-shell simulations, alongside the line of best fit in log scale (in blue, solid). In addition to the present study [4], we compare results to other two `PROMPI` simulations for the carbon [3] and oxygen [2] shells of massive stars. In the legend of the figure, we give the estimates of parameters A and n for each study.

Comparing the different tracks, we conclude that turbulent entrainment has a similar behaviour in different burning shells of late-phase massive stars. In particular, fitting the en-

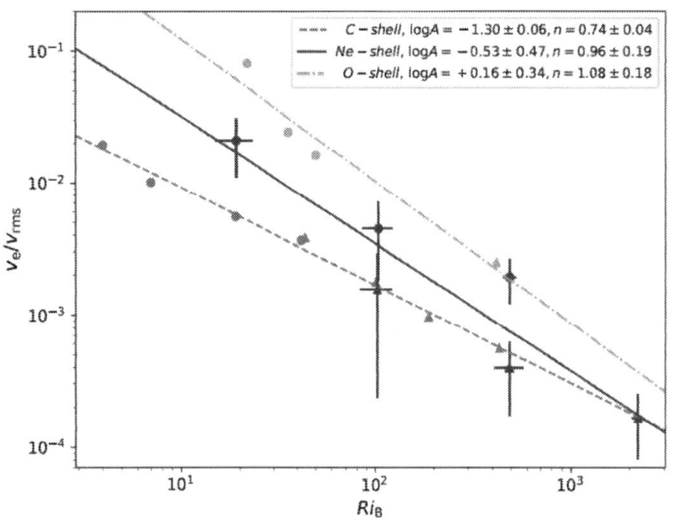

Figure 3. Entrainment rate vs. bulk Richardson number and linear regressions: Ne-shell from this study (blue, solid), C-shell from [3] (green, dashed), O-shell from [2] (orange, dot-dashed). Triangles are for the lower convective boundaries, circles for the upper ones. The boosting factor is inversely proportional to Ri_B, equal to 100, 10, 1 for the two triplets of blue circles and triangles. Error bars are standard deviations, computed in this work for the first time. Figure taken from [4].

trainment law to the data from our new set of Ne-shell simulations, the estimated parameter $n = 0.96 \pm 0.19$ is compatible with the value of 1 expected from geophysics studies [9]. Estimating A is more difficult, considering the fitting in log scale and the dispersion of the data: a reasonable estimate from our data is $A \sim 0.1 - 1.0$. This is in agreement with other results from stellar hydrodynamics simulations, but disagrees with 1D stellar evolution simulations, that predict a much smaller amplitude of about $\sim 10^{-4}$ for turbulent entrainment in the main sequence core [10, 11]. This divergence motivates the need for a more detailed investigation of convection and entrainment in 1D models, especially in the late phases of massive stars.

4 Conclusions

We presented a new set of 3D hydrodynamics simulations of the neon-burning shell in a 15 $M_\odot$ star, run with the PROMPI code. We study in detail the interplay between nuclear reactions and turbulent convection, focusing on the impact of the nuclear energy generation on the entrainment of stable material into the convective zone.

The key result of this study, as described in the work of [4], is that a strong entrainment is present in the simulations also when employing the unaltered input conditions from a state-of-the-art 1D stellar model, and in particular when the nuclear energy generation rate is not modified (nominal-luminosity simulation). We see that turbulent entrainment in our simulations follows the entrainment law in agreement with previous hydrodynamics studies [2, 3], showing this is an accurate and reliable way of modelling entrainment that can also be used for 1D stellar evolution models.

The discrepancy between 3D hydrodynamics and 1D stellar models on entrainment, especially concerning the estimate of parameter A in the entrainment law, can be traced back to a different behaviour during the advanced phases, or to not having implemented entrainment in the 1D input models. In both cases, investigating the effects of entrainment in stellar evolution models is crucial for understanding the origin of this discrepancy, and eventually solving it. To date, few studies [10, 11] have been implementing the entrainment law in a 1D stellar model, and only for the main sequence convective core.

A complete understanding of the complex phenomena that takes place in the stellar interiors can only be achieved by synergy and convergence between one- and multi-dimensional stellar models, and thanks to the implementation of the nuclear reactions that fuel convection in these environments. This will have an important impact on many different fields in astrophysics, like supernova studies, nucleosynthesis and chemical enrichment, but also the physics of neutron stars and black holes.

Acknowledgements

RH acknowledges support from the World Premier International Research Centre Initiative (WPI Initiative), MEXT, Japan and the IReNA AccelNet Network of Networks (National Science Foundation, Grant No. OISE-1927130). CG has received funding from the European Research Council (ERC) under the European Union's Horizon 2020 research and innovation program (Grant No. 833925). WDA acknowledges support from the Theoretical Astrophysics Program (TAP) at the University of Arizona and Steward Observatory. CG, RH, and CM acknowledge ISSI, Bern, for its support in organising collaboration. This article is based upon work from the ChETEC COST Action (CA16117) and the European Union's Horizon 2020 research and innovation programme (ChETEC-INFRA, Grant No. 101008324). The authors acknowledge PRACE for awarding access to the resource MareNostrum 4 at Barcelona Supercomputing Center, Spain, and the STFC DiRAC HPC Facility at Durham University, UK (Grants ST/P002293/1, ST/R002371/1, ST/R000832/1, ST/K00042X/1, ST/H008519/1, ST/ K00087X/1, ST/K003267/1). The University of Edinburgh is a charitable body, registered in Scotland, with Registration No. SC005336.

References

[1] E. Böhm-Vitense, **46**, 108 (1958)
[2] C.A. Meakin, D. Arnett, **667**, 448 (2007), astro-ph/0611315
[3] A. Cristini, R. Hirschi, C. Meakin, D. Arnett, C. Georgy, I. Walkington, **484**, 4645 (2019), 1901.10531
[4] F. Rizzuti, R. Hirschi, C. Georgy, W.D. Arnett, C. Meakin, A.S. Murphy, **515**, 4013 (2022), 2207.03223
[5] P. Eggenberger, G. Meynet, A. Maeder, R. Hirschi, C. Charbonnel, S. Talon, S. Ekström, **316**, 43 (2008)
[6] S. Ekström, C. Georgy, P. Eggenberger, G. Meynet, N. Mowlavi, A. Wyttenbach, A. Granada, T. Decressin, R. Hirschi, U. Frischknecht et al., **537**, A146 (2012), 1110.5049
[7] A. Maeder, **47**, 389 (1976)
[8] J.P. Zahn, **252**, 179 (1991)
[9] H.J.S. Fernando, Annual Review of Fluid Mechanics **23**, 455 (1991)
[10] E.I. Staritsin, Astronomy Reports **57**, 380 (2013)
[11] L.J.A. Scott, R. Hirschi, C. Georgy, W.D. Arnett, C. Meakin, E.A. Kaiser, S. Ekström, N. Yusof, **503**, 4208 (2021), 2103.06196

News on the slow neutron capture process in AGB stars

Diego Vescovi[1,*]

[1]Goethe University Frankfurt, Max-von-Laue-Strasse 1, Frankfurt am Main 60438, Germany

Abstract. Asymptotic giant branch (AGB) stars are responsible for the production of the main component of the solar s-process distribution. Despite enormous progress in the theoretical modeling of these objects over the last few decades, many uncertainties remain. The still-unknown mechanism leading to the production of ^{13}C neutron source is one example. The nucleosynthetic signature of AGB stars can be examined in a number of stellar sources, from spectroscopic observations of intrinsic and extrinsic stars to the heavy-element isotopic composition of presolar grains found in meteorites. The wealth of available observational data allows for constraining the processes occurring in AGB interiors. In this view, we discuss recent results from new AGB models including the effects of mixing triggered by magnetic fields, and show comparisons of the related s-process nucleosynthesis with available observations.

1 Introduction

Asymptotic giant branch (AGB) stars are known to be one of the main producers of carbon, nitrogen, fluorine, and about half of the elements heavier than strontium, thus contributing greatly to the chemical enrichment of the Universe [1]. These stars are characterized by recurrent thermal instabilities of the He-burning shell, known as thermal pulses (TPs), which interrupt the quiescent shell-H burning that dominates the nuclear energy generation for thousands of years during the interpulse phase. H- and He-burning products can be carried up to the stellar surface by episodes of convective mixing, possibly occurring after each thermal pulse, known as third dredge-up (TDU) events. These are responsible for the formation of carbon stars and the surface enrichment of heavy elements, produced in stellar interiors by the slow neutron capture process (s-process) in low-mass AGB stars ($M \lesssim 3M_{\odot}$) [2].

In these objects, the main neutron source is the ^{13}C(α, n)^{16}O reaction [3, 4], while the ^{22}Ne(α, n)^{25}Mg reaction is only marginally activated during the TPs [5]. For the ^{13}C(α, n)^{16}O reaction to effectively operate, some ^{13}C must be present in the He-rich zone located below the H-rich convective envelope. Its synthesis requires the penetration of a small quantity of protons below the envelope itself, where they are captured by the abundant ^{12}C thus forming a ^{13}C pocket. To date, the mechanism responsible for the creation of the ^{13}C pocket remains one of the major unknowns in the modeling of AGB stars and the main uncertainty in the relative nucleosynthesis of heavy elements through the s-process [6]. Over the past few years, many physical mechanisms for the partial mixing of H-rich material from the convective envelope have been investigated, as induced by convective overshoot [7, 8], rotation [9, 10], internal gravity waves [11, 12], and magnetic fields [13–15].

*e-mail: vescovi@iap.uni-frankfurt.de

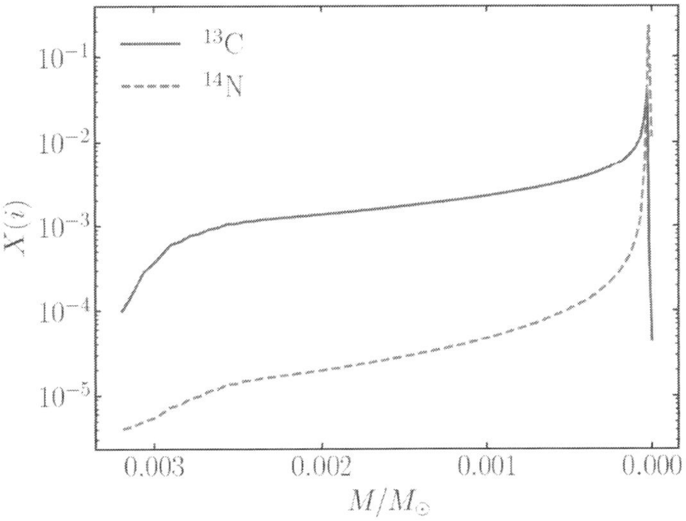

Figure 1. ^{13}C and ^{14}N abundance in the ^{13}C pocket region for FRUITY Magnetic models.

Because the s-process affects a large number of observable and is dependent on only a few fundamental quantities (such as neutron density, total neutron flux, and the cross sections of nuclei with a magic number of neutrons) [16], observational constraints for AGB stars are useful in distinguishing between suitable mechanisms. Here, we consider new theoretical predictions from FRUITY stellar models accounting for a magnetically-induced ^{13}C pocket and show their capability of reproducing observational constraints, as provided by the analysis of presolar stellar dust grains, spectroscopic observations of intrinsic and extrinsic AGB stars, and the chemical properties of Galactic open clusters.

2 Stellar magnetic models and observations

Magnetic fields may play an important role in the transport of angular momentum and chemical elements during stellar evolution [17]. In this regard, the buoyancy of material supported by magnetic pressure was shown to be an efficient physical mechanism for the transport of material from radiative regions to the convective envelope in low-mass giants stars [18] and provide a sufficient mass transport rate to explain the formation of the ^{13}C pocket in TP-AGB stars [13]. The inclusion of magnetic mixing in the FUNS evolutionary code was recently done by [19] to generate a new set of FRUITY stellar models [20]. According to these models, the action of magnetic mixing results in deep penetration of a relatively small number of protons, that are almost entirely consumed for the synthesis of ^{13}C. As a consequence, the local generation of ^{14}N through the p-captures on ^{13}C nuclei is greatly suppressed (see Figure 1). A ^{13}C reservoir with these characteristics, i.e. extended (a few 10^{-3} $M_{\odot}$) and with low concentrations of ^{13}C, was shown to be suitable in reproducing the distributions of s-process isotopes and elements, as measured in presolar silicon carbide (SiC) grains [21]. In particular, the mainstream (MS) ones, which account for about 90% of all SiC grains, are believed to have originated from carbon AGB stars and show their nucleosynthetic signature [22]. Therefore, isotope ratios of heavy elements measured in MS SiC grains are commonly used to study the s-process occurring in AGB stars and provide constraints on AGB stellar models [23–26]. Most of the presolar MS SiC grains have been shown to originate from AGB stars with $M \sim 2$ $M_{\odot}$ and close-to-solar metallicity [27, 28]. FRUITY Magnetic models

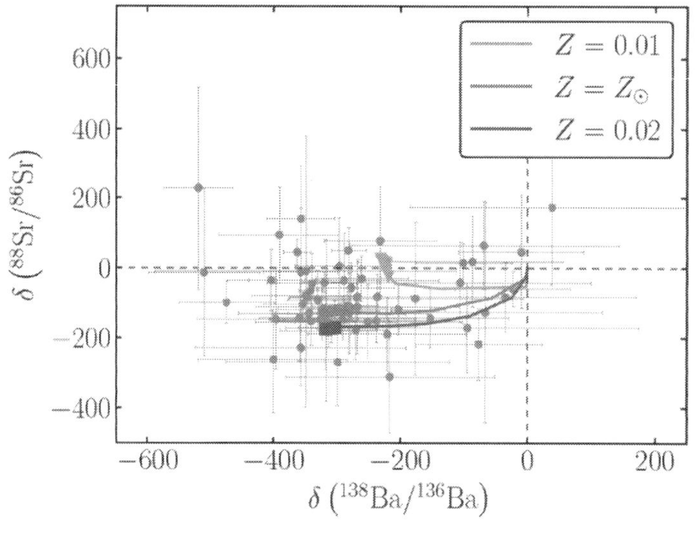

Figure 2. Comparison between FRUITY Magnetic 2 $M_\odot$ models at different metallicities and presolar grain isotope ratios. Isotope ratios are reported in the conventional δ notation, defined as the deviation of the ratio measured in a grain relative to the terrestrial value in parts per thousand. Grain data are from the PGD database [29].

within this mass and metallicity range, once properly calibrated, can accurately reproduce MS SiC grain isotope ratios for heavy elements [19] and, in particular, the negative δ-values of ^{88}Sr/^{86}Sr and ^{138}Ba/^{136}Ba ratios showed by the bulk of the data (see Figure 2).

The shortage of primary ^{14}N predicted by FRUITY Magnetic models has significant implications on the synthesis of light elements whose production involves nitrogen, including fluorine. In AGB stars, fluorine is synthesized via the reactions chain $^{14}\mathrm{N(n,p)}^{14}\mathrm{C}(\alpha,\gamma)^{18}\mathrm{O(p},\alpha)^{15}\mathrm{N}(\alpha,\gamma)^{19}\mathrm{F}$ [30]. Nitrogen and free neutrons are therefore simultaneously required to start the chain. The latter are provided by the activation of the $^{13}\mathrm{C}(\alpha,\mathrm{n})^{16}\mathrm{O}$ reaction. During the radiative burning of the ^{13}C pocket, the freshly synthesized neutrons are consumed together with the ^{14}N present in the He-intershell to produce ^{15}N, which is then burnt to create ^{19}F in the following convective TP. A second source of fluorine descends from the amount of ^{13}C and ^{14}N left in H-burning ashes (plus some eventual un-burnt ^{13}C in the pocket [8]) engulfed and burnt in the convective TP. Fluorine is then transported to the surface by convective processes during the following TDU. As a consequence, the abundance of ^{19}F in the envelope is expected to correlate with that of carbon and s-process elements. Direct evidence of fluorine production in AGB stars is provided by spectroscopic observations of photospheric [F/Fe] enhancements in both intrinsic and extrinsic stellar objects(see [31] and references therein). As shown in Figure 3, FRUITY Magnetic models predictions for low-mass AGB stars provide a good agreement for the F-enhancement trend with the metallicity. Even more interesting, these models are able to reproduce the observed spread for [F/s] vs. [s/Fe] ratios. Due to the shortage of ^{14}N in the ^{13}C pocket region, the fluorine production is dominated by the amount of ^{13}C and ^{14}N in the H-burning ashes [20]. As a result, the predicted fluorine envelope abundance is significantly reduced with respect to models exhibiting a large production of fluorine from primary ^{14}N [31].

Stellar yields from AGB stars are key ingredients to model the Galactic chemical evolution (GCE) and for interpreting the solar s-process distribution [1]. Additional precious information, regarding the chemical properties of the Galactic disk and its evolution, comes

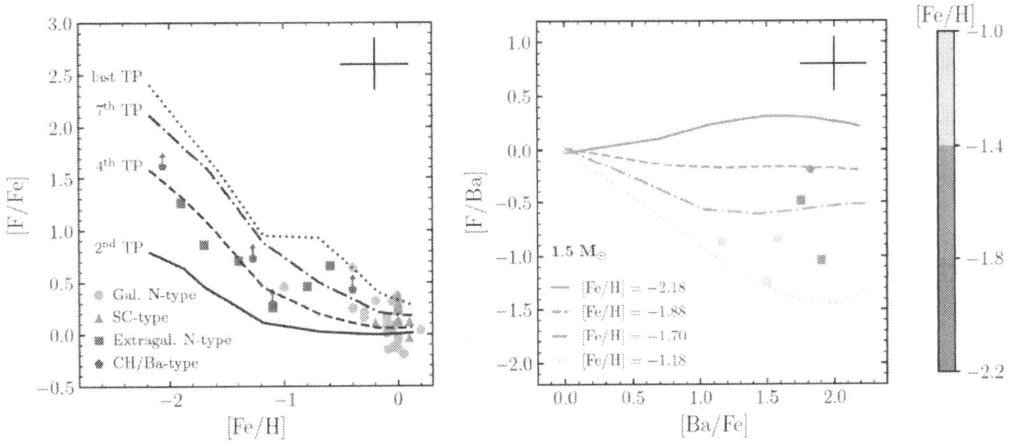

Figure 3. (Left panel) Comparison between observed [F/Fe] ratios as a function of the metallicity and FRUITY Magnetic models. Symbols refer to four data groups: circles, galactic (N-type) carbon stars; triangles, SC-type stars; squares, extragalactic carbon stars; pentagons, extrinsic CH/Ba stars. Lines represent theoretical predictions for 2 (for [Fe/H] ≥ -0.7) and 1.5 $M_{\odot}$ (for [Fe/H] < -0.7) AGB stars at different thermal pulses (TPs). (Right panel) [F/Ba] vs. [Ba/Fe] in the sample stars with [Fe/H] ≤ -1.0. Data points and theoretical lines are color-coded by [Fe/H]. Typical error bars are indicated.

from the study of open clusters. The latest observational data from the final release of the Gaia-ESO survey, show that there is an increasing, albeit weak, trend in the production of s-process elements as the cluster age decreases, but with radial variations, with respect to the center of the Galaxy [32]. These are related to the metallicity dependence of the stellar yields of the s-process and the radial dependence of the star formation history [33]. Specifically, in the inner part of the Galactic disk, the average [s/Fe] gradient does not vary significantly with age, thus indicating a substantial decrease of the s-process efficiency at the super-solar metallicity characteristic of the innermost part of the disk, as expected by the secondary nature of the s-process [2]. Nonetheless, GCE computations adopting AGB stellar yields from past FRUITY models overestimate the production of the s-process elements for clusters located at Galactocentric distances $R_{GC} < 7$ kpc [34]. Conversely, FRUITY Magnetic models are characterized by a lower neutron-to-seed ratio, due to the low ^{13}C concentration in the pocket (see Figure 1), so that lower neutron fluxes and s-process enhancements are attained at relatively high metallicity. In this regard, the behavior of neutron-capture elements in the inner Galactic disk is well explained by the current FRUITY Magnetic models [33].

3 Conclusions

In recent years, significant progress has been made in determining the evolution and nucleosynthesis of AGB stars. One of the lingering questions is the physical mechanism leading to the formation of the ^{13}C pocket in these stars. Among the various theories, mixing caused by magnetic instabilities has been shown to be a viable solution. In this work, we discussed a new set of FRUITY models that account for magnetic-buoyancy-induced mixing. These models provide a useful theoretical tool for explaining several observable constraints, such as heavy-element isotope ratios measured in presolar MS SiC grains, fluorine enhancements of AGB stars, and s-element abundances in galactic open clusters. These findings support the magnetic mixing as the potential mechanism behind the activation of the ^{13}C neutron source in AGB stars. Nonetheless, further analyses are required for a definite claim, including more comparisons with s-process enriched objects such as barium stars and post-AGB stars, as well as targeted multidimensional magnetohydrodynamic simulations.

DV acknowledges financial support from the German-Israeli Foundation (GIF No. I-1500-303.7/2019).

References

[1] N. Prantzos, C. Abia, Limongi et al., MNRAS **476**, 3432 (2018), 1802.02824
[2] M. Busso, R. Gallino, G.J. Wasserburg, ARA&A **37**, 239 (1999)
[3] O. Straniero, R. Gallino, M. Busso et al., ApJL **440**, L85 (1995)
[4] S. Cristallo, M. La Cognata, C. Massimi et al., ApJ **859**, 105 (2018), 1804.10751
[5] O. Straniero, A. Chieffi, M. Limongi et al., ApJ **478**, 332 (1997)
[6] F. Herwig, ARA&A **43**, 435 (2005)
[7] F. Herwig, T. Bloecker, D. Schoenberner et al., A&A **324**, L81 (1997)
[8] S. Cristallo, O. Straniero, R. Gallino et al., ApJ **696**, 797 (2009), 0902.0243
[9] F. Herwig, N. Langer, M. Lugaro, ApJ **593**, 1056 (2003), astro-ph/0305491
[10] L. Siess, S. Goriely, N. Langer, A&A **415**, 1089 (2004)
[11] P.A. Denissenkov, C.A. Tout, MNRAS **340**, 722 (2003)
[12] U. Battino, M. Pignatari, C. Ritter et al., ApJ **827**, 30 (2016), 1605.06159
[13] O. Trippella, M. Busso, S. Palmerini et al., ApJ **818**, 125 (2016), 1512.06777
[14] D. Vescovi, M. Busso, S. Palmerini, et al., ApJ **863**, 115 (2018), 1807.01058
[15] M. Busso, D. Vescovi, S. Palmerini et al., ApJ **908**, 55 (2021), 2011.07469
[16] F. Käppeler, R. Gallino, S. Bisterzo et al., Reviews of Modern Physics **83**, 157 (2011)
[17] A. Maeder, *Physics, Formation and Evolution of Rotating Stars* (Springer-Verlag Berlin Heidelberg, 2009)
[18] M. Busso, G.J. Wasserburg, K.M. Nollett et al., ApJ **671**, 802 (2007), 0708.2949
[19] D. Vescovi, S. Cristallo, M. Busso, N. Liu, ApJL **897**, L25 (2020), 2006.13729
[20] D. Vescovi, S. Cristallo, S. Palmerini et al., A&A **652**, A100 (2021), 2106.08241
[21] N. Liu, M.R. Savina, R. Gallino et al., ApJ **803**, 12 (2015), 1501.05883
[22] L.R. Nittler, F. Ciesla, ARA&A **54**, 53 (2016)
[23] S. Palmerini, M. Busso, D. Vescovi et al., ApJ **921**, 7 (2021), 2107.12037
[24] N. Liu, S. Cristallo, D. Vescovi, Universe **8**, 362 (2022), 2206.13721
[25] S. Taioli, D. Vescovi, M. Busso et al., ApJ **933**, 158 (2022), 2109.14230
[26] N. Liu, T. Stephan, S. Cristallo et al., arXiv e-prints arXiv:2209.10347 (2022)
[27] H.P. Gail, S.V. Zhukovska, P. Hoppe, M. Trieloff, ApJ **698**, 1136 (2009)
[28] S. Cristallo, A. Nanni, G. Cescutti et al., A&A **644**, A8 (2020), 2010.08268
[29] T. Stephan, M. Bose, A. Boujibar et al., *The Presolar Grain Database Reloaded - Silicon Carbide*, in *51st Annual Lunar and Planetary Science Conference* (2020), Lunar and Planetary Science Conference, p. 2140
[30] M. Forestini, S. Goriely, A. Jorissen, M. Arnould, A&A **261**, 157 (1992)
[31] C. Abia, S. Cristallo, K. Cunha, P. de Laverny, V.V. Smith, A&A **625**, A40 (2019)
[32] L. Magrini, C. Viscasillas Vazquez, L. Spina et al., arXiv e-prints arXiv:2210.15525 (2022), 2210.15525
[33] L. Magrini, D. Vescovi, G. Casali et al., A&A **646**, L2 (2021), 2101.04429
[34] G. Casali, L. Spina, L. Magrini et al., A&A **639**, A127 (2020), 2006.05763

Stellar s-process neutron capture cross sections on ^{A}Se and ^{A}Ce

Rudra N. Sahoo[1], *Moshe* Tessler[1,2,*], *Shlomi* Halfon[2], *Dani* Kijel[2], *Arik* Kreisel[2], *Michael* Paul[1], *Asher* Shor[2], and *Leo* Weissman[2]

[1]The Hebrew University of Jerusalem, Jerusalem, Israel 91904
[2]Soreq Nuclear Research Center, Yavne, Israel 81800

Abstract. We report on experiments at the Soreq Applied Research Accelerator Facility - Liquid-Lithium Target (SARAF-LiLiT) laboratory dedicated to the study of s-process neutron capture reactions. The kW-power proton beam at 1.92 MeV (1-2 mA) from SARAF Phase I yields high-intensity 30 keV quasi-Maxwellian neutrons ($3\text{-}5\times10^{10}$ n/s). The high neutron intensity enables Maxwellian averaged cross sections (MACS) measurements of samples with short-lived decay products. Neutron capture reactions on ^{nat}Se and ^{nat}Ce were investigated by activation in the LiLiT neutron beam and γ-spectrometry measurements of their decay products.

1 Introduction

Except for the lightest nuclei (H, He, Li) that were created during the nucleosynthesis era of the Big Bang, the nuclei of the elements are produced in stars. Neutrons generated in-situ, and in particular their captures on nuclear seeds, are responsible for the production of the large majority of the heavier nuclides ($A \geq 60$) [1]. Neutron-induced reactions remain at the forefront of experimental investigations for the understanding of stellar nucleosynthesis and chemical evolution of the Galaxy in the region of medium- and heavy-mass nuclides [2]. One of the neutron-induced nucleosynthesis paths, namely the so-called slow neutron capture process (s-process) consists of a succession of neutron captures, each occurring over a mean time longer than or commensurate with the typical mean life of β-decaying nuclides close to the valley of stability. The s-process is believed to occur in stars during their He core and shell burning stages, where neutrons are available from (α, n) reactions. In contrast, the rapid neutron capture process (r-process) consists of the sequential captures of a large number ($\approx$ 20) of neutrons on a fast time scale, much shorter than β-decaying times, occurring in an astrophysical site with an extreme neutron density ($\approx 10^{20}$ - 10^{22} cm^{-3}) [2].

We report on measurements of the cross section of neutron capture reactions ^{74,80,82}Se(n,γ) and ^{136,138,140,142}Ce(n,γ) relevant, respectively, to the weak and main s-processes. The ^{A}Se data pursues our recent study of the ^{69,71}Ga stellar (n,γ) reactions in the weak s-process regime [3]. The proton rich isotope ^{74}Se is a p nuclide, shielded from the s and r processes by stable nuclei in the region. The disentanglement of the different heavy-nuclide synthesis modes (s-, r- and p-processes) requires reliable and precise stellar

*e-mail: moshe.tessler@mail.huji.ac.il

neutron-capture cross sections. Such is the case also for the Ce isotopes in the main s-process [4]. In particular, ^{140}Ce is found to be one of the most important nuclides in the network of *s*-process reactions, affecting the abundances of a large number of isotopes [5]. The experiments were performed by the activation technique using a high-intensity (3-5$\times 10^{10}$ n/s) quasi-Maxwellian neutron beam that mimics conditions of stellar *s*-process nucleosynthesis. The neutron field was produced by a mA proton beam at E_p=1925 keV (beam power of 2–3 kW) as part of our experiment campaign at the Phase I of Soreq Applied Research Accelerator Facility (SARAF) [6], bombarding the Liquid-Lithium Target (LiLiT) [7, 8]. The cross sections were measured by counting the activity of neutron capture products via γ spectrometry with a high-purity germanium (HPGe) detector.

2 The Soreq Applied Research Accelerator Facility (SARAF) and the Liquid-Lithium Target (LiLiT): activation experiments

The SARAF accelerator [6] is a superconducting linear RF accelerator designed to produce a beam of light ions (m/q $\leq$ 2) with a continuous-wave (CW) beam intensity up to 5 mA. The layout of Phase I is shown in Fig. 1. Light-ion H+ or D+ are extracted at 20 keV/u from

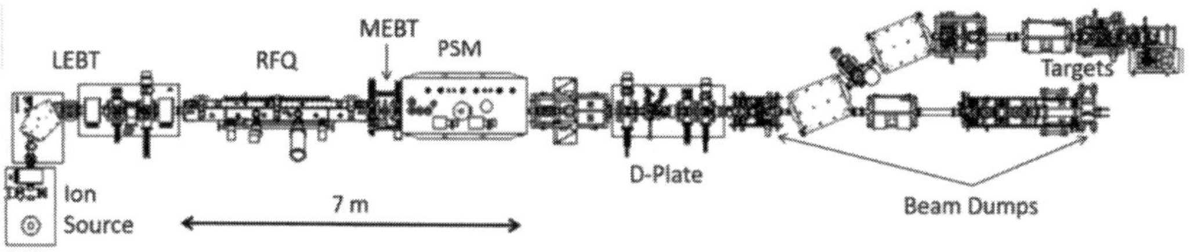

Figure 1: Layout of SARAF Phase I; see text for acronyms.

an Electron Cyclotron Resonance (ECR) ion source, mass analyzed and injected via a Low-Energy Beam Transport (LEBT) line to a 176 MHz four-rod Radio-Frequency Quadrupole (RFQ) with a final energy of 1.5 MeV/u. A short Medium-Energy Beam (MEBT) line transports the beam to a Prototype Superconducting Module (PSM) housing six Half-Wave Resonators (HWR) followed by the High Energy Beam (HEBT) line towards the LiLiT target at Station 1 (Fig. 1) via two 45° bending magnets.

The Liquid-Lithium Target (LiLiT) [7] was specifically designed and built to dissipate the high power of the CW proton beam from SARAF (up to $\approx$ 5 kW) and produce quasi-Maxwellian neutrons for nuclear astrophysics experiments (Fig. 2).

Liquid lithium is forced-flown from the top into a 15 mm wide, 1.8 mm thick film supported by a thin stainless steel wall (concave as viewed in the proton beam direction). The proton beam impinges directly (windowless) onto the lithium surface emitting neutrons mainly in the forward direction. The liquid lithium acts both as a (thick) target producing neutrons near the surface ($\approx$ 4 μm deep) and as a beam dump. The high power ($\approx$ 0.7 MW/cm^3) at the Bragg peak depth is mechanically transported by the fast flow (2-5 m/s) to a reservoir and heat exchanger.

An activation measurement consists of irradiation of a secondary target in the neutron field generated by the ^{7}Li(p,n) reaction in LiLiT. An experimental cross section for the neutron-induced reaction is extracted from the offline decay counting of the activated target or by atom counting of the residual nuclide. The target must be positioned as close as possible to the Li film to maximize the neutron dose. The experimental setup is illustrated in Fig. 2. The secondary target is located in a so-called experimental chamber separated

https://doi.org/10.1051/epjconf/202327906002

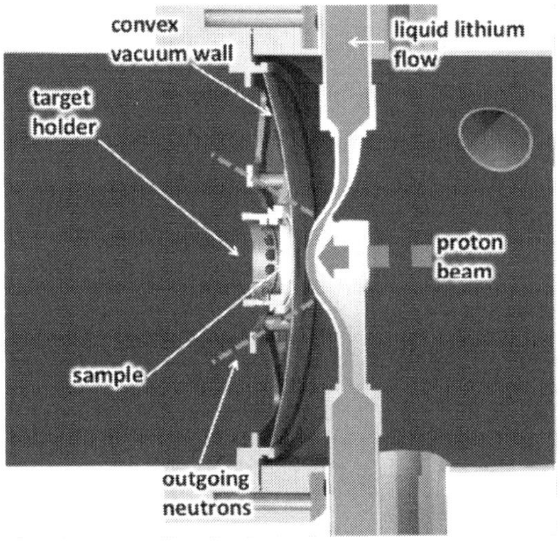

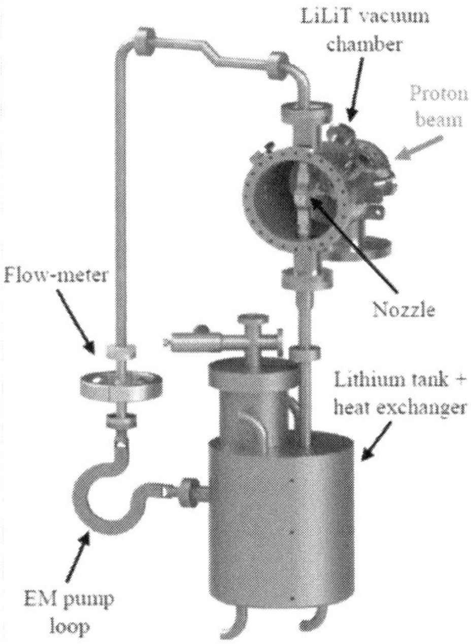

Figure 2: (right) Schematic drawing of the Liquid-Lithium Target (LiLiT). (left) Diagram of the Liquid-Lithium Target (LiLiT) and activation target assembly. The (1-2 mA, ≈9 mm full width) proton beam (dashed red arrow) impinges directly on the windowless liquid-lithium film. The light blue shows the liquid-lithium circulating flow (see [7] for details). The activation samples are mounted at the center of a ring target holder made of Al and positioned in the outgoing neutron cone (green dashed lines) at a distance of ≈6 mm from the liquid-lithium film surface. The targets are in a vacuum chamber separated from the LiLiT chamber by a 0.5 mm stainless steel concave vacuum wall.

from the LiLiT chamber by a curved vacuum wall. The proton beam (full width of ≈ 9 mm) impinges on the free surface of the liquid lithium film, resulting in outgoing forward-directed neutrons. The wall curvature allows the secondary target within its holder to be at a distance of 6±1 mm from the liquid-Li surface. The secondary target is sandwiched within the holder between two Au foils serving as neutron fluence monitors. The setup and simulations were carefully benchmarked in [8, 9] through measurements of the $^{94,96}\mathrm{Zr}(n,\gamma)^{95,97}\mathrm{Zr}$ cross sections where a method of extrapolation of experimental cross sections to Maxwellian averaged cross sections (MACS) is described.

3 ^{A}Se and ^{A}Ce measurements

^{nat}Se and ^{nat}Ce targets were irradiated at SARAF-LiLiT for MACS measurements. After the irradiation, the induced activities are measured with a High-Purity Ge (HPGe) detector for each sample separately. Gamma spectra from the activated targets showing the major activation lines are displayed in Fig. 3.

When using a thick liquid-lithium target with a high-intensity proton beam, there is an ample production of high-energy γ rays (17.6 MeV and 14.6 MeV) from the radiative capture $^{7}\mathrm{Li}(p,\gamma)^{8}\mathrm{Be}$ reaction. In cases where there are a stable A-1 isotope, an unstable A isotope and a stable A+1 isotope, the nuclide A can be produced through the A-1(n,γ) or A+1(γ,n) reactions. In order to measure the contribution of the A+1(γ,n) reaction to the production of

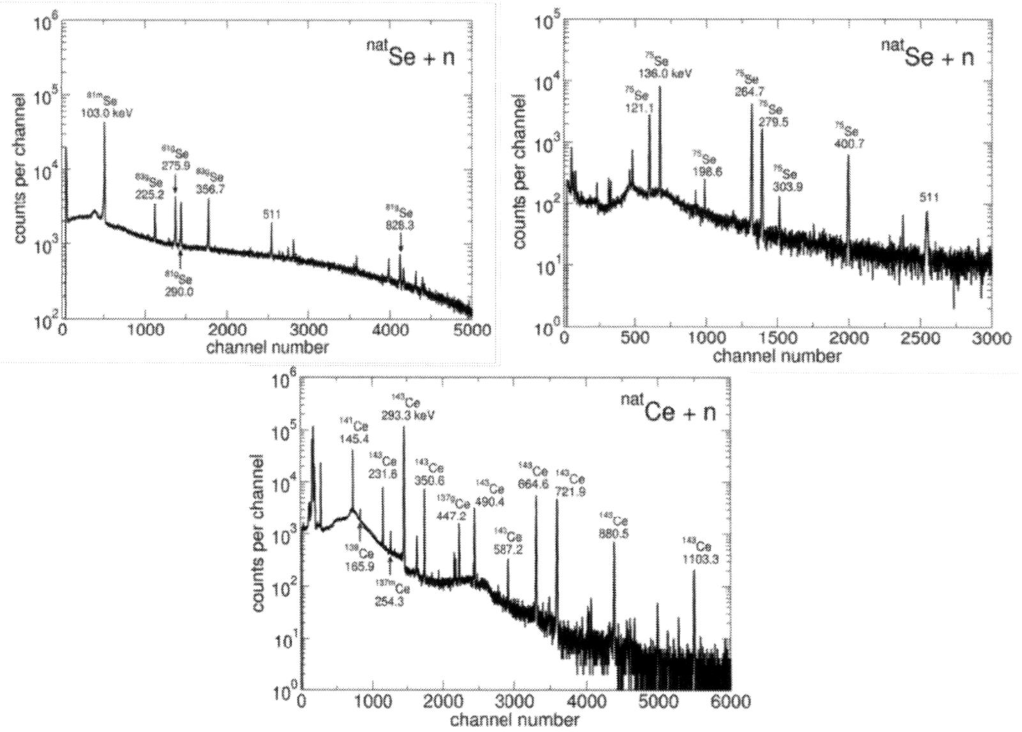

Figure 3: Gamma spectra accumulated with a shielded High-Purity Ge detector for targets of ^{nat}Se (top) and ^{nat}Ce (bottom) activated in the neutron spectrum from LiLiT. The identified lines of the respective (n, γ) reaction products are indicated. The top left spectrum is for the short-lived ^{81,83}Se isotopes and the top right spectrum is for the longer-lived ^{75}Se isotope.

A, an irradiation of protons on LiLiT is conducted below the neutron production threshold. This results in the irradiation of the sample with γ-rays, but without neutrons. The (γ, n) counts are then subtracted to obtain only the (n, γ) cross sections. The thick-target intensity of the 17.6 MeV and 14.6 MeV γ rays at the SARAF-LiLiT setup from the ^{7}Li$(p, \gamma)^{8}$Be reaction is estimated to be $\approx 7 \times 10^{8}\ \gamma s^{-1}$ per mA of proton beam at our relevant proton energy [8, 9]. Gamma spectra from the γ activated targets are displayed in Fig. 4.

From the isotope activity relative to the Au activity (using the ^{197}Au(n, γ) cross section [12, 13]), the experimental cross section is obtained. The Maxwellian Average Cross Section (MACS) at a given thermal energy kT is then obtained by applying a correction to the experimental cross section (see [8–10] for details).

Preliminary results for the MACS are generally in good agreement with published values; Table 1 lists preliminary results. Further analysis is in progress to reach final values that will be published when ready.

4 Acknowledgments

This work was supported by the Pazy Foundation (Israel). M.P. acknowledges support by the European Union (ChETEC-INFRA, project no. 101008324) and the German-Israeli Foundation (GIF), Grant Nr. 3013004592.

 https://doi.org/10.1051/epjconf/202327906002

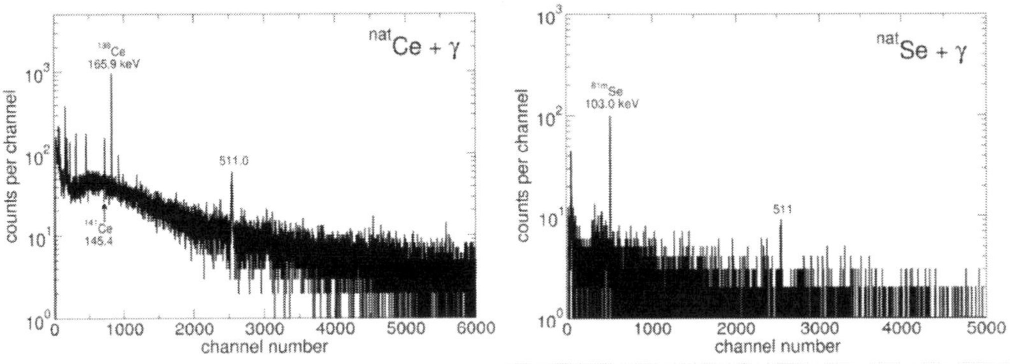

Figure 4: Gamma spectra accumulated with a shielded High-Purity Ge detector for targets of ^{nat}Ce (left) and ^{nat}Se (right) activated below the neutron production threshold resulting in only gammas and no neutrons. The ^{139,141}Ce and ^{81m}Se counts are then subtracted to obtain only the (n, γ) cross sections.

Table 1: Preliminary values (in mb) for MACS at 30 keV for the Se and Ce isotopes.

Nuclide	Recommended, KADoNiS-1.0 [11]	This work (preliminary)
^{74}Se	302(17)	306(11)
^{80}Se $\rightarrow{}^{81m}$Se	8.6(9)	9.3(4)
^{82}Se $\rightarrow{}^{83g}$Se	2.0(2)	2.1(2)
^{136}Ce	328(21)	340(20)
^{136}Ce $\rightarrow{}^{136m}$Ce	28.2(16)	28.9(14)
^{138}Ce	179(5)	201(11)
^{140}Ce	11.73(44)	8.5(5)
^{142}Ce	29.9(10)	26.4(7)

References

[1] E. M. Burbidge, G. R. Burbidge, W. A. Fowler, F. Hoyle, Rev. Mod. Phys. **29**, 547 (1957).
[2] F. Kaeppeler *et al.*, Rev. Mod. Phys. **83**, 157 (2011).
[3] M. Tessler *et al.*, Phys. Rev. C **105**, 035801 (2022).
[4] F. Kaeppeler, *et al.*, Phys. Rev. C **53**, 1397 (1996).
[5] A. Koloczek, *et al.*, At. Data Nucl. Data Tables **108**, 1 (2016).
[6] I. Mardor, *et al.*, Eur. Phys. J. A **54**, 91 (2018).
[7] S. Halfon, *et al.*, Rev. Sci. Instrum. **85**, 056105 (2014).
[8] M. Paul, *et al.*, Eur. Phys. J. A **55**, 44 (2019).
[9] M. Tessler, *et al.*, Phys. Lett. B **751**, 418 (2015).
[10] M. Tessler, *et al.*, Phys. Rev. C **104**, 015806 (2021).
[11] KADoNiS v1.0, Available online: https://exp-astro.de/kadonis1.0/.
[12] C. Massimi, *et al.*, Phys. Rev. C **81**, 044616 (2010).
[13] C. Lederer, *et al.*, Phys. Rev. C **83**, 034608 (2011).

The impact of n_TOF data on s-process modelling

Samuele Lanzi[1], *Sergio* Cristallo[2,3], *Francesco* Giacomini[4], *Cristian* Massimi[1,3,*], *Alberto* Mengoni[5], and *Diego* Vescovi[6]

[1]Università di Bologna, Dipartimento di Fisica e Astronomia, Italy.
[2]Istituto Nazionale di Astrofisica (INAF) - Osservatorio Astronomico d'Abruzzo, Teramo, Italy
[3]Istituto Nazionale di Fisica Nucleare (INFN), Sezione di Perugia, Perugia, Italy
[4]Istituto Nazionale di Fisica Nucleare (INFN), CNAF, Bologna, Italy
[5]Agenzia nazionale per le nuove tecnologie, l'energia e lo sviluppo economico sostenibile (ENEA), Bologna, Italy
[6]Goethe University Frankfurt, Max-von-Laue-Strasse 1, Frankfurt am Main 60438, Germany.

Abstract. We show examples of the impact of the Maxwellian averaged capture cross sections determined at n_TOF over the past 20 years on AGB stellar nucleosynthesis models. In particular, we developed an automated procedure to derive MACSs from evaluated data libraries, which are subsequently used as input to stellar models computed by means of the FuNS code. In this contribution, we present a number of s-process abundances obtained using different data libraries as input to stellar models, with a focus on the role of n_TOF data.

1 Introduction

The slow neutron capture process (the *s* process [1, 2]) accounts for the formation of approximately half of elements heavier than iron. It is the result of a sequence of neutron radiative captures and β-decays. More in detail, starting from iron seeds, the *s* process produces heavy elements up to bismuth. Contrary to the rapid process (*r* process), β-decay rates are faster than neutron capture rates, and therefore the reaction flow proceeds along the β stability valley of nuclei. The *s* process can take place in two different scenarios: during He-shell flashes in the asymptotic giant branch (AGB) phase [3–6] of low- and intermediate-mass stars (also referred to as the main component, and responsible for the formation of isotopes with A>90) and in core-He burning and shell-C burning of massive stars [7] (also referred to as the weak component, and responsible for the formation of isotopes with 60<A<90).

Together with β-decay rates, neutron capture cross sections are the basic nuclear physics input to the slow neutron capture process. Since 2001, the neutron time-of-flight facility n_TOF at CERN has provided accurate cross section data for the *s* process. More in detail, (n,γ) cross sections were measured as a function of energy, using the time-of-flight method. Moreover, improved detection systems, innovative ideas, and collaborations with other neutron facilities have led to a considerable contribution of the n_TOF collaboration to *s* process studies. Results have been reported for stable and radioactive samples, i.e., ^{24,25,26}Mg, ^{54,57}Fe, ^{58,59,62,63}Ni, ^{70,72,73}Ge, ^{90,91,92,93,94,96}Zr, ^{139}La, ^{140}Ce, ^{147}Pm, ^{151}Sm, ^{154,155,157}Gd, ^{171}Tm, ^{186,187,188}Os, ^{197}Au, ^{203,204}Tl, ^{204,206,207}Pb and ^{209}Bi isotopes (see [8] for

*e-mail: massimi@bo.infn.it

more details), and others are being studied or planned to be studied in the near future. In some cases, these results were then used to improve Evaluated Nuclear Data File such as ENDF/B-VIII.0 [9], JENDL-5.0 [10] and JEFF-3.3 [11] or TENDL-2021 [12] and other international projects. The same data were used to derive the neutron-induced cross section averaged over the stellar neutron-energy distribution, typically referred to as Maxwellian averaged cross section (MACS) - more details in Sec. 2. This quantity represents the key nuclear physics input to stellar models for the study of nucleosynthesis in Red Giants stars.

We have developed an automated procedure to derive pointwise MACS for temperatures between 10 and 1000 MK, using the data from evaluated nuclear data files (in ENDF format), in the format required by FuNS [13]. This system allowed us to study the impact of n_TOF data on AGB nucleosynthesis models, as well as to compare nuclear data in international libraries.

2 From the ENDF format to stellar reaction rates

The astrophysical reaction rate is a function of the number density of interacting particles times the reaction rate per particle pair $\langle\sigma v\rangle$. This latter term describes the probability of nuclear reactions between two particles, moving at relative velocity v. In a stellar plasma, the interacting particles are in thermodynamic equilibrium, and consequently, their kinetic energy is linked to their thermal motion (i.e. the relative velocity can be described by a Maxwell-Boltzmann distribution). More in detail, its maximum occurs at the velocity $v_T = \sqrt{\frac{2kT}{\mu}}$, μ being the reduced mass of the system formed by the interacting particles. By expressing the velocity distribution as energy distribution:

$$\langle\sigma v\rangle = \left(\frac{8}{\pi\mu}\right)^{1/2} \frac{1}{(kT)^{3/2}} \int_0^\infty E\sigma(E)e^{-\frac{E}{kT}dE}, \tag{1}$$

the reaction rate can be expressed in terms of the MACS:

$$\text{MACS} = \frac{\langle\sigma v\rangle}{v_T} = \frac{2}{\sqrt{\pi}(kT)^2} \int_0^\infty E\sigma(E)e^{-\frac{E}{kT}dE}. \tag{2}$$

Stellar models, as for example FuNS [13], require as input the reaction rate evaluated at temperatures between 10^7 K and 10^{10} K in intervals of $10^{0.1}$ Kelvin. With the aim of producing these input files, we have developed an automatic procedure running over all the isotopes contained in a library, consisting of these 4 steps:

1. download of the preferred evaluation from the IAEA webpage, each evaluation containing about 1000-3000 isotopes;

2. interpretation of the ENDF file, from which the pointwise (n,γ) cross section $\sigma(E)$ is reconstructed in the laboratory system;

3. numerical integration to obtain the MACS from Eq. 2, after transformation to the center of mass system;

4. calculation of the astrophysical reaction rate using as MACS times v_T times the number density of particles.

In the same procedure, we have also included the option to multiply the reaction rate for the stellar enhancement factor from the Karlsruhe Astrophysical Database of Nucleosynthesis in Stars (KADoNiS) [14].

3 Results

The procedure described above has allowed us to perform systematic studies of the impact of different databases on the *s* process nucleosynthesis calculations.

Firstly, we have compared the MACS at $kT = 8$ keV obtained from the latest versions of the major libraries, namely ENDF/B-VIII.0, JEFF-3.3, JENDL-5 and TENDL-2021. Their ratios are shown in Fig. 1 for isotopes heavier than iron (A>60). This study includes

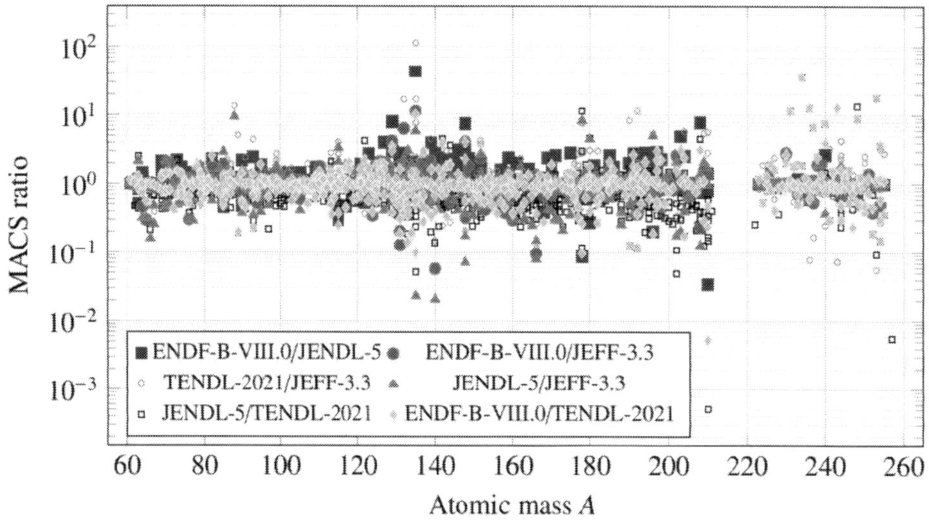

Figure 1. Ratio of MACS at $kT = 8$ keV obtained from major nuclear data libraries. Large deviations are observed for metastable states and short-lived radioisotopes.

metastable states as well as short-lived radioisotopes and therefore it should not surprise if the ratios span over 4 orders of magnitudes. Excluding these latter cases, large deviations between libraries are still present. For instance deviations of 20% for ^{63}Cu and ^{77}Se, and of 50% for ^{82}Se or ^{122}Sn are observed, to mention a few.

Another interesting study is related to s-only isotopes, whose production is entirely ascribed to the *s* process due to the presence of stable isobars. For those nuclei, a semi-empirical approach to the *s* process results in the so-called "local-equilibrium approximation". More in detail, for nuclei far from closed shell configuration, the abundance builds up until the destruction rate approaches the production rate. Consequently, a steady flow is reached along the *s*-process path, and therefore the product of the abundance times the MACS is approximately constant [1]. To date at n_TOF, 4 of the 33 *s*-only isotopes were studied: ^{58}Ni [16], ^{70}Ge [17], ^{154}Gd [18] and ^{186}Os [19, 20]. The MACS obtained experimentally at n_TOF, and reported in Fig. 2 are in very good agreement.

As a last example of the application of the described procedure, Fig. 3 shows the results of a refined stellar model for a 2 $M_\odot$ asymptotic giant branch star with metallicity Z = 0.01 to derive *s*-process abundances. (n,γ) cross sections from ENDF/B-VIII.0, JEFF-3.3, JENDL-5, and TENDL-2021 evaluations were given as input to the stellar model, and the obtained information was then used to interpret presolar grain isotopic abundances of barium, strontium, nickel, molybdenum, and zirconium isotopes (see, e.g., [21, 22]). In Fig. 3 the results obtained with data from KADoNiS 0.3 & n_TOF data, considered as a reference (REF), are also reported. It looks evident how much strontium, barium, and zirconium isotopic ratios are sensitive to the adopted library.

 https://doi.org/10.1051/epjconf/202327906003

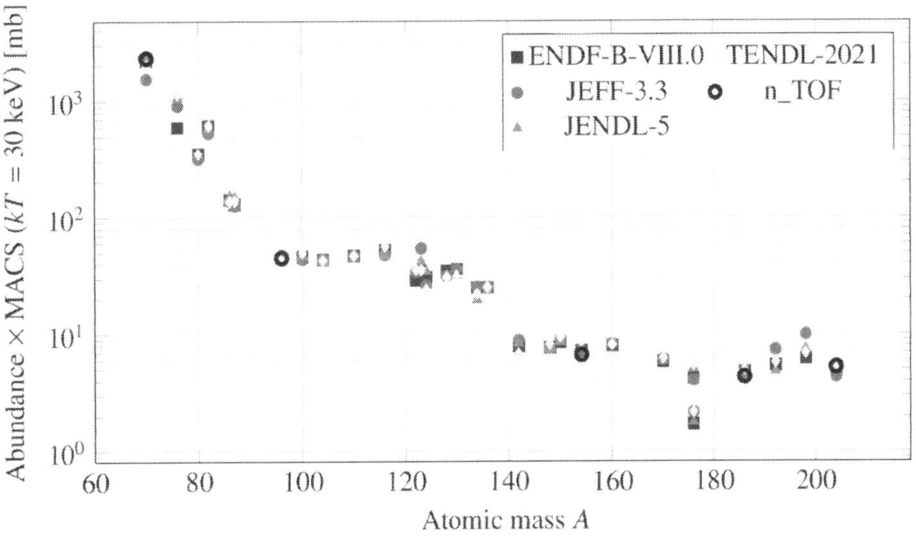

Figure 2. The product of the cross section (derived from nuclear data libraries) times the solar s-process abundances for s-only isotopes [15]. In the region between magic neutron numbers, the local equilibrium approximation is clearly visible. The four cases studied at n_TOF are also reported for comparison.

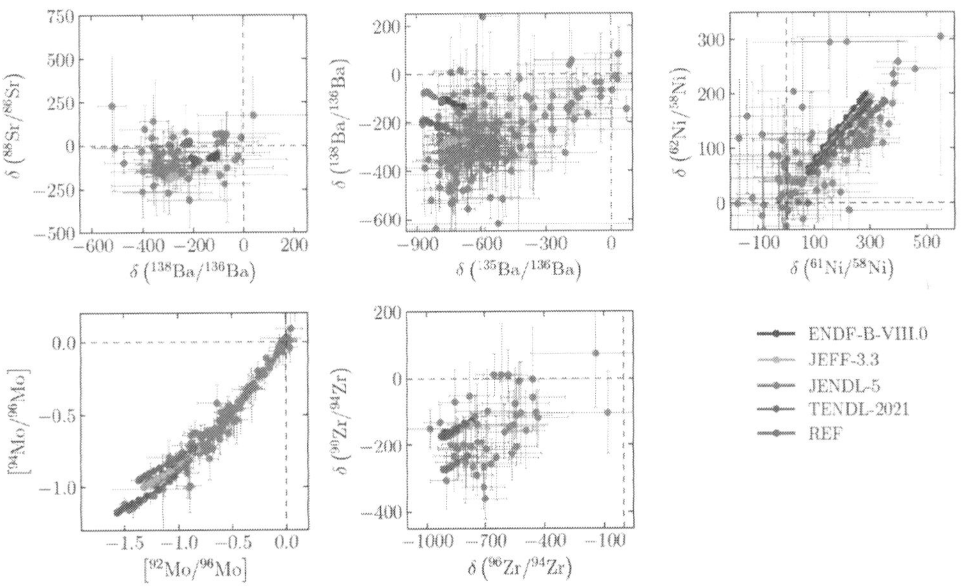

Figure 3. Comparison between AGB model calculations [23, 24] using different nuclear data input and isotopic abundances in presolar grains. The results obtained with data from KADoNiS 0.3, taken as a reference (REF), are shown in red. Grain data are from the PGD database [25].

4 Conclusions

The automated procedure described in this contribution seems to be a useful tool for systematic studies of the impact of cross section data on stellar models. In addition, it can be used to assess the impact of n_TOF cross section data on *s*-process nucleosynthesis calculations.

References

[1] E.M. Burbidge, G.R. Burbidge, W.A. Fowler, F. Hoyle, Reviews of Modern Physics **29**, 547 (1957)
[2] F. Käppeler, R. Gallino, S. Bisterzo, W. Aoki, Reviews of Modern Physics **83**, 157 (2011)
[3] M. Busso, R. Gallino, G.J. Wasserburg, ARA&A **37**, 239 (1999)
[4] F. Herwig, Annual Review of Astronomy and Astrophysics **43**, 435 (2005)
[5] O. Straniero, R. Gallino, S. Cristallo, Nuclear Physics A **777**, 311 (2006)
[6] A.I. Karakas, J.C. Lattanzio, PASA **31**, e030 (2014)
[7] M. Pignatari, R. Gallino, M. Heil et al., The Astrophysical Journal **710**, 1557 (2010)
[8] C. Massimi, S. Cristallo, C. Domingo-Pardo, C. Lederer-Woods, Universe **8** (2022)
[9] D. Brown, M. Chadwick, R. Capote, A. Kahler, A. Trkov, M. Herman, A. Sonzogni, Y. Danon, A. Carlson, M. Dunn et al., Nuclear Data Sheets **148**, 1 (2018), special Issue on Nuclear Reaction Data
[10] O. Iwamoto, N. Iwamoto, K. Shibata, A. Ichihara, S. Kunieda, F. Minato, S. Nakayama, *Status of JENDL*, in *European Physical Journal Web of Conferences* (2020), Vol. 239 of *European Physical Journal Web of Conferences*, p. 09002
[11] A.J.M. Plompen, O. Cabellos, C. De Saint Jean, M. Fleming, A. Algora, M. Angelone, P. Archier, E. Bauge, O. Bersillon, A. Blokhin et al., European Physical Journal A **56**, 181 (2020)
[12] A. Koning, D. Rochman, J.C. Sublet, N. Dzysiuk, M. Fleming, S. van der Marck, Nuclear Data Sheets **155**, 1 (2019), special Issue on Nuclear Reaction Data
[13] S. Cristallo, L. Piersanti, O. Straniero et al., ApJS **197**, 17 (2011), 1109.1176
[14] I. Dillmann, F. Käppeler, R. Plag, T. Rauscher, *KADoNiS v0.3 – The third update of the "Karlsruhe Astrophysical Database of Nucleosynthesis in Stars"*, in *EFNUDAT Fast Neutrons - Proceedings of the Scientific Workshop on Neutron Measurements, Theory and Applications - Nuclear Data for Sustainable Nuclear Energy - 28-30 April 2009, Geel, Belgium . EUR 23883 EN. Luxembourg (Luxembourg): Publications Office of the European Union; 2010. JRC56548. https://publications.jrc.ec.europa.eu/repository/handle/JRC56548 (accessed on January 2022)* (2010), p. 123
[15] K. Lodders, Space Science Reviews **217**, 44 (2021)
[16] P. Žugec, M. Barbagallo, N. Colonna, D. Bosnar, S. Altstadt, J. Andrzejewski, L. Audouin, V. Bécares, F. Bečvář, F. Belloni et al., Physical Review C **89**, 014605 (2014)
[17] A. Gawlik, C. Lederer-Woods, J. Andrzejewski, U. Battino, P. Ferreira, F. Gunsing, S. Heinitz, M. Krtička, C. Massimi, F. Mingrone et al., Physical Review C **100**, 045804 (2019), 1910.09360
[18] A. Mazzone, S. Cristallo, O. Aberle, G. Alaerts, V. Alcayne, S. Amaducci, J. Andrzejewski, L. Audouin, V. Babiano-Suarez, M. Bacak et al., Physics Letters B **804**, 135405 (2020), 2002.02322
[19] M. Mosconi, K. Fujii, A. Mengoni, C. Domingo-Pardo, F. Käppeler, U. Abbondanno, G. Aerts, H. Álvarez-Pol, F. Alvarez-Velarde, S. Andriamonje et al., Physical Review C **82**, 015802 (2010)
[20] K. Fujii, M. Mosconi, A. Mengoni, C. Domingo-Pardo, F. Käppeler, U. Abbondanno, G. Aerts, H. Álvarez-Pol, F. Alvarez-Velarde, S. Andriamonje et al., Physical Review C **82**, 015804 (2010)
[21] N. Liu, R. Gallino, S. Cristallo et al., ApJ **865**, 112 (2018), 1808.03614
[22] N. Liu, T. Stephan, S. Cristallo et al., ApJ **881**, 28 (2019), 1906.10776
[23] D. Vescovi, S. Cristallo, M. Busso, N. Liu, ApJL **897**, L25 (2020), 2006.13729
[24] D. Vescovi, S. Cristallo, S. Palmerini et al., A&A **652**, A100 (2021), 2106.08241
[25] T. Stephan, M. Bose, A. Boujibar et al., *The Presolar Grain Database Reloaded - Silicon Carbide*, in *51st Annual Lunar and Planetary Science Conference* (2020), Lunar and Planetary Science Conference, p. 2140

First measurement of the ^{94}Nb(n,γ) cross section at the CERN n_TOF facility

J. Balibrea-Correa[1*], V. Babiano-Suárez[1], J. Lerendegui-Marco[1], C. Domingo-Pardo[1], I. Ladarescu[1], A. Tarifeño-Saldivia[1], V. Alcayne[2], D. Cano-Ott[2], E. González-Romero[2], T. Martínez[2], E. Mendoza[2], C. Guerrero[3], F. Calviño[4], A. Casanovas[4], U. Köster[5], N. M. Chiera[6], R. Dressler[6], E. A. Maugeri[6], D. Schumann[6], O. Aberle[7], S. Altieri[8,9], S. Amaducci[10], J. Andrzejewski[11], M. Bacak[7], C. Beltrami[8], S. Bennett[12], A. P. Bernardes[7], E. Berthoumieux[13], R. Beyer[14], M. Boromiza[15], D. Bosnar[16], M. Caamaño[17], M. Calviani[7], F. Cerutti[7], G. Cescutti[18,19], E. Chiaveri[7,12], P. Colombetti[20,21], N. Colonna[22], P. Console Camprini[23,24], G. Cortés[4], M. A. Cortés-Giraldo[3], L. Cosentino[10], S. Cristallo[25,26], S. Dellmann[27], M. Di Castro[7], S. Di Maria[28], M. Diakaki[29], M. Dietz[30], E. Dupont[13], I. Durán[17], Z. Eleme[31], S. Fargier[7], B. Fernández[3], B. Fernández-Domínguez[17], P. Finocchiaro[10], S. Fiore[24,32], F. García-Infantes[33], A. Gawlik-Ramięga[11], G. Gervino[20,21], S. Gilardoni[7], F. Gunsing[13], C. Gustavino[32], J. Heyse[34], W. Hillman[12], D. G. Jenkins[35], E. Jericha[36], A. Junghans[14], Y. Kadi[7], K. Kaperoni[29], G. Kaur[13], A. Kimura[37], I. Knapová[38], M. Kokkoris[29], M. Krtička[38], N. Kyritsis[29], C. Lederer-Woods[39], G. Lerner[7], A. Manna[23,40], A. Masi[7], C. Massimi[23,40], P. Mastinu[41], M. Mastromarco[22,42], A. Mazzone[22,43], A. Mengoni[24,23], V. Michalopoulou[29], P. M. Milazzo[18], R. Mucciola[25,44], F. Murtas[45], E. Musacchio-Gonzalez[41], A. Musumarra[46,47], A. Negret[15], A. Pérez de Rada[2], P. Pérez-Maroto[3], N. Patronis[31], J. A. Pavón-Rodríguez[3], M. G. Pellegriti[46], J. Perkowski[11], C. Petrone[15], E. Pirovano[30], J. Plaza[2], S. Pomp[48], I. Porras[33], J. Praena[33], J. M. Quesada[3], R. Reifarth[27], D. Rochman[6], Y. Romanets[28], C. Rubbia[7], A. Sánchez-Caballero[2], M. Sabaté-Gilarte[7], P. Schillebeeckx[34], A. Sekhar[12], A. G. Smith[12], N. V. Sosnin[39], M. E. Stamati[31], A. Sturniolo[20], G. Tagliente[22], D. Tarrío[48], P. Torres-Sánchez[33], E. Vagena[31], S. Valenta[38], V. Variale[22], P. Vaz[28], G. Vecchio[10], D. Vescovi[27], V. Vlachoudis[7], R. Vlastou[29], A. Wallner[14], P. J. Woods[39], T. Wright[12], R. Zarrella[23,40], and P. Žugec[16]
and the n_TOF Collaboration

[1]Instituto de Física Corpuscular, CSIC - Universidad de Valencia, Spain
[2]Centro de Investigaciones Energéticas Medioambientales y Tecnológicas (CIEMAT), Spain
[3]Universidad de Sevilla, Spain
[4]Universitat Politècnica de Catalunya, Spain
[5]Institut Laue Langevin (ILL), Grenoble, France
[6]Paul Scherrer Institut (PSI), Villigen, Switzerland
[7]European Organization for Nuclear Research (CERN), Switzerland
[8]Istituto Nazionale di Fisica Nucleare, Sezione di Pavia, Italy
[9]Department of Physics, University of Pavia, Italy
[10]INFN Laboratori Nazionali del Sud, Catania, Italy
[11]University of Lodz, Poland
[12]University of Manchester, United Kingdom
[13]CEA Irfu, Université Paris-Saclay, F-91191 Gif-sur-Yvette, France
[14]Helmholtz-Zentrum Dresden-Rossendorf, Germany

*e-mail: javier.balibrea@ific.uv.es

[15]Horia Hulubei National Institute of Physics and Nuclear Engineering, Romania
[16]Department of Physics, Faculty of Science, University of Zagreb, Zagreb, Croatia
[17]University of Santiago de Compostela, Spain
[18]Istituto Nazionale di Fisica Nucleare, Sezione di Trieste, Italy
[19]Department of Physics, University of Trieste, Italy
[20]Istituto Nazionale di Fisica Nucleare, Sezione di Torino, Italy
[21]Department of Physics, University of Torino, Italy
[22]Istituto Nazionale di Fisica Nucleare, Sezione di Bari, Italy
[23]Istituto Nazionale di Fisica Nucleare, Sezione di Bologna, Italy
[24]Agenzia nazionale per le nuove tecnologie (ENEA), Italy
[25]Istituto Nazionale di Fisica Nucleare, Sezione di Perugia, Italy
[26]Istituto Nazionale di Astrofisica - Osservatorio Astronomico di Teramo, Italy
[27]Goethe University Frankfurt, Germany
[28]Instituto Superior Técnico, Lisbon, Portugal
[29]National Technical University of Athens, Greece
[30]Physikalisch-Technische Bundesanstalt (PTB), Bundesallee 100, 38116 Braunschweig, Germany
[31]University of Ioannina, Greece
[32]Istituto Nazionale di Fisica Nucleare, Sezione di Roma1, Roma, Italy
[33]University of Granada, Spain
[34]European Commission, Joint Research Centre (JRC), Geel, Belgium
[35]University of York, United Kingdom
[36]TU Wien, Atominstitut, Stadionallee 2, 1020 Wien, Austria
[37]Japan Atomic Energy Agency (JAEA), Tokai-Mura, Japan
[38]Charles University, Prague, Czech Republic
[39]School of Physics and Astronomy, University of Edinburgh, United Kingdom
[40]Dipartimento di Fisica e Astronomia, Università di Bologna, Italy
[41]INFN Laboratori Nazionali di Legnaro, Italy
[42]Dipartimento Interateneo di Fisica, Università degli Studi di Bari, Italy
[43]Consiglio Nazionale delle Ricerche, Bari, Italy
[44]Dipartimento di Fisica e Geologia, Università di Perugia, Italy
[45]INFN Laboratori Nazionali di Frascati, Italy
[46]Istituto Nazionale di Fisica Nucleare, Sezione di Catania, Italy
[47]Department of Physics and Astronomy, University of Catania, Italy
[48]Uppsala University, Sweden

Abstract. One of the crucial ingredients for the improvement of stellar models is the accurate knowledge of neutron capture cross-sections for the different isotopes involved in the *s*-,*r*- and *i*- processes. These measurements can shed light on existing discrepancies between observed and predicted isotopic abundances and help to constrain the physical conditions where these reactions take place along different stages of stellar evolution.

In the particular case of the radioactive ^{94}Nb, the ^{94}Nb(n,γ) cross-section could play a role in the determination of the *s*-process production of ^{94}Mo in AGB stars, which presently cannot be reproduced by state-of-the-art stellar models. There are no previous ^{94}Nb(n,γ) experimental data for the resolved and unresolved resonance regions mainly due to the difficulties in producing high-quality samples and also due to limitations in conventional detection systems commonly used in time-of-flight experiments.

Motivated by this situation, a first measurement of the ^{94}Nb(n,γ) reaction was carried out at CERN n_TOF, thereby exploiting the high luminosity of the EAR2 area in combination with a new detection system of small-volume C6D6-detectors and a high quality ^{94}Nb-sample. The latter was based on hyper-pure ^{93}Nb material activated at the high-flux reactor of ILL-Grenoble. An innovative ring-configuration detection system in close geometry around the capture

sample allowed us to significantly enhance the signal-to-background ratio. This set-up was supplemented with two conventional C_6D_6 detectors and a high-resolution $LaCl_3$(Ce)-detector, which will be employed for addressing reliably systematic effects and uncertainties.
At the current status of the data analysis, 18 resonance in ^{94}Nb+n have been observed for the first time in the neutron energy range from thermal up to 10 keV.

1 Introduction

The slow neutron capture process, or s process, is responsible for about half of the isotopic abundances of chemical elements heavier than iron [1]. This process takes place in Massive Stars ($>10M_\odot$) and in Asymptotic Giant Branch (AGB) stars of $M<4M_\odot$ at low or moderated neutron fluxes, mainly produced by the $^{13}C(\alpha, n)$ and $^{22}Ne(\alpha, n)$ reactions[2]. In such stellar conditions, the neutron capture time scale is slower than for β-decay process, thus populating isotopes close to the stability-valley. In order to model and understand the s process and the corresponding stellar media one of the crucial ingredients comes from the nuclear data, in particular from the neutron capture cross-sections of the involved isotopes [2, 3]. This experimental information serves as input for state-of-the-art stellar models [2, 3], which can be progressively improved and refined by comparing their predictions with astronomical observations of elemental abundances and with isotopic compositions of pre-Solar meteorites [4, 5].

In this context, the isotopic abundances measured in presolar SiC grains represent a long standing puzzle [5]. The predicted relative isotopic abundance of ^{94}Mo is of 1% or less, whereas isotopic analysis of pre-solar SiC grains yield a factor of five more (5%) [6]. This problem has been discussed recently by different groups [6, 7]. One of the possible solutions to this discrepancy could come from the uncertain nuclear data involved in the abundance calculations [6, 7]. In the particular case of the long-lived ^{94}Nb isotope, the ^{94}Nb(n,γ) cross section could play a role in the s-process production of ^{94}Mo in AGB stars if the β-decay to ^{94}Mo would be higher than expected or, correspondingly, the cross section would be smaller [7]. As of today, there are no previous measurements of the ^{94}Nb(n,γ) reaction for the resolved and unresolved resonance regions. This situation was mainly ascribed to difficulties in the production of high-quality ^{94}Nb samples, as well as limitations in conventional detection systems commonly used in time-of-flight experiments. With the present work, the latter two difficulties have been significantly alleviated, as it is described below.

2 Experimental challenges and setup

The first challenge in the measurement of the ^{94}Nb(n,γ) cross-section was related to the production of a suitable sample for a TOF experiment. The conventional methodology of producing ^{94}Nb from activation of commonly available ^{93}Nb materials is not suitable for an accurate neutron capture cross-section determination based on the time-of-flight technique. Although ^{93}Nb is naturally mono-isotopic, there is no supplier that can certificate a sample containing less than about 100 ppm of Ta. This quantity is far too much because it would take many years to sit out the ^{182}Ta produced during a neutron irradiation in a nuclear reactor. To avoid such problematic, a hyper-pure sample of ^{93}Nb, originally produced at the Institute of Solid State and Materials Research of Dresden [8], was made available for this measurement thanks to a collaboration with the Institute Laue-Langevin (ILL) in France. The synthesis of the ultra-high-purity Nb was based on a refined molten salt electrolysis, zone melting and high and ultra-high vacuum heat treatments. The final ^{93}Nb material available, 304 mg with <1 ppm of Ta, had a shape of two thin wires of 45 and 47 mm long and 1 mm diameter. This high purity material was afterwards activated at the high-flux nuclear ILL-Grenoble reactor for 51 days and a power weighted fluence of 42 full-power days. A careful characterization of the activated sample was performed at the Paul Scherrer Institute (PSI), in Switzerland, by means of a customized HPGe gamma-ray spectroscopy set-up. In this study an activity of 10.1 MBq of ^{94}Nb was found, without any trace of contaminants, thereby yielding $9.24 \cdot 10^{18}$ atoms of ^{94}Nb. This represents ~1% of the total number of atoms present in the bulk of the sample.

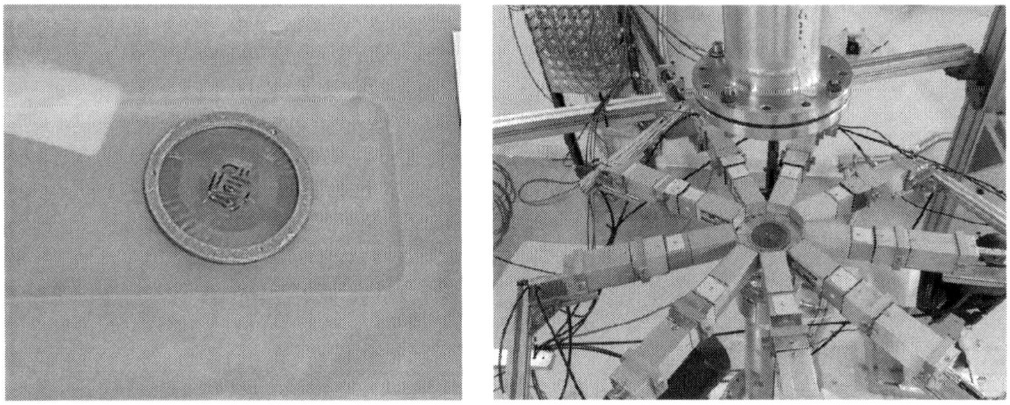

Figure 1. Left panel: Picture of the ^{94}Nb sample manufactured at PSI. Right panel: Picture of the experimental setup used during the experimental campaign at the n_TOF facility in April 2022.

After the irradiation at ILL and characterization at PSI, the ^{94}Nb sample for the capture experiment was prepared in a hotcell at PSI. Its final shape is shown in the left panel of Fig 1. The initial aim of producing an spiral-shaped sample was hindered by the radiation damage produced in the Nb material during the neutron irradiation at ILL.

The other main challenge in this measurement was related to the decay activity of the Nb sample itself. The β-decay of ^{94}Nb is followed by a two-step γ-ray cascade of 702 keV and 871 keV. With an activity of 10 MBq, the TOF measurement of this sample is still quite challenging owing to the gamma-ray contribution of the sample activity to the overall background and the relatively small concentration (1%) of ^{94}Nb in the sample. Due to the aforementioned arguments, the neutron-capture measurement of ^{94}Nb was carried out at the vertical experimental area (EAR2) of the CERN n_TOF facility [9]. EAR2 is specially well suited for this particular type of samples because of its high instantaneous neutron flux and its still high energy resolution [10]. A full detailed report on the proposal can be found in [11]. Additionally, during the last years an effort has been made at CERN n_TOF in order to develop new detection systems capable of coping with the high count-rate requirements of EAR2 and with a fast response to recover quickly after the so-called gamma-flash [9, 10]. These requirements have been achieved, to a large extent, with the new segmented Total Energy Detector (s-TED). Their small active volume, which is about 1/9 liter of C_6D_6, together with new PMTs (Hamamatsu-R11265U) designed for fast-response and high count-rate conditions, mitigate the effects of the gamma-flash and enlarge the neutron energy range where they can operate under stable and well controlled performance conditions [12]. A full paper detailing the detector construction, performance and first results is in preparation [13]. The new s-TED detectors were arranged in an innovative compact-ring configuration, designed to maximize the sensitivity for the (n,γ) reactions in the capture sample itself. A picture of the complete experimental setup during the ^{94}Nb campaign is shown in the right panel of Fig. 1. Thus, the main detection system consists of an array of 9 s-TED detectors arranged as-close-as-possible around the capture sample. Additionally, two C_6D_6 and one $LaCl_3$(Ce) were placed in the setup with the goal of addressing systematic effects and uncertainties, detecting possible angular anisotropies in the prompt gamma-ray cascades and extract spectroscopic information.

The experimental campaign took place between March and April of 2022, including also dedicated runs for background determination and capture-yield normalization. For the latter aim also hyperpure ^{93}Nb samples and ^{197}Au samples with specific geometries were used.

3 First preliminary results

In addition to the main runs with a ^{94}Nb sample an ancillary measurement of an hyper-pure ^{93}Nb sample was carried out. The latter had the main aim of determining the contribution of ^{93}Nb(n,γ) within the ^{94}Nb(n,γ) measurement, and also to address corrections for multiple scattering and neutron flux intersection. The counts per proton pulse registered with the s-TED array using the ^{94}Nb- and ^{93}Nb-samples in the neutron energy range from thermal up to 10 keV are displayed in both panels of Fig. 2. It is worth to mention that, while for the stable ^{93}Nb-sample the main background component is due to the contribution from the sample backing (blue component in right panel of Fig. 2), in the case of the ^{94}Nb measurement the background is fully dominated by the β-decays in the radioactive sample itself (black line in the left-panel of Fig. 2).

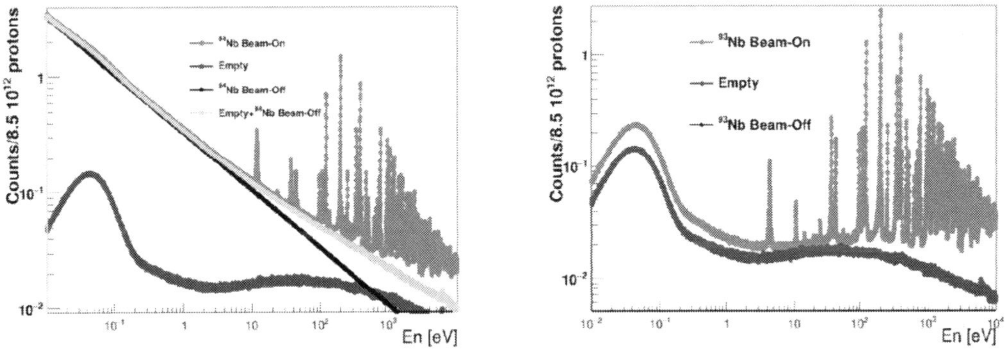

Figure 2. Counts per proton pulse registered with the s-TED array during different configurations. The left panel shows the results for the ^{94}Nb-sample measurement, whereas the pure 93Nb sample measurement is shown on the right panel.

Some preliminary results after background subtraction are shown in both panels of Fig. 3.

The left panel in Fig. 3 shows the first two candidates for ^{94}Nb+n resonances at low neutron energy. The count-rate of the ancillary ^{93}Nb-sample measurement is also shown for comparison. Some Ta impurities, labelled as C, are apparent in the latter measurement which, however, do not interfer with the determination of the ^{94}Nb(n,γ) capture yield. The right-panel in Fig. 3 shows a higher neutron-energy range, between 260 eV and 330 eV. From the three resonances visible in the figure resonances, only the big resonance near 330 eV can be ascribed to capture on ^{93}Nb, whereas the other two levels at lower energy belong to capture events on ^{94}Nb. The observed first neutron resonances up to 200 eV are in very good agreement with the unpublished transmission data from M. R. Serpa performed in 1970 [14]. It is worth to remind that the ^{94}Nb/^{93}Nb ratio is of only ~1%, which shows the overall high sensitivity attained with s-TED detectors for the capture-channel of interest and the large

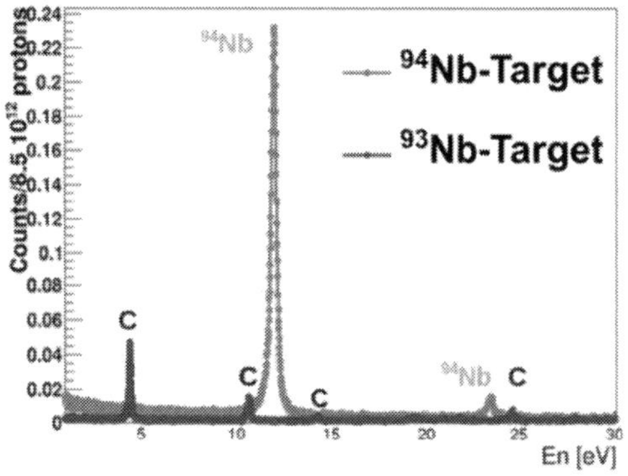

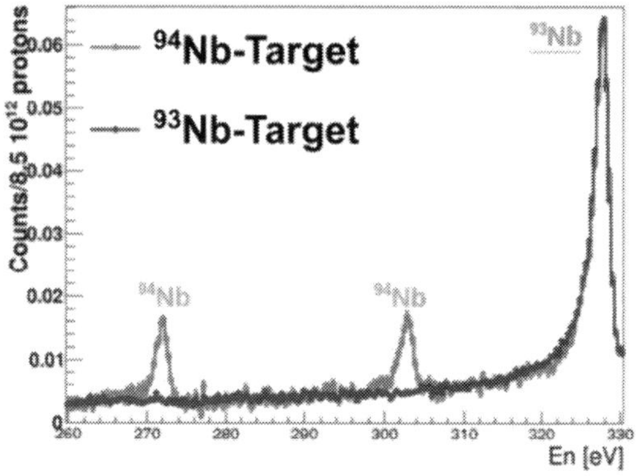

Figure 3. Comparison between experimental yield after background subtraction for ^{94}Nb and ^{93}Nb-spiral configurations for two selected neutron energy ranges. The ^{94}Nb, ^{93}Nb and contaminant resonances are tagged by orange, light-blue and black colors, respectively.

reduction of sample-related background contribution achieved thanks to the high neutron luminosity of EAR2.

At the current preliminary status of the analysis, a completely new capture 18 resonances in the neutron energy range from thermal up to 10 keV have been identified as possible candidates for ^{94}Nb(n,γ) reaction. The next steps in the data analysis comprise the implementation of a Bayesian-based algorithm for reliably addressing the different background components, the determination of the capture-yield by applying the pulse-height weighting technique[15] and the final R-matrix analysis of the resolved resonances. Additionally it is foreseen to perform a complementary measurement at the n_TOF NEAR station [16] via activation, once the facility is fully characterized and operational, for validating the results from this experimental campaign at several stellar temperatures via activation.

4 Summary and conclusions

A first measurement of the ^{94}Nb(n,γ) reaction has been carried out at the CERN n_TOF EAR2 facility in order to determine the neutron-capture cross-section over a broad neutron-energy range, from thermal energy up to several tens of keV [11]. This experiment was carried out with the aim of resolving discrepancies between AGB-model predictions [6] and measured isotopic abundances of Mo-isotopes in SiC-grains[5]. The final impact of the measured capture data will be evaluated once the final results of the measurement are obtained.

Until now, there were no previous experimental data on ^{94}Nb(n,γ) owing to the difficulties to produce a suitable sample for a TOF experiment, and the stringent requirements of the experimental set-up for such a TOF measurement.

In this work, a sample containing 304 mg of hyperpure ^{93}Nb and ^{94}Nb with an ^{94}Nb/^{93}Nb isotopic ratio of only ~1%, was produced thanks to materials and resources from the Institute of Solid State and Materials Research of Dresden, ILL Grenoble and PSI Villigen.

The ^{94}Nb(n,γ) time-of-flight experiment was carried out at the CERN n_TOF facility between March and April of 2022 using a new generation of neutron-capture detectors designed to overcome difficulties related to the use of conventional C_6D_6 detectors [17] in the high instantaneous-flux conditions of EAR2. A compact-ring array-configuration around the capture-sample under study allowed one to achieve an excellent signal-to-background ratio and optimize the overall measuring time of 40 days. This set-up was complemented with two conventional C_6D_6 detectors and one $LaCl_3$(Ce) detector. The latter were aimed at obtaining additional information, that can be relevant to address systematic effects and to derive also spectroscopic information.

The preliminary results reported here show that in this measurement it was possible to cover the neutron energy range from thermal up to several tens of keV. In the present status of the analysis, about 18 capture resonances have been identified as candidates for the ^{94}Nb(n,γ) reaction.

Acknowledgments

This work has been carried out in the framework of a project funded by the European Research Council (ERC) under the European Union's Horizon 2020 research and innovation programme (ERC Consolidator Grant project HYMNS, with grant agreement No. 681740). This work was supported by grant ICJ220-045122-I funded by MCIN/AEI/ 10.13039/501100011033 and by European Union NextGenerationEU/PRTR. The authors acknowledge support from the Spanish Ministerio de Ciencia e Innovación under grants PID2019-104714GB-C21, FPA2017-83946-C2-1-P, FIS2015-71688-ERC, CSIC for funding PIE-201750I26. In line with the principles that apply to scientific publishing and the CERN policy in matters of scientific publications, the n_TOF Collaboration recognises the work of V. Furman and Y. Kopatch (JINR, Russia), who have contributed to the experiment used to obtain the results described in this paper.

References

[1] F. Käppeler, A. Mengoni, Nuclear Physics A **777**, 291 (2006), special Isseu on Nuclear Astrophysics
[2] F. Käppeler, R. Gallino, S. Bisterzo, W. Aoki, Rev. Mod. Phys. **83**, 157 (2011)
[3] F. Herwig, Annual Review of Astronomy and Astrophysics **43**, 435 (2005), `https://doi.org/10.1146/annurev.astro.43.072103.150600`
[4] E. Zinner, Annual Review of Earth and Planetary Sciences **26**, 147 (1998), `https://doi.org/10.1146/annurev.earth.26.1.147`
[5] C. Burkhardt, T. Kleine, F. Oberli, A. Pack, B. Bourdon, R. Wieler, Earth and Planetary Science Letters **312**, 390 (2011)
[6] M. Lugaro, A.M. Davis, R. Gallino, M.J. Pellin, O. Straniero, F. Kappeler, The Astrophysical Journal **593**, 486 (2003)
[7] G. Cescutti, R. Hirschi, N. Nishimura, J.W.d. Hartogh, T. Rauscher, A.S.J. Murphy, S. Cristallo, Monthly Notices of the Royal Astronomical Society **478**, 4101 (2018), `https://academic.oup.com/mnras/article-pdf/478/3/4101/25096994/sty1185.pdf`
[8] A. Koethe, J.I. Moench, Materials Transactions, JIM **41**, 7 (2000)
[9] C. Weiß, E. Chiaveri, S. Girod, V. Vlachoudis, O. Aberle, S. Barros, I. Bergström, E. Berthoumieux, M. Calviani, C. Guerrero et al., Nuclear Instruments and Methods in Physics Research Section A: Accelerators, Spectrometers, Detectors and Associated Equipment **799**, 90 (2015)
[10] M. Sabaté-Gilarte, M. Barbagallo, N. Colonna, F. Gunsing, P. Žugec, V. Vlachoudis, Y.H. Chen, A. Stamatopoulos, J. Lerendegui-Marco, M.A. Cortés-Giraldo et al., The European Physical Journal A **53**, 210 (2017)
[11] J.B. Correa, C.D. Pardo, Tech. rep., CERN (2021), `http://cds.cern.ch/record/2731959?ln=es`

[12] J. Lerendegui-Marco, et al., **15th International Conference on Nuclear Data for Science and Technology (ND2022)** (2022)

[13] V. Alcayne, et al., **15th International Conference on Nuclear Data for Science and Technology (ND2022)** (2022)

[14] T.S. et al., Tech. rep., IAEA (1979), `https://www-nds.iaea.org/exfor/servlet/X4sX4arc?op=se&entry=10404&fid=1670&reqx=3246`

[15] U. Abbondanno, G. Aerts, F. Alvarez-Velarde, H. Álvarez-Pol, S. Andriamonje, J. Andrzejewski, G. Badurek, P. Baumann, F. Bečvář, J. Benlliure et al. (n_TOF Collaboration), Phys. Rev. Lett. **93**, 161103 (2004)

[16] G. Gervino, O. Aberle, A.P. Bernardes, N. Colonna, S. Cristallo, M. Diakaki, S. Fiore, A. Manna, C. Massimi, P. Mastinu et al., Universe **8** (2022)

[17] R. Plag, M. Heil, F. Käppeler, P. Pavlopoulos, R. Reifarth, K. Wisshak, Nuclear Instruments and Methods in Physics Research Section A: Accelerators, Spectrometers, Detectors and Associated Equipment **496**, 425 (2003)

Direct reactions for nuclear astrophysics

Philipp Erbacher[1,*], *Brückner* Benjamin[1], *Sophia Florence* Dellmann[1], *Kathrin* Göbel[2], *Tanja* Heftrich[1], *Markus* Reich[1], *Cem Deniz* Kurtulgil[1], *Rene* Reifarth[1], *Meiko* Volknandt[1], *Mario* Weigand[1], *Alexandra* Härth[1], *Marcel* Bennedik[1], *Tabea Marie* Kuttner[1], *Janina* Strahl[1], and *Ernest* Ademi[1]

[1]Goethe University Frankfurt
[2]Facility for Antiproton and Ion Research in Europe GmbH

Abstract. The neutron activation technique is a well established method to measure neutron capture cross sections relevant for the s-process. The ^{7}Li(p,n) reaction at $E_p = 1912$ keV is often used as a neutron source since the energy distribution of the emitted neutrons closely resembles a Maxwell-Boltzmann spectrum of $k_B T = 25$ keV, mimicking the ^{22}Ne(α,n) phase in TP-AGB stars. The weak s-process, which takes place in massive stars, can reach energies up to $k_B T = 90$ keV. Neutron spectra corresponding to a Maxwell-Boltzmann distribution with $k_B T > 25$ keV cannot be produced by the ^{7}Li(p,n) reaction directly. We developed a method to obtain quasi-Maxwellian neutron capture cross sections over a wide energy range by combining a set of spectrum average cross sections measured at six different proton energies and distances between the lithium target and the sample. The measured spectrum averaged cross section can be used to calculate the Maxwellian-Averaged cross-section (MACS) from $k_B T = 25$ keV to $k_B T = 90$ keV. Over the last two years neutron capture cross sections on over 20 isotopes have been measured at Goethe University Frankfurt using this methodology. An overview of the current experimental method, challenges during data analysis and the first results are presented.

1 Introduction

The neutron activation technique is a well established methodology to measure neutron capture cross sections relevant for the s-process. The ^{7}Li(p,n) reaction at $E_p = 1912$ keV is often used as a neutron source since the energy distribution of the emitted neutrons closely resembles a Maxwell-Boltzmann spectrum of $k_B T = 25$ keV, mimicking the ^{22}Ne(α,n) phase in TP-AGB stars. However, for network simulations of the s-process, which takes place in massive stars at temperatures of up to $k_B T = 90$ keV, experimental neutron capture cross sections at higher neutron energies are needed. Neutron spectra corresponding to a Maxwell-Boltzmann distribution with $k_B T > 25$ keV cannot be produced by the ^{7}Li(p,n) reaction directly (see Fig. 1), but this can be achieved by a combination of multiple different neutron spectra, the so-called 90 keV method.

*e-mail: erbacher@iap.uni-frankfurt.de

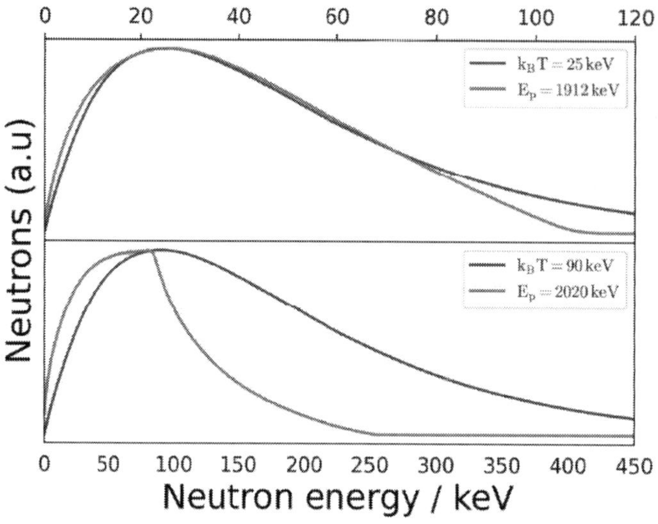

Figure 1: Angle integrated energy distribution of neutrons hitting the sample resulting from the ^{7}Li(p,n) reaction (red) for proton energies of 1912 keV and 2020 keV compared to a Maxwell-Boltzmann distribution (blue) at $k_{\mathrm{B}}T = 25$ keV (top) and $k_{\mathrm{B}}T = 90$ keV (bottom), respectively. While the 1912 keV proton beam can reproduce a $k_{\mathrm{B}}T = 25$ keV distribution quite well, higher proton energies do not directly yield a Maxwell-Boltzmann distribution for higher temperatures.

2 The 90 keV method

The neutrons are produced via the ^{7}Li(p,n) reaction. The neutron energy spectra depend on the incident photon energy and the angle with respect to the beam direction.

Hence the energy distribution observed by the sample can either be changed by varying the distance between the lithium target and the sample, or changing the size and shape of the sample under study. During the experiment, we used an up to 100 µm thick lithium target to ensure that a possible degradation of the lithium target during irradiation does not affect the neutron energy spectrum significantly.

A quasi-Maxwellian $S_{\mathrm{qMB}}(E)$ spectrum can be approximated via the linear combination of six different Non-Maxwellian spectra $S_i(E)$ where each spectrum is scaled by a factor α_i:

$$S_{\mathrm{qMB}}(E) = \sum_{i=1}^{6} \alpha_i S_i(E) \tag{1}$$

For each sub-spectrum the spectrum average cross section $\sigma_{i,SACS}$ (SACS) is measured via the activation method and can be described by:

$$\sigma_{i,SACS} = \frac{\int_0^\infty S_i(E)\sigma(E)dE}{\int_0^\infty S_i(E)dE} \qquad i = 1, 2, 3, 4, 5, 6. \tag{2}$$

Thus, a Maxwellian-Averaged cross section at a given stellar temperature can be obtained by combining the standard equation with Eq. 1 and Eq. 2:

	E_p keV	ΔE_p keV	d_{Li} µm	r_{beam} mm	Distance mm	r_{sam} mm	$k_{\mathrm{B}}T = 60\,\mathrm{keV}$ α	β	$k_{\mathrm{B}}T = 90\,\mathrm{keV}$ α	β
S_1	1912	2.0	100	3.0	2.0	10.0	-0.29782	-0.11306	-0.47561	-0.12444
S_2	1961	2.0	100	3.0	3.0	10.0	0.22154	0.12086	0.09198	0.03458
S_3	1979	2.0	100	3.0	3.0	10.0	0.31224	0.19835	0.10361	0.04536
S_4	1990	2.0	100	3.0	4.0	10.0	0.24089	0.18188	0.15319	0.07972
S_5	2091	2.0	100	3.0	5.0	10.0	0.39403	0.49935	0.62589	0.54663
S_6	2238	2.0	100	3.0	3.0	10.0	0.05539	0.11260	0.29849	0.41815

Table 1: Activation and scaling parameters for $k_{\mathrm{B}}T = 60\,\mathrm{keV}$ and $k_{\mathrm{B}}T = 90\,\mathrm{keV}$. The scaling factors α are used to scale the neutron spectra supplied by PINO [1]. The measured cross sections are scaled by the weighting factors β and can be used to calculate the MACS using Eq. 6.

$$\sigma_{\mathrm{qMB}}(E) = \frac{2}{\sqrt{\pi}} \frac{\int_0^\infty S_{qMB}(E)\sigma(E)dE}{\int_0^\infty S_{qMB}(E)dE} \tag{3}$$

$$\underset{\mathrm{Eq.(1)}}{=} \frac{2}{\sqrt{\pi}} \frac{\sum_{i=1}^{6} \alpha_i \int_0^\infty S_i(E)\sigma(E)dE}{\int_0^\infty \sum_{j=1}^{6} \alpha_j S_j(E)dE} \tag{4}$$

$$\underset{\mathrm{Eq.(2)}}{=} \sum_{i=1}^{6} \alpha_i \underbrace{\frac{\frac{2}{\sqrt{\pi}} \int_0^\infty S_i(E)\sigma(E)dE}{\int_0^\infty \sum_{i=1}^{6} \alpha_i S_i(E)dE}}_{k_i} \sigma_{i,SACS} \tag{5}$$

$$= \sum_{i=1}^{6} \alpha_i k_i \sigma_{i,SACS} = \sum_{i=1}^{6} \beta_i \sigma_{i,SACS} \tag{6}$$

In the last step α_i was combined with the normalization constant k_i to yield the weighting factor β_i used to calculate the MACS values for each stellar temperature from the measured spectrum average cross section. Examples for the weighting factors for a $k_{\mathrm{B}}T = 60\,\mathrm{keV}$ and $k_{\mathrm{B}}T = 90\,\mathrm{keV}$ distribution are tabulated in Tab. 1.

2.1 Optimization

We used the Monte-Carlo based tool PINO [1] to simulate the neutron spectra resulting from the ^{7}Li(p,n) reaction for proton energies ranges spanning from 1912 keV and 2400 keV and sample distances between 2 mm and 9 mm.

A set of six spectra was chosen from these simulation results via chi-squared optimization to reproduce the Maxwell-Boltzmann distribution for different temperatures between 25 keV and 90 keV by only tuning the scaling factors α (see Eq. 6). Two examples for $k_{\mathrm{B}}T = 60\,\mathrm{keV}$ and $k_{\mathrm{B}}T = 90\,\mathrm{keV}$ are shown in Fig. 2. The optimized parameters for both stellar temperatures are tabulated in Tab. 1. It is not possible to reproduce a Maxwell-Boltzmann distribution perfectly using a limited number of subspectra, especially for the high energy tail of the distribution. We estimated the discrepancy between our optimization to the real MACS by folding both the quasi-Maxwellian Spectrum and the ideal one with the well known neutron capture cross section of gold and a 1/v trend. The discrepancies obtained are below 6% and 3% for the ^{197}Au(n,γ) reaction and 1/v, respectively.

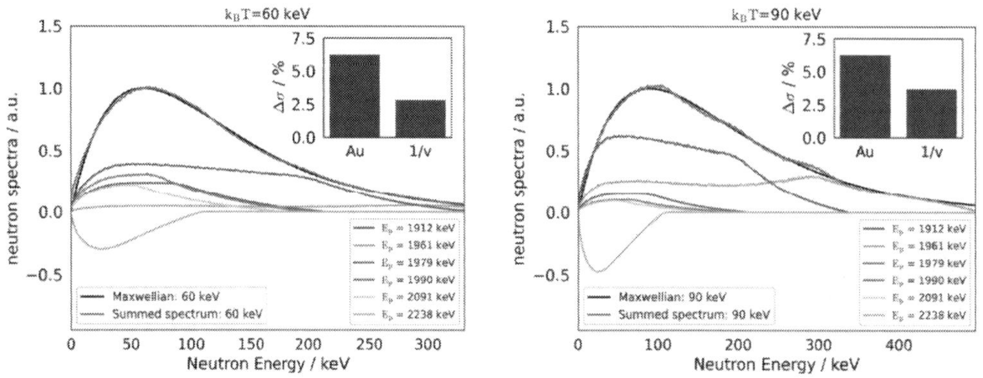

Figure 2: Optimized spectra for $k_BT = 60\,\text{keV}$ (**left**) and $k_BT = 90\,\text{keV}$ (**right**) compared to the corresponding Maxwell-Boltzmann distribution. An estimation of the possible discrepancies between measured cross section using the optimized spectrum and a ideal Maxwell-Boltzmann distribution is shown in the upper right corner of each graph based on the neutron capture cross section of gold and a 1/v trend (see Sec. 2.1).

3 Uncertainties

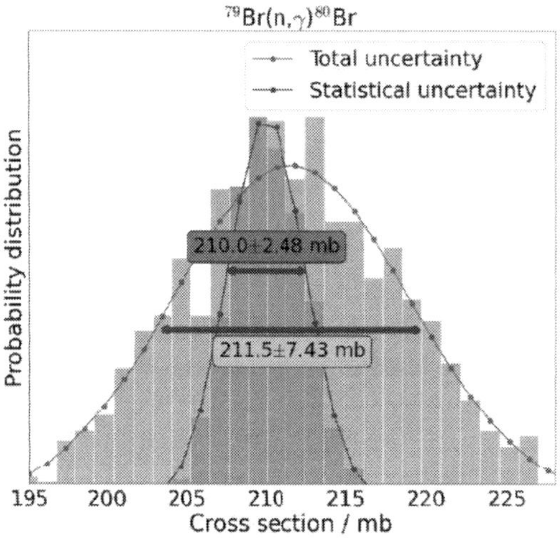

Figure 3: Uncertainties of the measured ^{79}Br(n,γ) cross section at $k_BT = 90\,\text{keV}$ obtained via the monte-carlo method. The total uncertainty and the partial uncertainty resulting only from statistical uncertainties e.g. counting events are shown in gray and red, respectively.

Multiple different sources of uncertainties need to be considered to calculate the uncertainties of the SACS yielding from e.g. the counting statistics, neutron capture cross sections of the neutron flux monitor, detector efficiencies, half-lifes and γ-ray intensities used in the calculation. The uncertainties of the calculated SACS, which are combined to calculate the MACS via the aforementioned methodology, are highly correlated. Thus, we used a Monte-Carlo based method to obtain the uncertainties for the MACS. An example is shown Fig. 3.

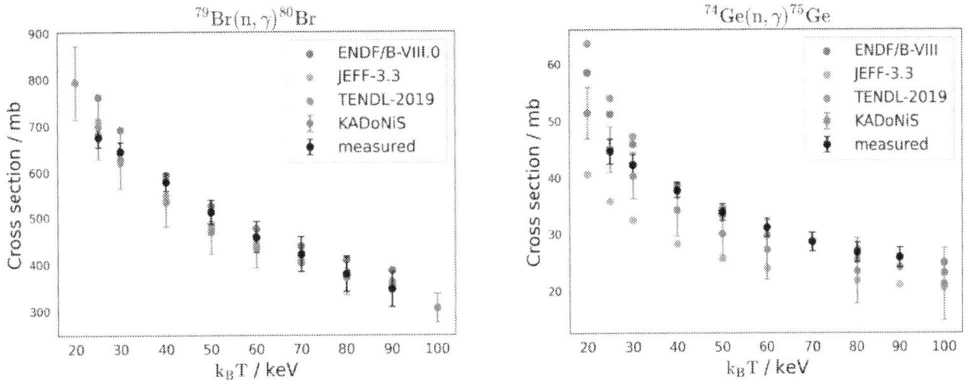

Figure 4: Preliminary averaged cross sections for the neutron capture reaction ^{79}Br(n,γ)^{80}Br (**left**) and ^{74}Ge(n,γ)^{75}Ge isotopes (**right**) for stellar temperatures spanning from $k_BT = 25$ keV and $k_BT = 90$ keV. For comparison the convolutions of evaluated cross sections (ENDF, JEFF and TENDL) [4–6] with the Maxwell-Boltzmann distributions are also shown.

All input values were sampled, 10000 times according to their uncertainties distributions and the MACS were calculated for each set of input parameters. A Gaussian distribution was assumed for all uncertainties.

4 Preliminary results

We measured over 20 different isotopes, including partial cross sections for the ground and isomeric state. Fig. 4 shows our preliminary results for Maxwellian-Averaged cross-section for the ^{79}Br(n,γ) and ^{74}Ge(n,γ) reactions compared to evaluated cross section of ENDF, JEFF and TENDL and the KADoNiS database [3–6]. Our data for the ^{79}Br(n,γ) reaction are in good agreement with the cross section found in the KADoNiS database over the whole measured temperature range. For the ^{74}Ge(n,γ) reaction our results are in good agreement with the KADoNiS data at $k_BT = 25$ keV, but starts to diverge for higher temperatures. In this case, MACS values tabulated in KADoNiS for temperature above $k_BT > 30$ keV are not directly based on measurements, but rely on theoretical model predictions which were scaled to measured cross sections at $k_BT = 30$ keV. This could explain the discrepancies with our result for higher temperatures.

5 Outlook

We are currently investigating the possibility to extend this methodology to lower energies between 5 keV and 25 keV. The first results of this innovative methodology are presented in the proceeding of Tanja Heftrich, included in this conference [2].

References

[1] R. Reifarth et al., Nuclear Instruments and Methods in Physics Research Section A: Accelerators, Spectrometers, Detectors and Associated Equipment, 139-143 (2009)

[2] T. Heftrich et al., Activations for lower s-process temperatures, NPA 2022

[3] I. Dillmann, R. Plag, F. Käppeler, and T. Rauscher, KADoNiS v0.3 - the third update of the Karlsruhe Astrophysical Database of Nucleosynthesis in Stars

[4] A.J. Koning, D. Rochman, J. Sublet, N. Dzysiuk, M. Fleming and S. van der Marck, TENDL: Complete Nuclear Data Library for Innovative Nuclear Science and Technology, Nuclear Data Sheets 155 (2019) 1

[5] Plompen, A.J.M., Cabellos, O., De Saint Jean, C. et al. The joint evaluated fission and fusion nuclear data library, JEFF-3.3. Eur. Phys. J. A 56, 181 (2020)

[6] D.A.Brown et al., ENDF/B-VIII.0: The 8th major release of the nuclear reaction data library with CIELO-project cross sections, new standards and thermal scattering data, Nucl. Data Sheets 148(2018)1

Presolar grain isotopic ratios as constraints to nuclear physics inputs for s-process calculations

Sara Palmerini[1,2,3,*], *Maurizio* Busso[1,2], *Diego* Vescovi[4,5], *Sergio* Cristallo[5,2], *Alberto* Mengoni[6,7], *Stefano* Simonucci[8,2], and *Simone* Taioli[9,10]

[1]Department of Physics and Geology, University of Perugia, via A. Pascoli s/n, Perugia, Italy
[2]INFN Sezione di Perugia, via A. Pascoli s/n, Perugia, Italy
[3]INAF Astronomical Observatory of Rome, Via Frascati 33, Monte Porzio Catone (Rome), Italy.
[4]Goethe University Frankfurt, Max-von-Laue-Strasse 1, Frankfurt am Main D-60438, Germany
[5]INAF, Osservatorio Astronomico d'Abruzzo, Via Mentore Maggini snc, I-64100 Collurania, Teramo, Italy
[6]ENEA, Agenzia Nazionale per la nuove Tecnologie, l'Energia e lo Sviluppo Economico Sostenibile, Via Martiri di Monte Sole 4, Bologna, Italy
[7]INFN Sezione di Bologna, Viale Berti Pichat 6/2, Bologna, Italy
[8]Physics Division, School of Science and Technology, University of Camerino, Via Madonna delle Carceri 9B, Camerino, Macerata, Italy
[9]European Centre for Theoretical Studies in Nuclear Physics and Related Areas, Str. delle Tabarelle, 286, Trento, Italy
[10]Trento Institute for Fundamental Physics and Applications, Via Sommarive, 14,Trento, Italy

Abstract. The isotopic abundances in presolar SiC grains of AGB origin provide important and precise constraints to those star nucleosynthesis models. By comparing the values of the s-element abundances resulting from calculations with the ones measured in these dust grains, it turns out that new measurements of weak-interaction rates in ionized plasmas, as well as of neutron-capture cross sections, are needed, especially in the region near the neutron magic numbers 50 and 82.

1 Introduction

The Asymptotic Giant Branch (AGB) phase is a late stages of stellar evolution undergone by low- (≤ 3 M$_\odot$) and intermediate (≤ 7 M$_\odot$) mass stars. AGB stars have being studied for decades not because of their large number, nor because they are quite simple objects, but because they are site of the so called $s-process$, the slow neutron capture nucleosynthesis mechanism responsible for the production of about half part of the nuclei heavier than Fe in the Galaxy. In particular, stars with mass ≤ 3 M$_\odot$ are responsible for the synthesis of the main component of the s-process, namely of the elements from Sr to Pd. The flux of neutrons available for the captures crucially affects the effectiveness of the s-process nucleosynthesis. In low mass AGBs, neutrons are mainly delivered by the $^{13}C(\alpha,n)^{16}O$ reaction, which is burnt in the ^{13}C reservoirs called *pocket* in the He-rich layers just below the H-burning shell (for more details about AGB stellar structure see e.g. [1, 2] or [3]).

*e-mail: sara.palmerini@pg.infn.it

The formation mechanism of the ^{13}C-pocket determines its chemical composition, its extension (in mass and depth) and then it is the main source of uncertainties in modeling AGB star nucleosynthesis, more that the precision by which the $^{13}C(\alpha,n)^{16}O$ is know at astrophysical energies [4].

In addition to stellar physics, nuclear physics is also source of uncertainties in s-process studies. Indeed the nuclear inputs used for our calculations do not always have the needed precision. By its nature the s process runs along the $\beta-$stability valley, synthesizing increasingly heavier isotopes of an element through (n,γ) processes, when an unstable nucleus is produced it decays and the neutron capture chain continues along the stable isotopes of the element of the daughter nucleus. By definition the s-process is slow, i.e. the neutron capture rates are on average lower than the decay ones of the unstable nuclei produced. However, there are cases in which the rates of the two reactions are comparable and the processes compete, forming a branching of the nucleosynthesis path. The branching points of the s-process are key points and the distribution of the neutron flux between the 2 channels affects the abundance distribution of the synthesized isotopes. From the observed abundances it is then possible to obtain information about the neutron density and the physical conditions of the nucleosynthesis environment. However, it must be taken into account that given a branching point/nucleus a more intense neutron flux, a larger neutron capture cross section or a slower decay rate can produce the same effect on the nucleosynthesis yields. This is the reason why knowing with high precision the nuclear inputs is essential to account for the observations and to decide what are the most reliables ones among the several hypothesis and free parameters introduced by the stellar physics. In this note we want to analyzed relevant issues concerning the nuclear input parameters (cross sections and weak-interaction rates), using as a test-bench for nucleosynthesis calculations the isotopic ratios of heavy elements measured in presolar SiC. In doing that we adopt the MHD-based mixing scheme for the ^{13}C pocket formation presented by [7] (and references therein), which was shown to shape a pocket suitable to reproduce isotopic ratios of heavy elements measured in presolar SiC grains[1] by [5, 6].

2 MACS and $\beta-$ decay rates

The Maxwellian average cross sections (MACS) of (n, γ) reactions, employed in nucleosynthesis calculations, are usually known with sufficient precision for stable nuclei, for which they are relatively easy to be measured. Technical limits make instead difficult, or impossible, to measure neutron capture cross sections and decay rates of unstable nuclei in stellar conditions. For this reason the nuclear physics input uncertainties are concentrated in the proximity of the s-process branching points, which occur close to the magic neutron numbers where unstable nuclei and isomeric states concentrate too.

Nuclear physics input for astrophysical calculations are usual provided by databases, MACS for our AGB models are taken from the KADONIS on-line repository, in its version 1.0 [10], which collects the status of the art of experimental results and whose data are in general agreement with those reported by the National Nuclear Data Center of the Brookhaven National Laboratory [11]. The advent of radioactive ion beam facilities and the use of indirect measurement techniques (see e.g. [13] and references therein) are allowing new experiment to remedy the lack of data on neutron capture reactions of unstable nuclei,

[1]Presolar SiC grains are dusts that condense in the atmospheres of various stars including AGBs. Among these those belonging to the MainStream (MS) group are believed to have AGB origins due to their high content of s elements. The grains have come down to us included in primordial meteorites and by analyzing them it is possible to obtain their isotopic composition and the one of the envelope of the stars where they formed. Such analysis provide relevant constraints on the s-process nucleosynthesis [8, 9].

but data are still missing in several cases, therefore, when experimental data are not available, we used theoretical Hauser-Feshbach computations from the TALYS2008 [14].

A pivotal information for s-process calculations are the rates for weak interactions as a function of temperature and density. Half-lives of many nuclei are essentially unchanged for varying temperatures, but in some case the physical conditions (i.e. plasma temperature and degree of ionization) can modify β−decay (or electron capture) rates of order of magnitude. Nowadays, measurements of decay rates in plasmas under conditions of pressure, temperature, and degree of ionization typical of the stellar interior are either missing or very rare. New precious data could be provided in the near future by the PADORA plasma trap, which is currently under commissioning at INFN LNS [16]. Up to now the most extended and complete compilation of weak interaction rates, as a function of temperature, is the one by Takashi and Yokoi 1987 [17]. In this paper, we refer as standard (ST) case to the nucleosynthesis calculations we run employing the ensemble of the nuclear physics input from [10, 14] and [17].

To identify crucial points in the s-process pathway for which input data changes might be important, Palmerini et al. 2021 [23] considered a long series of possible local changes of the nuclear data affected by considerable uncertainties. The results of nucleosynthesis calculations performed using ad hoc modified parameter sets are referred to as version 2 (V2) models. The authors performed an extensive analysis including the branchings of the s-process that affect the nucleosynthesis of Kr, Sr, Zr, Nb, Mo, Cs and Ba. For sake of brevity, we deal here only about the analysis carried out for the elements belonging to the light-s (Kr and Sr) and heavy-s (Cs and Ba) peaks.

3 Light−s vs heavy−s

The ^{85}Kr branching point is perhaps the most famous one among the branchings of the s-process. Here the neutron capture on ^{84}Kr (whose recommended cross section at 30 keV is 33.1 mb [10]) populates both the ground and the isomeric state of ^{85}Kr, according with the more recent data collected in our ST case the branching ratio turns out to be of about 60% in favor of the isomer (which has $t_{1/2} = 4.5h$), while the estimate results to be just the 40% repeating the calculations using the ^{84}Kr MACS from the previous release of Kadonis, the 0.3 [18].

The nucleosynthesis flow through the ^{85}Kr isomeric state feeds the production of $^{86-87}$Sr passing through $^{85-86}$Rb, only ^{88}Sr is instead produced if neutron captures on ^{85}Kr ground state populate ^{86}Kr and ^{87} Rb. Therefore, Sr isotopic ratios shown by presolar grains depend on cross sections of Sr as well as those of $^{84-85}$Kr and on the decay rates of ^{85}Kr^{m} and ^{86}Rb.

For sakes of comparison we adopt the ^{84}Kr(n,γ)^{85}Kr cross section reported in [18] in our set V2 of nuclear physics input. However, smaller values of the ^{88}Sr/^{86}Sr isotopic ratio could be obtained also considering MACS from K1, but assuming different values of the ^{85}Kr half-life in stellar conditions (about 10.5 yr) or of its cross section, which has only theoretical estimate with uncertainty of the order of 50% (73 ± 34 mb [10]).

Available data for neutron capture cross sections of Ba isotopes are generally affected by rather small uncertainties, but the s-process path that leads to the Ba synthesis passes through a long series of branching points. Indeed, around N = 82, the four Cs isotopes with mass from 134 to 137 form four branching points along the nucleosynthesis path. After ^{133}Cs, the flux proceeds through ^{134}Cs, where n-captures compete with β−decay (laboratory half-life of 2 yr), to excited states of ^{134}Ba and, in a smaller part to via electron captures to ^{134}Xe (half life of $6.8 \cdot 10^5$ yr). From ^{134}Cs, neutron captures feed the longer-lived ^{135}Cs and then ^{136}Cs (half-life of 13.16 d) and ^{137}Cs (half-life of 30.07 yr), whose decay rates remain essentially unchanged for varying temperatures. This is not so for ^{134}Cs and ^{135}Cs. According to [17] the

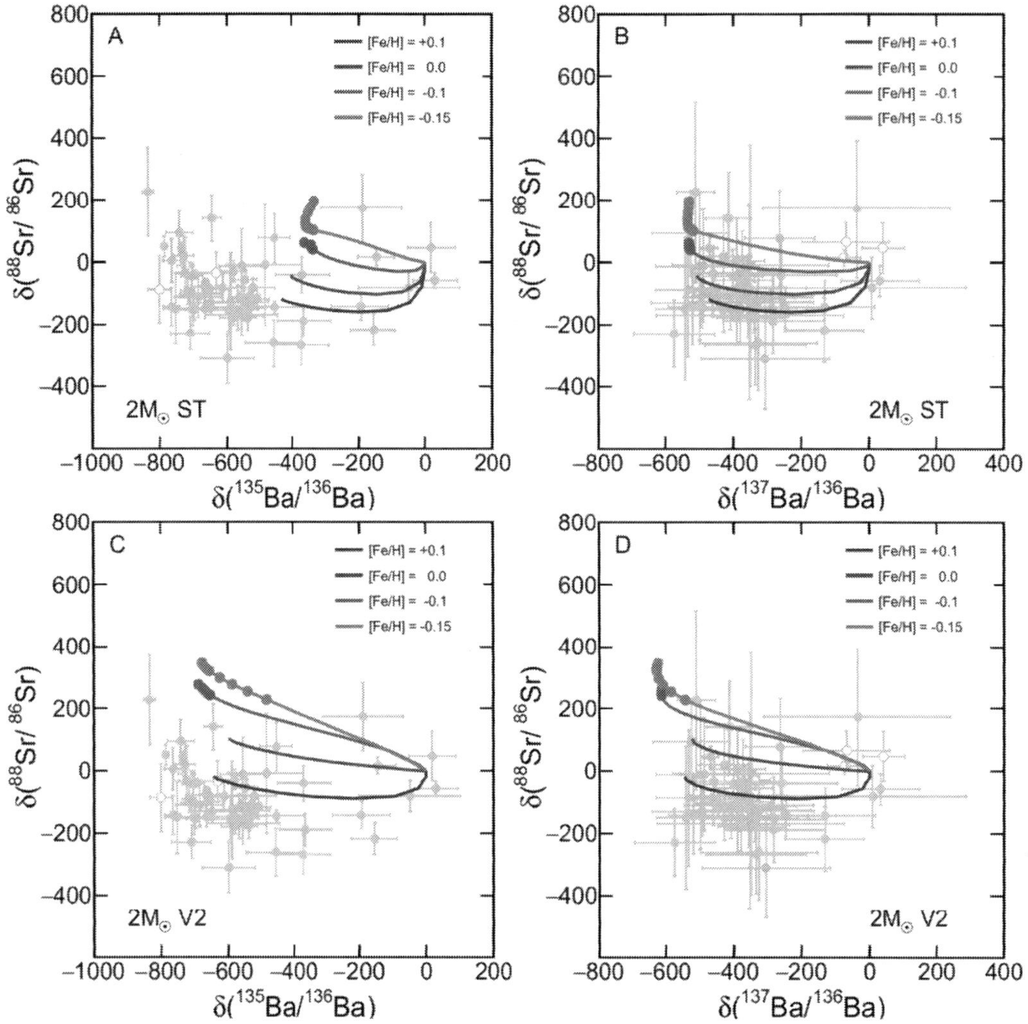

Figure 1. Predictions from $2M_{\odot}$ AGB models of different metallicities (red curves) with SiC grains data (from [19], gray dots with error bars). The case of the isotopic ratios ^{88}Sr/^{86}Sr versus ^{135}Ba/^{136}Ba (panels A and C) and ^{137}Ba/^{136}Ba (panels B and D) are shown. Panels A and B show the envelope composition resulting form calculations run by employing nuclear physics input from set ST, while panels C and D show the output of the same models run with input from set V2 (see the text for details). The full red dots represent model abundance ratios after third dredge-up episodes once the envelope has reached a carbon rich composition [see 7].

half-life of ^{134}Cs is reduced with respect to the laboratory by a factor of about 200. However to reproduce the galactic abundances of the 2 s-only nucleus ^{135}Ba and ^{136}Ba the temperature dependence of the decay rate for ^{134}Cs has to be assumed about 8 times less steep than suggested by [17] or by changing upward the cross section of ^{134}Cs by more than a factor of two [7] . Recently, thanks to an approach based on the solution of the Dirac-Hartree Fock equations, Taioli et al. (2022) [20] found the temperature dependence of the ^{134}Cs and the ^{135}Cs β−decay rates to smaller than so far assumed. In particular, the half-lives of ^{134}Cs estimated by the authors is longer than the one by [17] by a factor ranging from 2.5 (near 8-10 keV) to more than 30 (at 30 keV). Similar effects were found also for ^{135}Cs, whose branching importance is increased by the reduction of the efficiency of the ^{134}Cs decay. Such results confirm the findings of Li et al. (2021) [21], which also obtained an improved fit to Ba isotopic abundances in SiC grains thanks to a new and independent estimate of ^{134}Cs half-life. In our computations, we consider the reduced rates of both ^{134}Cs ^{135}Cs decays in our V2 cases.

4 Conclusion

In Figure 1 s-process predictions of AGB models with $2M_{\odot}$ are compared with measurements in presolar SiC grains by considering the relative behavior of the isotopes of Sr and Ba. Those are elements belonging to the light-s and the heavy-s peak, respectively, and their isotopic abundances are defined by the neutron flow in the region with $N = 50$ and $N = 82$. In particular, the case of the isotopic ratios ^{88}Sr/^{86}Sr as a function of ^{135}Ba/^{136}Ba and of ^{137}Ba/^{136}Ba are illustrated. The area covered by the grain data and the one covered by the models generally overlap, but in panel A.

The fit to the grains improves by changing the nuclear physics input from panel A to panel C, this is so thanks to the longer half-lives of ^{134}Cs and ^{135}Cs, which allow the models to cover a wider range of values of the ^{135}Ba/^{136}Ba ratios. Moreover, the distributions of the predictions for the ^{137}Ba/^{136}Ba and the ^{88}Sr/^{86}Sr isotopic ratios are larger when computed by adopting nuclear input from set V2 instead of those from ST.

In any case more improvements are possible and needed, because models in the C-rich phases (namely when C/O ratio in the stellar envelope is equal or larger than 1) reproduce the abundances of just a small portion of the grain sample. This is shown by the red full dots in Figure 1, which deal with the stellar abundances, after each thermal pulse, once the envelopes have became carbon rich. The fit do not improve significantly even considering $3M_{\odot}$ stars. Such a problem is pretty common in the classical approaches to account for s-isotope distribution of SiC grains in the literature [see e.g. 22]. A possible solution has been presented by [23] that suggested the magnetic winds responsible for the formation of the ^{13}C pocket to generate also a further enrichment in s elements of the stellar envelope. In this scenario, SiC grains can start forming in domains (spots) rich in C and s elements when the stellar envelopes are still O-rich.

We can conclude that very precise esperimetal data about unstable nuclei are mandatory to validate the extra-mixing scenario suggested by [23] to account for SiC grain formation and composition as well as all the other hypothesis that have been already formulated, or will be formulated, on this subject.

References

[1] Busso, M., Gallino, R., Wasserburg, G. J., ARA&A 37 239 (1999)
[2] Herwig, F., ARA&A *43* 435 (2005)
[3] Karakas, A.I, Lattanzio, J.C., PASA *31* 030 (2014)
[4] Cristallo, S., La Cognata, M., Massimi, C., ApJ 859 105 (2018)
[5] Palmerini, S., Trippella, O., Busso, M., et al., GeCoA **221**, 21 (2018)
[6] Vescovi, D.; Cristallo, S.; Palmerini, S., et al., A&A 652 100 (2021)
[7] Busso, M., Vescovi, D., Palmerini, S., et al., ApJ 908 55 (2021)
[8] Zinner, E., *Meteoritics and Planetary Science* (ed. A. M. Davis, New York, Elsevier, 2014) 181
[9] Lugaro, M., Karakas, A. I.. Pető, M., Plachy, E., GCA, 221, 6 (2018)
[10] Dillmann, I., Szücs , T., Plag, R. et al., Nuclear Data Sheets 120 171 (2014)
[11] Pritychenko, B. and Mughabghab, S. F., NDS 113 3120 (2012)

[12] Reifarth, R., Erbacher, P., Fiebiger, S., et al., EPJP 133 56 (2018)
[13] Spartá, R., La Cognata, M., Guardo, G. L., Frontiers in Physics 10 896011(2022)
[14] Goriely, S., Hilaire, S., Koning, A. J., A&A **487** (2008)
[15] Prantzos, N., Abia, C., Cristallo, S., et al., MNRAS **491** 1832 (2020)
[16] Mascali, D., Busso, M., Mengoni, A., et al., EPJWC 22701013 (2022)
[17] Takahashi, K. and Yokoi, K., Atomic Data and Nuclear Data Tables **36** 375 (1987)
[18] Dillmann, I.,Heil, M., Kaeppeler, F. et al., AIPCS 819 123 (2006)
[19] Stephan, T., Bose, M., Boujibar, A., et al., LPSC 2140 (2020)
[20] Taioli, S., Vescovi, D., Busso, M., et al., apJ 933 158 (2022)
[21] Li, K.-A.; Qi, C., Lugaro, M., et al., ApJ 919 19L (2021)
[22] Vescovi, D., Cristallo, S., Busso, M., Liu, N., ApJ, 897 L25 (2020)
[23] Palmerini, S., Busso, M., Vescovi, D., et al., ApJ 921 7 (2021)

A new approach to β-decays studies impacting nuclear physics and astrophysics: the PANDORA setup

D. Mascali[1*], D. Santonocito[1], M. Busso[2,3], L. Celona[1], A. Galatà[4], M. La Cognata[1], G. S. Mauro[1], A. Mengoni[5,6], E. Naselli[1], F. Odorici[5], S. Palmerini[2,3], A. Pidatella[1], R. Ràcz[7], S. Taioli[8,9,10], G. Torrisi[1]

[1]National Institute for Nuclear Physics (INFN), Laboratori Nazionali del Sud, 95123 Catania, Italy
[2]National Institute for Nuclear Physics (INFN), Sez. di Perugia, 06123 Perugia, Italy
[3] Department of Physics and Geology, University of Perugia, 06123 Perugia, Italy
[4] National Institute for Nuclear Physics (INFN), Laboratori Nazionali di Legnaro, 35020, Italy
[5] National Institute for Nuclear Physics (INFN), Sez. di Bologna, 40127 Bologna, Italy
[6] Italian National Agency for New Technologies, Energy and Sustainable Economic Development (ENEA), 40129 Bologna, Italy
[7] Institute for Nuclear Research (ATOMKI), 4026 Debrecen, Hungary
[8] Trento Institute for Fundamental Physics and Applications, TIFPA-INFN, 38123 Trento, Italy
[9] European Centre for Theoretical Studies in Nuclear Physics and Related Areas, Fondazione Bruno Kessler, 38123 Trento, Italy
[10] Faculty of Applied Physics and Mathematics, Gdańsk University of Technology, Gdańsk, Poland

Abstract. Theory predicts that lifetimes of β-radionuclides can change dramatically as a function of their ionization state. Experiments performed in Storage Rings on highly ionized atom have proven nuclei can change their beta decay lifetime up to several orders of magnitude. The PANDORA (Plasmas for Astrophysics, Nuclear Decay Observation and Radiation for Archaeometry) experiment is now conceived to measure, for the first time, nuclear β-decay rates using magnetized laboratory plasma that can mimic selected stellar-like conditions in terms of the temperature of the environment. The main feature of the setup which is based on a plasma trap to create and sustain the plasma, a detector array for the measurement of the gamma-rays emitted by the daughter nuclei after the decay process and the diagnostic tools developed to online monitor the plasma will be presented. A short list of the physics cases we plan to investigate together with an evaluation of their feasibility will be also discussed.

1 Introduction

Stellar nucleosynthesis in massive stars proceeds through subsequent fusion reactions until iron, where it stops because the fusion of still heavier nuclei needs energy instead of providing it. Heavier nuclei are created by the interplay between neutron captures and β-decays; these two processes represent the driving mechanism allowing to increase nuclear mass and charge up to the limits imposed by the nuclear stability [1,2], according to the so-called s-process and r-process. This interplay determines the pathways of stellar nucleosynthesis and, thus, the abundances of atomic nuclei observed in the Universe. However, a major difference exists between terrestrial and stellar conditions: stellar nucleosynthesis proceeds in a hot and dense environment that affects the degree of ionization of the atoms involved in the stellar nucleosynthesis and allows the population of atomic excited states that can induce significant effects on the nuclear beta decay constant or make the beta decay possible in nuclei stable in terrestrial conditions [3-5]. Actually, our knowledge of the effects of the external environment on the nuclear β-decay rates is rather limited. Early attempts showed variations lower than about 0.05% (see, e. g. [6]) as a function

* Corresponding author: davidmascali@lns.infn.it

of pressure and temperature, whilst near the end of the '60s a chemically induced change of about 3.5% in the half-life of ^{7}Be (see, e. g. [6]) was observed. An important breakthrough was achieved using storage rings. The so-called "bound-state β-decay" [7] was observed for the first time on many isotopes. For example, fully stripped $^{187}Re^{75+}$ ions decayed by 9 orders of magnitude faster than neutral ^{187}Re atoms that have a half-life of 42 Gyr [8]. Moreover, bare $^{163}Dy^{66+}$ nuclei, being stable as neutral atoms, become radioactive, with a half-life of 33 days, thus allowing the occurrence of a branching reaction in the s-process path [9]. In order to get a deeper knowledge of the effect of the environment on the beta decay main features and evaluate the possible implications in s-process nucleosynthesis we are developing a new experimental setup, named PANDORA [10,11]. This facility will be dedicated to the measurement of the β-decay process in stellar-like conditions where, due to the high temperature of the plasma, a charge state distribution (CSD) of the ions is established in non-local thermodynamic equilibrium (non-LTE) conditions, and its effect on the decay can be investigated as a function of the temperature. Such a measurements can be extremely relevant for those radionuclides involved in nuclear-astrophysics processes (BBN, s-process, CosmoChronometers, early solar system formation).

2 Pandora experimental setup

The PANDORA project aims to build a compact minimum-B magnetic trap (with B_{max} = 3.0 T at the injection and extraction sides) [12], where plasmas are heated by microwaves via Electron Cyclotron Resonance (ECR) mechanism at 18-21 GHz, up to densities $n_e \sim 10^{11}$-10^{13} cm^{-3}, and temperatures $T_e \sim 0.1$-30 keV. The predicted variation of decay rates will be measured as a function of the thermodynamic plasma properties (density and temperature), whose combination determines the CSD of the in-plasma ions. This paper will describe the overall setup including the array of 14 HPGe detectors dedicated to the measurement of the γ-rays emitted after β-decays, supported by a plasma multi-diagnostics system consisting of RF polarimeters, optical and X-ray spectroscopy, X-ray imaging and space resolved spectroscopy for the simultaneous measurement of plasma density and temperature.

2.1 The magnetic system

A plasma trap that uses a magnetic field for plasma confinement will be employed in PANDORA [12]. It consists of a group of 3 axial solenoids and an hexapole nested inside, coaxial with the solenoids (SEXT-IN-SOL configuration). This configuration produces a superposition of an axial magnetic field (from the axial coils) and a radial magnetic field (from the hexapole coils) called *minimum-B* in which the field grows in every direction, both axially and radially starting from the plasma chamber center. The whole superconducting magnetic system is enclosed in a single cryostat, surrounded by an iron yoke. Its warm bore allows to place a cylindrical plasma chamber with inner radius $R_{CH\ IN}$= 140 mm and a length L = 700 mm. Due to the need to place the gamma detection array and some diagnostic tools around the trap, 16 tapered holes has been created in the cryostat, yoke and properly positioned in the interspace of the six hexapole (as shown in fig. 1). Such a choice allows the detection of gamma-rays without any significant absorption effect due to the crossing of the material except the plasma chamber walls that, in correspondence to the holes, will be properly reduced in thickness. The magnetic system will be able to operate at any point inside the operative ranges shown in Table 1, while ensuring plasma stability. The plasma trap confines the plasma in a Magneto-Hydrodynamical stable mode. The plasma is heated using microwaves at 18 + 21 GHz via ECR. Three klystrons (see fig 2) will generate an RF power of up to 6.5 kW, scalable up to a value of 10 kW in full-power operations. The trap procurement started in May 2021. The tender for procurement (a special type called

"Competitive Dialogue", according to Italian regulations) is expected to end by March. 2023, then the construction will start.

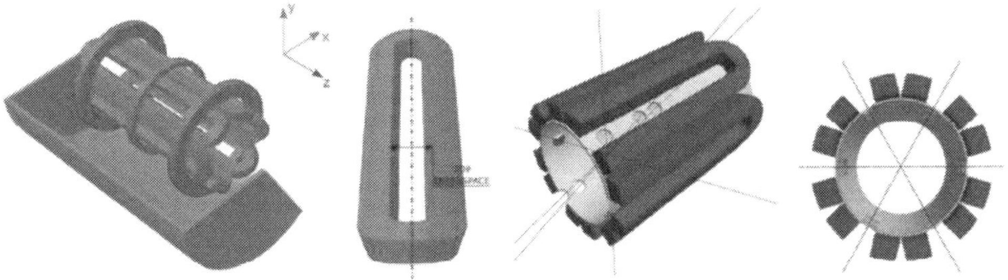

Fig. 1 – Final design of the magnetic superconducting trap (on the left), with details about the hexapole structure and the radial holes in the plasma chamber made to allow gamma-rays to be detected by the HPGe detectors surrounding the trap.

Tab. 1 – List of requirements for the superconducting magnetic trap of PANDORA.

MAGNETIC FIELD REQUIREMENTS	
B_{inj} max @ z = - 350 mm	3 T
B_{inj} operative range	1.7 T – 3 T
B_{ext} max @ z = 350 mm	3 T
B_{ext} operative range	1.7 T – 3 T
B_{min} @ z = 0 mm	0.4 T
B_{hex} @ R_{CH_IN} = 140 mm	1.6 T
Liquid He	Free
Warm Bore radius	150.5 mm
Distance between mirrors	700 mm
Stray field	Less than 0.02 T

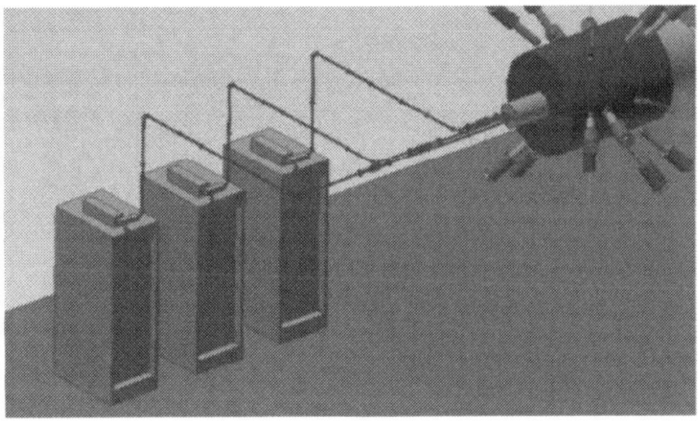

Fig. 2 – Render view of the PANDORA trap surrounded by HPGe detectors, showing the three Klystrons used to produce the high-power RF (18+21 GHz) and the corresponding microwave branching (waveguides) needed to generate the plasma via ECR.

2.2 The detectors array

The gamma-ray detectors array of PANDORA [13,14] (see figure 3) plays a fundamental role in the project. It will be used to measure the in-plasma decay rate of selected radionuclides, β emitters, through the detection of the γ-rays emitted by the daughter nuclei populated in the β decay process. The presence of the coils and the hexapole poses important limitations on the number of detectors that can be placed around the trap without being shielded by the magnetic system. Therefore, the design of the γ array represents a compromise between the best detection efficiency achievable and the mechanical constraints imposed by the feasibility of creating many apertures in the cryostat and the yoke without altering the magnetic field shape. The results of the design study suggest that the best configuration for the γ array is made of 14 large volume HPGe detectors, twelve placed radially in correspondence to dedicated apertures, created between the warm bore radius and the external iron yoke through the cryostat and the inner cold mass, and two axially, as illustrated in figure 3. They will "look" into the plasma through aluminum windows made in the walls of the plasma chamber. The whole array has a photopeak efficiency which reaches its maximum value of 1.8×10^{-3} around 200 keV and then progressively decreases down to 8×10^{-4} in the range 1400-2000 keV, where it is rather constant. Such a value is sufficient, due to the high number of decaying ions present in the plasmoid, to measure the predicted variations in the decay rates of the first physics cases selected in PANDORA such as ^{176}Lu, ^{134}Cs, ^{94}Nb. In these cases preliminary simulations indicate that a 3-sigma confidence level is achievable in few months (Lu), days or weeks (Cs and Nb), depending on the activity of the specific nuclide under investigation. For the initial experimental campaign, it will be possible to use 16 HPGe detectors of the GALILEO experiment [15] thanks to a collaboration agreement between PANDORA and GAMMA collaborations signed in Oct. 2021. The gamma detection array will work under harsh conditions due to the self-emitted background produced by electron Bremsstrahlung which will lead to a counting rate of about 50 kHz background in each detector. For this reason, dedicated pre-amplifiers, developed at the LNL, showing a limited worsening in detector energy resolution at high rates will be used [14].

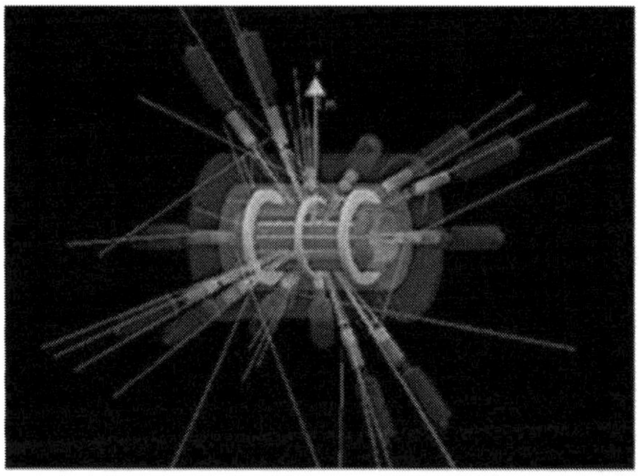

Fig. 3 – Render view of the experimental setup showing the magnetic trap, the HPGe detectors and the GEANT4 ray tracing simulations of the γ-rays.

However, detector performances can deteriorate with time especially considering the harsh conditions to which they will be exposed. Therefore, it was decided to build, at the LNS, a dedicated detector laboratory to cope with these needs with the support of the GAMMA collaboration and the LNL DetLab in order to be able to repair, test and run ordinary maintenance procedures on the HPGE detectors.

2.3 The plasma diagnostics

The main aim of the PANDORA project is to map the evolution of β decay nuclear lifetime as a function of plasma density and temperature. Therefore, an online monitoring on the plasma parameters during the measurement is fundamental to correlate the variation of the decay lifetime to plasma properties. A simultaneous measurement of density and temperature resolved in space and time is thus needed to reconstruct the spatial structure of the plasma and its temporal behavior, in stable and turbulent regimes. ECRIS plasmas are in non-LTE conditions, with three different electron populations: hot, at kT~100 keV or more; warm from kT~100 eV up to tens of keV; cold, at kT~1 eV ÷ 100 eV. Only the warm component of the electron energy distribution function has the right energy to ionize the atoms up to high charge states, since the cold electrons have insufficient energy to remove electrons from the inner shells of atoms (although they are highly involved in the confinement dynamics) and in the hot electron domain the ionization cross section is inversely proportional to the electron energy. A complete characterization of ECR plasmas, which can emit radiation in a broad range of frequencies, requires a multi-diagnostic setup [16] able to measure the density and temperature of the different components. PANDORA diagnostics will consist of:

- A Silicon Drift Detector (SDD) for plasma density and temperature measurements in the warm electron domain [17];
- Two X-rays pin-hole cameras for 2D space resolved spectroscopy [18,19];
- A High-purity Germanium detector for hard X-rays time resolved spectroscopy;
- A spectrometer for the plasma-emitted visible light characterization (resolution of R=13900, $\Delta\lambda$= 0.035nm at λ=486 nm);
- A microwave Polarimeter for line integrated total density measurement [20];
- RF probes and fast scopes for time-resolved measurements and monitoring of plasma instabilities (via a 80 Gs/s-20 GHz bandwidth oscilloscope) [21];
- RF probes connected to a Spectrum Analyzer (SA) for plasma radio-emission in GHz ranges [22].

Some other tools are under development to improve the performances of the setup. In particular, the collaboration is working on the implementation of the Microwave Imaging Profilometry (MIP), to achieve an online measurement of local values of the electron density that can be synergic with the microwave polarimeter already described [23], as well as on the Incoherent Thomson Scattering (ITS) [24], which is an alternative technique for investigating the absolute plasma density, the electron energy distribution functions, and even the electron global drift velocity.

2.4 Injection of radioisotopes in the plasma

The short list of physics cases for PANDORA phase 1 is made of three solid elements. In order to inject them inside the ECR plasma it is necessary to produce a neutral vapor first, or an "ejection" of neutral particles towards the plasma chamber. Two techniques are planned to be used: the resistive oven (for the ^{176}Lu and ^{134}Cs) and the sputtering (for ^{94}Nb). In the first case, the efficiency is estimated to be larger than 30% and ion density is expected to amount to 1-10% of the plasma buffer's. In the second case (sputtering), the efficiency drops to 10% or lower, with a relative concentration around 1%. As it concerns the case of ^{134}Cs lifetime investigation, a dedicated protocol is planned. Such an isotope is commercially available in liquid solutions of HCl: Cesium dissociates after the drying of the compound, at T<500 °C, making it available for ionization by plasma electrons. The results of these feasibility tests are fundamental to plan the next experiments campaign focused on the evaporation of isotopes like ^{176}Lu and ^{133}Cs.

3 Physics cases: perspectives

Tens of physics cases of potential interest for astrophysics have been identified, with a shortlist of priorities which includes ^{94}Nb ($t_{1/2}$ ~ 2 x10^{4} y), ^{134}Cs ($t_{1/2}$ ~ 2 y) [25], ^{176}Lu ($t_{1/2}$~ 3.8 x 10^{10}). Theoretical calculations in fact suggest that an important variation in β-decay lifetime is expected for all these isotopes, as a function of plasma temperature, which could have important implications in s-process nucleosynthesis branchings. For example, the half-life dependence on plasma temperature is reported in fig. 4 for ^{176}Lu.

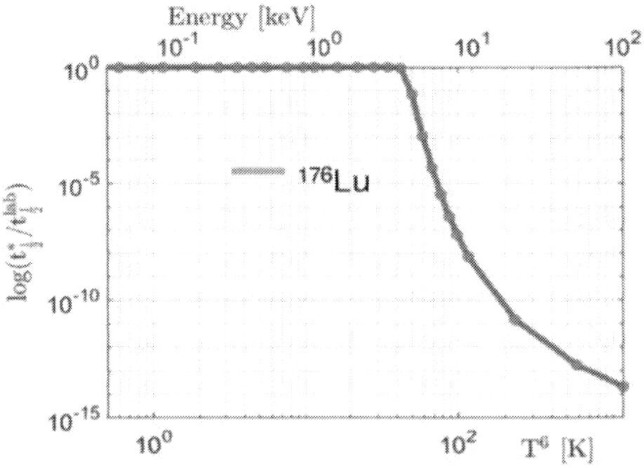

Fig. 4 – Theoretical prediction of stellar half-life to terrestrial half-life ratio for ^{176}Lu nucleus as a function of plasma temperature expressed in T^6 [10^6 K] (Energy [keV]). Calculations were performed at plasma densities of 10^{12} cm^{-3}.

^{176}Lu nucleus, whose role as a cosmo-thermometer or a cosmo-chronometer is not fully defined yet, plays a major role in s process nucleosynthesys as it determines the abundance of ^{176}Hf, an s-only nucleus. The ^{176}Lu/^{176}Hf ratio is largely affected by the branch at ^{176}Lu, which is strongly sensitive to the temperature during the thermal pulses. Therefore the measurement of the dependence of the beta half-life dependence on the temperature might shed light on this issue. The decay rate of ^{134}Cs influences the abundances of two s-only isotopes ^{134}Ba and ^{136}Ba that are not well reproduced by the present-day nucleosynthesis models, the uncertainty affecting the ^{134}Ba/^{136}Ba ratio being mainly associated to the β-decay rate of ^{134}Cs. The ^{94}Nb nucleus provides the main production channel for ^{94}Mo in conditions typical of the s process.
Simulations were performed on all physics cases foreseen, considering the expected lifetime variation and, therefore, the effective activity in the plasma, assuming a plasma volume of 1500 cm^{3} with a realistic concentration of the radioactive isotope to be studied. Taking into account the HPGe array detection efficiency, to reach a 3-sigma confidence level the experimental run should last from about one day up to about three months depending on the physics case investigated and the lifetime variation observed [13]. Numerical simulations [26, 27] are now ongoing to estimate the expected density and temperature distribution in PANDORA, and then to perform a kind of "virtual experiment" addressing and supporting the data taking that will start at the end of 2024.

The possibility to investigate of β decay rates in stellar-like conditions is expected to show a strong dependence on the temperature T, while a very weak dependence on plasma density [5] is foreseen. At a given T, not only the CSD, but also a complex configuration of atomic/electronic excited states in the various ions is achieved in plasma. Hence, ECR plasmas that are in non-LTE conditions give the unique opportunity to investigate the actual effect of electron temperature on the decay rate. An equivalent CSD can be then extrapolated to LTE scenarios applying to many astrophysical plasmas allowing for a direct comparison with stellar model predictions. This new approach has, therefore, all the potential to provide a more complex but more coherent and "realistic" picture of the decay, closer to astrophysical contexts than any result obtained from storage rings experiments.
Within PANDORA we also intend to provide, in the forthcoming years, critical new nuclear and atomic physics inputs from experiments and advanced theoretical models, in particular, the in-plasma atomic physics knowledge on the opacity of several metallic r-process elements. These data will allow us to extend our understanding of the microphysics of kilonovae (KN). Preliminary work has been already carried out in the framework of the PANDORA collaboration to support first-of-its-kind experimental measurements of plasma opacity with in-laboratory plasmas resembling these KN-stage conditions. In this view, the results of recently performed experiments at the INFN-LNS to reproduce stable early-stage ejecta thermodynamical conditions of under-dense and low-temperature plasmas are reported in [28]. These are the grounds to tackle the astrophysical problem in laboratory plasmas with an interdisciplinary approach.

4 Acknowledgements

Authors are grateful to the INFN 3rd Nat. Comm. of INFN for the Financial Support under the Grant PANDORA_Gr3.

References

1. R. Spartà et al. Front. Phys., 26 May 2022 Sec. Nuclear Physics
2. M. Busso et al., Front. Astron. Space Sci., 9 (2022) 956633
3. K. Takahashi and K. Yokoi, Nuclear Physics A 404(3):578-598 · August 1983.
4. K. Takahashi and K. Yokoi, Atom. Data and Nuc. Data Tab. 36(3), 1987, 375-409
5. S. Palmerini S. et al., The Astrophysical Journal 2021, 921 (1), 7
6. G. T. Emery, Ann. Rev. Nucl. Sci. 1972, 22, 165-202
7. J. N. Bahcall, Phys. Rev. 124, 495 – 15 October 1961
8. M. Jung, et al., 1992, Phys. Rev. Lett. 69 2164
9. Yu. A. Litvinov, et al., 2007, Phys. Rev. Lett. 99 262501
10. D. Mascali et al., European Physical Journal A 03/2017; 53(7)
11. D. Mascali et al., Universe, 80 (2022) 8
12. G. Mauro et al., Front. Phys., 10 (2022) 931953
13. E. Naselli et al., Front. Phys., 10 (2022) 935728
14. A. Goasduff et al., Front. Phys., 10 (2022) 936081
15. A. Goasduff et al., NIM A 2021, 1015, 165753
16. E. Naselli et al., J Instrum., 14 (2019) 10008
17. B. Mishra et al., Phys. Plasmas. 28, 102509 (2021)
18. S. Biri et al., J. Instrum. 16, P03003 (2021)
19. E. Naselli et al., Condens. Matter 7, 5 (2022)
20. G. Torrisi et al., Front. Astron. Space Sci., 9, 949920 (2022)
21. D. Mascali et al., Rev. Scient. Instr. 93, 033302 (2022)

22. E. Naselli et al., Plasma Sources Sci. Technol. 28, 085021 (2019).
23. G. Torrisi et al. 2022 JINST 17 C01050
24. S. Tsikata, et al. (2022), Front. Astron. Space Sci. 9:936532.
25. S. Taioli et al. 2022 ApJ 933 158
26. A. Galatà, et al. Front. Phys. 10:947194. doi: 10.3389/fphy.2022.947194
27. B. Mishra, et al. Front. Phys. 10:932448. doi: 10.3389/fphy.2022.932448
28. A. Pidatella, Front. Astron. Space Sci. 9:931744. doi: 10.3389/fspas.2022.931744

Stellar neutron spectra at 28 keV thermal temperature

Elizabeth Musacchio-González[1,*], *Pierfrancesco* Mastinu[1], and *Guido* Martín-Hernández[2]

[1]INFN-Laboratori Nazionali di Legnaro, Viale dell'Università 2, 35020, Legnaro, Italy.
[2]Centro de Aplicaciones Tecnológicas y Desarrollo Nuclear, 5ta y 30, Miramar, La Habana, Cuba.

Abstract. To calculate the reaction rate in the neutron capture processes it is common to work with the Maxwellian Average Cross Section (MACS), defined as the reaction rate scaled by the most probable neutron velocity of the Maxwell-Boltzmann distribution. For MACS determination with lower uncertainties, the need of a neutron spectrum as similar as possible to the stellar one motivates this work. At the CN Van der Graaff accelerator of the LNL-INFN laboratories, an experimental measurement is performed to produce a Maxwell-Boltzmann neutron spectrum with 28 keV of thermal temperature. The neutron time-of-flight spectrometry is implemented to determine 0°-90° integrated neutron spectrum, employing the ^{7}Li(p,n)^{7}Be reaction as neutron source, an initial proton energy of 3.17 MeV and a 51 μm thickness aluminum foil, as proton energy shaper.

1 Introduction

Nuclei from carbon to iron were found to be produced by charged particle reactions during the evolutionary phases from stellar He to Si burning. Nucleosynthesis of elements beyond Fe (B=8.8 MeV/A) is produced in stars by successive (n,γ) reactions and β-decays. The stellar velocity neutron spectrum is a Maxwell-Boltzmann distribution. Depending on the stellar site and the evolutionary stage of the star the most important temperatures (kT) are in the range of 8-90 keV, being 30 keV the standard temperature of reference [1].

The possibility of reproducing the abundance of the elements in the universe depends on stellar reaction rate calculations. For a given temperature and neutron density and for the s-process mainly, the Maxwellian-Averaged Cross Section (MACS) directly describes the reaction rate inside the stars. The calculated MACS uncertainties of several stable and most of the unstable isotopes are higher than the requested accuracy i.e., for the s-process is 3-5%. Hence, the importance of measuring the MACS with the least possible uncertainty. To achieve this goal it is mandatory to have a neutron beam as similar as possible to the Maxwell-Boltzmann neutron spectrum (MBNS), and this is the main purpose of this work.

2 The method

A possible method to obtain a Maxwell-Boltzmann neutron spectrum (MBNS) at 30 keV was published by Mastinu *et al.* [2], in 2009. The work proposed to modify or shape the neutron energy spectrum by shaping the projectile energy distribution in accelerator-based

*e-mail: musacchio@lnl.infn.it

neutron sources. In case the projectiles were protons coming from the ^{7}Li(p,n)^{7}Be reaction, the method is based on the idea that if the proton beam distribution is wider, with protons that cover higher energies, the production of neutrons with higher energies will be obtained. This is done by means of a foil, placed before the neutron producing target, intercepting the projectile beam. This method improves the neutron flux at the sample position and avoid the use of moderators. As proposed by Martín-Hernández *et al.* [3], by choosing different thicknesses, different materials, or combinations of both, different thermal temperatures can be obtained with this method. Figure 1 shows a schematic representation of this method.

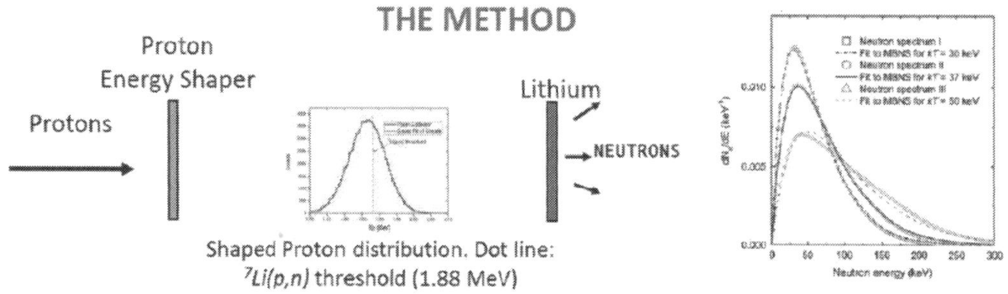

Figure 1. Method to produce a Maxwell-Boltzmann neutron spectrum (MBNS), proposed by Mastinu *et al.* [2].

In the present work, the mentioned method was employed. Different materials and their isotopes were studied as proton energy shaper. The energy threshold for neutron production, the value of the (p,n) cross section, the neutron yield, and the gamma yield were taken into consideration for the analysis. The software SRIM 2013 [4] was employed to obtain the proton energy distributions for various incident monochromatic proton energies impinging on different thicknesses of the studied materials. The theoretical integral neutron spectrum was obtained using the kinematics of the reaction and then fitted to a Maxwell-Boltzmann distribution, determining the kT and the R-square coefficient of determination for that fit. The neutron spectrum that fits best the Maxwell-Boltzmann distribution with kT = 30 keV was determined with a fitting code that was developed employing as parameters the proton energy and the FWHM of the proton beam after passing the shaper. After the best fit program calculations, an aluminum foil of 51 μm thickness as proton shaper and an initial proton energy of 3.17 MeV were found to be the best set of parameters to obtain a MBNS with a kT=28 keV.

3 Experimental measurement

The experimental measurement of the Maxwell-Boltzmann neutron spectrum was carried out at the CN Van de Graaff accelerator of the National Laboratories of Legnaro of the Italian National Institute for Nuclear Physics (LNL-INFN), in Padua, Italy. The experimental setup employed in the experiment is shown in Figure 2. The neutron time-of-flight spectrometry (nTOF) was implemented to determine the neutron spectrum, using the ^{7}Li(p,n)^{7}Be nuclear reaction as neutron source and a 600 kHz proton pulsed beam. In the experiment, a proton energy of 3.17 MeV was set in the accelerator and, the 51 μm thickness Al foil was placed before the Li target.

A natural lithium metallic target of 100 μm thickness in a 300 μm copper backing was produced and placed at the end of the accelerator beamline. Three ^{6}Li-glass detectors

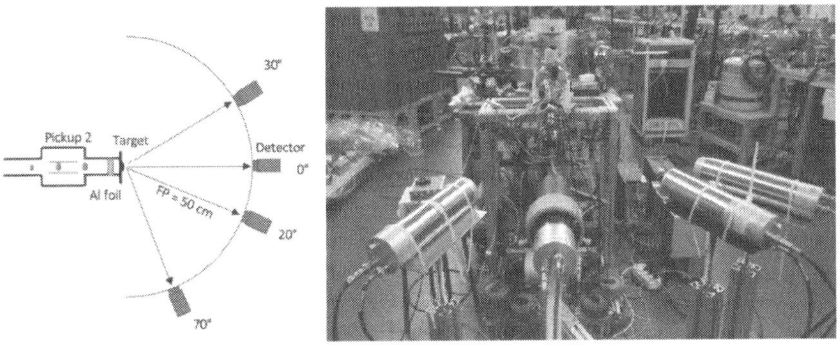

(a) (b)

Figure 2. a) Schematic representation of the experimental setup. b) Experimental setup. Detectors were located at 50 cm from the target at different angles to measure the neutron time-of-flight.

purchased from the Scionix company were used for TOF measurements. Two one inch thick detectors and one half inch thick were used. A fourth 3 mm thick Li-glass detector was also used, but as neutron counter monitor for normalization. This detector was also employed to check the target and beam stability. The four Li-glass detectors were placed precisely at 50 cm from the lithium target (flight path) with a low mass designed goniometer. Differential angular neutron TOF distributions from 0 to 90 degrees in steps of 10° were measured in order to obtain the 0°-90° integrated neutron spectra.

4 Results and discussion

The angle-integrated neutron time-of-flight (nTOF) spectrum is shown in Figure 3 a) and was obtained by summing the weighted TOF spectra from 10 up to 90 degrees for each detector. The differences in the spectrum are due to differences in the efficiency of each detector. A new approach to transform TOF spectrum into energy spectrum was also implemented, using the detector response matrix. This method is long and complex and deserves a separate paper that it is being written. The proposed conversion method considers not only the mean moderation time of neutrons in the detector but also its distribution in time, making an exact deconvolution of the time spectra. This detector response matrix was calculated with the MCNPX [5] simulation code, including the detector geometry and materials composition and determining the neutron time distribution inside the detector for monoenergetic neutron sources. It is important to know precisely the geometry and materials of the detector since the conversion method relies on the response matrix. The geometry and materials composition of each type of detector were directly provided by the Scionix company.

The FWHM of the Gaussian fit for the gamma flash distribution gave a value of 3.43±0.05 ns (σ=1.46±0.02 ns). Because of this, 4 ns was taken as the time resolution of the beam. To mimic the beam time structure in the MCNPX simulation, the source was sampled as a Gaussian-shaped time-dependent distribution with FWHM=4 ns.

Even in the case of monoenergetic neutron beams, different glass thickness in the detectors means different moderation times, and therefore different TOFs. Because of this, two separated analyses were done: one for the half inch thickness detector and the other for the one inch thickness detectors. Figure 3 b) shows the angle-integrated neutron energy spectrum after the new conversion method was applied to the nTOF spectrum for each detector type. As expected, the emitted neutron spectrum is the same even when it was measured with different detectors.

For the data analysis, the time bins were chosen to have almost the same counting statistic per bin (about 1% relative error). With the detector acquired statistics, it was possible to obtain the neutron energy spectrum with 34 nonlinear energy bins, from 1 keV to 322 keV. The higher part of the energy spectrum was obtained using 4 ns temporal bins, because of the already mentioned time resolution of the measurement. At lower energies, the spectrum was obtained with energy bins of 2 keV. The energy resolution goes from 2.6% at 1.85 keV to 28.1% at 300 keV.

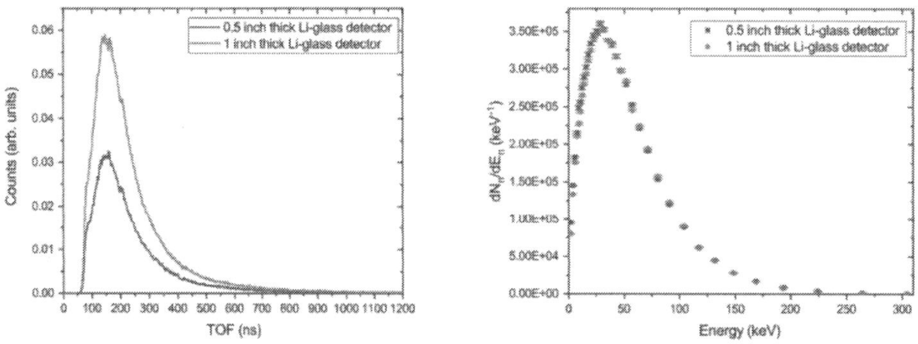

(a) (b)

Figure 3. a) Angle-integrated neutron time-of-flight (TOF) spectrum, obtained by summing the weighted TOF spectra from 10 up to 90 degrees for each detector. b) Angle-integrated neutron energy spectrum obtained with the new conversion method.

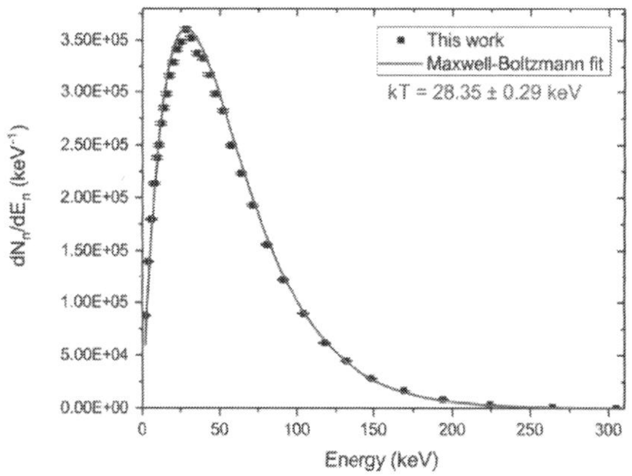

Figure 4. Final angle-integrated, from 0° to 90°, neutron energy spectrum and the least squares fit to a Maxwell-Boltzmann distribution.

The final Maxwell-Boltzmann neutron spectrum (MBNS) is shown in Figure 4, calculated as the average between the values for each energy bin from the two detectors type (Figure 3 b)). The least squares fit to a Maxwell-Boltzmann distribution is also shown, giving a value of thermal temperature of 28.35±0.29 keV. It is observed in the figure the good agreement between the experimental data and the Maxwell-Boltzmann fit, even in the high energy part of the spectrum, validating the method proposed by Mastinu *et al.* [2].

5 Conclusions

In the experimental measurement, neutron time-of-flight spectrometry was implemented to determine the neutron spectrum. As source of neutrons, the ^{7}Li(p,n)^{7}Be reaction was employed, using a lithium metallic target and a 600 kHz proton pulsed beam from the Van de Graaff accelerator of the LNL-INFN laboratories, in Legnaro (Padua), Italy. To produce a Maxwell-Boltzmann neutron spectrum, the method of using a proton energy shaper was employed and validated. The most important result of the present work is that a MBNS with a kT=28 keV was measured and well reproduced. Irradiating a sample with this neutron field, produced with 3.17 MeV protons and the 51 μm Al foil, a very accurate measurement of the MACS at said thermal temperature can be performed. This is precisely the next experiment to be performed: to measure the "gold standard" for MACS with the activation technique.

References

[1] P.F. Mastinu, J. Praena, G. Martín Hernández, N. Dzysiuka, G. Prete, R. Capote, M. Pignatarie, A. Ventura, Physics Procedia **26**, 261 – 273 (2012).

[2] P.F. Mastinu, G. Martín Hernández, J. Praena, Nuclear Instruments and Methods in Physics Research, Section A: Accelerators, Spectrometers, Detectors and Associated Equipment, **601**, 333-338 (2009).

[3] G. Martín-Hernández, P.F. Mastinu, J. Praena, N. Dzysiuk, R. Capote Noy, M. Pignatari, Applied Radiation and Isotopes, **70**(8), 1583–1589 (2012).

[4] J. F. Ziegler, SRIM - The Stopping and Range of Ions in Matter, http://www.srim.org (2013).

[5] D.B. Pelowitz, LA-CP-11-00438, MCNPX Users Manual Version 2.7.0 (2011).

Shedding light on $^{17}O(n,\alpha)^{14}C$ reaction at astrophysical energies with Trojan Horse Method and Asymptotic Normalization Coefficient

Giovanni Luca Guardo[1*], *Livio* Lamia[1,2,3] for the ASFIN collaboration

[1]INFN-Laboratori Nazionali del Sud, Catania, Italy
[2]Dipartimento di Fisica e Astronomia 'Ettore Majorana', Università di Catania, Catania, Italy
[3]Centro Siciliano di Fisica Nucleare e Struttura della Materia (CSFNSM), Catania, Italy

Abstract. Indirect methods have been established in the past as a complementary way of increasing our knowledge about nuclear structure and low-energy cross section measurements. Among these, the neutron induced reaction cross sections appear to be of particular interest because of their role both for unstable and stable beams. In view of this, we report here the combined study of the $^{17}O(n,\alpha)^{14}C$ reaction accomplished by the Trojan Horse Method (THM) and the Asymptotic Normalization Coefficient (ANC) method. The low lying resonances 8038, 8125, 8213, and 8282 keV in ^{18}O are studied and Γ_n are derived. A comparison with direct data and recent THM experimental data is presented. The independent ANC investigation corroborates our previous THM results, confirms the consistence of the two indirect investigation and shows new frontiers also in view of neutron induced reactions with RIB's.

1 Introduction

The $^{17}O(n,\alpha)^{14}C$ reaction is considered in astrophysical models for its role in the relevant "s(slow)-process" since it could act as a possible "neutron poison" for the neutron-induced nucleosynthesis thus influencing the final stellar abundances of some elements such as Ni or Sr [1]. Thus, its reaction rate must be known in the energy region of interest for astrophysics, going from few keV up to about 400 keV. At such energies, the intermediate $^{17}O+n\rightarrow^{18}O$ nucleus presents four excited levels (8038, 8125, 8213 and 8282 keV) affecting the magnitude of the $^{17}O+n$ cross section at astrophysical energies because of the 8044 keV threshold for the neutron emission of ^{18}O. Besides the role of the two 8213 keV ($J^\pi=2^+$) and 8282 keV ($J^\pi=3^-$) states, already investigated by the direct measurements [2], up-to-now poor information is available about the 8125 keV ($J^\pi=5^-$) resonant level, thus requiring a detailed experiment.
However, neutron-induced reaction cross section measurements are difficult to be performed due to the practical difficulties of having "easily available" neutron beams. For such a reason,

* Corresponding author: guardo@lns.infn.it

the Trojan Horse Method (THM) has been applied to the $^{17}O(n,\alpha)^{14}C$ case and discussed in [3,4]. The THM measurements [3,4] show an evident contribution of the 8213 keV and 8282 keV ^{18}O excited states with an additional contribution of the 8125 keV resonant level, naturally suppressed in the direct measurements because of the large l=3 angular momentum required for its population in the ^{17}O+n channel. Thanks to our THM investigation, information about the corresponding neutron (Γ_n) or alpha partial widths (Γ_α) is now available. To complement such information, mainly deduced by the (modified) R-matrix formalism for THM described in [5], we proposed a new experiment dedicated to the study of these resonant ^{18}O levels, with particular regard to the measurement of the 8125 keV level by means of the corresponding ANC.

2 Method and experiment

The ANC is a fundamental nuclear characteristic of importance in nuclear reactions since it corresponds to the amplitude of the overlap function of the bound-state wave functions of the initial and final nuclei. Hence, the ANC method allows one to obtain the cross section of a peripheral direct capture reaction A(a,γ)B in terms of the radial overlap integral of the X and B nuclei of the A(X,Y)B transfer reaction , where the nuclei X and A can be considered as X = Y+a and B = A+a [6]. This method has been widely used in the last 20 years to investigate proton, neutron and α direct captures in stellar environments [7-9]. In the present case, in order to extract the ANC for the ^{18}O resonant states, the $^{17}O(d,p)^{18}O$ reaction has been studied, using the deuterium for its simple p+n configuration. The experiment was performed at the isochronus cyclotron U-120M of the Nuclear Physics Institute of the Czech Academy of Science (NPI-ASCR) in Rez (Prague). A 16.3 MeV beam was delivered onto a 90% pure ^{17}O gas target, connected to the beam line with 3μm-thick havar entrance and exit windows. The beam energy was chosen in order to maximize the peripherical contribution of the reaction (see [10] and reference therein). The detection setup, already used in many experiments [11], consisted of 8 standard point-like ΔE-E telescopes, with thicknesses of 500μm and 5000μm, respectively. Three telescopes were placed at fixed angles of 17°, 27° and 37° with respect to the beam line and have been used also as the monitors. Five telescopes were placed on the turntable plate on the opposite side with a relative angular step of 10°. The whole set of five telescopes could measure reaction products in the range of θ_{LAB} from 6° to 67°.

3 Data analysis and conclusions

After the performed detector calibrations and uncertainties evaluations, the next stage of the present analysis has been the extraction of the angular distribution for both, the elastic scattering ^{17}O+d and the reaction $^{17}O(d,p)^{18}O$. This is a standard procedure followed elsewhere [12] and focuses first on extracting the optical parameters from DWBA calculation on the ^{17}O+d elastic scattering angular distributions and then the ANC for the resonant ^{18}O levels populated in the $^{17}O(d,p)$ reaction.

Once extracted, each ANC value ($|C_l|^2$) can be connected to the corresponding Γ_n of the ^{18}O level state by the following formula [13]:

$$\Gamma \approx \frac{k_r \hbar c^2 C_l^2}{\mu_{ab}} = \frac{\sqrt{2\mu_{ab}E_r}\hbar c C_l^2}{\mu_{ab}} = \sqrt{\frac{2E_r}{\mu_{ab}}}\hbar c C_l^2 \qquad (1)$$

where E_r is the resonance energy (in MeV), μ is the reduced mass (in MeV), $\hbar c$=197 MeV fm and $|C_l|^2$ is in fm^{-1}.

Then, considering also the Γ_α as reported in [4], it is possible to calculate the cross section and compare it with the direct and inverse data [2,14,15]. This comparison is reported in Fig. 1 where the red line represents the cross section calculated as described before while the red band is the error associated to the Γ_n of about 20%. Light blue points are data taken from the work of [2], blue points from [14] while black points are the inverse data of [15] converted using the detailed balance. From Fig.1 it is clear the presence of a quite good agreement between the ANC calculation and the direct data at the energy region above 150 keV. Below this region the direct measurements present a very big scatter distribution, while the ANC calculation decrease a lot due to the absence of a proper evaluation of the subthreshold level.

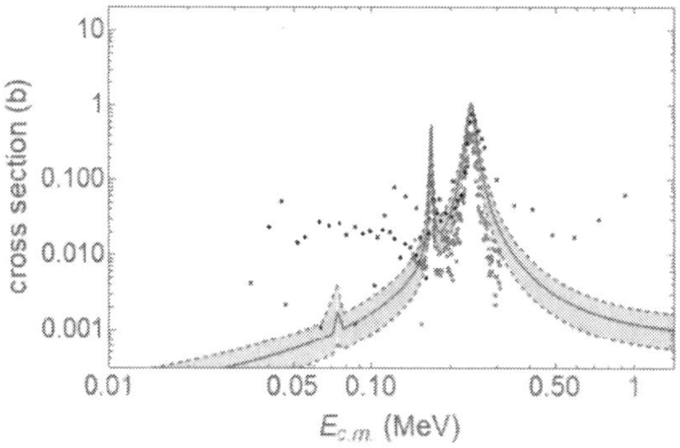

Figure 1 - ANC cross section compared with available direct data (see text for details)

Moreover, with the same parameters it is possible to compare also the THM data [4], as reported in Fig. 2. Again, red line and red band are used to highlight the cross section calculated from the Γ_n derived from the ANC data while the black points represent the THM data.

Not yet included is the subthreshold level centered at -7 keV since for this level the analysis is still ongoing. In both cases the results are in good agreement, confirming the possibility to use independent ANC investigation to corroborate THM results and thus opening new frontiers also in view of neutron induced reactions with RIBs, where there is still a lack of direct measurements due to practical problems in the simultaneous production of neutron and radioactive beams.

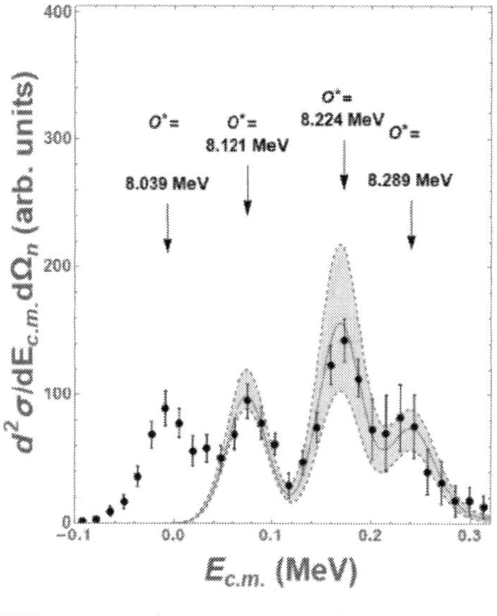

Figure 2 - ANC cross section compared with available THM data

As a final remark, Table I compare the Γ values for the ^{18}O level state available in the literature, as a final cross-check of the good agreement between the methods and to underline once more that the combined study presented here allows us to measure for the first time the resonance parameters of the 8125 keV ($J^{\pi}=5^{-}$) level state, suppressed in direct measurement due to the centrifugal barrier.

Table 1 - Total level width available in literature compared with the present measurement.

$E_{c.m.}$ (keV)	Γ (eV) Ref. [2]	Γ (eV) Ref. [4]	Γ (eV) This work
75	----	36±5	33±6
178	2258±135	2260±300	2150±420
244	14739±590	14700±3800	16670±3350

References

[1] M. Pignatari et al., AstroPhys. J., **710** (2010) 1557
[2] J. Wagemans et al., Phys. Rev. C **65** (2002) 034614
[3] M. Gulino et al., Phys. Rev. C **87** (2013) 012801
[4] G.L. Guardo et al., Phys. Rev. C **95** (2017) 025807
[5] A.M. Mukhamedzhanov et al., J. of Phys. G **35** (2008) 014016
[6] L.D. Blokhintsev, I. Borbely, and E. I. Dolinskii, Sov. J. Part. Nucl. **8** (1977) 485
[7] V. Burjan et al., Eur. Phys. J. A **55** (2019) 114
[8] A. M. Mukhamedzhanov et al., Phys. Rev. C **84** (2011) 024616
[9] G. G. Kiss et al., Phys. Lett. B **807** (2020) 135606
[10] V. Burjan, J. Mrazek and G. D'Agata, Frontiers in Science **7** (2020) 562466
[11] G. D'Agata et al., Phys. Rev. C **103** (2021) 015806
[12] R. E. Tribble et al., Rep. Prog. Phys. **77** (2014) 106901
[13] A. M. Mukhamedzhanov and R. E. Tribble, Phys. Rev. C **59** (1999) 3418
[14] P. E. Koehler and S. M. Graff, Phys. Rev. C **44** (1991) 2788
[15] R. M. Sanders, Phys. Rev. **104** (1956) 1434

Analysis methods to determine the bound-state beta-decay half-life of Thallium-205

Guy Leckenby[1,2,*], *Ragandeep Singh* Sidhu[3,4,5], *Rui Jiu* Chen[4,5], *Yuri* Litvinov[4], *Jan* Glorius[4], *Christopher* Griffin[1], *Iris* Dillmann[1,6], and the E121 Collaboration

[1]TRIUMF, Vancouver, BC V6T 2A3, Canada
[2]Dept. Physics & Astronomy, University of British Columbia, Vancouver, BC V6T 1Z1, Canada
[3]School of Physics and Astronomy, University of Edinburgh, EH9 3JZ, UK
[4]GSI Helmholtzzentrum für Schwerionenforschung GmbH, D-64291 Darmstadt, Germany
[5]MPI für Kernphysik, Saupfercheckweg 1, 69117 Heidelberg, Germany
[6]Dept. Physics & Astronomy, University of Victoria, Victoria, British Columbia V8P 5C2, Canada

Abstract. Bound-state β^--decay is an exotic decay mode that produces temperature-dependent stability in nuclei. A striking example is ^{205}Tl, in part because of its impact on the ^{205}Pb/^{204}Pb cosmochronometer—a short-lived radionuclide clock that can provide unique constraints on *s* process material in the early solar system. The bound-state β^--decay of ^{205}Tl was measured at GSI, where fully stripped ^{205}Tl^{81+} ions were produced and stored in the Experimental Storage Ring. Decay occurred during storage producing increased ^{205}Pb daughters with increased storage time. This contribution briefly outlines the experiment and describes analytical corrections required to extract the half-life.

1 Introduction

Bound-state β^--decay is a rare of example of how altering the electron density can change nuclear properties. Despite being characterised in 1947 [1], it was not measured experimentally until 1992 [2] with the development of the Experimental Storage Ring at GSI Helmholtzzentrum für Schwerionenforschung. The decay mode differs from normal β^--decay because the β-electron is created directly in a bound orbital of the daughter nucleus, thus increasing the Q-value of the reaction by the binding energy of the bound electronic orbital. Thallium-205 is particularly interesting because neutral ^{205}Tl is stable, with ^{205}Pb decaying via electron capture with a Q-value of 50.6 keV [3]. However, since the binding energy of a K-electron for Tl is 85.5 keV, the bound-state β^--decay of fully ionised or hydrogen-like ^{205}Tl can occur, primarily to the first excited state of ^{205}Pb as shown in Figure 1. This makes bare ^{205}Tl^{81+} radioactive with a half-life first predicted to be 124 days [4].

The bound-state β^--decay of ^{205}Tl has a large impact on the termination of the *s* process. ^{205}Pb is one of 10 Short-Lived Radioactive (SLR) nuclei (short-lived on an astronomical timescale; 1-100 My) that decay into isotopic ratios that can preserve the abundance patterns of the Early Solar System (ESS). ^{205}Pb is particularly unique as it is the singular *s*-only SLR (only being produced by the *s* process) and also probes the final branching point of the *s* process at ^{204}Tl (half-life only 3.8 years). However, the first excited state of ^{205}Pb at just

*e-mail: gleckenby@triumf.ca

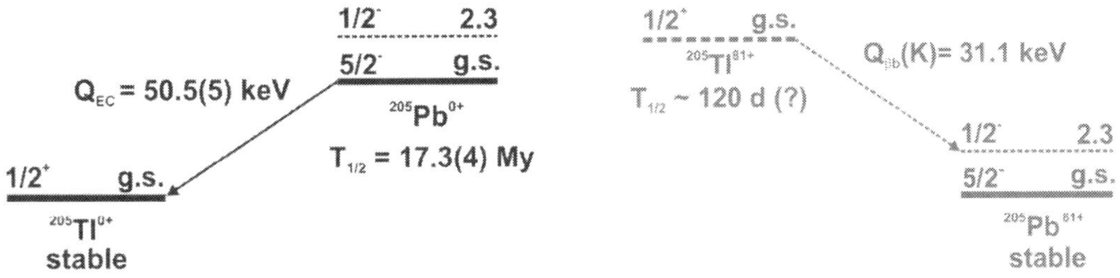

Figure 1. The low-lying nuclear structure for ^{205}Tl and ^{205}Pb shows the interplay between the neutral and fully stripped states. Note that because the bound-state β^--decay proceeds overwhelmingly to the first excited state of ^{205}Pb, it's half-life provides access to the nuclear matrix element of the two spin-$\frac{1}{2}$ states, which is a crucial astrophysical transition.

2.3 keV can be significantly thermally populated in stellar environments, drastically reducing the effective half-life due to the reduced spin difference of the EC decay. Takahashi and Yokoi [5, 6] identified that the bound-state β^--decay of ^{205}Tl could provide a mechanism to counterbalance the ^{205}Pb thermal destruction. The predicted temperature at which bound-state β^--decay would become competitive was ~ 180 MK (depending on electron density), which is reached both in the He-shell flashes of Thermally Pulsating AGB stars and the C-shell of massive stars. This idea was supported by the confirmation of live ^{205}Pb during the ESS, but inconsistent meteorite measurements have made it difficult to predict exactly what site (if any) could have provided this ^{205}Pb [7–9]. Thus an experimental value was required to provide clarity on the exact mechanism.

This contribution will briefly outline the experimental method used, but will focus on the analytic corrections required to extract the half-life. The intention is to outline the analysis methods used to derive the half-life, and thus will not comment on the result or its impact.

2 Experimental Method

The measurement of the bound-state β^--decay of ^{205}Tl was conducted at GSI using the full accelerator system [10] and the Experimental Storage Ring (ESR) [11]. Thallium cannot be used directly as a source as its vapours are highly toxic, so the primary beam was provided by a specifically-developed enriched-^{206}Pb source. The $^{206}Pb^{67+}$ charge state was accelerated by the UNILAC and SIS-18 to 678 MeV/u with an intensity of 10^9 particles per second. This primary beam was impinged on a ^{9}Be target producing fully ionised $^{205}Tl^{81+}$ at 592 MeV/u by projectile fragmentation, amongst a cocktail of contaminant species. $^{205}Tl^{81+}$ was selected by the FRagment Separator (FRS) using $B\rho$-ΔE-$B\rho$ separation with an Al energy degrader and highly selective focal plane slits suppressing the amount of contaminant $^{205}Pb^{81+}$ to $<$ 0.2%. An ion optical simulation showing the separation of species along different trajectories through the FRS is shown in Figure 2. Each injection from the FRS produced $\sim 5 \times 10^5$ $^{205}Tl^{81+}$ ions at 400 MeV/u.

To achieve the required intensity of $\sim 10^6$ $^{205}Tl^{81+}$ ions, up to 100 injections were stacked and cooled inside the ESR. The $^{205}Tl^{81+}$ ions were then stored for periods of up to 10 hours to accumulate $^{205}Pb^{81+}$ daughter ions. Because the mass difference of $^{205}Tl^{81+}$ and $^{205}Pb^{81+}$ ions is just 53 keV and bound-state β^--decay daughters have the same charge state as the parent ions, both species have the same mass-to-charge ratio and were stored on the same orbit in the ESR. To count the number of ^{205}Pb daughters, the bound β-electron was stripped away by interacting the stored ions with an Ar gas target for 10 minutes producing $^{205}Pb^{82+}$

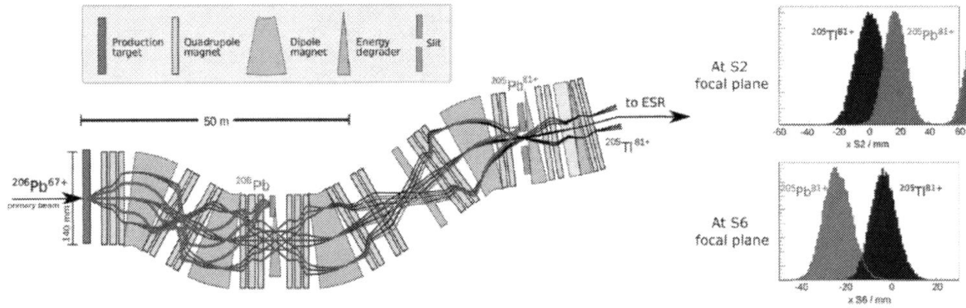

Figure 2. A MOCADI ion-optical simulation shows the trajectories of relevant projectile fragments through the FRS. Primary $^{206}Pb^{81+}$ beam is well separated at the focal plane, whilst an energy degrader is required to separate the isobar $^{205}Pb^{81+}$ from the desired $^{205}Tl^{81+}$. Image courtesy of *Helmut* Weick.

ions. Because the storage times were small compared to the half-life of the decay, daughter products grow linearly with storage time. Background contribution by $^{205}Pb^{81+}$ contamination from the FRS and nuclear charge exchange reactions in the gas target was determined from the $t_s = 0$ intercept, whilst the growth of $^{205}Pb^{81+}$ intensity was directly attributable to bound-state β^--decay as all other parameters were maintained between storage runs.

Throughout the experiment, ions in the ring were monitored by a non-destructive 245 MHz resonant Schottky noise pickup [12] that measures the Schottky noise power from the revolution of ions. As the revolution frequency in the storage ring is determined by the mass-to-charge ratio, each ion in the ring can be identified and its intensity monitored. Ions that left the acceptance of the ring were monitored with particle detectors: the CsI–Silicon Particle detector for Heavy ions Orbiting in Storage rings (CsISiPHOS) [13] on the inside of the ring and a Multi-Wire Proportional Chamber (MWPC) [14, 15] on the outside of the ring.

Each storage run can be characterised by 4 key stages:

1. the storage period after the accumulation of $^{205}Tl^{81+}$ ions
2. the stripping stage where the gas target is active for 10 minutes
3. the electron cooling stage where ions were cooled into well-resolved peaks
4. the measurement window where the intensities of ions were taken

Whilst the Pb and Tl Schottky intensities were measured during stage 4, the ratio that is required to determine the half-life is the value at the end of storage (stage 1), so a number of corrections need to be made before the ratios of different storage times can be compared.

3 Analysis Corrections

There are four corrections that need to be applied to the raw Schottky power intensities.

3.1 Resonance correction

The first is the Resonance Correction (RC), which accounts for the resonant response of the Schottky detector. Whilst the detector is sensitive between 241.6–248.4 MHz, a strong resonance response is centred at 243.6 MHz associated with the fundamental resonance frequency of the cavity. This amplifies the Schottky power of species closer to the resonance peak, and this amplification needs to be corrected for. Extensive details on this effect will be provided in an upcoming publication.

 https://doi.org/10.1051/epjconf/202327906010

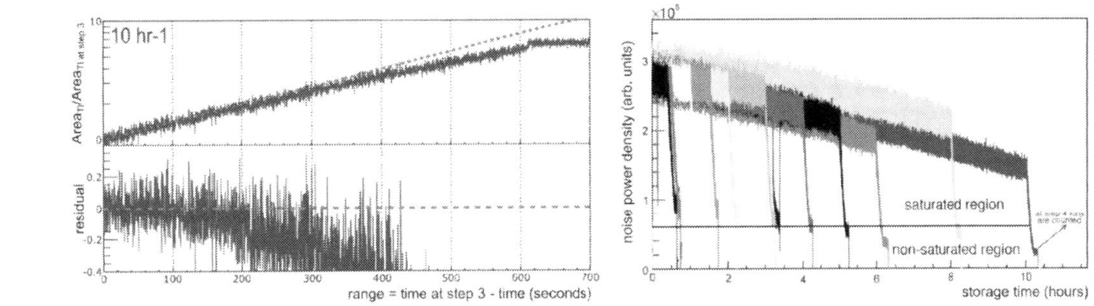

Figure 3. Due to an incorrectly applied amplifier in the NTCAP-DAQ [16], intense peaks (mostly Tl) showed a saturation effect visible because the peaks do not decay exponentially. The onset of saturation was determined by the departure of exponential decay (left figure), where the decay rate was measured independently by an MWPC detector. This threshold is then compared to the Tl intensity for all storages (right figure, colours correspond to different storages). This figure is adapted from [17].

3.2 Schottky saturation correction

The second is the Saturation Correction (SC), which accounts for a saturation effect observed in the recorded Schottky data. An unexpected interaction between updated accelerator control systems and the recent New Time Capture (NTCAP) Schottky data acquisition [16] resulted in the adaptive amplifier on the NTCAP-DAQ being permanently activated. This caused a saturation effect above a threshold that meant noise power density was not proportional to ion number, causing Schottky lines to no longer decay exponentially. To identify the intensity threshold at which saturation set in, the decay constant of the beam measured by the MWPC was used in a fit to the Schottky noise power data such that the only free parameter was the final intensity at the end of stripping. This fit is shown in Figure 3, where the time coordinate is inverted to look at time from the end of stage 2. It is clear that the data deviates from exponential behaviour for noise power intensities over 600×10^3 (arb. units for the Tl peak), which was chosen to mark the saturated and non-saturated regions shown in Figure 3.

This analysis revealed that for 12 shortest storage times, the Schottky noise power data was saturated for the $^{205}Tl^{81+}$ peak at the measurement stage (note that the $^{205}Pb^{82+}$ peak had a much smaller intensity, well below the saturation threshold). The Tl peak intensity during stage 2 (stripping) spans all values of the Tl peak at stage 4 (measurement). The first 10 hour storage (10hr-1) was best suited as a calibrant as it had the lowest final intensity and thus the largest portion of stage 2 in the non-saturated region. Two models were applied to the 10hr-1 data:

$$N_{\text{exp}}(t) = N_{S3} \exp(-\lambda_{MWPC} t) \tag{1}$$

$$N_{\text{sat}}(t) = N_{S3} \exp(-\lambda_0 t)(C_1 + C_2 t) \tag{2}$$

where $N_{\text{exp}}(t)$ was the expected behaviour using the MWPC decay rate whilst $N_{\text{sat}}(t)$ was a phenomenological model that replicated the saturation effect. The ratio between these two models defined a Saturation Correction factor $SC = \frac{N_{\text{exp}}}{N_{\text{sat}}} > 1$ for any given noise power density at stage 4. Statistical errors on this factor were determined by using a Monte Carlo calculation to propagate the fit errors.

3.3 Charge-changing cross-section ratio

The third correction factor has been termed the Cross-Section Ratio Correction (CSRC). This correction accounts for the fact that whilst $^{205}Pb^{82+}$ ions stripped by the gas target are recorded, $^{205}Pb^{80+}$ ions from electron recombination will leave the acceptance of the ring. Since the beam and target are identical for these two reaction channels, the correction factor is the ratio of the charge-changing reaction cross-sections:

$$\frac{\sigma_{\text{ioniz}} + \sigma_{\text{rec}}}{\sigma_{\text{ioniz}}} = \frac{N_{\text{CsISi}} + N_{\text{MWPC}}}{N_{\text{CsISi}}} = \frac{R_{\text{DCCT}}}{R_{\text{DCCT}} - R_{\text{MWPC}}}, \tag{3}$$

where N_X is the number of ions counted in a detector X and R_X is the rate in that detector. To measure this cross-section ratio, H-like primary beam $^{206}Pb^{81+}$ was reacted with the gas target, and the CsISiPHOS and MWPC particle detectors captured the ionisation and recombination products respectively.

Due to inexplicable effects in the CsISiPHOS data, the ratio was instead determined from the total ion loss rate, with ionisation to bare Pb being determined by the subtraction of the recombination rate from the total loss rate, as in the last equality of Equation 3. Because the Schottky data was saturated, a DC beam Current Transformer (DCCT) was used to measure the total ion current in the ring. The DCCT was recorded with a scaler counter, and the high event rate meant that the scalar counter didn't accrue many counts per event, so a stable derivative was very challenging. In particular, the derivative did not stabilise until a window of 50 seconds was used. The DCCT also has a zero-current offset to ensure non-zero scalar counting even if the ring is empty. This offset was measured in two ways, from 124 'stray' events when the ring was empty and from the offset of 14 exponential + constant fits from the 14 injections of primary beam during this measurement.

In the first, and most direct approach, the ions reacted (measured by the DCCT) over these 50 second windows were used to calculate 70 independent values for CSRC, each with uncertainty values from the underlying counting statistics from the scaler and MWPC events and the DCCT offset calibration. The correlation in the DCCT offset was handled by propagating errors using a Monte Carlo calculation, which confirmed the statistical errors were Gaussian. However, the resulting normalised residuals indicated the raw error bars underestimated the statistical spread around the mean, which was attributed to unquantified missing statistical uncertainty given that the DCCT was not intended for precision measurements. A Bayesian analysis was used to allow the data to constrain this missing uncertainty, which was assumed to be Gaussian from the Central Limit Theorem with a Jeffrey's prior. Integrating out the nuisance parameters lead to a Gaussian posterior distribution for the CRSC of $R_1 = 1.433(9)$.

In the second approach, the DCCT data was fit with an exponential decay function, circumventing the problem of an uncertain local derivative with a global fit. However, the decay rates determined from the DCCT were systematically lower than from the MWPC, despite measuring the same interaction with the gas target. A potential explanation was found in the Schottky spectrum where broad peaks next to the main beam were observed that grew over time, despite only having primary beam in the ring. Throughout this part of the experiment, the stochastic cooler [18] could not be turned off, so it was proposed that the stochastic cooler was providing a RF-kick to ions throughout the storage, building a side band over time which would reduce the measured decay constant of the DCCT rate explaining the behaviour. Whilst this theory wasn't able to be definitively confirmed, the decay rate from the MWPC was more reliable. Thus two values were calculated, $R_2 = 1.425(5)$ from explicitly fitting a 'side band', and $R_3 = 1.418(4)$ from simple exponential decay. This yielded a spread of analysis models producing a modelling systematic uncertainty of 0.005, taken from the variance of a uniform distribution. The side band model was chosen as the most accurate analysis.

Additionally, ~ 5% of MWPC events demonstrated anomalous behaviour despite only primary beam being in the ring. A systematic error of 0.012 from the difference in including vs excluding these events was added. This resulted in a final value for the charge-changing cross-section ratio of $R = 1.425(5)_{\text{stat}}(13)_{\text{syst}}$.

3.4 Gas Stripping Efficiency

The fourth correction factor accounts for the interaction with the gas target. In particular, $^{205}Tl^{81+}$ ions are lost to electron recombination whilst $^{205}Pb^{82+}$ daughters are uncovered by stripping (recombination is already accounted for by the CSRC), so the intensity of the two species throughout gas stripping is given by:

$$N_{\mathrm{Tl}^{81+}}(t) = N_{\mathrm{Tl},2} \exp(-\lambda_{\mathrm{str,Tl}} t) \tag{4}$$

$$N_{\mathrm{Pb}^{82+}}(t) = N_{\mathrm{Pb},2}(1 - \exp(-\lambda_{\mathrm{str,Pb}} t)) \tag{5}$$

where $N_{\mathrm{X},2}$ is the ion number at the start of stage 2 and $\lambda_{\mathrm{str,X}}$ is the stripping decay constant, which is specific to individual species.

An MWPC was inserted during most storage times, so individual values for $\lambda_{\mathrm{str,Tl}}$ were determined for 11 of the 18 storage times by fitting an exponential to the event rate measured in the detector. For $^{205}Tl^{81+}$ ions at 400 MeV/u into a gas target of $\sim 3 \times 10^{12}$ atoms/cm^{-2}, the average decay constant was $\lambda_{\mathrm{str,Tl}} = 0.00312(9)$ s^{-1}. The value of $\lambda_{\mathrm{str,Pb}}$ was determined from the MWPC rate during the $^{206}Pb^{81+}$ primary beam measurements, ensuring the same systematics. For $^{206}Pb^{81+}$ ions at 400 MeV/u into a gas target of $\sim 3 \times 10^{12}$ atoms/cm^{-2}, the average decay constant was $\lambda_{\mathrm{str,Pb}} = 0.00757(2)$ s^{-1}. This means that for a 10 minute stripping duration, the stripping efficiency for revealing $^{205}Pb^{82+}$ ions was $\sim 99\%$.

These four correction factors can be applied to the measured raw Schottky intensities at stage 4 to provide a ratio at the end of storage:

$$\frac{N_{\mathrm{Pb},2}}{N_{\mathrm{Tl},2}} = \frac{N_{\mathrm{Pb},4}}{N_{\mathrm{Tl},4}} \frac{RC \cdot CSRC}{SC} \frac{\exp(-\lambda_{\mathrm{str,Tl}} t_{\mathrm{str}})}{1 - \exp(-\lambda_{\mathrm{str,Pb}} t_{\mathrm{str}})}. \tag{6}$$

4 Half-life Fit

With the ratios at the end of storage determined, the increase in $^{205}Pb^{82+}$ daughters can be fit using linear regression. Because the storage losses due to electron recombination in the electron cooler and collisions with the rest gas in the ring determine the ion intensity throughout storage, they need to be incorporated when solving the differential growth equation. The full solution to this differential equation is given by:

$$R(t_s) = \left(R_0 + \frac{\lambda_{\beta_b}/\gamma}{\lambda_{\beta_b}/\gamma + \lambda^{cc}_{\mathrm{Tl}} - \lambda^{cc}_{\mathrm{Pb}}}\right) \exp\left((\lambda_{\beta_b}/\gamma + \lambda^{cc}_{\mathrm{Tl}} - \lambda^{cc}_{\mathrm{Pb}}) t_s\right) - \frac{\lambda_{\beta_b}/\gamma}{\lambda_{\beta_b}/\gamma + \lambda^{cc}_{\mathrm{Tl}} - \lambda^{cc}_{\mathrm{Pb}}}, \tag{7}$$

where $R(t_s) = N_{Pb}(t_s)/N_{Tl}(t_s)$ is the ratio, R_0 the background contribution, λ_{β_b} the bound-state β^--decay decay rate, γ the Lorentz factor for conversion into the laboratory frame, and λ^{cc} the respective loss rates during storage. Using a Taylor series, the exponential can be expanded, and noting that $\lambda_{\beta_b} \ll \lambda^{cc}$, this expression can be simplified to:

$$R(t_s) = \frac{\lambda_{\beta_b}}{\gamma} t_s \left(1 + \tfrac{1}{2}(\lambda^{cc}_{Tl} - \lambda^{cc}_{Pb}) t_s + O(t_s^2)\right) + R_0 \exp\left((\lambda^{cc}_{Tl} - \lambda^{cc}_{Pb}) t_s\right). \tag{8}$$

The approximation in Equation 8 differs from the full result in Equation 7 by $< 0.15\%$.

The four corrections described in Section 3 consist of a mix of individual and average values depending on measurement constraints and the plausibility of varying experimental conditions. The uncertainties on the average values used will thus be correlated across different storage times, making a simple χ^2-minimisation very challenging to correctly handle the correlations. To solve this a Monte Carlo calculation was done to propagate the correlated errors consistently, where 10^5 linear regressions were applied to resampled data producing a distribution in the determined β_b^--decay rate λ_{β_b}.

At present, the collaboration is still finalising how to handle the background contamination consistently, so the final half-life result is not ready at this stage. However, the analysis is almost finalised and important experimental clarification on the role of ^{205}Pb produced during the *s* process will be published soon.

Acknowledgements

This research has been conducted in the framework of the SPARC, ILIMA, LOREX, and NucAR collaborations, experiment E121 of FAIR Phase-0 supported by GSI. This work was supported by the the European Research Council (ERC) under the EU's Horizon 2020 research and innovation program [Grant Agreement No. 682841 "ASTRUm"]; the Natural Sciences and Engineering Research Council of Canada (NSERC) [NSERC Discovery Grant No. SAPIN-2019-00030]; and the Excellence Cluster ORIGINS from the German Research Foundation DFG [Excellence Strategy EXC-2094–390783311].

References

[1] R. Daudel, M. Jean, M. Lecoin, J. Phys. Radium **8**, 238 (1947)

[2] M. Jung, F. Bosch, K. Beckert, H. Eickhoff, H. Folger, B. Franzke, A. Gruber, P. Kienle, O. Klepper, W. Koenig et al., Phys. Rev. Lett. **69**, 2164 (1992)

[3] F.G. Kondev, M. Wang, W.J. Huang, S. Naimi, G. Audi, Chin. Phys. C **45**, 030001 (2021)

[4] K. Takahashi, R.N. Boyd, G.J. Mathews, K. Yokoi, Phys. Rev. C: Nucl. Phys. **36**, 1522 (1987)

[5] K. Takahashi, K. Yokoi, Nucl. Phys. A **404**, 578 (1983)

[6] K. Yokoi, K. Takahashi, M. Arnould, Astron. Astrophys. **145**, 339 (1985)

[7] J.M. Huey, T.P. Kohman, Earth Planet. Sci. Lett. **16**, 401 (1972)

[8] E. Palk, R. Andreasen, M. Rehkämper, A. Stunt, K. Kreissig, B. Coles, M. Schönbächler, C. Smith, Meteorit. Planet. Sci. **53**, 167 (2018)

[9] R.G.A. Baker, M. Schönbächler, M. Rehkämper, H.M. Williams, A.N. Halliday, Earth Planet. Sci. Lett. **291**, 39 (2010)

[10] M. Steck, Y.A. Litvinov, Prog. Part. Nucl. Phys. **115**, 103811 (2020)

[11] B. Franzke, Nucl. Instrum. Methods Phys. Res., Sect. B **24-25**, 18 (1987)

[12] M.S. Sanjari, P. Hülsmann, F. Nolden, A. Schempp, J.X. Wu, D. Atanasov, F. Bosch, C. Kozhuharov, Y.A. Litvinov, P. Moritz et al., Phys. Scr. **T156**, 014088 (2013)

[13] M. Najafi, I. Dillmann, F. Bosch, T. Faestermann, B. Gao, R. Gernhäuser, C. Kozhuharov, S. Litvinov, Y. Litvinov, L. Maier et al., Nucl. Instrum. Methods Phys. Res., Sect. A **836**, 1 (2016)

[14] H. Stelzer, Nucl. Instrum. Methods Phys. Res., Sect. A **310**, 103 (1991)

[15] O. Klepper, C. Kozhuharov, Nucl. Instrum. Methods Phys. Res., Sect. B **204**, 553 (2003)

[16] C. Trageser, C. Brandau, C. Kozhuharov, Y.A. Litvinov, A. Müller, F. Nolden, S. Sanjari, T. Stöhlker, Phys. Scr. **2015**, 014062 (2015)

[17] R.S. Sidhu, PhD, Heidelberg University, Germany (2021), `https://doi.org/10.11588/heidok.00030275`

[18] F. Nolden, K. Beckert, F. Caspers, B. Franczak, B. Franzke, R. Menges, A. Schwinn, M. Steck, Nucl. Instrum. Methods Phys. Res., Sect. A **441**, 219 (2000)

Neutron activations for lower s-process temperatures

Tanja Heftrich[1,*], *Kafa* Al-Khasawleh[2], *Sandra* Borzek[1], *Benjamin* Brückner[1], *Sophia Florence* Dellmann[1] *Ozan Can* Dogan[1], *Asmaa* El Mard[1], *Philipp* Erbacher[1], *Madeleine* Gail[1], *Kathrin* Göbel[3], *Fabian* Habermehl[1] *Benedict* Heybeck[1], *Timo* Kisselbach[1], *Deniz* Kurtulgil[1], *Rim* Mourad[1], *Markus* Reich[1], *René* Reifarth[1], *Silas* Sheriff[1], *Tom* Staab[1], *Meiko* Volknandt[1], and *Mario* Weigand[1]

[1]Goethe University Frankfurt, Frankfurt, Germany
[2]Jordan Atomic Energy Commission/Jordan Research and Training Reactor
[3]Facility for Antiproton and Ion Research in Europe GmbH (FAIR GmbH), Darmstadt, Germany

Abstract. The slow neutron capture process produces heavy elements in different stellar sites at different temperatures. Neutron capture cross sections for stellar temperatures between $k_BT = 5$ keV and $k_BT = 100$ keV are crucial for a quantitative understanding of the s-process abundance distribution. Over the last decade activation measurements were performed at the Goethe University Frankfurt to study cross sections at a temperature of $k_BT = 25$ keV. We developed a new method to measure neutron capture cross sections at $k_BT = 6$ keV.

1 Introduction

Heavy elements beyond iron are mainly synthesised by neutron capture reactions. The slow neutron capture process (s-process) accounts for about 50 %. The corresponding neutron energies in the different astrophysical sites range from a few to about hundred keV [1, 2]. By using the activation technique, neutron capture reactions have been studied over the last decades at $k_BT = 25$ keV. With a modified approach, we measured neutron capture cross sections for lower s-process temperatures. With ring-shaped samples, a specific range of neutron emission angles is covered such that the neutron energy distribution corresponds to a Maxwell-Boltzmann distribution of $k_BT = 6$ keV. We did first proof-of-principle measurements using nickel and tantalum samples. The neutron yield is significantly higher than in previous approaches using the ^{18}O(p,n) reaction as a neutron source [3]. While the neutron yield using the ^{18}O(p,n) approach is about 10^{-4} lower than for the standard 25 keV setting, our approach reduces the neutron yield only by a factor of 10^{-2}.

2 Activations for lower s-process temperatures

In the activation method, neutrons are produced by protons impinging on a Lithium target. The sample of interest is sandwiched by gold foils, which act as monitors for the neutron flux (Fig.1). The neutrons are produced via the reaction ^{7}Li(p,n). A lithium glass detector is placed behind the samples. The neutron flux fluctuations are monitored using the lithium

*e-mail: heftrich@iap.uni-frankfurt.de

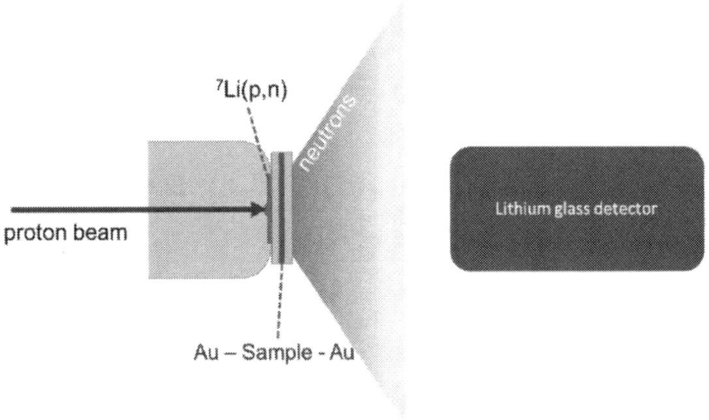

Figure 1. The proton beam impings on a lithium target where neutrons are produced via the reaction ^{7}Li(p,n). The neutrons pass the sample sandwich consisting out of the sample of interest which is surrounded by two gold foils. The gold foils acts as neutron monitors. A lithium glass detector is placed at zero degree in order to take account for beam fluctuations.

glass detector. The working horse method for the last decades delivers quasi stellar neutrons at $k_B T = 25$ keV for proton energies of $E_P = 1912$ keV, if the sample covers the whole neutron cone. The simulation of the emitted neutrons was done using PINO [4] (Fig.2, left panel). The radioactive reaction product in the sample was characterized by γ-spectroscopy. Our new method relies on a partial angular coverage of the neutron cone. We simulated different geometries varying position and size of ring-shaped samples as well as the proton energy.

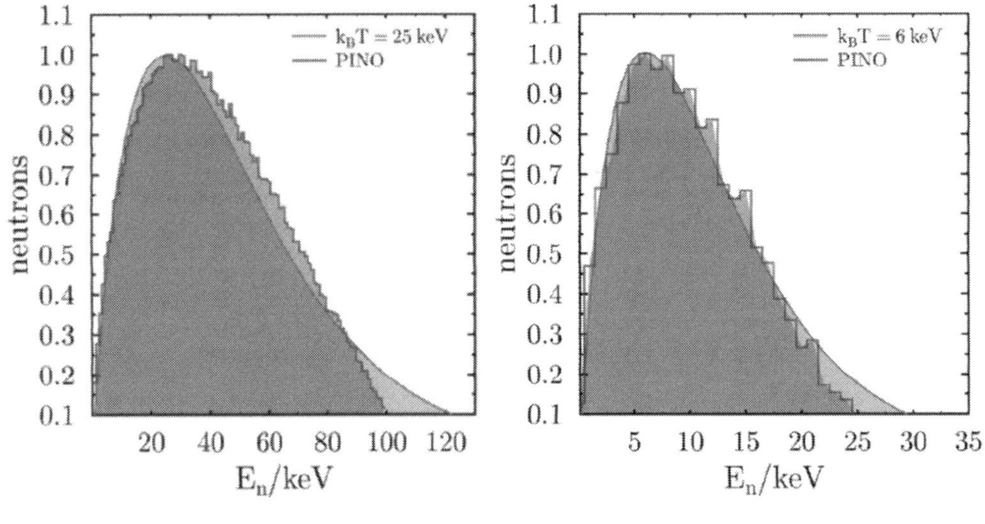

Figure 2. PINO-Simulation for neutrons passing the sample placed 3.2 mm after the lithium target (blue) compared with a Maxwellian distribution (green).
Left: $k_B T = 25$ keV, disk-shaped sample with 10 mm radius, E_{proton}=1912 keV.
Right: $k_B T = 6$ keV, ring-shaped sample with an inner radius of 6 mm and outer radius of 10 mm, E_{proton}=1914 keV.

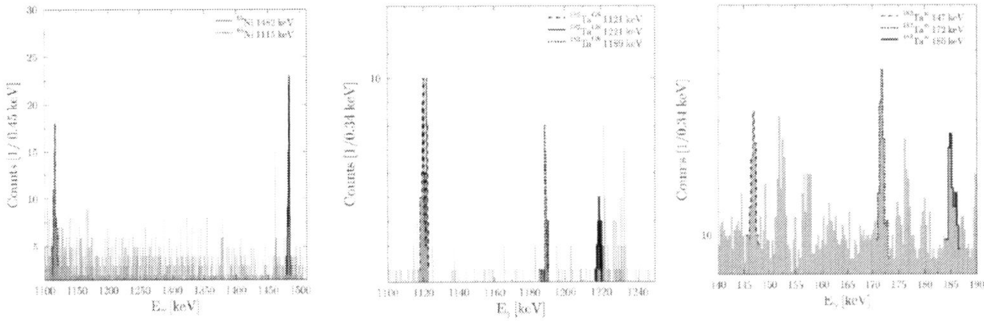

Figure 3. Measurements using γ-spectroscopy for the reactions for low s-process temperatures at $k_B T = 6$ keV for the reactions **Left:** ^{64}Ni(n,γ)^{65}Ni, **Middle:** ^{181}Ta(n,γ)^{182}Ta, and **Right:** ^{181}Ta(n,γ)^{182}Ta^m.

Using a proton beam energy of $E_P = 1914$ keV and a ring-shaped sample placed 3.2 mm after the lithium target with an inner radius of 6 mm and an outer radius of 10 mm covers the neutron cone such, that the resulting neutron spectrum closely resembles a Maxwellian distribution of $k_B T = 6$ keV (Fig.1, right panel).

3 Proof of principle

We measured the reaction ^{64}Ni(n,γ)^{65}Ni at the energy of $k_B T = 6$ keV and $k_B T = 25$ keV. We identify the produced radioactive nuclei using γ-spectroscopy with two Broad Energy Germanium (BEGe) detectors in head-to-head geometry. Fig.3 shows a spectrum of the ring measurements. The left panel shows the identification of the produced radioactive ^{65}Ni using the γ-emission lines at E_γ = 1482 keV and 1115 keV. Preliminary results are listed in Tab. 1.

As a second proof-of-principle, we performed a measurement of the reaction ^{181}Ta(n,γ)^{182}Ta and ^{181}Ta(n,γ)^{182}Ta^m. The identification of the radioactive ^{182}Ta nuclei as well as its isomeric state ^{182}Ta^m was done using γ-spectroscopy. Fig.3 shows the measurement for the ground state of ^{182}Ta (middle panel) and for the isomeric state of ^{182}Ta (right panel). The identification of the ground state ^{182}Ta was done using the γ-emission lines at E_γ = 1121 keV, 1221 keV and 1189 keV. The isomeric state could be resolved by the γ-emission lines at E_γ = 147 keV, 172 keV and 185 keV. The preliminary cross sections are given in Tab. 1.

4 Summary

To perform activation measurements at lower energies, it is possible to vary the angular coverage of the neutron cone. This can be achieved using ring-shaped of the samples. We

Table 1. Preliminary neutron capture cross sections of the reactions ^{64}Ni(n,γ)^{65}Ni, ^{181}Ta(n,γ)^{182}Ta and ^{181}Ta(n,γ)^{182}Ta^m at $k_B T = 6$ keV and $k_B T = 25$ keV.

preliminary	^{64}Ni(n,γ)^{65}Ni	^{181}Ta(n,γ)^{182}Ta^{GS}	^{181}Ta(n,γ)^{182}Ta^m
$\sigma_{k_B T=6\,\mathrm{keV}}$[mb]	12	2136	3
$\sigma_{k_B T=25\,\mathrm{keV}}$ [mb]	4	454	0.8

successfully performed first measurements using a Nickel and a Tantalum sample to study the cross sections for the reactions ^{64}Ni(n,γ)^{65}Ni, ^{181}Ta(n,γ)^{182}Ta, and ^{181}Ta(n,γ)^{182}Ta^{m} at $k_{\mathrm{B}}T = 6\,\mathrm{keV}$. Preliminary results are shown in Tab. 1. The preliminary neutron capture on ^{64}Ni was determined to $\sigma_{k_{\mathrm{B}}T=6\,\mathrm{keV}} = 12\,\mathrm{mb}$ which fits quit well to the KADONIS value $\sigma_{k_{\mathrm{B}}T=5\,\mathrm{keV}} = 11.8\,\mathrm{mb}$ [5]. The neutron capture on ^{181}Ta into the ground state of ^{182}Ta was preliminary determined to $\sigma_{k_{\mathrm{B}}T=6\,\mathrm{keV}} = 2136$ mb compared to KADONIS with $\sigma_{k_{\mathrm{B}}T=5\,\mathrm{keV}} = 2765$ mb [5]. The neutron yield is reduced by a factor of 100 compared to the standard 25 keV setting, but is still a factor of 100 higher than previous attempts using the ^{18}O(p,n) reaction.

References

[1] R. Reifarth, C. Lederer, F. Käppeler, Journal of Physics G Nuclear Physics **41**, 053101 (2014)

[2] R. Reifarth, P. Erbacher, S. Fiebiger, K. Göbel, T. Heftrich, M. Heil, F. Käppeler, N. Klapper, D. Kurtulgil, C. Langer et al., Eur. Phys. J. Plus **133**, 424 (2018)

[3] M. Heil, S. Dababneh, A. Juseviciute, F. Käppeler, R. Plag, R. Reifarth, S. O'Brien, Phys. Rev. C **71**, 025803 (2005)

[4] R. Reifarth, M. Heil, F. Käppeler, R. Plag, Nucl. Inst. Meth. A **608**, 139 (2009)

[5] I. Dillmann, R. Plag, F. Käppeler, T. Rauscher, *KADoNiS v0.3 - The third update of the "Karlsruhe Astrophysical Database of Nucleosynthesis in Stars*, in *Proceeding of the workshop EFNUDAT Fast Neutrons - scientific workshop on neutron measurements, theory & applications* (2009), Geel, Belgium

EPJ Web of Conferences **279**, 07001 (2023) https://doi.org/10.1051/epjconf/202327907001
NPA-X 2022

The intermediate neutron-capture process in AGB stars

Arthur Choplin[1,*], *Lionel* Siess[1], and *Stephane* Goriely[1]

[1]Institut d'Astronomie et d'Astrophysique, Université Libre de Bruxelles (ULB), CP 226, B-1050 Brussels, Belgium

Abstract. The intermediate **neutron-capture** process is thought to arise when protons are mixed in a convective helium-burning zone. This can happen during the early Thermally-Pulsing (TP) Asymptotic Giant Branch (AGB) phase of low-mass, low-metallicity stars. After discussing the differences between the s- and i-processes in AGB stars, we highlight some critical (n, γ) reactions for i-process nucleosynthesis that may be experimentally constrained by the β-Oslo method. We then compare our s- and i-process nucleosynthesis predictions to the abundances of the Carbon-Enhanced Metal-Poor star HE2258-6358, which shows a composition pattern midway between the s- and r-process.

1 Introduction

Most of the elements heavier than **iron** were synthesized by **neutron-capture** processes. In addition to the slow and rapid **neutron-capture processes**, an intermediate **neutron-capture** process (i-process) is thought to arise when protons are mixed in a convective helium-burning zone (proton ingestion event or PIE). One possible astrophysical site is the early Thermally-Pulsing (TP) Asymptotic Giant Branch (AGB) phase of low-mass, low-metallicity stars ([1] and references therein). Very few AGB models have consistently followed **the possible i-process nucleosynthesis.** In this **proceedings paper**, after discussing the differences between the s- and i-processes in AGB stars, we highlight critical (n, γ) rates, which are unknown experimentally, and compare our AGB nucleosynthesis predictions to the Carbon-Enhanced Metal-Poor star HE2258-6358 whose abundance pattern is in-between s and r (CEMP-r/s star).

2 Differences between the s- and the i-process in AGB stars

The **Thermally-Pulsing** AGB star are evolved low-and intermediate mass stars and the site of a very rich nucleosynthesis [2–5]. Their degenerate carbon-oxygen core is surrounded by an unstable helium-burning shell, surrounded by a hydrogen-burning shell and an extended convective envelope. The two burning shells are separated by the intershell region. The energy is alternatively produced by hydrogen burning (during interpulses) and helium-burning (during thermal pulses).

*e-mail: arthur.choplin@ulb.be

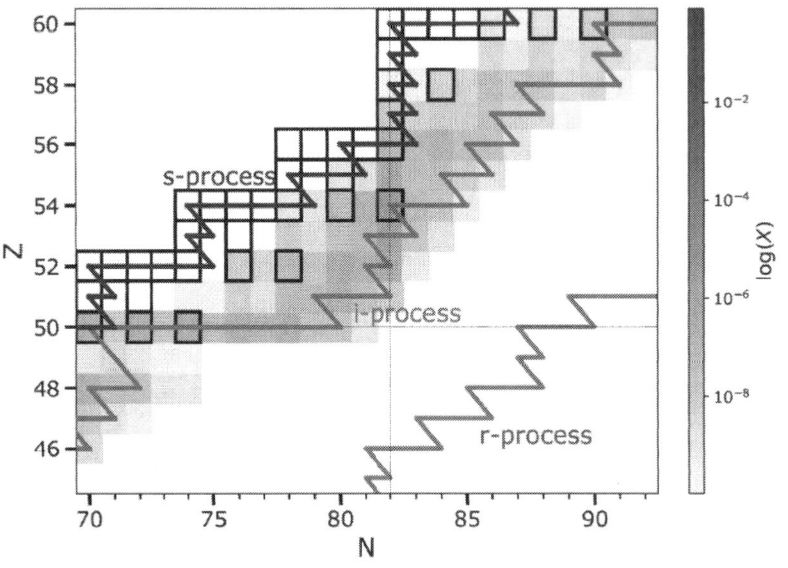

Figure 1. The main i-process path (red line) in a 1 $M_\odot$, [Fe/H] = −2.3, AGB model, at the bottom of the second thermal pulse, at the time of maximum neutron density (N_n = 2.2×10^{15} cm^{-3}). The chemical abundances are shown by the red colour scale in mass fraction. **The black squares are stable nuclei.** Typical s- and r-process paths (in blue and **green, respectively**) are shown for comparison.

2.1 The radiative s-process

After a thermal pulse, the inner boundary of the convective envelope moves inwards (third dredge up). Depending on mixing parameters (e.g., **those controlling the overshoot** below the envelope), protons from the envelope can be partially mixed in the intershell region, react with ^{12}C via ^{12}C$(p,\gamma)^{13}$N$(\beta^+)^{13}$C and form the so-called ^{13}C-pocket. Neutrons are then released by the reaction ^{13}C$(\alpha,n)^{16}$O and the s-process takes place during the interpulse in radiative conditions. The temperature is about 100 MK, the timescale of the order of 10^4 yr (duration of an interpulse) and the neutron density $\sim 10^8$ cm^{-3}. The radiative s-process is usually thought to be the main mechanism in AGB stars to synthesize trans-iron elements, and in particular the heavy (Ba, La, Ce, Pr, Nd) and very heavy (Pb) s-elements.

2.2 The convective s-process

When the H-burning ashes are engulfed in the convective pulse, ^{14}N is quickly converted to ^{22}Ne by ^{14}N$(\alpha,\gamma)^{18}$F$(,e^+\nu_e)^{18}$O$(\alpha,\gamma)^{22}$Ne. If the temperature at the bottom of the pulse is **greater than** 300 **MK** (*i.e.* for an initial stellar mass $M_{\rm ini} \gtrsim 3$ M$_\odot$), the ^{22}Ne(α,n) reaction is activated and leads to a short burst of neutron with neutron density of $10^{10} - 10^{14}$ cm^{-3} during typically one year [6]. Due to its convective nature, the s-process in such conditions is not expected to contribute as much as the radiative s-process but can significantly affect the light s-elements (Sr, Y, Zr), as well as the relative abundances around branching points.

2.3 The i-process

It was shown that during a thermal pulse, the entropy barrier at the bottom of the H-burning shell in low metallicity stars can be surmounted by the energy released by the pulse [7, 8]. In

Table 1. Most uncertain (n, γ) reactions (from Ba to Pb) along the i-process path (at a neutron density of $\sim 10^{15}$ cm^{-3}), according to the three different nuclear models A, B and C. **Reactions are included when at least one of the ratios of the Maxwellian-averaged cross sections from different models (at T = 250 MK, given in the table) is higher than 10.**

reaction	$\langle\sigma\rangle_B/\langle\sigma\rangle_A$	$\langle\sigma\rangle_C/\langle\sigma\rangle_A$	reaction	$\langle\sigma\rangle_B/\langle\sigma\rangle_A$	$\langle\sigma\rangle_C/\langle\sigma\rangle_A$
^{150}Pr(n,γ)	2.13e+0	1.42e+1	^{176}Er(n,γ)	1.87e+0	1.58e+1
^{158}Pm(n,γ)	5.91e+0	1.53e+1	^{173}Tm(n,γ)	8.61e-1	1.40e+1
^{159}Sm(n,γ)	4.61e+0	2.86e+1	^{174}Tm(n,γ)	1.98e+0	1.07e+2
^{163}Eu(n,γ)	9.86e-1	1.88e+1	^{185}Lu(n,γ)	1.64e-1	4.34e+0
^{164}Eu(n,γ)	8.53e-1	1.10e+1	^{190}Ta(n,γ)	1.47e-1	2.92e+0
^{165}Eu(n,γ)	1.16e+0	2.83e+1	^{191}Re(n,γ)	1.27e-1	1.51e+0
^{161}Gd(n,γ)	1.95e+0	1.80e+1	^{192}Re(n,γ)	6.32e-2	1.83e+0
^{163}Gd(n,γ)	2.03e+0	1.45e+1	^{216}Pb(n,γ)	1.57e-2	9.71e-1
^{167}Dy(n,γ)	2.40e+0	1.01e+1			

these conditions, the top of the helium-convective zone reaches the hydrogen-rich layers, so that **protons can be engulfed (in a PIE)** and mixed down to high temperatures in a convective timescale of typically 1 hr. The protons burn on the fly by $^{12}C(p,\gamma)^{13}N$ and do not reach the bottom of the pulse. The ^{13}N isotope beta-decays to ^{13}C in 10 min and $^{13}C(\alpha,n)^{16}O$ is activated. At the bottom of the pulse, the temperature is typically 250 MK and the $^{13}C(\alpha,n)^{16}O$ reaction leads to neutron densities of 10^{15} cm^{-3} during $\sim$ 1 day. During the peak neutron emission, the **neutron-capture** timescale τ_n is of the order of 1 min. For instance, in the i-process model shown in Fig. 1, τ_n for the ^{129}Sn isotope is 46 s.

The s- and -i-processs use the same neutron source $^{13}C(\alpha,n)^{16}O$ but they operate at very different temperatures ($\sim$ 100 vs 250 MK, respectively) and environment (radiative vs convective). Also due to the large amount of nuclear energy released during the proton injection by $^{12}C(p,\gamma)^{13}N$, the structure, surface composition and evolution of the AGB star can be substantially impacted [1].

3 Nuclear uncertainties

The abundances resulting from i-process nucleosynthesis are affected by nuclear uncertainties. About 70 % of the **neutron-capture** rates involved during i-process nucleosynthesis are not known experimentally. To study the impact of nuclear uncertainties, we calculated the i-process nucleosynthesis during a PIE using three different sets of theoretical (n,γ) rates, as detailed in Ref. [9]:

- Model A : the fiducial rates obtained with TALYS reaction code [10].
- Model B : same as model A but considering the photon strength functions **from [11], instead of [12] as in model A**.
- Model C : same as model A but adding the direct capture mechanism [13].

We chose models B and C out of the ten sets included in [9] since they give the lower and upper limits for the heavy element production. **According to these three nuclear models, we identified the most critical and uncertain** (n,γ) **reactions for the nuclei between Ba and Pb located along the i-process path (or nearby) at** $N_n \sim 10^{15}$ **cm**$^{-3}$ **(Table 1). This value of the neutron density is reached at the bottom of the thermal pulse of our 1** $M_\odot$**, [Fe/H] =** -2.5 **model star. The different nuclear models give rise to differences of typically 0.4 dex in the surface abundance enrichment of the AGB star. The impact is**

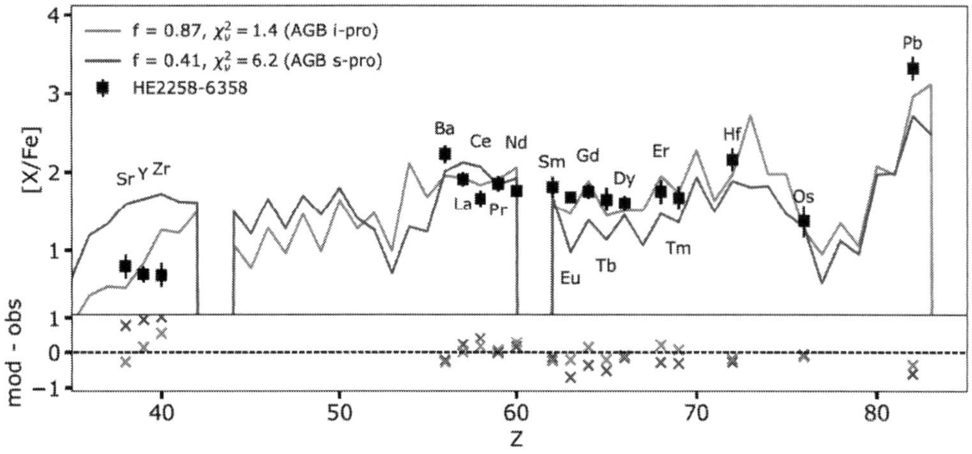

Figure 2. Best fits to the CEMP-r/s star HE2258-6358 using a 2 $M_\odot$ AGB model with [Fe/H] = −2.5 that experienced s-process only (blue) and using a 1 $M_\odot$ AGB model with [Fe/H] = −2.5 that experienced i-process (red). The method of fitting is described in the Section 6.2 of [1]. The dilution factor f (Eq. 8 and 10 in [1]) and the smallest χ^2_ν value (Eq. 7 in [1]) are indicated. The abundance data is from [15].

the highest in the $75 \lesssim Z \lesssim 80$ **region with differences of up to** $\sim$ **1 dex (Figure 6 in [9])**. An accurate rate determination of these reaction rates is therefore desired to reduce the uncertainties affecting i-process nucleosynthesis in low-metallicity AGB stars. Although the involved isotopes are generally away from the valley of stability, it may be possible to constrain experimentally some of these reactions using **the** β**−Oslo method [14]**.

4 The i-process and the CEMP-r/s stars

The CEMP-r/s stars are main-sequence or red giant stars that show abundance patterns midway between the s- and r-processes. Because they are not sufficiently evolved, they could not have enriched their surface by themselves. Instead, one (or more) external stellar source(s) should have polluted their surface. A possibility is that these stars have acquired their heavy element pattern from at least two distinct astrophysical sources that produced the s- and r-processes separately. A difficulty with this scenario is to explain the correlation that seems to exist between Ba (mostly formed by s-process) and Eu (mostly formed by r-process) [6]. An alternative paradigm is that the CEMP-r/s stars acquired their chemical pattern by mass accretion from an AGB companion **that experienced the i-process [16]**. This scenario is supported by the fact that most CEMP-r/s stars are in binary systems [17].

As shown in Fig. 2, the heavy element abundance pattern of the CEMP-r/s star HE2258-6358 is well reproduced by the yields of an AGB star that experienced i-process nucleosynthesis (red pattern). By contrast, a standard radiative s-process in an AGB cannot account for the high abundances in the Eu region and overestimates Sr, Y and Zr (blue pattern).

Another issue with the radiative s-process models is that the level of enrichment **may be** too small to explain the CEMP-r/s stars, **although it strongly depends on the efficiency of the third dredge-up, which is very uncertain (e.g. section 4 in [5]).** When fitting the abundances of a CEMP-r/s star, the dilution factor f is freely varied so as to minimize the χ^2 value. A dilution factor of 0.87, for instance, means that 13 % of the AGB material was mixed with 87 % of the envelope of the accreting star (details can be found in Sect. 6.2 of [1] and in [18]). The dilution factor can also be linked to physical quantities related to the binary system properties:

$$f = \frac{M_{\rm env}}{M_{\rm env} + M_{\rm acc}},$$

where $M_{\rm env}$ is the envelope mass of the r/s star before the accretion episode and $M_{\rm acc}$ the mass accreted by the r/s star from the AGB companion.

For the s-process AGB model, the dilution factor is $f = 0.41$, which gives $M_{\rm acc} = 0.54\ M_\odot$. Accreting this much mass is likely not realistic and if happening, will possibly boost the evolution of the r/s-star to the white dwarf stage, thereby making it unobservable today. In the case of the i-process AGB model, the dilution factor of $f = 0.87$ leads to the more realistic value of $M_{\rm acc} = 0.06\ M_\odot$. Noticing that $M_{\rm acc} = \beta M_{\rm wind}$ with $M_{\rm wind}$ the mass lost by the AGB star during its lifetime ($\sim 0.5\ M_\odot$ in this case), and β the accretion efficiency parameter, we obtain $\beta \sim 0.1$, which is compatible with 3D hydrodynamical simulations of wind mass transfer in AGB binary systems [19].

5 Conclusions

We discussed the i-process in low-mass, low-metallicity AGB stars. Although the i-process in AGB shows similarities with the radiative s-process in AGB, the i-process arises at much higher temperatures and in convective conditions. Using three different nuclear models, we extracted 17 critical (n, γ) reactions for i-process nucleosynthesis (between Ba and Pb) **whose rates are the most uncertain** (Table 1). Finally, we have shown the ability of i-process AGB models to explain the CEMP-r/s star HE2258-6358. This suggests that low-mass, low-metallicity AGB stars are a viable site for the i-process.

Acknowledgments

This work was supported by the Fonds de la Recherche Scientifique-FNRS under Grant No IISN 4.4502.19. L.S. and S.G. are senior FRS-F.N.R.S. research associates.

References

[1] A. Choplin, L. Siess, S. Goriely, A&A, **648**, A119 (2021), 2102.08840
[2] I. Iben, Jr., A. Renzini, ARA&A, **21**, 271 (1983)
[3] M. Busso, R. Gallino, G.J. Wasserburg, ARA&A, **37**, 239 (1999)
[4] F. Herwig, ARA&A, **43**, 435 (2005)
[5] A.I. Karakas, J.C. Lattanzio, PASA, **31**, e030 (2014)
[6] M. Lugaro, A.I. Karakas, R.J. Stancliffe, C. Rijs, ApJ, **747**, 2 (2012), 1112.2757
[7] M. Schwarzschild, R. Härm, ApJ, **150**, 961 (1967)
[8] M.Y. Fujimoto, Y. Ikeda, I. Iben, Jr., ApJ, **529**, L25 (2000)
[9] S. Goriely, L. Siess, A. Choplin, A&A, **654**, A129 (2021), 2109.00332
[10] A.J. Koning, D. Rochman, Nuclear Data Sheets **113**, 2841 (2012)
[11] S. Goriely, E. Khan, M. Samyn, Nucl. Phys. A **739**, 331 (2004)
[12] S. Goriely, P. Dimitriou, M. Wiedeking, T. Belgya, R. Firestone, J. Kopecky, M. Krticka, V. Plujko, R. Schwengner, S. Siem et al., Eur. Phys. J. A **55**, 172 (2019)
[13] K. Sieja, S. Goriely, Eur. Phys. J. A **57**, 110 (2021)
[14] A. Spyrou, S.N. Liddick, A.C. Larsen, M. Guttormsen, K. Cooper, A.C. Dombos, D.J. Morrissey, F. Naqvi, G. Perdikakis, S.J. Quinn et al., Phys. Rev. Lett., **113**, 232502 (2014), 1408.6498

[15] V.M. Placco, A. Frebel, T.C. Beers, A.I. Karakas, C.R. Kennedy, S. Rossi, N. Christlieb, R.J. Stancliffe, ApJ, **770**, 104 (2013), 1304.7869
[16] M. Hampel, R.J. Stancliffe, M. Lugaro, B.S. Meyer, ApJ, **831**, 171 (2016), 1608.08634
[17] D. Karinkuzhi, S. Van Eck, S. Goriely, L. Siess, A. Jorissen, T. Merle, A. Escorza, T. Masseron, A&A, **645**, A61 (2021), 2010.13620
[18] A. Choplin, L. Siess, S. Goriely, A&A, **662**, C3 (2022)
[19] M.I. Saladino, O.R. Pols, C. Abate, A&A, **626**, A68 (2019), 1903.04515

Nuclear physics in the N≈126 region relevant for the *r* process

Zsolt Podolyák[1,2,*]

[1]Department of Physics, University of Surrey, Guildford GU2 7XH, UK
[2]ISOLDE, EP Department, CERN, CH-1211 Geneva, Switzerland

Abstract. Understanding how the heavy chemical elements are made in the astrophysical *r*-process requires information on both the properties of the nuclei involved and that of the environment. The availability of experimental information on the neutron-rich N∼126 nuclei is discussed, with emphasis on phenomena specifically relevant to this region: the large role of first-forbidden β decays, and that of the Δn=0 selection rule in Gamow-Teller decays. The development of nuclear data bases by combining different theoretical approaches is suggested.

1 Introduction

The rapid neutron-capture (*r*) process is known to synthesise about half of the nuclei heavier than iron [1]. The observational data on the *r*-process is scarce. Only the solar system provides us with elemental and isotopic abundances. Here the *r*-process abundances are determined as the difference of the total observed, minus the slow neutron-capture (*s*) process abundances. Since in the case of the *s*-process both the astrophysical site and the majority of the nuclear data are known, these calculations are robust. The such obtained *r*-process abundances, exhibit wide peaks at around masses A∼80, 135, 190, connected to the magic neutron numbers at N=50, 82, 126. This abundance pattern became a *de facto* definition of the *r*-process. This is what individual *r*-process calculations aim to reproduce, independently whether this originates from a single process, from a single site, or is a mixture of these. The single event scenario seems rather unlikely. It was suggested that the long-lived ^{182}Hf and ^{26}Al isotopes have different origins in the early solar system [2]. Another illustration of the (in)probability of a single event: it was advocated that the plasma in Local Bubble in which the solar system is embedded originated from 14-20 supernovae explosions from the last 13M years [3]. Despite all these, the fact that a single dataset is available, justifies the present approach, with single astrophysical event adopted in the *r*-process abundance calculations.

Observational information is available also from a large number of stars, for which the abundances of (some) elements were determined. Around a dozen of them, all very old, as judged by their very low metallicity, exhibit similar abundance patterns than that of the solar system *r*-process [1]. This is considered as proof of the universality of the *r*-process. However, there are other low metallicity stars in which the abundance decreases drastically with atomic number, pointing toward an incomplete *r*-process [4], with large number of intermediate cases; or the opposite of this, stars which have too much actinide nuclei (Th, U) [5].

*e-mail: z.podolyak@surrey.ac.uk

These observational data suggest that the *r*-process abundance distribution is dependent on the environment (neutron density, temperature), and possibly also on the site (SNe, merger); conclusion supported by the astrophysical calculations. Most likely, different sites, with different neutron density conditions, contribute to the production of *r*-process elements [1]. Additional information on the *r*-process can be gained from the chemical evolution of elements (changes in abundances as function of metallicity), pre-solar grains, meteorites carrying long lived isotopes (^{129}I, ^{247}Cm [6]), or the observation of long lived isotopes on Earth and Moon (^{60}Fe, ^{244}Pu, ^{53}Mn), see e.g. [7].

The largest uncertaintiy in the *r*-process is due to the uncertain site(s) where it happens. The site(s) needs to provide high neutron densities, with the most discussed candidate being supernovae of several types and mergers. The recent observation in 2017 of the kilonova remnant of the first observed gravitational wave event attributed to a neutron-star merger provided evidence for this scenario. The evolution of the remnant's luminosity in different wavelengths is consistent with the synthesis of heavy neutron-rich nuclei [8]. While no individual heavy elements could be observed, some properties of the spectrum were linked to strontium [9].

2 Nuclear data

The abundance calculations, as input, need information about environment, as well as nuclear data. The nuclear data clearly has an effect an the *r*-process yield calculations, as attested by the numerous studies accompanying the reporting of new lifetime and mass values (e.g. [10, 11]). As a general statement, shorter lifetimes for the N$\sim$126 (82) nuclei shift the predicted A$\sim$195 (130) *r*-process peak to higher masses. At the same time β-delayed neutron emission moves the peak to lighter masses, widening it (more exotic nuclei emit more neutrons) and smoothening it (largely removing the odd-even mass staggering) [12].

Sensitivity tests are used to identify nuclei and their properties which are particularly important for the quantitative understanding of the *r*-process yields. These are often biased towards the progenitors of final products with high abundance [13, 14]. It seems that there are no really special nuclei, but the understanding of whole regions are needed. This conclusion seems to be supported by the fact that the $\sim$130 abundance peak is not better understood than the $\sim$195 one, despite that the mass and lifetimes of several N=82 *r*-process path nuclei were already measured [15] (uncertainties due to fission recycle provide an alternative explanation).

3 Nuclear structure in the N$\sim$126 region

The neutron-rich N$\sim$126 region is different in two ways when compared to lighter nuclei. First-forbidden (FF) β decays are expected to compete against allowed ones, and even to dominate. In addition, this is the region where the effect of the so called Δn=0 β decay selection rule for Gamow-Teller transitions becomes important. We discuss these two aspects one by one.

3.1 Competition between first-forbidden and allowed beta decays

Global β-decay calculations show that allowed beta decay (with selection rules ΔI=0, $\pm$1 and $\Delta\pi$=No) dominate over first-forbidden ones (ΔI=0,$\pm$1,$\pm$2 and $\Delta\pi$=Yes) on the large part of the nuclide chart [16, 17]. But in the case of heavy nuclei, starting around N=126, first-forbidden transition compete and can become dominant, in stark contrast with the N$\sim$82 and N$\sim$50 regions. Since the calculation of FF transitions is notoriously difficult, the differencies in the predicted lifetimes are large. Table I, comparing lifetimes and FF contributions obtained from recent ISOLDE experiments on ^{207,208}Hg [18] demonstrates this.

Table 1. Experimentally determined halflives and first-forbidden beta decay contribution for the β decays of ^{208}Hg and ^{207}Hg, compared with theoretical predictions.

Nucleus		Exp	Theory						
			[19]	[20]	[21]	[22]	[17]	[16]	[23]
^{208}Hg	$T_{1/2}$ (s)	135(10) [24]	168.9	8.1	12.1	5.3	0.9	70	
	FF/total	≈1 [24]				0.040	0.87	1	
^{207}Hg	$T_{1/2}$ (s)	174(12) [25]	40.1	9.1		313.2	4.0	61	
	FF/total	≈1 [26]				0.986	0.77	1	1

3.2 The Δn=0 selection rule

A less well-known selection rule for the otherwise allowed β-decays is that the number of nodes (n) in the radial wave functions of the initial and final states has to be the same. This Δn=0 requirement plays a major role in the β decay of heavy neutron-rich nuclei as there are several pairs of single particle Δn=0 orbitals for N>126 and Z<82. If the selection rule is strictly obeyed, β decay between them is forbidden, resulting in longer lifetimes than otherwise. The greatest impact is on nuclei where the Fermi level lies high above N=126 and/or much below Z=82, e.g. nuclei on the astrophysical r-process pathway, influencing the nucleosynthesis of heavy elements. This selection rule has little effect on isotopes which are proton-rich or close to the stability line, making the experimental investigation of its validity difficult. The most stringent test was provided by the β decay of ^{207}Hg, where the level of forbideness of the Δn=1 $\nu 1g_{9/2} \rightarrow \pi 0g_{7/2}$ transition was recently investigated at ISOLDE. An upper limit of 3.9x10^{-5} was obtained for the probability of this decay, corresponding to a logft>8.8 (95% confidence limit) [27]. This selection rule is based on the fact that the overlap integral between the Δn=0 orbitals vanishes, which is valid only if the proton and neutron feel the same potential. This is not stricly true due to the Coulomb force and the different spin-orbit couplings (parallel and antiparallel). While in ^{207}Hg the selection rule was upheld, global lifetime calculations suggest that in heavy nuclei about 20% of the decay is via these Δn=0 transitions [17]. In order to really understand the relevance of this selection rule, experimental information from much more exotic nuclei is needed, as the two available examples (^{207}Hg and ^{209}Tl) are strongly relyant on small components of the wave functions, obtained from shell model calculations. For example, in the case of the ^{207}Hg decay, the probability that there are two holes in the $\pi g_{7/2}$ orbital in its ground-state is rather low at ~10^{-4} [27].

4 Which experimental data is relevant for the r process?

The nuclear data needed for the abundance calculations are mainly masses (used to calculate Q values and, crucially, reaction cross sections) and lifetimes. Properties of excited states, unless these are long lived isomers which are included as separate nuclides, are not used in the network calculations. Figure 1 summarises the data available in the neutron-rich N≥126 region. Lifetimes along N=126 are only known for four nuclei below ^{208}Pb, down to ^{204}Pt [28]. Mass values are published only down to ^{206}Hg [29]. These properties are the ones directly compared with the global nuclear calculations. Information on ground-state poperties which are less used to constrain calculations are β-delayed neutron-emission probabilities [30] and charge radii [31]. However, there is a much wider knowledge on properties of excited states. Information on them, such as energies and spin-parities of yrast states are available down to ^{203}Ir [32], five protons from ^{208}Pb. Often there are transition strength information as well (because of the high sensitivity of the isomeric decay technique), which are used to obtain information on the wave functions. These considerable knowledge on excited states usually

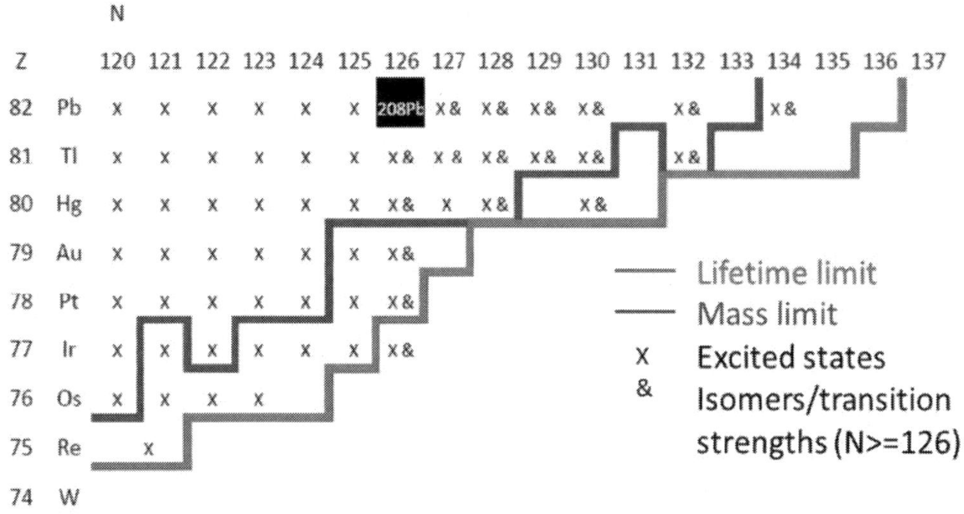

Figure 1. Status of experimental knowledge on the neutron-rich N 126 nuclei.

is not taken advantage of when testing nuclear theories. This is a shame, as excited states in less exotic nuclei, especially transition strength information, can provide knowledge on the ground state of the more exotic ones. The reason for this omission is that global calculations are somewhat simplified when compared to the more microscopic ones, and cannot be used to predict the details of the nuclear structure. On the other hand, while the shell model could predict all properties of ground- and excited states, these are computationally prohibitive for nuclei with large number of valence nucleons. Models such as the particle-vibrational coupling try to provide the bridge between these two, global and localised, type of calculations. For now, when creating nuclear databases it would be advantageous to combine (for lifetimes and masses) predictions from different theoretical frameworks, always using the one considered more reliable for a given region. In a sense such mixing is what is done, when measured values (like in the N∼82 region) are combined with theoretical ones. One would need to check and smoothen discontinuities at the boundaries of different calculations. The such obtained database would be more reliable than the existing ones relying on individual global calculations.

5 Outlook

Multimessenger astronomy observations, similar to the pioneering one which followed the GW17017 event [8], will help to understand how similar/different such events are and therefore provide information on the universality of the *r* process. There is a possibility that light-curve studies will prove the production of ^{254}Cf, whose decays has a large impact after 100 days [13]. More data on the difficult to determine abundances of Th or U, the only two elements produced exclusively in the *r* process would be important for similar reasons. From experimental point of view, the radioactive beam facilities coming online will allow to measure directly the properties of some of the N=126 *r*-process path nuclei for the first time. This will help to determine how much material goes into the fission region, and get better understanding of the importance of fission recycling (for a given astrophysical scenario). This, together with experimental nuclear data might clarify the origin of the rare-earth *r*-process peak, weather it is related to discontinuities of nuclear properties (mass, neutron capture rate) or is the result of fission.

References

[1] J.J. Cowan *et al.*, Rev. Mod. Phys. **93**, 015002 (2021)
[2] J.C. Holst *et al.*, PNAS **110**, 8891 (2013)
[3] D. Breitschwerdt *et al.*, Nature **532**, 73 (2016)
[4] S. Honda *et al.*, Astrophys. J. **666**, 1189 (2007)
[5] E.M. Holmbeck *et al.*, Astrophys. J. **859**, L24 (2018)
[6] B. Cote *et al.*, Science **371**, 945 (2021)
[7] A. Wallner et al., Science **372**, 742 (2021)
[8] E. Pian *et al.*, Nature **551**, 67 (2017)
[9] D. Watson *et al.*, Nature **574**, 497 (2019)
[10] J. Wu *et al.*, Phys. Rev. Lett. **118**, 072701 (2017)
[11] H.F. Li *et al.*, Phys. rev. Lett. **128**, 152701 (2022)
[12] N. Nishimura, Zs. Podolyák, D.-L. Fang, T. Suzuki, Phys.Lett. B **756**, 273 (2016)
[13] I. Kullmann *et al.*, arXiv:2207.07421v1
[14] M.R. Mumpower *et al.*, Progr. Part. Nucl. Phys. 86, 86 (2016)
[15] Meng Wang *et al.*, Chin. Phys. C **45**, 030003 (2021)
[16] E.M. Ney, J. Engel, T. Li, N. Schunck, Phys.Rev. C **102**, 034326 (2020)
[17] T. Marketin, L. Huther, G. Martinez-Pinedo, Phys. Rev. C **93**, 025805 (2016)
[18] Zs. Podolyák, EPJ Web of Conf. **260**, 03005 (2022)
[19] P. Moller, B.Pfeiffer, K.-L.Kratz, Phys. Rev. C **67**, 055802 (2003)
[20] H. Koura, T. Tachibana, M. Uno, M. Yamada, Prog. Theor. Phys. **113**, 305 (2005)
[21] I. Borzov, Nucl. Phys. A **777**, 645 (2006)
[22] D.L. Fang, B.A. Brown, T. Suzuki, Phys. Rev. C.**88**, 034304 (2013)
[23] A. Kumar, P.C. Srivastava, Nucl. Phys. A **1014**, 122255 (2021)
[24] R.J. Carroll *et al.*, Phys. Rev. Lett. **125**, 192501 (2020)
[25] B.Jonson, O.B.Nielsen, L.Westgaard, J.Zylicz, Proc.Int.Conf.Nuclei Far from Stability, Helsingor, Denmark, Vol.2, p.640 (1981); CERN-81-09 (1981)
[26] T. Berry *et al.*, Phys. Rev. C **101**, 054311 (2020)
[27] T. Berry *et al.*, Phys. Lett. B **793**, 271 (2019)
[28] A.I. Morales *et al.*, Phys.Rev.Lett. **113**, 022702 (2014)
[29] L. Chen *et al.*, Phys. Rev. Lett. **102**, 122503 (2009)
[30] R. Caballero-Folch *et al.*, Phys. Rev. Lett. **117**, 012501 (2016); Phys. Rev. C **95**, 064322 (2017)
[31] T. Day Goodacre *et al.*, Phys. Rev. Lett. 126, 032502 (2021)
[32] S.J. Steer *et al.*, Phys. Rev. C **84**, 044313 (2011)

EPJ Web of Conferences **279**, 08002 (2023) https://doi.org/10.1051/epjconf/202327908002
NPA-X 2022

Constraining nucleosythesis in neutrino-driven winds using the impact of (α, xn) reaction rates

Athanasios Psaltis[1,*], *Almudena* Arcones[1,2], *Melina L.* Avila[3], *Maximilian* Jacobi[1], *Camilla Juul* Hansen[4], *Linda* Lombardo[5], *Zach* Meisel[6], *Peter* Mohr[7], *Fernando* Montes[8,9], *Wei Jia* Ong[10], and *Hendrik* Schatz[8,9]

[1]Institut für Kernphysik, Technische Universität Darmstadt, Schlossgartenstr. 2, Darmstadt 64289, Germany
[2]GSI Helmholtzzentrum für Schwerionenforschung GmbH, Planckstr. 1, Darmstadt 64291, Germany
[3]Physics Division, Argonne National Laboratory, Argonne, IL 60439, USA
[4]Institute for Applied Physics, Goethe University Frankfurt, Max-von-Laue-Str. 12, Frankfurt am Main 60438, Germany
[5]GEPI, Observatoire de Paris, Université PSL, CNRS, 5 place Jules Janssen, 92195 Meudon, France
[6]Institute of Nuclear & Particle Physics, Department of Physics & Astronomy, Ohio University, Athens, OH, 45701 USA
[7]Institute for Nuclear Research (ATOMKI), H-4001 Debrecen, Bem tér 18/c, Hungary
[8]National Superconducting Cyclotron Laboratory, East Lansing, MI, 48824, USA
[9]Joint Institute for Nuclear Astrophysics – Center for the Evolution of the Elements, Michigan State University, East Lansing, Michigan 48824, USA
[10]Nuclear and Chemical Sciences Division, Lawrence Livermore National Laboratory, Livermore CA USA 94550

Abstract. The lighter heavy elements of the first r-process peak, between strontium and silver, can be synthesized in the moderately neutron-rich neutrino–driven ejecta of either core–collapse supernovae or neutron star mergers via the weak r–process. This nucleosynthesis scenario exhibits uncertainties from the absence of experimental data from (α, xn) reactions on neutron–rich nuclei, which are currently based on statistical model estimates. We have performed a new impact study to identify the most important (α, xn) reactions that can affect the production of the lighter heavy elements under different astrophysical conditions using new, constrained (α, xn) reaction rates based on the Atomki-V2 αOMP. Our results show how when reducing the nuclear physics uncertainties, we can use abundance ratios to constrain the astrophysical conditions/environment. This can be achieved in the near future, when the key (α, xn) reaction rates will be measured experimentally in radioactive beam facilities.

1 Introduction

The astrophysical *r*-process can produce around half of the elements heavier than iron in explosive environments, such as the merging of a binary neutron star system [1]. Spectroscopic observations of galactic halo metal-poor stars show a large scatter in the light heavy elements of the first *r*-process peak, between strontium and silver (Z= 38-47), compared to a rather

*e-mail: psaltis@theorie.ikp.physik.tu-darmstadt.de

robust production of the lanthanides [2]. For this reason, many other processes have been suggested to produce the lighter heavy elements.

A possible scenario is nucleosynthesis in neutrino-driven outflows of core-collapse supernovae or neutron star mergers, via the weak *r*-process (also known as α-process) [3]. This nucleosynthesis process operates for $T \approx 2{-}5$ GK and creates heavy elements mainly through α- and proton-induced reactions on neutron-rich nuclei. Refs. [4, 5] have demonstrated that the main nuclear physics uncertainty in weak *r*-process nucleosynthesis is caused by the (α, n) reaction rates, and more specifically from the α-nucleus potential (αOMP), since these rates are calculated through the statistical Hauser-Feshbach model.

2 Impact study of the (α, xn) reaction rates

We recently performed an impact study of the (α, xn) reaction rates to the weak *r*-process [6], employing the Atomki-V2 αOMP [7]. We explored the phase space of the astrophysical conditions, based on Ref. [8] and used a Monte Carlo technique to systematically vary the (α, xn) reaction rates.

Instead of finding the impact on *elemental* abundances, as Ref. [9], we explored the impact on *abundance ratios*, which can be better constrained through observations. Figure 1 shows results of the predicted abundance ratios of Sr/Y and Y/Zr for various astrophysical conditions. We also show the recent observations by Lombardo *et al.* [10], which includes 52 giant metal-poor stars ([Fe/H] < −1.5), using high resolution spectra from the UVES instrument of the Very Large Telescope (VLT).

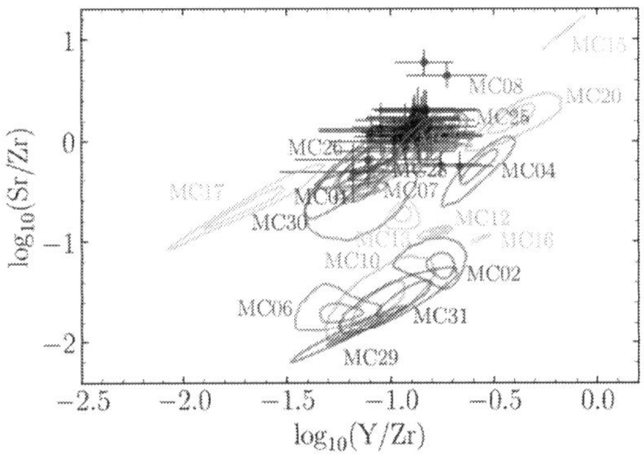

	Y_e	s (k_B/nuc)	τ (ms)		Y_e	s (k_B/nuc)	τ (ms)
MC02	0.45	113	11.9	MC16	0.49	126	15.4
MC04	0.44	66	19.2	MC17	0.46	132	12.4
MC06	0.40	56	63.8	MC20	0.41	42	59.3
MC07	0.47	96	11.6	MC22	0.40	40	46.7
MC08	0.43	78	35.0	MC25	0.46	96	20.9
MC10	0.40	54	31.0	MC26	0.40	84	36.2
MC12	0.48	85	9.7	MC28	0.46	113	11.9
MC13	0.43	64	35.9	MC29	0.41	66	41.4
MC15	0.48	103	20.4	MC31	0.43	71	11.4

Figure 1. Two-dimensional Kernel Density Estimates (KDEs) of the Sr/Y and Y/Zr elemental ratios from Ref. [6] with the addition of stars from the sample of Ref. [10]. In the right, the main properties of each model (electron fraction Y_e, entropy per nucleon s and expansion timescale τ) are presented. The contours show the 1 and 2σ uncertainties for each calculation. See the text for details.

Based on the analysis above, we have identified 35 (α, xn) reactions which affect the production of the lighter heavy elements in the weak *r*-process (see Ref. [6] for the complete list). In the following, we discuss experimental plans to measure these important reactions at radioactive ion beam facilities.

3 Experimental Plans

Measuring the relevant (α, xn) reaction rates will help us better constrain the relevant astrophysical conditions that produce the lighter heavy elements between strontium and silver via the weak *r*-process. There is an intense interest in the nuclear astrophysics community regarding these reactions, and many measurements already have either been proposed or recently been performed in various stable and radioactive ion beam facilities worldwide, such as in ATOMKI, Argonne, FRIB and TRIUMF. In the following, we shall briefly discuss a recently approved proposal by the ATLAS PAC at Argonne to perform the first measurement of the ^{93}Sr$(\alpha, n)^{96}$Zr reaction using MUSIC. This reaction has been shown to affect many elemental ratios under many astrophysical conditions in the impact study of Ref. [6].

The MUSIC (Multi-Sampling Ionization Chamber) detector [11] is a a highly-efficient, self-normalizing active-target system that is capable of measuring cross sections of reactions relevant for astrophysics. It is sensitive to the energy losses of the particles that travel through the gas (in our case helium) and has the ability to measure excitation functions using a single beam energy. For the proposed experiment, the ^{93}Sr radioactive beam will be provided by the nuCARIBU neutron-induced fission source.

Recently, MUSIC successfully measured the ^{100}Mo(α, xn) reaction, which demonstrated its capability for (α, xn) reaction cross sections up to $A = 100$ [12]. The detector can also be transported to other facilities, such as FRIB, to study important reactions that were identified in our impact study [6].

4 Summary & Discussion

Explaining the origin of the lighter heavy elements between strontium and silver is an exciting open question in nuclear astrophysics. Nucleosynthesis in neutrino-driven ejecta through the weak r-process provides a possible solution, but there are still uncertainties, both in the astrophysical conditions and the underlying nuclear physics input.

We recently performed an impact study of the (α, xn) reaction rates to the weak r-process using the Atomki-V2 αOMP and identified key reactions that affect the abundance ratios that are observed in metal-poor stars. Future experimental and observational efforts will help to further reduce the relevant reaction rate uncertainties and shed light to the production of the lighter heavy elements in the Cosmos.

Acknowledgements

This work was supported by the Deutsche Forschungsgemeinschaft (DFG, German Research Foundation)—Project No. 279384907—SFB 1245, the European Research Council Grant No. 677912 EUROPIUM, and the State of Hesse within the Research Cluster ELEMENTS (Project ID 500/10.006)

References

[1] J.J. Cowan et al., Rev. of Mod. Phys. **93**, 015002 (2021)
[2] C. Sneden, J.J. Cowan, R. Gallino, Annu. Rev. Astron. Astrophys. **46**, 241 (2008)
[3] A. Arcones, F. Montes, Astrophys. J **731**, 5 (2011)
[4] J. Pereira, F. Montes, Phys. Rev. C **93**, 034611 (2016)
[5] J. Bliss et al., J. Phys. G **44**, 054003 (2017)
[6] A. Psaltis et al., Astrophys. J **935**, 27 (2022)
[7] P. Mohr et al., At. Data Nucl. Data Tables **142**, 101453 (2021)
[8] J. Bliss et al., Astrophys. J **855**, 135 (2018)
[9] J. Bliss et al., Phys. Rev. C **101**, 055807 (2020)
[10] L. Lombardo et al., A&A **665**, A10 (2022)
[11] M. Avila et al., Nucl. Instrum. Methods. Phys. Res. A **859**, 63 (2017)
[12] W.J. Ong et al., Phys. Rev. C **105**, 055803 (2022)

Understanding the cosmic abundance of ^{22}Na: lifetime measurements in ^{23}Mg

C. Fougères[1,2,*], F. de Oliveira Santos[1,**], N. A. Smirnova[3], C. Michelagnoli[1,4], and GANIL-E710 / AGATA collaborations

[1]Grand Accélérateur National d'Ions Lourds (GANIL), CEA/DRF-CNRS/IN2P3, Caen, France
[2]Physics Division, Argonne National Laboratory, Lemont, USA
[3]LP2IB, Université de Bordeaux, CNRS/IN2P3, Gradignan, France
[4]Institut Laue-Langevin, Grenoble, France

Abstract. Simulations of explosive nucleosynthesis in novae predict the production of ^{22}Na, a key astronomical observable to constrain nova models. Its gamma-ray line at 1.275 MeV has not yet been observed by the gamma-ray space telescopes. The ^{20}Ne/^{22}Ne ratio in presolar grains, a possible tool to identify nova grains, also depends on ^{22}Na produced. Uncertainties on its yield in classical novae currently originate from the rate of the ^{22}Na(p, γ)^{23}Mg reaction. At peak novae temperatures, this reaction is dominated by a resonance at E_R=0.204 MeV, corresponding to the E_x=7.785 MeV excited state in ^{23}Mg. The resonance strengths measured so far disagree by one order of magnitude. An experiment has been performed at GANIL to measure the lifetime and the proton branching ratio of this key state, with a femtosecond resolution for the former. The reactions populating states in ^{23}Mg have been studied with a high resolution detection set-up, i.e. the particle VAMOS, SPIDER and gamma tracking AGATA spectrometers, allowing the measurements of lifetimes and proton branchings. We present here a comparison between experimental results and shell-model calculations, that allowed us to assign the spin and parity of the key state. Rather small values obtained for reduced $M1$ matrix elements, $|M(M1)| \lesssim 0.5\ \mu_N$, and proton spectroscopic factors, $C^2S_p<10^{-2}$, seem to be beyond the accuracy of the shell model. With the reevaluated ^{22}Na(p, γ)^{23}Mg rate, the ^{22}Na detectability limit and its observation frequency from novae are found promising for the future space telescopes.

Many nuclei are synthesized in explosive stellar environments, like classical novae, supernovae and neutron star mergers. Novae are transient astronomical events and the most frequent explosive events in our galaxy, after X-ray bursts. These explosions happen in a close binary stellar system consisting of a white dwarf accreting hydrogen-rich matter from its companion. This matter is progressively compressed at the white dwarf surface until the conditions required for hydrogen combustion are reached. Nucleosynthesis takes place during this explosive stage while a part of the new material is being ejected in the interstellar medium. Beyond this well understood picture, large uncertainties remain in our knowledge of novae, for example the white dwarf initial conditions (mass, luminosity), the accretion rate and its composition, the amount of admixed white dwarf material with the accreta, the

*e-mail: cfougeres@anl.gov
**e-mail: oliveira@ganil.fr

ejected mass, to list a few [1]. After the explosion, gas is crystallized, imprinting the ejecta composition into grains. The isotopic composition of presolar grains of a putative nova origin shed some lights into these stellar explosions [1–3]. Furthermore, novae are thought to end as supernovae of type Ia [4] which are used as standard candles to determine distances and, so, to estimate the cosmic expansion. Therefore, astronomical observables are required to improve our understanding of novae: ^{22}Na has revealed itself as a key candidate.

The lifetime of the ^{22}Na radioisotope (τ=2.6 yr) makes it a good radioactive tracer of novae [7] since: *(i)* the lifetime is longer than the duration of the opaque phase following the explosion (on timescales of hours), *(ii)* the decay time is short enough to ensure space-time correlation. Hence, ^{22}Na is a promising candidate for observing γ-ray emissions from novae [8]. The emitted γ rays at E_γ=1.275 MeV (and 0.511 MeV) are expected from hours to months after the explosion [9]. They were searched for by the missions CGRO/COMPTEL [10] and INTEGRAL/SPI [11], leading only to upper limits for the ejected amount of ^{22}Na (left Fig. 1). New space telescopes are in preparation, namely COSI [12] and e-ASTROGAM [13], with

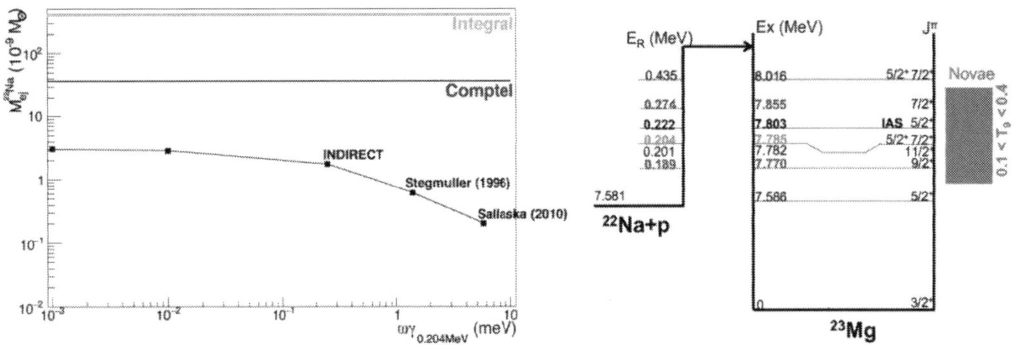

Figure 1. Left: The expected ^{22}Na mass ejected from a nova is given as a function of the strength of the E_R=0.204 MeV resonance, from simulations with the MESA code [5]. Colored lines mark the Integral and Comptel space telescope lower sensitivity limits. **Right:** Levels scheme of ^{23}Mg and resonance energies E_R of the proton capture reaction. The Gamow window of novae is shown with the red zone.

sensitivities 20 times higher than INTEGRAL. The radioisotope ^{22}Na can be also measured, indirectly, by the over-abundance of the daughter nucleus ^{22}Ne in presolar grains. Neon noble gas is not expected to easily condense into dust grains and, so, the presence of ^{22}Ne implies in-situ β^+ decays of ^{22}Na. A ^{22}Ne excess was observed [14]. Therefore, it is critical to well quantify ^{22}Na nucleosynthesis in novae.

ONe novae are predicted to be the main site of ^{22}Na production [6]. In these sites, the matter is composed of a mixture between the transferred proton rich gas and the white dwarf envelope with mainly oxygen and neon. The nucleosynthesis involves H to Ca isotopes, proton capture reactions and β^+ decays [1]. The production pathway of ^{22}Na is ^{20}Ne(p,γ)^{21}Na(p,γ)^{22}Mg(β^+)^{22}Na at high temperatures and ^{20}Ne(p,γ)^{21}Na(β^+)^{21}Ne(p,γ)^{22}Na otherwise. The thermal window of novae is [0.05, 0.5] GK. The destruction of ^{22}Na proceeds via β^+ decays and ^{22}Na(p,γ)^{23}Mg. The latter reaction is the least known, resulting in a factor of 10 uncertainty in the predicted amount of ^{22}Na ejected (left Fig. 1). Since the direct capture contribution was found to be negligible [15, 16], the total reaction rate is determined by several narrow resonances in the Gamow window $<\sigma v>_{\rm tot}= \sum_i (\frac{2\pi}{\mu_{(^{22}{\rm Na,p})} k_B T})^{\frac{3}{2}} \hbar^2 (\omega\gamma)_i \exp(-\frac{E_{Ri}}{k_B T})$ (right Fig. 1). A direct measurement at astrophysical energies showed that a resonance at E_R = 0.204 MeV dominates the ^{22}Na(p,γ)^{23}Mg rate [17]. However, its contribution disagrees by a factor 3 or even more with other measurements [16, 18]. Such a direct measurement is complex due to the radioactive nature of the target. The resonance strength can be determined indirectly from the spectroscopic properties of the E_x=7.785 MeV state in ^{23}Mg, i.e. $\omega\gamma = \frac{2J_{^{23}Mg}+1}{(2J_{^{22}Na}+1)(2J_p+1)} \frac{\Gamma_p \times \Gamma_\gamma}{\Gamma_{tot}}$. The total and proton widths can be experimentally accessed via the state lifetime $\tau = \hbar/\Gamma_{tot}$ and proton branching ratio $BR_p = \Gamma_p/\Gamma_{tot}$. But the measured lifetime, τ=10(3) fs [19], is in contradiction with shell-model predictions, as explained hereafter.

An experiment has been performed at GANIL facility, France, to strongly reduce the rate uncertainties of ^{22}Na(p,γ)^{23}Mg through the measurements of τ and BR_p. It is briefly presented here. Deduced spectroscopic properties are then discussed in relation with results from shell-model calculations.

1 Experimental approach

The state of interest has been populated with a ^{24}Mg beam of E=4.6 MeV/u onto a ^{3}He-implanted gold target. The reactions ^{3}He(^{24}Mg, ^{4}He)^{23}Mg* have been selected and measured with: the spectrometer VAMOS [20] for ^{4}He identification and momenta measurement, the γ-ray tracking spectrometer AGATA [21], and the silicon telescope SPIDER for proton momenta measurements. Momenta reconstruction has been improved thanks to a small gas chamber placed in front of VAMOS. The γ-ray transitions and lifetimes have been measured for about 20 identified states in ^{23}Mg*. The branching ratios have been measured for the measured protons from states close to the proton emission threshold. Details can be found in Ref. [15] and results will be more extensively discussed in forthcoming publications.

2 Nuclear structure considerations

Insights from theory are now discussed in relation with the performed experimental study of ^{23}Mg. The shell-model calculations are first briefly explained. Then, the measured and theoretical reduced M1 matrix elements of the γ-ray transitions are compared for several levels in ^{23}Mg. We finally focus on the spectroscopic properties of the E_x=7.785 and 7.803 MeV states, the latter being the Isobaric Analog State (IAS) of the ^{23}Al ground state ($J^\pi = \frac{5}{2}^+$). Among the calculated levels in ^{23}Mg, the E_x=7.785 MeV state has not yet been clearly identified because of its uncertain spin value [19, 22]. While identifying the best shell-model candidate for this state, its spin assignment and possible isospin-mixing with the IAS are discussed.

Shell-model calculations

We present here a theoretical analysis of the partial ^{23}Mg level scheme, based on the *sd*-shell calculations with charge-dependent versions of USDA and USDB [25, 26] (USDAcpn and USDBcpn) using the `NushellX` code [27]. In general, those interactions are known to provide a very good description of *sd*-shell nuclei. Assessment of shell-model states with observations is more challenging at higher energies because of the increasing level density. This has still been achieved for the E_x~7.5-8 MeV states in ^{23}Mg (details in [15]). Excited states in ^{23}Mg, including resonant ones, have thus been identified by comparing experimental energies, gamma decay branching ratios and ^{23}Al β^+ feeding with theoretical quantities. Let us recall that a partial electromagnetic decay width Γ_γ is related to the reduced γ-ray transition strength B_γ via $\Gamma_{\gamma,(L,\, i\to f)} = \frac{8\pi(L+1)}{L[(2L+1)!!]^2}(\frac{E_{\gamma,0}}{\hbar c})^{2L+1} B_\gamma$. Then, the reduced M1 matrix elements $|M(M1)|$ can be derived with $|M(M1)| = \sqrt{(2J_i+1)B(M1)}$. The proton decay width can be expressed as $\Gamma_p = \sum_{\ell,j} C^2S_p(\ell j) \times \Gamma_{s.p.}(\ell j)$ where $C^2S_p(\ell j)$ refer to proton spectroscopic factors, $\Gamma_{s.p.}(\ell j)$ to the single-particle proton decay widths which are calculated from proton scattering cross-sections on a Woods-Saxon potential.

Reduced M1 matrix elements of the ^{23}Mg states

Experimental $B(M1)$ values have been determined from *(i)* measured (τ, $E_{\gamma,0}$), *(ii)* identified M1 transitions [29] among observed γ rays and *(iii)* the assumption of small E2 contributions. The comparison of experimental and theoretical values of reduced M1 matrix elements is presented in Fig. 2 for transitions in ^{23}Mg and ^{21,23}Na [30]. Optimum effective g-factors USDA(6) and USDB(6) from Table I of Ref. [28] have been used to parameterize the M1 operator. The results for the state of interest are emphasized by the red contours. At the bottom, the ratios of shell-model to experimental reduced matrix elements highlight differences by about a few times for small values of $|M(M1)|$. Although the accuracy in reproduction of $|M(M1)|$ for selected isotopes close to ^{23}Mg is consistent with the rms deviations of about 0.22 - 0.20 μ_N reported for USDA and USDB [28], we see that the M1 transition from the key state of interest is among the most challenging for theory.

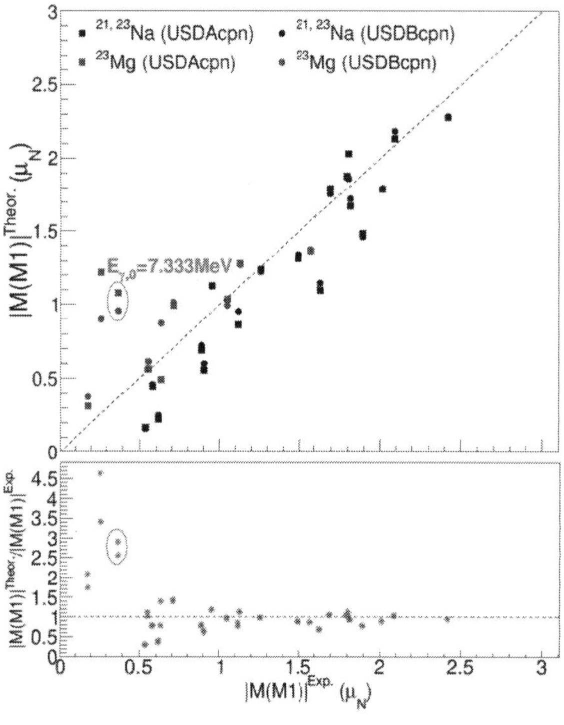

Figure 2. Correlation between measured and theoretical reduced matrix elements $|M(M1)|$ values for transitions in ^{23}Mg and ^{21,23}Na [30]. Theoretical values were obtained from *sd*-shell calculations with USDAcpn and USDBcpn [25, 26], experimental ones from measured (τ, $E_{\gamma,0}$), resulting in rms=0.322 μ_N (USDAcpn) and rms=0.273 μ_N (USDBcpn). It follows that the largest relative deviations are observed for the lowest $|M(M1)|$ values, as for the E_x=7.785 MeV state shown in the red contours.

Spectroscopy of the E_x=7.785 and 7.803 MeV states

The two states measured at E_x=7.785 and 7.803 MeV (IAS) are now discussed in more detail. The γ-ray width, never measured for the IAS, is 3.0(2) eV for the mirror state, in agreement with the shell-model results of 2 eV (USDAcpn) and 3 eV (USDBcpn). From the previous experiment on BR_p [31], the proton width and spectroscopic factor are deduced for this isospin-forbidden proton emission, i.e. Γ_p=5(2) meV and C^2S_p=0.0056(20) versus C^2S_p=0.0073 (USDAcpn) and 3×10^{-5} (USDBcpn).

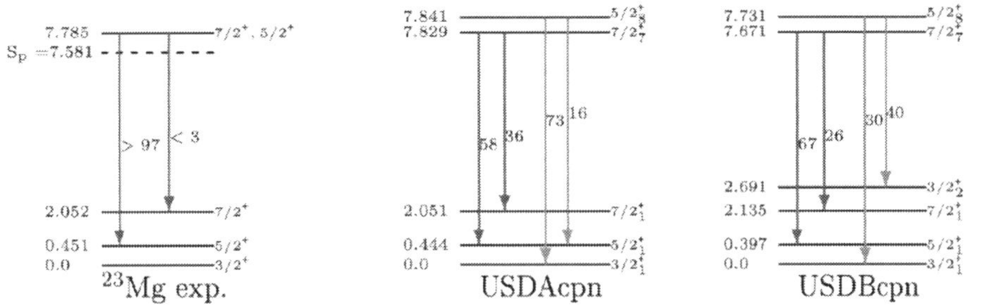

Figure 3. Comparison between γ-ray transitions from the E_x=7.785 state and from shell-model candidates: (n_{SM}=7, $J^\pi=\frac{7}{2}^+_7$) and (n_{SM}=8, $J^\pi=\frac{5}{2}^+_8$). A good agreement is observed for the blue transitions, strengthening the n_{SM}=7 candidate as the E_x=7.785 MeV state (intensities are given in %.)

Three different spins were proposed for the E_x=7.785 MeV state, either $\frac{7}{2}^+$[19] or ($\frac{3}{2}^+$, $\frac{5}{2}^+$)[22]. The value $\frac{3}{2}^+$[22] is unlikely since this state was well observed in the beta-delayed proton emission of ^{23}Al ($\frac{5}{2}^+$)[23, 24]. The spin of ^{22}Na ground state being 3^+, the transferred angular momentum should be at least ℓ = 2, that is too high to be competitive with the γ emission. As shown in Fig. 3 (in red), the theoretical γ-decay pattern of the $\frac{5}{2}^+_8$ state (eighth predicted state, n_{SM}=8) strongly differs from the observed one. On the contrary, the $J^\pi=\frac{7}{2}^+_7$ state (in blue) agrees well with the measured decay pattern. This is true regardless of the interaction used. Moreover, a spin assignment $\frac{5}{2}^+$ of the key state would imply a strong isospin-mixing with the IAS (as predicted by the shell model) which is just 18 keV above. However, the measured $\log_{10}$(ft)=3.305(23) of the IAS [32] is in excellent agreement with no mixing hypothesis, and the one of the E_x=7.785 MeV state agrees with the calculated $\frac{7}{2}^+_7$ state. Consequently, $J^\pi=\frac{7}{2}^+$ is assigned for the key state.

This conclusion of no mixing is supported by the small gamma width of the key state, $\Gamma_\gamma \lesssim 0.1$ eV, if compared to the gamma width of the IAS. Regarding proton emission, from the measured BR_p, a small spectroscopic factor $C^2S_p \approx 10^{-3}$ was obtained. Theoretical predictions are C^2S_p=0.003 (USDAcpn) and 0.05 (USDBcpn), assuming $7/2^+$ assignment. Although the former result captures the order of magnitude, we remark that similar to M1 observable, very small values of spectroscopic factors may not be perfectly described.

3 Summary

In the present work, a detailed comparison has been drawn between experimentally observed and shell-model states in ^{23}Mg. It appears that shell-model predictions may have large relative uncertainties, up to a factor of 10, for very small ($<10^{-2}$) spectroscopic factors and M1 transition probabilities. This is apparently the case of the key state. Nevertheless, shell model has brought helpful arguments favoring $\frac{7}{2}^+$ spin assignment of the E_R=0.204 MeV resonant state. The measured total and partial widths of this state have led to a reevaluation of the ^{22}Na(p,γ)^{23}Mg rate at novae temperatures. According to the future COSI and eASTROGAM sensitivities, it allows us to estimate the maximum detectability distance and a conservative observation frequency of ^{22}Na based on the detected ONe novae [1, 33]. These predictions, illustrated in Fig. 4, are encouraging the search for ^{22}Na with the new space telescopes.

Acknowledgments

The authors thank the GANIL accelerator staff for their beam delivery and their support, the AGATA collaboration, and the Normandie region.

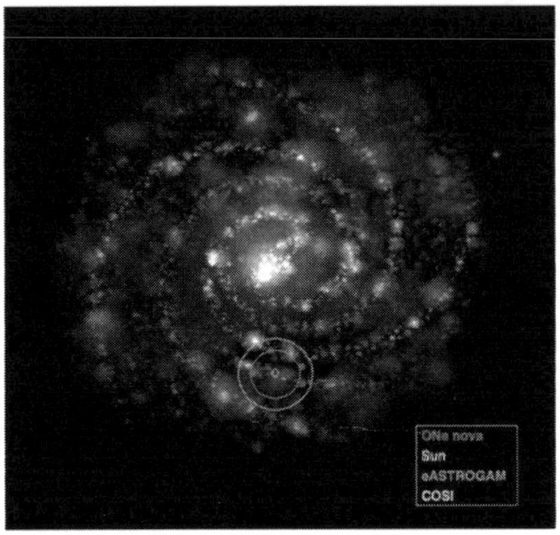

Figure 4. Map of the Milky Way with the expected flux of ^{22}Na, represented by the red dots, from the ONe novae observed for the last 60 years. Circles represent the limit distance of COSI and eASTROGAM to detect γ-rays from ^{22}Na. The former should detect such emission once every 20 yr.

References

[1] José, J., CRC Press, (2020)
[2] Amari, S. et *al.*, ApJ, **551**, 1065-1072 (2001)
[3] Bose, M. et *al.*, ApJ, **873**, 14 (2019)
[4] Bravo, E. et *al.*, MNRAS **517**, L31–L35 (2022)
[5] Paxton, B. et *al.*, Astrophys. J. Suppl. Ser., **208**, 4 (2013)
[6] José, J., and Coc, A. and Hernanz, M., ApJ, **520**, 347–360 (1999)
[7] Diehl, R., Astronomical Review, **8**, 4-52 (2013)
[8] Clayton, D.D. et *al.*, ApJ, **187**, 101 (1974)
[9] Hernanz, M, 2*nd* eASTROGAM Workshop (2017).
[10] Schoenfelder, V. et *al.*, Astrophys. J. Suppl. Ser., **86**, 657 (1993)
[11] Vedrenne, G. et *al.*, A&A, **411**, (2003)
[12] Tomsick, J.A. et *al.*, Astro2020 APC White Paper (2020)
[13] De Angelis, A. et *al.*, JHEAp, **19**, 1-106 (2018)
[14] Black, D.C., Geochimica et Cosmochimica Acta, **36**, (1972)
[15] Fougères, C., Normandie Université Ph.D. thesis, HAL **tel-03768524**, (2022)
[16] Stegmüller, F. et *al.*, Nucl. Phy. A., **601**, 168-180 (1996)
[17] Sallaska, A. L. et *al.*, Phy. Rev. Lett., **105**, 152501 (2010)
[18] Seuthe, S. et *al.*, Nucl. Phy. A., **514**, 471-502 (1990)
[19] Jenkins, D. G. et *al.*, Phy. Rev. Lett., **92**, 031101 (2004)
[20] Rejmund, M. et *al.*, Nucl. Instr. Meth. A, **646**, 184-191 (2011)
[21] Clément, E. et *al.*, Nucl. Instr. Meth. A, **855**, 1-12 (2017)
[22] Kwag, M. S. et *al.*, Eur. Phys. J. A, **56**, 108 (2020)
[23] Saastamoinen, A. et *al.*, Phys. Rev. C, **83**, 045808 (2011)
[24] Friedman, M. et *al.*, Phys. Rev. C, **101**, 052802 (2020)
[25] Ormand, W. E. and Brown, B. A., Nuclear Physics A **491**, 1-23 (1989)
[26] Brown, B. A. and Richter, W. A., Phys. Rev. C **74**, 034315 (2006)
[27] Brown, B. A. and Rae, W. D. M., Nuclear Data Sheets **120**, 115-118 (2014)
[28] Richter, W. A. and Mkhize, S. and Brown, B. A., Phys. Rev. C **78**, 064302 (2008)

[29] Shamsuzzoha Basunia, M. and Chakraborty, A., Nuclear Data Sheets, **171**, 1-252 (2021)
[30] Hasan, A. K. et *al.*, Ukrainian Journal of Physics **65**, 3 (2020)
[31] Peräjärvi, K. et *al.*, Phys. Let. B, **492**, 1-7 (2000)
[32] Zhai, Y., Ph.D. thesis (2007)
[33] Hachisu, I. et *al.*, Astrophys. J. Suppl. Ser., **242**, 18 (2019)

The γ-process nucleosynthesis in core-collapse supernovae

Lorenzo Roberti[1,2,3,*], *Marco* Pignatari[1,3,4], *Maria* Lugaro[1,5,6], *Zsolt* Fülöp[7], and *Gyorgy* Gyurky[7]

[1]Konkoly Observatory, Research Centre for Astronomy and Earth Sciences, Eötvös Loránd Research Network (ELKH), Konkoly Thege Miklós út 15-17, H-1121 Budapest, Hungary; MTA Centre of Excellence
[2]INAF – Osservatorio Astronomico di Roma Via Frascati 33, I-00040, Monteporzio Catone, Italy
[3]NuGrid Collaboration, http://nugridstars.org
[4]E. A. Milne Centre for Astrophysics, University of Hull, Hull HU6 7RX, UK
[5]School of Physics and Astronomy, Monash University, VIC 3800, Australia
[6]Eötvös Loránd University, Institute of Physics, Budapest 1117, Pázmány Péter sérány 1/A, Hungary
[7]Institute for Nuclear Research (ATOMKI), H-4001 Debrecen, Hungary

Abstract. Neutron-capture processes made most of the abundances of heavy elements in the Solar System, however they cannot produce a number of rare proton-rich stable isotopes (p–nuclei) lying on the left side of the valley of stability. The γ–process, i.e., a chain of photodisintegrations starting on heavy nuclei, is recognized and generally accepted as a feasible process for the synthesis of p–nuclei in core collapse supernovae (CCSNe). However this scenario still leaves some puzzling discrepancies between theory and observations. We aim to explore in more detail the γ–process production from massive stars, using different sets of CCSNe models and the latest nuclear reaction rates. Here we show our preliminary analysis, by identifying the γ–process sites and focusing on progenitors of CCSNe that experience a C–O shell merger just before the collapse of the Fe core.

1 Introduction

The existence of 35 stable p–rich isotopes of elements above Fe[1] shows that not all the heavy nuclides are made by the slow [1] and rapid [2] neutron capture processes, and alternative processes must exist. These neutron-deficient nuclei have been thought to be produced through a chain of photodisintegrations on heavy **isotopes**, historically known as p– or γ–process [3–6]. This process is still widely accepted today as a main candidate for the origin of most of the p-rich heavy isotopes. It is activated in O/Ne rich layers of massive stars during core collapse supernova (CCSN) explosions on pre-existing local and heavier nuclides.

Although the γ–process nucleosynthesis in CCSNe has been explored for many decades, there are still some fundamental problems with it. The main two historical limitations of

*e-mail: lorenzo.roberti@csfk.org

[1]namely: ^{74}Se, ^{78}Kr, ^{84}Sr, ^{92,94}Mo, ^{96,98}Ru, ^{102}Pd, ^{106,108}Cd, ^{112,114,115}Sn, ^{113}In, ^{120}Te, ^{124,126}Xe, ^{130,132}Ba, ^{136,138}Ce, ^{138}La, ^{144}Sm, ^{152}Gd, ^{156,158}Dy, ^{162,164}Er, ^{168}Yb, ^{174}Hf, ^{180}Ta, ^{180}W, ^{184}Os, ^{190}Pt, and ^{196}Hg.

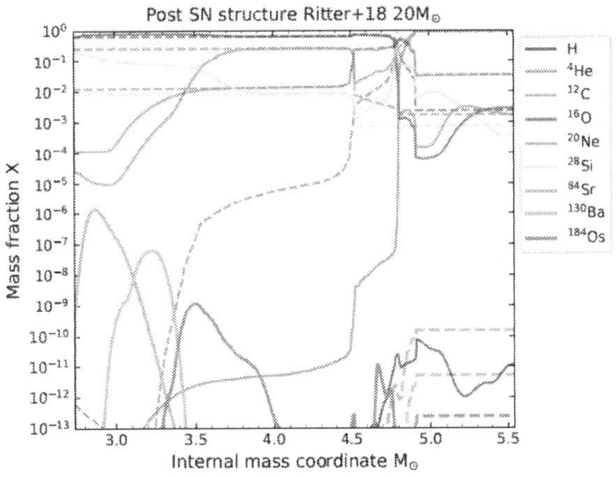

Figure 1. Abundance distribution before (dashed) and after (solid) the CCSN explosion in a 20 $M_\odot$ at Z=0.02 [14]. The *p*–nuclei ^{84}Sr, ^{130}Ba and ^{184}Os are taken as representative for the γ–process nucleosynthesis.

the γ–process in massive stars are that (1) the average *p*–process yields result in an under-production by about a factor of three relative to the amount required to explain the Solar System distribution [7, 8]; and (2) ^{92,94}Mo and ^{96,98}Ru are systematically under-produced by more than an order of magnitude compared to the other *p*–process species [9]. To overcome these problems, the γ–process from thermonuclear supernovae (SNIa), from the explosion of white dwarfs [10], has been proposed as an additional astrophysical source [8]. However, large uncertainties are affecting also this alternative γ–process astrophysical source, for example how many SNIa are generated from the Chandrasekhar mass progenitors (see discussion in [11]). Recently, **[12]** suggested that the efficiency of the γ–process could be enhanced in fast rotating massive stars at sub-solar metallicity, due to an increased production of *s*–process seeds. However, the frequency of such spinstars in the latest stellar generations is still unclear. Therefore, **it becomes of paramount importance** to explore in more detail possible solutions that could solve the problems of the γ–process production from massive stars, in particular for Mo-Ru region.

Our new project is a collaboration with scientists at the Institute for nuclear research (ATOMKI) in Debrecen and with an international research team of nuclear astrophysics experts, with the aim of studying in detail the γ–process nucleosynthesis in CCSNe. The main goals of this project are the analysis of γ–process yields in existing sets of CCSN models, the update of the nuclear reaction rates for the γ–process nucleosynthesis, and finally the production of γ–process stellar yields in new sets of CCSNe, using the NuGrid nucleosynthesis codes [13].

2 The γ–process nucleosynthesis in our CCSN models

In the preliminary analysis presented here we consider massive star explosions for progenitor with an initial mass of 15, 20, and 25 $M_\odot$ [14–17]. The typical production site for the γ–process nucleosynthesis shown in Figure 1 corresponds to the O/Ne–rich layers that the shock wave crosses after the CCSN explosion. This region of the star reaches a peak temperature between 2.5 and 3.3×10^9 K, which is the characteristic value at which photodisintegrations can efficiently convert the heavy material into neutron deficient nuclides. A useful

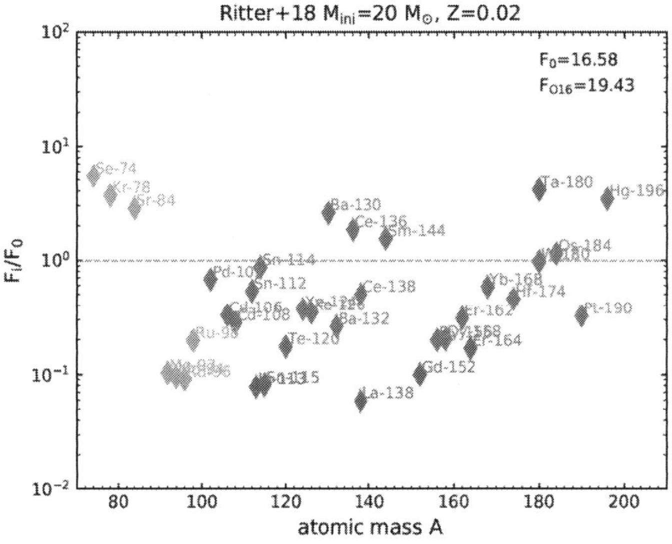

Figure 2. Normalized p–nuclide overproduction factors of the total ejected yields for the model showed in Figure 1. The orange symbols represent the p–nuclei that may have an additional explosive contribution, while the green symbols represent the p–nuclei that have also s–, r–process or neutrino-capture contributions.

parameter to evaluate the efficiency of the γ–process nucleosynthesis is the normalized p–nuclide overproduction factor F_i/F_0 (Figure 2), where the overproduction factor F_i represents the production of the i_{th} p–nuclide with respect to solar, while F_0 is the average overproduction factor for all the γ–process isotopes. The average overproduction factor F_0 is often compared to the overproduction factor of ^{16}O, one of the main products of CCSNe. It should be noted that not all the p–nuclei have a γ–process only origin. In fact, the lighter nuclei may have an additional contribution from other explosive nucleosynthetic components, e.g., α-process [18] or neutrino winds [19]. Some other p–nuclei may also have a s–, r–process and neutrino-capture contribution [9, 20]. Therefore, F_0 **must be considered** carefully: it could be enhanced by an isotope having also an additional component, or it could be reduced if some nucleosynthesis components are missing.

On top of the γ–process in the O/Ne–rich layers, massive stars may experience the ingestion of some C in the O convective shell in the latest stages of their evolution [14, 15, 21, 22]. This leads to the formation of an extended mixed convective zone in which the C shell material is brought to O burning temperature triggering an efficient γ–process nucleosynthesis (Figure 3). In this case, a large fraction of the CCSN yields of γ–process isotopes is dominated by the production in this convective C–O shell before the explosion, and their abundance might increase by orders of magnitude as compared to standard massive star models. Therefore, the C–O shell mergers may provide an additional production site for the γ–process nucleosynthesis in massive stars, leaving different fingerprints on the overproduction factors that will need to be studied in detail (Figure 4).

3 Conclusions and outlook

Although much progress has been made in the framework of the computation of stellar models and nucleosynthesis, the interpretation of the p–nuclei abundance pattern in the Solar

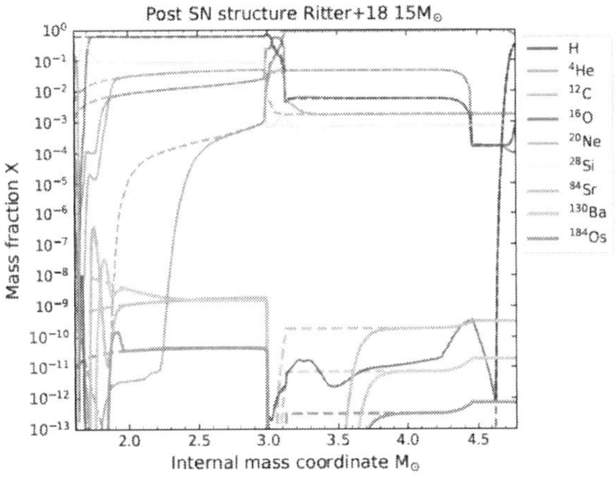

Figure 3. Abundance distribution before (dashed) and after (solid) the CCSN explosion in a 15 $M_\odot$ at Z=0.02 [14]. In this particular case, the star experiences a C–O shell merger just before the explosion and the γ–process nuclei abundances are dominated by the pre-supernova production during the merger.

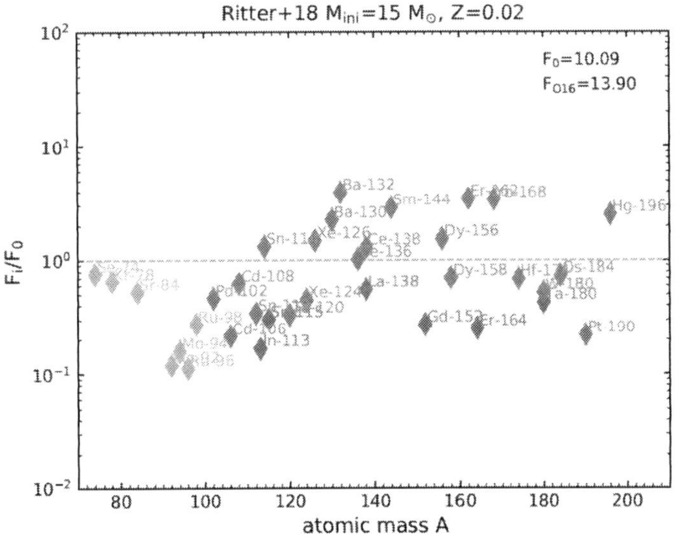

Figure 4. Same as Figure 2 but in the case of the star showed in Figure 3. In this case, there is a larger relative production of the isotopes that have an only γ–process contribution.

System is unclear and severe discrepancies still exist. Our upcoming more detailed study of the γ–process nucleosynthesis in massive stars is required to shed light on their production and to explore alternative scenarios. For example, the C–O shell merger in the latest stage of the evolution of massive stars needs to be analyzed. In this case, the production of γ–process isotopes is significantly increased before the explosion. Our goal in the next future is to compute a new set of CCSN models, taking into account models with C–O shell mergers and with an updated γ–process nuclear network.

Acknowledgements

We thank the support from the OTKA grant 138031 and the ERC Consolidator Grant (Hungary) programme (RADIOSTAR, G.A. n. 724560). LR and MP acknowledge the support to NuGrid from JINA-CEE (NSF Grant PHY-1430152) and STFC (through the University of Hull's Consolidated Grant ST/R000840/1), and ongoing access to `viper`, the University of Hull High Performance Computing Facility. MP and ML acknowledges the support from the "Lendület-2014" Programme of the Hungarian Academy of Sciences (Hungary). We thank the ChETEC COST Action (CA16117), supported by the European Cooperation in Science and Technology. This work was supported by the European Union's Horizon 2020 research and innovation programme (ChETEC-INFRA – Project no. 101008324), and the IReNA network supported by US NSF AccelNet (Grant No. OISE-1927130).

References

[1] F. Käppeler, R. Gallino, S. Bisterzo, W. Aoki, Reviews of Modern Physics **83**, 157 (2011), `1012.5218`

[2] J.J. Cowan, C. Sneden, J.E. Lawler, A. Aprahamian, M. Wiescher, K. Langanke, G. Martínez-Pinedo, F.K. Thielemann, arXiv e-prints arXiv:1901.01410 (2019), `1901.01410`

[3] A.G.W. Cameron, AJ**62**, 9 (1957)

[4] E.M. Burbidge, G.R. Burbidge, W.A. Fowler, F. Hoyle, Rev. Mod. Phys. **29**, 547 (1957)

[5] M. Arnould, A&A**46**, 117 (1976)

[6] S.E. Woosley, W.M. Howard, ApJS**36**, 285 (1978)

[7] M. Rayet, M. Arnould, M. Hashimoto, N. Prantzos, K. Nomoto, A&A**298**, 517 (1995)

[8] C. Travaglio, T. Rauscher, A. Heger, M. Pignatari, C. West, ApJ**854**, 18 (2018), `1801.01929`

[9] M. Arnould, S. Goriely, Phys. Rep.**384**, 1 (2003)

[10] W. Hillebrandt, M. Kromer, F.K. Röpke, A.J. Ruiter, Frontiers of Physics **8**, 116 (2013), `1302.6420`

[11] U. Battino, M. Pignatari, C. Travaglio, C. Lederer-Woods, P. Denissenkov, F. Herwig, F. Thielemann, T. Rauscher, MNRAS**497**, 4981 (2020), `2007.13677`

[12] Choplin, A., Goriely, S., Hirschi, R., Tominaga, N., Meynet, G., A&A **661**, A86 (2022)

[13] M. Pignatari, F. Herwig, R. Hirschi, M. Bennett, G. Rockefeller, C. Fryer, F.X. Timmes, C. Ritter, A. Heger, S. Jones et al., ApJS**225**, 24 (2016), `1307.6961`

[14] C. Ritter, F. Herwig, S. Jones, M. Pignatari, C. Fryer, R. Hirschi, MNRAS**480**, 538 (2018), `1709.08677`

[15] T. Rauscher, A. Heger, R.D. Hoffman, S.E. Woosley, ApJ**576**, 323 (2002), `arXiv:astro-ph/0112478`

[16] T.V. Lawson, M. Pignatari, R.J. Stancliffe, J. den Hartogh, S. Jones, C.L. Fryer, B.K. Gibson, M. Lugaro, MNRAS(2021)

[17] A. Sieverding, G. Martínez-Pinedo, L. Huther, K. Langanke, A. Heger, ApJ**865**, 143 (2018)

[18] S.E. Woosley, R.D. Hoffman, ApJ**395**, 202 (1992)

[19] A. Arcones, F. Montes, ApJ**731**, 5 (2011), `1007.1275`

[20] M. Pignatari, K. Göbel, R. Reifarth, C. Travaglio, International Journal of Modern Physics E **25**, 1630003-232 (2016), `1605.03690`

[21] C.A. Meakin, D. Arnett, ApJ**637**, L53 (2006), `astro-ph/0601348`

[22] R. Andrassy, F. Herwig, P. Woodward, C. Ritter, MNRASp. 2556 (2019), `1808.04014`

Neutron-induced charged-particle reaction studies in nuclear astrophysics with a Micromegas based gaseous detector

Chandrabhan Yadav[1,*], *Akiva* Green[1], and *Moshe* Friedman[1]

[1]Racah Institute of Physics, Hebrew University of Jerusalem, 91904 Jerusalem, Israel

Abstract.
Neutron-induced charged-particle reaction studies on various unstable nuclei play an important role in understanding various nucleosynthesis processes occurring in explosive astrophysical scenarios. We are pursuing a novel experimental approach to study neutron-induced charge particle cross-sections for various unstable nuclei at effective temperatures of 1.5-3.5 GK using the $^7Li(p,n)^7Be$ reaction as a neutron source with three orders of magnitude higher neutron intensities with respect to currently available neutron time-of-flight facilities. We plan to perform our experiments with a 10-μA proton beam at the Physikalisch Technische Bundesanstalt facility (PTB, Germany), with a Micromegas based gaseous detector being developed.

1 Introduction

Neutron-induced charged-particle reaction studies on various unstable nuclei are imperative for a complete understanding of the nucleosynthesis occurring in many astrophysical sites and events. This information is used as input for network calculations that provide predictions on the expected abundances [1–3]. Comparing these predictions to the observed abundances, may, for example, provide an indication either on the accuracy of the models, the contribution of different astrophysical events to the galactic isotopic content, or the frequency of specific events such as supernovae.

A complete understanding of supernovae nucleosynthesis requires data on thousands of nuclear reaction rates of different types, most of them not yet experimentally accessible, at temperatures of 1.5-3.5 GK. Sensitivity studies have pointed out several (n, p) and (n, α) reactions to have relatively high impact on the final isotopic abundances [3–7]. Experimental cross section data on (n, p) and (n, α) reactions at the relevant energies for explosive nucleosynthesis is scarce or non-existent for most isotopes of interest. One reason is the need for an intense neutron source at energies of 10-2000 keV, corresponding to a Maxwellian flux distribution at temperatures of 1.5-3.5 GK. Currently, such experiments are performed using the neutron Time-of-Flight (TOF) technique. Today's state-of-the-art neutron TOF facility, produces neutrons at an intensity of about 10^6 n/s at the relevant energy range [8]. Another difficulty is the target size. Even for stable isotopes, the limited range of the outgoing charged particle impose a strong limit on the effective target size. For unstable isotopes the target size

*e-mail: chandrabhan.yadav@mail.huji.ac.il

is even a bigger problem, both in terms of production and target handling. Note that the activation approach is not feasible for most isotopes of interest, which naturally decay via electron capture or β^+, which are indistinguishable from the $(n,\ p)$ reaction by activation measurements. Our novel experimental approach to study this cross-section measurements at stellar temperatures using ^{7}Li$(p, n)^7$Be reaction as a neutron source with three-four orders of magnitude higher neutron intensities as compared to TOF facilities is described in detailed in a recent article [9]. We will use this method to study $(n,\ p)$ and $(n,\ \alpha)$ cross-section at explosive stellar temperature, for various nuclei for which experimental data is poor or non-existent. We plan to conduct our experiments at PTB, Germany, with a neutron intensity on the order of 10^9 n/s on the target, a three orders of magnitudes above existing neutron TOF facilities with a gaseous detector being developed as described in detail ahead.

2 Experimental details and plans

We aim to measure the neutron-induced charge-particle reaction at neutron energies, 1 - 2000 keV. We plan to conduct measurements important to supernovae nucleosynthesis with life times on the order of 10^5 years or above, which we consider feasible with the current experimental setup. Neutrons are produced via the ^{7}Li$(p,\ n)^7$Be reaction with a monoenergetic proton beam at various energies from 1.9 MeV - 3.6 MeV in a step of 0.1 MeV. The target of interest will be placed in a gaseous detector, and the outgoing charged particles are detected by ionization.

2.1 The experimental set-up

Fig. 1 (left) shows a schematic illustration of the experimental setup. A monoenergetic proton beam with varying energies from 1.9 MeV - 3.6 MeV in a step of 0.1 MeV impinges on a thick Li target to produce a neutron beam via ^{7}Li$(p,\ n)^7$Be reaction, which further impinges on a target place inside the Micromegas-based gaseous detector. The detector is located 2 cm downstream of the neutron source. Fig. 2 (a) shows the calculated neutron spectra at the target, for a thick lithium target, and 5 cm^2 target located 2 cm downstream the lithium target at proton energies from 1.9 MeV to 3.6 MeV and Fig. 2 (b) shows the total neutron intensity going through the sample as a function of proton energy. The neutron spectra were calculated using SimLiT [10]. Reaction products from the neutron-induced reaction are detected inside the gas-filled detector.

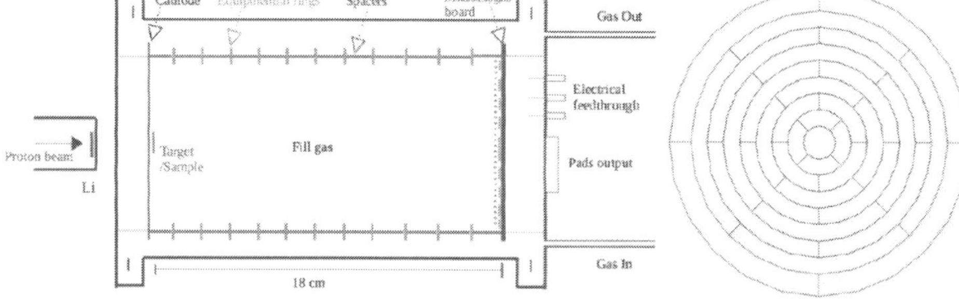

Figure 1. (Left) Conceptual design of the experimental setup and a Micromegas-based gas filled detector. (Right) Segmentation scheme of the anode pad.

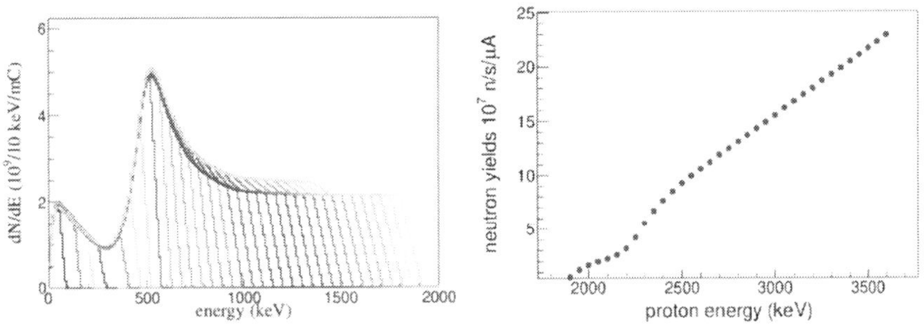

(a) (b)

Figure 2. (a) Calculated neutron spectra at the target, for a thick lithium target, and 5 cm^2 target located 2 cm downstream the lithium target. The different colors represents different proton beam energies from 1.9 MeV to 3.6 MeV. The neutron spectra were calculated using SimLiT [10]. (b) Total neutron intensity going through the target as function of proton energy. The values are the integration over the spectra in the left figure.

2.2 The detector

The detector is cylindrical with a cathode, equipotential rings, and a segmented Micromegas detector. The sample will be placed in the center of the cathode inside the chamber facing the Micromegas. Such a design of the detection system provides a large solid angle for charge-particle detection. The separation between the cathode and the Micromegas is 18 cm, and the active volume diameter is 16 cm. The segmentation scheme of the Micromegas board is shown in the right panel of Fig. 1.

2.3 Fill gas

We carried out detailed analysis for various gas mixtures for the detector being developed for neutron-induced charge particle reaction experiments. Fig. 3 shows electron drift velocity as a function of the field for various gas mixtures obtained using the Magboltz program [11]. We plan to use Ne(90%) and CF_4(10%) gas mixtures in the pressure range of 3 - 4 atmosphere as detection gas. The Ne/CF_4 gas mixture is chosen as the detection gas as mixtures containing argon and hydrogen had to be omitted to avoid background from neutron scattering and ^{41}Ar. The addition of CF_4 gas has many desirable merits viz; high drift velocity, helpful for fast detectors and reduced sensitivity to neutron background compared to hydrogenated molecules [12]. Pure CF_4 gas can also be used at low pressure but may have one demerit of lower gain as compared to Ne/CF_4 gas [12].

2.4 Read-out plane

The anode plane will be a segmented Micromegas detector. It is divided into nine circular rings with a central pad ring of diameter 3 cm, second, third, fourth, fifth, sixth, seventh, and eighth rings of diameter, 4.5 cm, 6 cm, 7.5 cm, 9 cm, 10.5 cm, 12 cm, 14 cm respectively, and ninth ring of diameter 16 cm. These outer rings are further segmented into four sections, thus giving 33 segmented pads on the anode plane, as shown in the right panel of Fig.1. The central pad faces the target; hence it will detect any heavy charged particle emitted from the reaction. It will serve as a trigger to exclude most of the background events. The outermost ring, shown in red, will serve as a veto pad to exclude events that do not originate in the

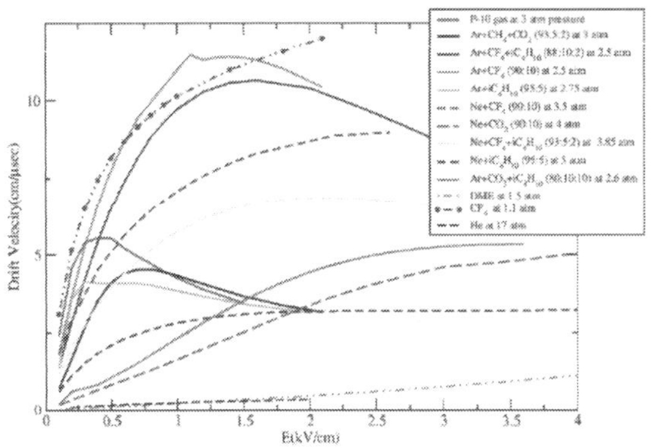

Figure 3. Electron drift velocity as a function of field for various gas mixtures obtained using [11].

detector. The purpose for the rest of the segmentation is to allow particle identification based on ΔE-E technique, and to reduce the background rates on each individual pad as discussed ahead.

2.5 Geant4 simulation and Background

One of the significant challenges of this experiment is the intense expected background. In order to investigate the expected background level and improve the design of the detector, we carried out a Geant4 [13] simulation of the designed experimental setup. The primary background sources are neutron and γ radiation. The neutron spectrum will mainly produce low-energy background due to hadron scattering in the gas. Similarly, an intense 478-keV γ radiation from the ^{7}Li(p, $p'\gamma$) reaction will cause low-energy background due to Compton scattering in the gas. This background can be reduced by adding a 1.6 cm lead shielding between the source and the detector. Other γ radiations background are the 14.6 and 17.6 MeV γ-rays from the ^{7}Li(p, γ) reaction. Although they are of lower intensity and have a low probability of interacting with the gas, they still contribute via (γ, p) and (γ, α) reactions. The β^+ decays from the target itself do not contribute much to the background. Fig. 4 (a) shows a graphical illustration of Geant4 simulation in which γ-rays being bombarded on detector. Fig. 4 (b) shows the estimated background from all sources as calculated by Geant simulation for 3.6 MeV incident proton energy on Li along with expected yields and energies from the ^{26}Al(n, p_1) channel, which is the most important channel for the total ^{26}Al(n, p) rate. The broad proton peak is due to the wide neutron energy distribution. The spectrum presented in Fig. 3 was obtained for 10^{10} primary particle events. Another issue is the background rate. Although it is possible to define a high threshold for this experiment and reduce the DAQ load, the actual ionization rate in the gas might hinder the operation of the detector. One of the goals of the segmentation on the anode plane we chose is to reduce the rate on a specific pad to a reasonable level. The simulation of the experimental setup implies that the proposed experiment is feasible despite the high background level.

2.6 Experiments

We plan to carry out, ^{26}Al(n, p)^{26}Mg, ^{26}Al(n, α)^{23}Na [$t_{1/2}$=7.2 x 10^5 Years], and ^{40}K(n, p)^{40}Ar, ^{40}K(n, α)^{37}Cl [$t_{1/2}$=1.25 x 10^9 Years] cross-section measurements. For the ^{26}Al case, the target thickness is limited by the availability of the unstable isotope and its price.

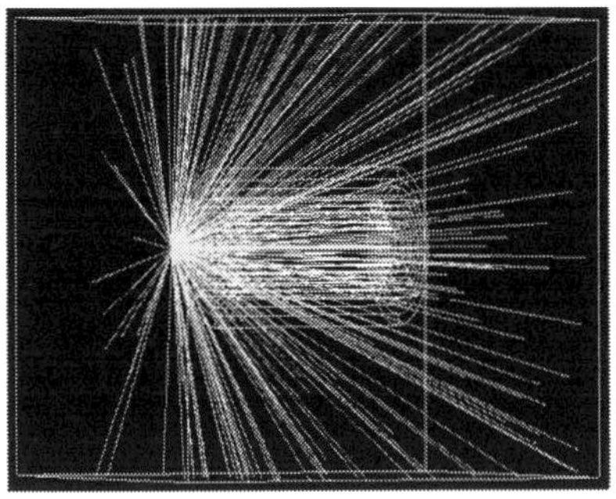

(a)

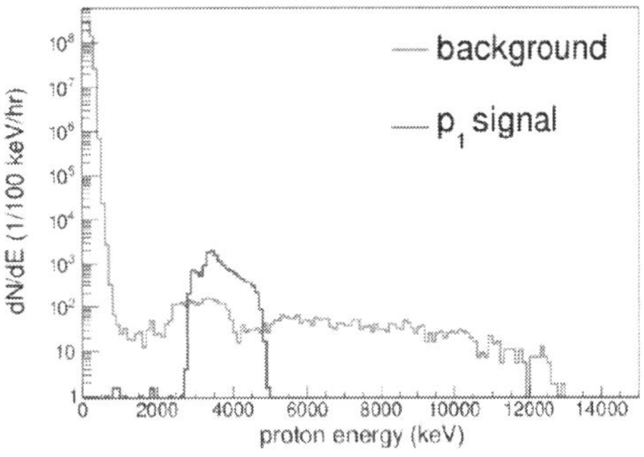

(b)

Figure 4. (a) A graphical illustration of Geant4 simulation of the detection setup where gamma rays are being bombarded on the detector. (b) Simulated background for proton energy of 3.6 MeV, compared to the expected yields from the ^{26}Al(n, p_1) channel.

Considering a 100 nCi target and 10-μA proton beam, and using TENDL-2019 cross section evaluation [14], a rate of 10^3 -10^4 detections per hour is calculated for the protons and the α-particles. For ^{40}K case, considering 0.25 mg/cm2 target thickness and assuming 10-μA proton beam, a rate of 10^4 -10^5 detections per hour is expected for both protons and α-particles.

3 Summary

Measurements of neutron-induced cross sections on unstable nuclei at explosive stellar temperatures require high-intensity neutron fluxes at the relevant energies. We plan to use the ^{7}Li$(p,n)^7$Be reaction with varying proton energies to produce a set of cross section measurements that can be combined with proper weights to obtain Maxwellian-Averaged Cross-Section (MACS) values at temperatures of 1.5–3.5 GK. Further work is being pursued with regard to detector building, testing and commissioning for the first experiment.

References

[1] J. José, *Stellar Explosions: Hydrodynamics and Nucleosynthesis* (CRC Press, Boca Raton, 2016).
[2] J. José and C. Iliadis, Reports Prog. Phys. **74 (9)**, 096901 (2011).
[3] C. Iliadis *et al.*, Astrophys. Journal, Suppl. Ser. **193** (1) (2011) 16.
[4] K. Hermansen *et al.*, Astrophys. Journal **901** (1) (2020) 77.
[5] R. D. Hoffman *et al.*, Astrophys. Journal **521** (2) (1999) 735–752.
[6] S. Wanajo, H. T. Janka, S. Kubono, Astrophys. Journal **729** (1) (2011) 46.
[7] A. Parikh *et al.*, Astron. Astrophys. **557** (2013) A3.
[8] C. Weiß *et al.*, Nucl. Instrum. Method A **A 799**, 0168–9002 (2015).
[9] M. Friedman, Eur. Phys. J. A. **56**, 155 (2020).
[10] M. Friedman *et al.*, Nucl. Instrum. Methods A **698**, 117–126 (2013).
[11] S. F. Biagi, Nucl. Instrum. Method A **421**, (1999) 234–240.
[12] F. Sauli, *Gaseous Radiation Detectors: Fundamentals and Applications*, Cambridge University Press; 1st edition (2014).
[13] S. Agostinelli *et al.*, Nucl. Inst. Methods Phys. Res. A **506**, 250-303 (2003).
[14] A. J. Koning *et al.*, Nuclear Data Sheets **155** (2019) 1-55.

Nucleon-nucleon interaction and the neutrino surface of neutron star mergers

Rajan Anderson Dornan[1,*] and *O. L.* Caballero[1,**]

[1]Department of Physics, University of Guelph, Guelph, ON, Canada

Abstract. Neutrinos act as an important cooling mechanism in an array of explosive astrophysical events. Modeling of their fluxes and spectra is key for the interpretation of their detection and the understanding of the synthesis of heavy elements. The neutrino surface is key in the modeling of neutrino spectra. In previous work, the points of last neutrino scattering were found assuming no interaction between nucleons. In this work we incorporate this interaction using a few different models and observe differences in where the neutrinos decouple from the system.

1 Introduction

Neutrinos are known to determine the fate of explosive environments such as core-collapse supernova, and mergers involving neutron stars. Neutrinos cool matter and can also provide energy to produce winds, in which heavy elements can be synthesized [1]. Compared to photons, neutrinos decouple from matter at much higher densities providing a window to stellar interiors. Thus, understanding neutrino scattering from highly dense matter is of paramount importance and as such has gotten a lot of attention (see e.g. [2–6] to name a few).

The neutrino surface is defined by the last points of neutrino scattering in a stellar environment. The local temperature at decoupling is determinant in the modeling of neutrino spectra from emitting sources. In previous work, we have studied the neutrino surface, fluxes, and detection, from neutron star mergers and accretion disks around black holes [7–9]. In those works, we assumed that matter was dissociated into neutrons, protons, and electrons, and we ignored in-medium effects. In this work, we present our first study of the neutrino surface when nucleonic interactions are introduced in the neutrino scattering off nucleons. We use the formalism of the virial equation of state (EoS) and Skyrme forces in the framework of density functional theory (DFT). Below we present the simulation used, our post-processing approach, and results that motivate future work.

2 Neutrinos from neutron star mergers

Our results are based on a post-processing study of 3D general relativistic simulations of a neutron star merger. These simulations included strong magnetic fields, neutrino cooling and different microscopic Equations of State (EoS)[8]. For this particular study we used a

*e-mail: rander05@uoguelph.ca
**e-mail: ocaballe@uoguelph.ca

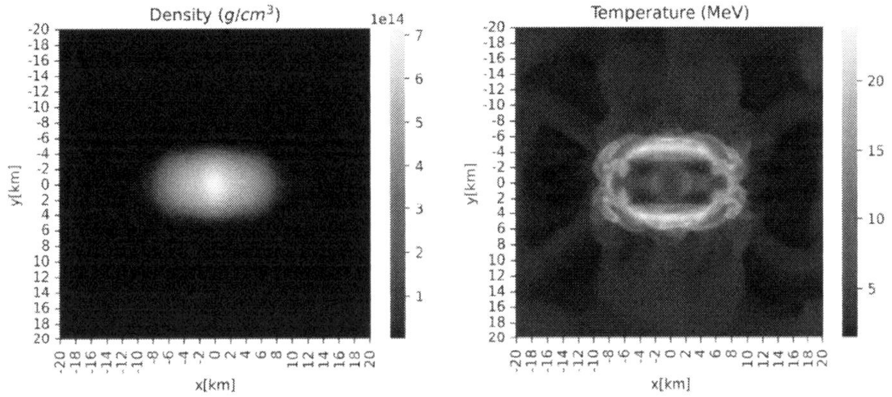

Figure 1: The left panel shows our density profile in g/cm^3, and the right panel shows the temperature profile in MeV

snapshot at approximately 3.5 ms after merger of a simulation with the DD2 EoS [10]. This EoS predicts a radius of 13 km for an isolated neutron star of mass 1.35 $M_{\odot}$. In ref. [8] this EoS was classified as intermediate in terms of stiffness. The density and temperature profile as seen in the equatorial plane of the merger are shown in Figure 1.

To determine the neutrino surface we must examine the optical depth which we defined in along a path in the z direction as

$$\tau = \int_{z_0}^{\infty} n_i \langle \sigma_i \rangle dz, \tag{1}$$

where n_i represents the density of species i, and $\langle \sigma_i \rangle$ represents the thermally averaged cross section of species i (we use Einstein summation). A blackbody at optical depth 2/3, will radiate approximately the same energy as our system [11], and thus we find z_0, the last point of neutrino scattering, such that $\tau = 2/3$.

In order to find the set of points z_0, we must include the cross sections of all interactions undergone by neutrinos in the system. In our previous work cross sections between neutrinos and nucleons did not account for nucleonic interaction [9], which modify the scattering cross sections, and may be significant due to the highly dense nature of the system. As a first pass, here we modified the cross sections of the neutral current reactions $\nu + p \to \nu + p$ and $\nu + n \to \nu + n$, and will include in-medium effects in the current reactions in future work.

3 Neutrinos and Nucleon-nucleon interactions

The study of neutrino scattering in a medium of strongly interacting nucleons is done by via the nuclear matter response functions χ. These quantify the change in expectation values in a given observable, in response to some perturbation. The dissipative component of these response functions relates to the structure factors S, that modify the free differential cross section. Using the temperature dependent fluctuation dissipation theorem we have [12]

$$-\frac{1}{\pi} \operatorname{Im} \chi(\mathbf{q}, \omega) = \left(1 - e^{-\beta\omega}\right) S(\mathbf{q}, \omega), \tag{2}$$

where $\beta = 1/k_B T$, T is the temperature, $\mathbf{q}$ is the momentum and ω the energy loss. In the absence of interactions, these response functions turn into the identity in position space, or delta functions centered around $\omega = 0$ in momentum space.

3.1 The Virial EoS

The virial EOS is a model independent method for describing finite-temperature low-density nuclear matter that takes into account interactions between species. Previous development has been done to study the response function of low density warm matter. We refer the reader to Ref. [13] for details. Here we summarize the main points.

In this formalism the pressure of a system of different species i in terms of the fugacity, to second order, can be written as [14, 15]

$$\frac{P}{T} = \sum_i \frac{s_i}{\lambda_i^3} z_i + \sum_{(i,j)} \frac{s_i}{\lambda_i^3} b_{ij} z_i z_j , \tag{3}$$

where T is the temperature, $z_i = e^{(\mu_i + E_i)/T}$ is the fugacity (with chemical potential μ_i and binding energy $E_i > 0$), s_i is the spin degeneracy of a nucleon and $\lambda_i = \sqrt{2\pi/m_i T}$ is the thermal wavelength, all these for species i. b_{ij}, are the second virial coefficients (note $b_i = b_{ii}$) and characterize interactions between particles of species i and j. It is possible to determine the second virial coefficients from nuclear scattering phase shifts (see [14]). For a system with one single species (for example neutrons) the vector response is written as [16, 17]

$$S_V = \frac{T}{(\partial P/\partial n)_T}. \tag{4}$$

Horowitz et al [13] found that for a mixture of neutrons and protons which densities are given by

$$n_{n(p)} = \frac{2}{\lambda^3}\Big(z_{n(p)} + 2b_n z_{n(p)}^2 + 2b_{pn} z_n z_p\Big), \tag{5}$$

the vector response S_V is

$$S_V = 1 + \frac{4}{\lambda^3}\frac{C_v^{n2} b_n z_n^2 + 2C_v^n C_v^p z_n z_p + C_v^{p2} z_p^2 b_n}{C_v^{n2} n_n + C_v^{p2} n_p}, \tag{6}$$

where $C_v^{n(p)}$ are weak charges of neutron and proton. Similarly, the axial or spin response S_A was found via the densities of spin polarized nuclear matter

$$n_{n(p)}^{\pm} = \frac{1}{\lambda^3}\Big(z_{n(p)}^{\pm} + 2b_+ z_{n(p)}^{\pm\,2} + 2z_{n(p)}^{\pm}(b_- z_{n(p)]}^{\mp} + b_{pn}^+ z_{p(n)}^{\pm} + b_{pn}^- z_{p(n)}^{\mp})\Big), \tag{7}$$

where the virial coefficients b_+ and b_-, describe the interaction between two particles of the same species (neutrons or protons) with spins aligned in the same direction and opposite direction, respectively. b_{pn}^+ and b_{pn}^- correspond to the second virial coefficients of a neutron and a proton with spins pointing in the same direction and opposite direction, respectively. $z_i^{\pm}$ is the fugacity of species i with spin up (+) or down (−). In this way,

$$S_A = 1 + \frac{4}{\lambda^3}\frac{\left(z_p^2 + z_n^2\right) b_a - 2z_p z_n b_{pn}^a}{n_n + n_p}, \tag{8}$$

where $b_a = b_+ - b_-$ and $b_{pn}^a = b_{pn}^+ - b_{pn}^-$. Finally, the in-medium differential cross section per unit volume V can be written in terms of the structure factors as [13]

$$\frac{1}{V}\frac{d\sigma}{d\Omega} = \frac{G_F^2 E_\nu^2}{16\pi^2}\left(g_A^2(3 - \cos\theta)(n_n + n_p)S_A + (1 + \cos\theta)n_n S_V\right), \tag{9}$$

where G_F is the Fermi constant, g_A is the axial charge of the nucleon, E_ν is the neutrino energy, S_A and S_ν are response functions derived from our virial equation of state, and θ is the angle of deflection. Note that by finding our cross section per volume we find our inverse mean free path, $n\sigma$.

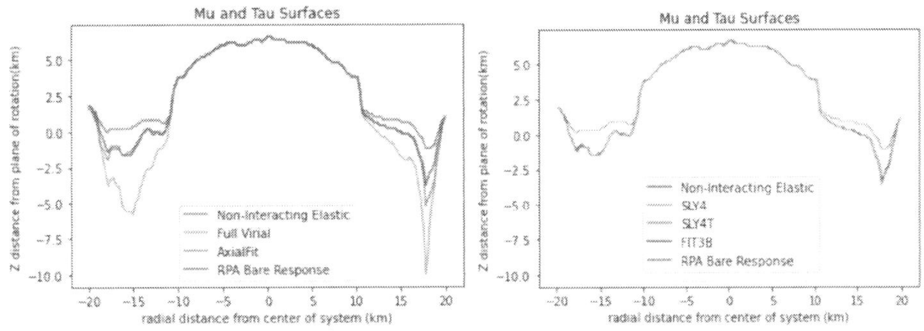

Figure 2: Muon(tau) neutrino surfaces using different neutral current cross sections. The left panel shows the differences between the structure factors from the virial equation of state, a bare RPA response and a fit between the two, whereas, the right panel focuses on various parameterizations of the Skryme Energy Density Functional.

3.2 Skyrme forces

Nuclear interactions have also been studied by using energy density functionals, which provide the energy of the system as a functional for a variety of different densities. In this way interaction energy is included by experimentally tuning different parameters. One common energy density functional is the Skyrme energy density functional, whose standard form looks as follows [18]

$$\mathcal{E}_{Skyrme} = C_T^s \mathbf{s}_T^2 + C_T^{\Delta s}\mathbf{s}_T \cdot \Delta \mathbf{s}_T + C_T^{sT}\mathbf{s}_T \cdot \mathbf{T}_T + C_T^{\nabla s}(\nabla \cdot \mathbf{s}_T)^2 + C_T^j \mathbf{j}_T^2 + C_T^{\nabla j}\mathbf{s}_T \cdot \nabla \times \mathbf{j}_T$$
$$+ C_T^\rho \rho_T^2 + C_T^{\Delta\rho}\rho_T \Delta\rho_T + C_T^\tau \rho_T \tau_T + C_T^J \mathbb{J}_T^2 + C_T^{\nabla J}\rho_T \nabla \cdot \mathbf{J}_T \quad (10)$$

where the C's represent parameters tuned to experimentally known quantities, T represents an isospin index, where T=0 is the density regardless of isospin, T=1 is the difference in densities between the two isospin densities. The other variables represent different types of densities (eg. Spin, kinetic, etc). We numerically calculate the structure factors from these functionals, following the work of A. Pastore et al. [19]. In terms of the Skyrme functionals the differential cross section is given by

where $E_\nu(E'_\nu)$ is the incoming(outgoing) neutrino energy, k' is the outgoing neutrino momentum, and $S^{(0,0)}$, $S^{(1,0)}$ and $S^{(1,1)}$ can be derived from response functions in different spin channels. Finally, we also make use of the fit provided by Horowitz et al [13] that matches the virial EoS at low density and follows Random Phase Approximation results at high density.

4 Results and Discussion

Figure 2 shows a slice of the neutrino surface corresponding to the simulation discussed in section 2. We have found differences when the neutrino scattering off nucleons includes the correlation between nucleons caused by nucleon-nucleon interactions. The effect is stronger on the edges of the merger, on the equatorial plane, away from the center of gravity where the density has decreased. The height at which neutrinos decouple is higher when the nucleon-nucleon interactions are ignored. Therefore, neutrinos decouple at higher temperatures due to nucleon correlations. We also note that surfaces using Skyrme energy density functionals are closer to the non-interacting case. However not much difference was seen between the different parameterizations of the Energy density functional. In future work we will study the impact of these differences in the neutrino detection as seen on neutrino observatories on Earth.

5 Acknowledgments

O. L. C and R. Anderson thank A. Pastore for providing a code to perform some of the calculations. O. L. C. acknowledges support from NSERC.

References

[1] A. Arcones and F. K. Thielemann, J. Phys. G **40**, 013201 (2013)
[2] I. Vidana, D. Logoteta and I. Bombaci, Phys. Rev. C **106**, 035804 (2022)
[3] P. T. P. Hutauruk, H. Gil, S. i. Nam and C. H. Hyun, Phys. Rev. C **106**, 035802 (2022)
[4] Z.Lin, A. W. Steiner, & J. Margueron, 2022, arXiv:2207.05927
[5] S. Reddy, M. Prakash and J. M. Lattimer, Phys. Rev. D **58**, 013009 (1998)
[6] G. I. Lykasov,, C. J. Pethick,, & A. Schwenk, Phys. Rev. C, **78**, 045803 (2008).
[7] L. Lehner, et al, Class. Quant. Grav. **33**, 184002 (2016)
[8] C. Palenzuela et al, Phys. Rev. D **92**, 044045 (2015)
[9] O. L. Caballero, G. C. McLaughlin, and R. Surman, Phys. Rev. D **80**, 123004 (2009)
[10] S. Typel, G. Ropke, T. Klahn, D. Blaschke and H. H. Wolter, Phys. Rev. C **81**, 015803 (2010)
[11] S. L. Shapiro, S. A. Teukolsky, *Black Holes, White Dwarfs, and Neutron Stars: The Physics of Compact Objects*, Wiley-VCH, (1983) 579-586
[12] A. Altland and B. D. Simons, *Condensed Matter Field Theory*. Cambridge University Press, 2010.
[13] C. J. Horowitz, O. L. Caballero, Z. Lin, E. O'Connor and A. Schwenk, Phys. Rev. C **95**, 025801 (2017)
[14] C.J. Horowitz and A. Schwenk, Nucl. Phys. A **776**, 55 (2006).
[15] C.J. Horowitz and A. Schwenk, Phys. Lett. B **638** 153 (2006).
[16] C. J. Horowitz and A. Schwenk, Phys. Lett. B **642** 326 (2006)
[17] L.D. Landau and E.M. Lifschitz, *Statistical Physics*, Pergamon N.Y., Second Edition (1969), Eq. (114.14).
[18] M. Bender, P. H. Heenen, P. G. Reinhard, Rev.of Mod. Phys. 121–180 (2003).
[19] A. Pastore, G. E. Brown, J. Navarro, Physics Reports, 563, (2015)

Evolution of Accreting Neutron Star Common Envelopes and Their Nucleosynthesis - Possible P-nuclide Production

Sophie E D Abrahams[1,2,*], *Chris* Fryer[2], *Alexander* Hall-Smith[1], *Alison* Laird[1], and *Christian* Diget[1]

[1]School of Physics, Engineering and Technology, University of York, Heslington, York, YO10 5DD, UK
[2]Center for Theoretical Astrophysics, Los Alamos National Laboatory, Los Alamos, NM 87545, USA

Abstract. In a binary system, a short lived phase called a common envelope can occur in which both stars share the envelope of the primary. During this phase, there is the opportunity for nucleosynthesis to occur at high temperatures and densities in the accretion disk which forms around the secondary star. The system of interest is one in which the secondary star is a neutron star. After introducing the system, our model is described and preliminary results are presented. Preliminary simulations show that light p-nuclides ^{92}Mo and ^{96,98}Ru are overproduced, and that the closer the material is to the neutron star prior to ejection, the fewer p-nuclides are produced.

1 Introduction

In a binary system, a short phase can occur during which the two stars orbit within a common stellar envelope. This common envelope (CE) phase occurs as a result of Roche Lobe Overflow - for example when the primary star is expanding into its red giant phase - and the secondary accreting star cannot incorporate material quickly enough. The accretor is overwhelmed by the envelope of the expanding star and the two objects share an envelope. During the common envelope phase, the orbital energy of the binary is lost through dynamical and tidal friction. This results in the ejection of the primary's envelope and the tightening of the binary's orbit. This evolutionary stage is thought to be a vital step in the creation of compact binary systems such as neutron star mergers and X-ray binary systems [1].

In the scenario of interest here, the accretor is a neutron star and the expanding star is entering its red giant phase. Once inside the CE, the neutron star will slowly inspiral towards the core of the red giant while accreting hydrogen rich material at hypercritical rates [1], [3]. The angular momentum of the accreting material will cause it to hang up in an accretion disk. If the accretion rate is high, the temperature of this accreting material will become high enough to drive nuclear burning in the accretion disk as material approaches the neutron star [3]. A proportion of this material will then be ejected from the system, current hydrodynamical simulations say up to 25% [2]. When the material is ejected, it moves into the companion red giant. When the red giant ends its life as a supernova, the material which has undergone nucleosynthesis in the neutron star accretion disk can be ejected into the interstellar medium (ISM) [3].

*e-mail: seda500@york.ac.uk

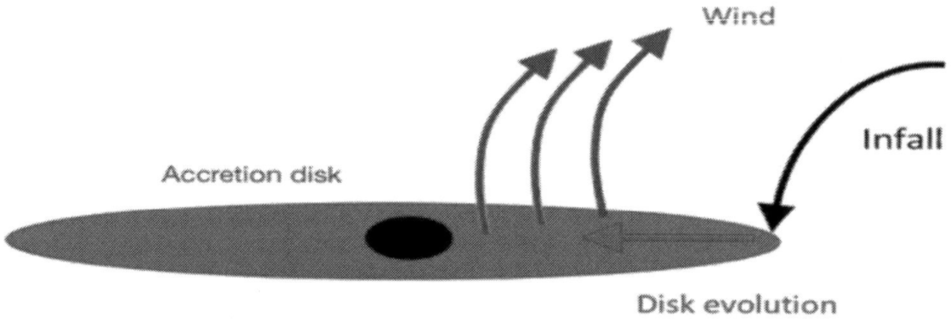

Figure 1. Schematic view of the path followed by material entering neutron star accretion disk.

Nucleosynthesis in the accretion disk of this system is not well understood. Keegans et al. (2019) conducted an initial investigation of the nucleosynthesis in accreting neutron star common envelopes using trajectories that assumed that the accreting material had no angular momentum. They find that this system could contribute to galactic chemical evolution (GCE) and find that in some simulations, some p-nuclides are overproduced. P-nuclides are a set of proton-rich isotopes ranging from ^{74}Se to ^{196}Hg - see Table 1 of [4]. In p-process models, p-nuclide isotopes are underproduced by a factor of four compared to Solar System abundances [5]. In particular, the isotopes ^{92,94}Mo and ^{96,98}Ru are underproduced by an additional order of magnitude compared to the other p-nuclides [5]. They are thought to be produced by the so-called p-process, where p-process is an umbrella term used to describe the processes which produce p-nuclides [6]. The possible sites hosting the p-process are debated.

Work presented here builds on the foundations laid by Keegans et al. (2019), improving upon them by using trajectories which account for the angular momentum needed to form an accretion disk.

2 Method

2.1 Accretion Disk Model

In order to model accretion around the neutron star, a modified Bondi-Hoyle-Lyttleton accretion model is used. This model is adapted from Popham et al. (1999), where they model hyperaccreting black holes [8]. In the model used here, the accretion rate is set to:

$$\dot{M}_{\rm B} \approx \lambda_{\rm BHL} 4\pi r_{\rm B}^2 \rho (v^2 + c_s^2)^{\frac{1}{2}}. \qquad (1)$$

Where $\lambda_{\rm BHL}$ is a non-dimensional scaling parameter to lower the accretion rate due to limiting factors such as angular momentum of the system or gradients in the accretion disk (see [9] for more details), $r_{\rm B}$ is the Bondi radius, v is the velocity of the material bound to the accreting object and $c_{\rm s}$ is the speed of sound in the accreting material.

Figure 1 shows a schematic view of the trajectory followed by the material. The material infalls towards the neutron star. The angular momentum of the accreting material causes the formation of an accretion disk around the neutron star. New material joins the accretion disk at the outer edge, and loses angular momentum as it moves towards the centre of the accretion disk. During this process, material is ejected from the disk by a wind and moves back into the companion star. All through this process, the material is exposed to temperatures and densities which allow for nucleosynthesis to occur. We model this process by post-processing

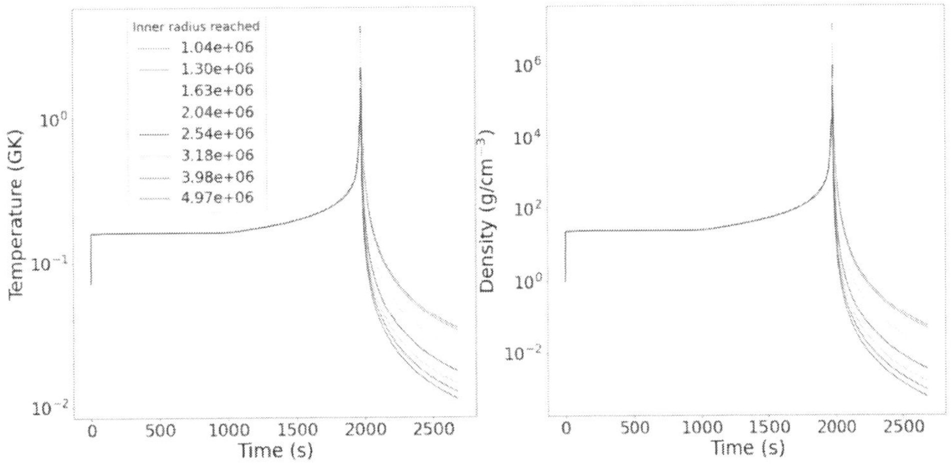

Figure 2. Left: Temperature-time plot for trajectories. Right: Density-time plot for trajectories. Material in trajectories is accreting around a $1.5M_\odot$ neutron star at an accretion rate of $8 \times 10^{-5}\ M_\odot s^{-1}$. Unit of inner radius is cm.

temperature-density trajectories representative of the material's trajectories. Figure 2 shows temperature time and density time plots for the eight trajectories explored. Material in the accretion disk is ejected at different radii. As the material moves inwards in the accretion disk towards the neutron star, it is exposed to higher temperatures and densities (see Figure 2). Therefore, the material's final composition depends on the innermost radius reached before ejection. We denote the point at which material is ejected from the accretion disk as the "inner radius", in units of cm. The smaller the inner radius is, the further into the accretion disk - closer to the neutron star - the material moves prior to ejection. In order to explore how the nucleosynthesis of p-nuclides is affected by the inner radius, we tested a set of eight trajectories with different inner radii. We can also explore the average production of p-nuclides over the eight trajectories, which taken together serve as an approximation of an accretion disk. To do this, we averaged the mass fraction of each p-nuclide over the trajectories. We then divided this average by the initial abundance of the p-nuclide to find the average overproduction from the approximated accretion disk.

2.2 Post-Processing Code - NuGrid

The post-processing code `NuGrid` was used to calculate the nucleosynthesis. The `NuGrid` run was a version of `Modular2ininet3` adapted so that `REACLIB` was used to replace outdated rates, such as those from Iliadis et al. (2001) (see [11] for more details) [10], [11], [12]. The network was composed of 5200 isotopes. The initial abundances were solar scaled, and have a metallicity of $Z_m = 0.02$. This is to reflect the high hydrogen abundance in the accreting material.

3 Preliminary Results

Figure 3 shows the impact of varying the inner radius reached by the material on the production of p-nuclides. As the material moves closer to the neutron star before ejection - and the peak temperature and density reached increases (see Figure 2) - higher mass p-nuclides

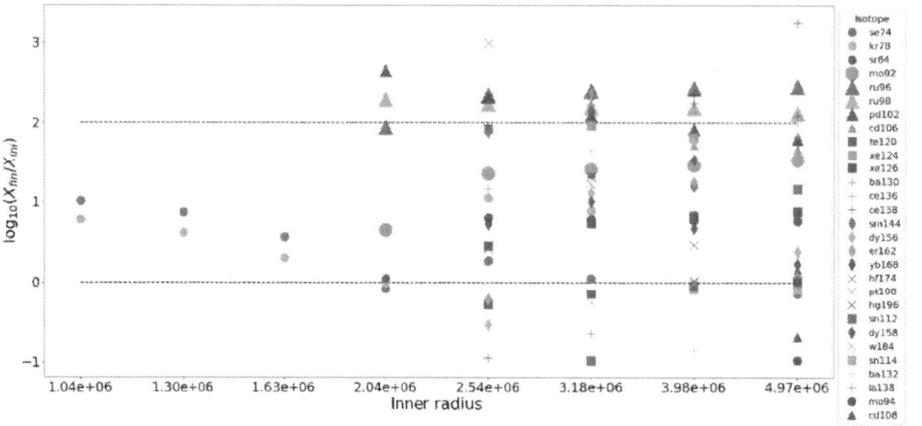

Figure 3. Effect of changing the inner radius on production of p-nuclides for a set of trajectories modelling accretion around a 1.5M$_\odot$ neutron star at an accretion rate of 8×10^{-5} M$_\odot$s^{-1}. Any isotopes which had an overproduction of less than −1 were not included.

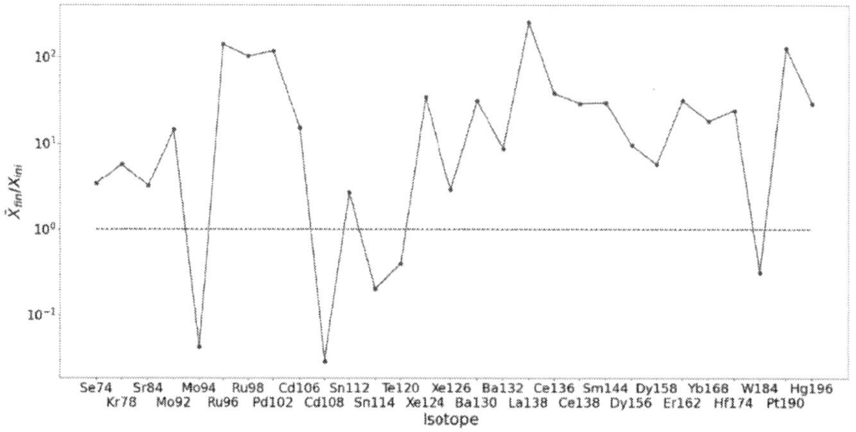

Figure 4. Average overproduction of p-nuclides for **an approximated** accretion disk composed of eight trajectories modelling accretion around a 1.5M$_\odot$ neutron star at an accretion rate of 8×10^{-5} M$_\odot$s^{-1}.

are not overproduced in the final abundances. For the three smallest inner radii, only ^{74}Se and ^{78}Kr are overproduced. For all but the three closest inner radii, the isotopes ^{96,98}Ru are consistently overproduced by 2 dex or more. The exception to this is ^{96}Ru at inner radius 2.04×10^6 cm, where it is overproduced by just under 2 dex. For all but the three closest inner radii, ^{92}Mo is overproduced and for all but the four closest inner radii it is overproduced by more than 1 dex. Exploration is ongoing to determine what prevents p-nuclides with A > 78 from being overproduced when the inner radius decreases and the material moves closer to the neutron star. A likely cause is that photodisintegration reactions are breaking down higher mass p-nuclides due to the higher temperatures reached at lower inner radii (see Figure 2 for more).

Figure 4 shows the average overproduction of p-nuclides for an approximated accretion disk. While ^{94}Mo is underproduced, other light p-nuclides - ^{92}Mo, ^{96,98}Ru are overproduced by dex 1 and dex 2 respectively, as is the light p-nuclide ^{102}Pd. Heavier p-nuclides are also overproduced, including the long lived radioisotope ^{138}La. Figure 4 shows evidence that an approximation of an accretion disk can overproduce light p-nuclides, which are underproduced in current p-process models. Further work needs to be done to understand the nucleosynthesis at work in this system.

4 Conclusions

Initial results show that p-nuclides are overproduced in accreting neutron star common envelopes. In particular light, p-nuclides - ^{92}Mo and ^{96,98}Ru - which are usually underproduced in p-process models. This is significant as nucleosynthesis in accreting neutron star common envelopes has not been explored in great detail. In the future, we will investigate the nucleosynthesis in accretion disks on larger grids, explore the impact of parameters such as the angular momentum of the accreting material and the mass of the accretor. Work will also involve better understanding the affect of varying the metallicity of the accreting material, as well as placing any overproduction in the context of the ISM and GCE.

5 Acknowledgements

SEDA thanks Andrés Yagüe López for support. The authors thank the anonymous referee whose comments strengthened these proceedings. LA-UR-22-33183.

References

[1] Ivanova N., Justham S., Chen X., De Marco O., Fryer C. L., Gaburov E., Ge H., et al., 2013, A&ARv, 21, 59. doi:10.1007/s00159-013-0059-2

[2] Fryer C. L., Herwig F., Hungerford A., Timmes F. X., 2006, ApJL, 646, L131. doi:10.1086/507071

[3] Keegans J., Fryer C. L., Jones S. W., Côté B., Belczynski K., Herwig F., Pignatari M., et al., 2019, MNRAS, 485, 620. doi:10.1093/mnras/stz368

[4] Arnould M., Goriely S., 2003, PhR, 384, 1. doi:10.1016/S0370-1573(03)00242-4

[5] Pignatari M., Göbel K., Reifarth R., Travaglio C., 2016, IJMPE, 25, 1630003-232. doi:10.1142/S0218301316300034

[6] Iliadis C., 2015, Nuclear Physics of Stars. doi:10.1002/9783527692668

[7] Battino U., Pignatari M., Travaglio C., Lederer-Woods C., Denissenkov P., Herwig F., Thielemann F., et al., 2020, MNRAS, 497, 4981. doi:10.1093/mnras/staa2281

[8] Popham R., Woosley S. E., Fryer C., 1999, ApJ, 518, 356. doi:10.1086/307259

[9] MacLeod M., Ramirez-Ruiz E., 2015, ApJ, 803, 41. doi:10.1088/0004-637X/803/1/41

[10] Pignatari M., Herwig F., 2012, NPNew, 22, 18. doi:10.1080/10619127.2012.710142

[11] Cyburt R. H., Amthor A. M., Ferguson R., Meisel Z., Smith K., Warren S., Heger A., et al., 2010, ApJS, 189, 240. doi:10.1088/0067-0049/189/1/240

[12] Iliadis C., D'Auria J. M., Starrfield S., Thompson W. J., Wiescher M., 2001, ApJS, 134, 151. doi:10.1086/320364

Tension between implications from PREX-2 data and gravitational tidal response on dense matter equation of state

Vivek Baruah Thapa[1,2,*] and *Monika* Sinha[1]

[1]Indian Institute of Technology Jodhpur, Jodhpur-342037, India
[2]National Institute of Physics and Nuclear Engineering (IFIN-HH), RO-077125, Bucharest, Romania

Abstract. Recently an improved value of neutron skin thickness of ^{208}Pb was reported in Lead Radius EXperiment-2 (PREX-2) to be $R_{\rm skin} = R_n - R_p = (0.283 \pm 0.071)$ fm which corresponds to high estimations of nuclear symmetry energy ($E_{\rm sym}$) and its slope ($L_{\rm sym}$). The updated values of $E_{\rm sym}$ and $L_{\rm sym}$ commensurating to the neutron star observable estimations lie exterior to the astrophysical observed range. The higher values of $L_{\rm sym}$ at n_0 deduced from recent PREX-2 data correlates to matter being easily deformable (yielding higher radius values) around intermediate matter densities leading to higher values of $\tilde{\Lambda}$ creating a tension between the terrestrial and astrophysical observations. In this study, we exploit this tension to constrain the Δ-scalar meson coupling parameter space.

1 Introduction

The neutron stars (NSs), highly compact stars, contain densest matter inside them. The matter density inside the NS core varies in the range of a few times nuclear matter density [1]. Most of the models of highly dense matter are formulated to reproduce the experimentally obtained range of the saturation property parameters, viz., nuclear saturation density (n_0), saturation energy (E_0), incompressibility (K_0), symmetry energy ($E_{\rm sym}$), its slope with density ($L_{\rm sym}$), curvature of symmetry energy ($K_{\rm sym}$), and effective nucleonic Dirac mass (m^*_N). There is recent update in obtaining the experimental values of $E_{\rm sym}$ and its density dependence from the nuclear physics experiment. The Lead Radius EXperiment (PREX) collaboration reported its first results (PREX-1) [2] with $R_{\rm skin} = R_n - R_p = 0.33^{+0.16}_{-0.18}$ fm and corresponding $L_{\rm sym}(n_0)$ based on strong correlations to be in the range (35 – 265) MeV. Later on the ranges of $E_{\rm sym}$ and $L_{\rm sym}$ based on experimental data from finite nuclei and heavy-ion collisions (HICs) with different microscopic model calculations were estimated to be 28.5 – 34.9 MeV and 30.6 – 86.8 MeV respectively [3]. A latest compilation has been reported in ref.-[4] for the isospin properties and that to be in the range of (30.6 – 32.8) MeV, (55.7 – 63.9) MeV for $E_{\rm sym}$ and $L_{\rm sym}$ at nuclear saturation density (n_0) respectively. But very recently an improved value of ^{208}Pb was reported in PREX-2 as $R_{\rm skin} = (0.283 \pm 0.071)$ fm [5] with $\sim 1\%$ precision. This however leads to an estimation of $E_{\rm sym}$ and $L_{\rm sym}$ at n_0 to be in the ranges (38.1 ± 4.7) MeV, (106 ± 37) MeV respectively (1σ interval) with correlation coefficient as 0.978 [6]. The dense matter model should be tested and constrained with the recent astrophysical observations from NSs along with the nuclear physics experimental data. In recent years, the pursuit to

*e-mail: thapa.1@iitj.ac.in

determine and constrain the dense matter equation of state (EOS) can be drawn from the massive NS ($M \geq 2\ M_\odot$) observations [7–9], radius measurements of NS candidates from NICER (Neutron star Interior Composition ExplorER) space mission [10–14] as well as tidal response from gravitational-wave (GW) events [15–18] by the LIGO-Virgo Collaboration. Very recently another pulsar, namely PSR J0952−0607 has been observed with mass reported in the range $2.18 - 2.52\ M_\odot$ at $1-\sigma$ confidence interval [19]. The large lower bound of maximum NS mass leads to consider the appearance of heavier baryons inside the inner core of the NS [20–25]. Also the constrained of matter to be soft at intermediate density opens the possibility of Δ resonances to appear with the increase of density.

In present work we model the dense matter inside NS making compatible with recently updated ranges of E_{sym} and its density dependence as well as recent astrophysical observations within the covariant density functional (CDF) model of dense matter. (*Conventions*: We implement the natural units $G = \hbar = c = 1$ throughout the work)

2 Hadronic CDF model

In this study, to compute the dense matter EOS we implement CDF framework model. In this model scheme, the coupling constants are based on the fact to reproduce the experimental quantities known at nuclear saturation. For this work, we consider the entire baryon octet ($b \equiv N, Y$) along with the spin-3/2 Δ-resonances in the composition of dense matter. Leptons (e^-, μ^-) are brought into the picture to maintain the beta-equilibrium condition. We consider the effective interactions between the baryons are mediated via the isoscalar-scalar meson σ, isoscalar-vector mesons ω, ϕ and the isovector-vector ρ-meson. The hidden strangeness ϕ-meson mediates the repulsive interactions between the strange baryons only. The total Lagrangian density in this formalism is given by [1]

$$\mathcal{L} = \sum_b \bar{\psi}_b(i\gamma_\mu D^\mu_{(b)} - m^*_b)\psi_b + \sum_l \bar{\psi}_l(i\gamma_\mu\partial^\mu - m_l)\psi_l + \sum_\Delta \bar{\psi}_{\Delta\nu}(i\gamma_\mu D^\mu_{(\Delta)} - m^*_\Delta)\psi^\nu_\Delta + \frac{1}{2}(\partial_\mu\sigma\partial^\mu\sigma - m^2_\sigma\sigma^2) - \frac{1}{4}\omega_{\mu\nu}\omega^{\mu\nu} + \frac{1}{2}m^2_\omega\omega_\mu\omega^\mu - \frac{1}{4}\boldsymbol{\rho}_{\mu\nu}\cdot\boldsymbol{\rho}^{\mu\nu} + \frac{1}{2}m^2_\rho\boldsymbol{\rho}_\mu\cdot\boldsymbol{\rho}^\mu - \frac{1}{4}\phi_{\mu\nu}\phi^{\mu\nu} + \frac{1}{2}m^2_\phi\phi_\mu\phi^\mu - \mathrm{U}(\sigma), \quad (1)$$

where the covariant derivative is given by $D_{\mu(j)} = \partial_\mu + ig_{\omega j}\omega_\mu + ig_{\rho j}\boldsymbol{\tau}_{j3}\cdot\boldsymbol{\rho}_\mu + ig_{\phi j}\phi_\mu$ with j denoting the baryon particle spectrum. The baryon octet and Δ baryon Schwinger-Rarita fields along with their respective masses are represented by ψ_b, m_b and ψ_Δ, m_Δ respectively. $\omega_{\mu\nu}$, $\boldsymbol{\rho}_{\mu\nu}$ and $\phi_{\mu\nu}$ are the anti-symmetric field tensors corresponding to vector meson fields. To reproduce the nuclear matter incompressibility at n_0, the self-interaction of σ meson is included [1] by the term $U(\sigma) = (1/3)g_2\sigma^3 + (1/4)g_3\sigma^4$ with g_2, g_3 denoting the self-interaction coefficients (non-linear (NL) model). Alternate to the NL model is the density dependent (DD) coupling model. In latter model, the incompressibility of nuclear matter at n_0 can be reproduced without self-interaction term but considering the coupling parameters density dependent. In the density-dependent approach, a rearrangement term is necessary to maintain the thermodynamic consistency contributing to matter pressure explicitly via the chemical potential and is given by $\Sigma^r = \sum_b \left[\frac{\partial g_{\omega b}}{\partial n}\omega_0 n_b - \frac{\partial g_{\sigma b}}{\partial n}\sigma n^s_b + \frac{\partial g_{\rho b}}{\partial n}\rho_{03}\boldsymbol{\tau}_{b3}n_b + \frac{\partial g_{\phi b}}{\partial n}\phi_0 n_b\right] + \sum_\Delta(\psi_b \longrightarrow \psi^\nu_\Delta)$. The scalar couplings with the Λ, Σ-hyperons are determined from the hypernuclear binding energy fits corresponding to $U^N_\Lambda(n_0) = -30$ MeV and $U^N_\Sigma(n_0) = +30$ MeV [26]. We have considered the $\sigma - \Xi$ coupling to be $U^N_\Xi(n_0) = -20$ MeV in symmetric nuclear matter reported recently in Ref.-[27]. As for the vector meson couplings with the hyperons, they are incorporated according to SU(6) symmetry [28]. As the Δ-baryons are resonant states of nucleons, they do not couple with isoscalar-vector ϕ-meson. Due to insufficient knowledge on meson-Δ couplings in nuclear matter, we treat these couplings as parameters. The Δ isoscalar

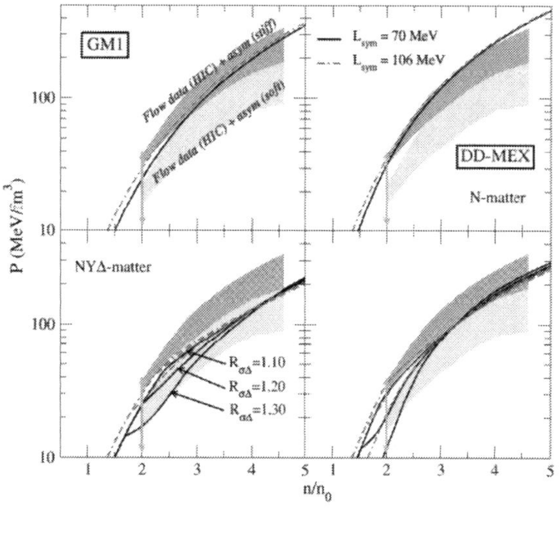

Figure 1. The matter pressure as a function of baryon number density for the nucleonic (upper panels), Δ-admixed hypernuclear (lower panels) dense matter EOSs considering variation in $L_{\rm sym}(n_0)$ and scalar meson-Δ coupling values. Re-calibrated $L_{\rm sym}(n_0) = 70,\ 106$ MeV EOSs are designated via the solid and dot-dashed curves respectively. The vertical bound at $2n_0$ is depicted from GW170817 event data via interpolation [16]. The shaded regions from density range $(2-4.5)\ n_0$ depict the flow data from HIC and modelled with stiffer as well as softer EOSs [29].

meson coupling gap is constrained in Ref.-[20] to be the range $0 \le R_{\sigma\Delta} - R_{\omega\Delta} \le 0.2$ where $R_{i\Delta} = g_{i\Delta}/g_{iN}$ and $i = \sigma,\ \omega,\ \rho$-mesons. In this study, we further constrain this parameter space on the basis of reconciling the tension between experimental nuclear and astrophysical observations. Here, we consider the isoscalar-vector and isovector-vector Δ-couplings to be $R_{\omega\Delta} = 1.1$, $R_{\rho\Delta} = 1.0$ respectively. The scalar Δ-coupling parameter space is varied in the range $1.1 \leqslant R_{\sigma\Delta} \leqslant 1.3$.

3 Results & Discussion

This section discusses the implications of exotic particles in reconciling the tension between the nuclear physics and astrophysical constraints. Here, we report the numerical results of various dense matter EOSs corresponding to variation in $R_{\sigma\Delta} - R_{\omega\Delta}$ coupling parameter space and isospin dependent saturation property.

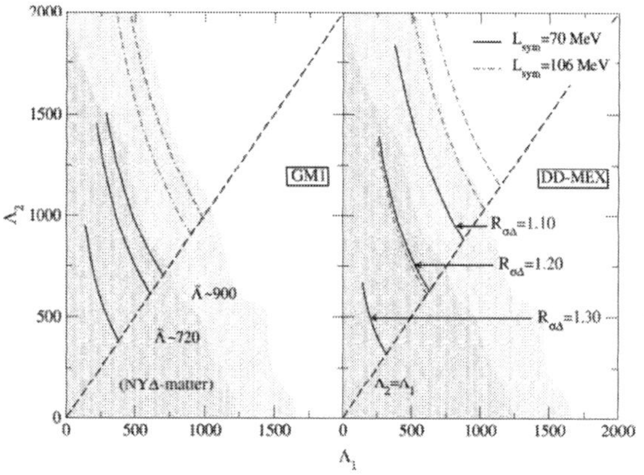

Figure 2. Tidal deformability parameters corresponding to the primary and secondary components involved in GW170817 event for various isospin dependent coupling in accordance to PREX-2 and alterations in $R_{\sigma\Delta} - R_{\omega\Delta}$. Here, we have considered the chirp mass to be $\mathcal{M} = 1.186\ M_\odot$. The shaded regions denote the tidal deformability upper bounds of $\tilde{\Lambda} \sim 900$ [15] (cyan) and $\tilde{\Lambda} \sim 720$ [17] (orange).

The pressure variation with baryon number density for the different EOS models are provided in fig.-1. It can be seen from the figure that the re-calibrated EOSs obtained via exploiting the $L_{\rm sym}$ and $\sigma - \Delta$ coupling parameter space for both the parametrization models lie within the constraints from GW170817 as well as the HIC data. Here, it is also evident that from the figure that considering the upper bound on $\sigma - \Delta$ coupling, $R_{\sigma\Delta} = 1.3$ satisfies the constraints from terrestrial as well as astrophysical observations. With further attractive optical potential for Δ-resonances in nuclear matter, the EOS models tend to be softer at the lower matter density regimes as evident from the figure. This result is in consistent with Ref.-[25]. The EOSs considered in this work fulfill the thermodynamic stability condition in addition to providing non-vanishing effective Dirac nucleon mass as pointed in Ref.-[30].

The tidal responses of both the components involved in GW170817 event are evaluated based on dense matter EOSs with variation in symmetry energy slope and scalar Δ-meson couplings. From the fig.-2, it can be seen that with higher values of $L_{\rm sym}$, the tidal responses also gradually increases. This is because of the increase in radius of intermediate mass NSs and $\tilde{\Lambda}$ is proportional to fifth-power of NS radius. And with higher values of $\sigma - \Delta$ coupling corresponding to further attractive potential, the tidal deformability is observed to decrease as invoking Δ^- at early matter densities induces more compactness. With the strict constraint of $\tilde{\Lambda} \sim 720$ obtained from recent reanalysis of GW170817 event data, it can be inferred that attractive potentials of Δ-resonances in dense matter is required to satisfy this constraint.

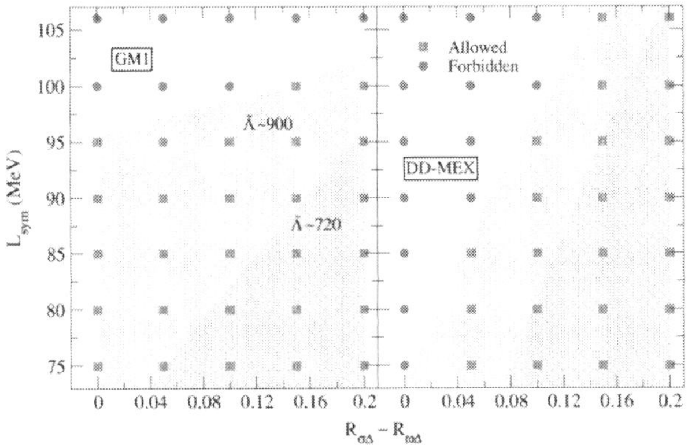

Figure 3. Coupling parameter space range of $R_{\sigma\Delta} - R_{\omega\Delta}$ for which constraints from nuclear physics experiments as well as astrophysical constraints are fulfilled. Here, we have varied $R_{\sigma\Delta}$ in the range of [1.10 − 1.30]. The shaded regions (cyan $\longrightarrow$ $\tilde{\Lambda} \sim 900$ and orange $\longrightarrow$ $\tilde{\Lambda} \sim 720$) are the allowed parameter space which fulfill both the terrestrial and astrophysical constraints.

The variation of Δ-coupling parameters with $L_{\rm sym}$ in dense matter EOS is shown in fig.-3. This figure provides the allowed range of $R_{\sigma\Delta}$ keeping $R_{\omega\Delta} = 1.10$ and $R_{\rho\Delta} = 1.0$ with changing $L_{\rm sym}$ scenarios. It can be seen that higher $L_{\rm sym}$ values demand the Δ potential in dense matter to be attractive in nature. In case of DD-MEX parameterization, for $L_{\rm sym} \leq 80$ MeV, the Δ coupling parameter space is unconstrained and all values of $R_{\sigma\Delta} - R_{\omega\Delta}$ are admissible following the former $\tilde{\Lambda} \sim 900$ constraint. While in case of GM1 coupling set, this $\tilde{\Lambda}$ constraint bounds the $R_{\sigma\Delta} - R_{\omega\Delta}$ parameter set for $L_{\rm sym} \leq 95$ MeV. Following strict upper bound of $\tilde{\Lambda} \sim 720$, the parameter space is further constrained leaving the coupling value of $R_{\sigma\Delta} = 1.30$ in case of DD-MEX set. However, it is noteworthy to mention that another analysis of the GW170817 event data [31] suggest much higher tidal deformability estimations ~ 1000 which would definitely put loose constraints on the poorly known Δ−coupling values.

In another recent experiment to measure $R_{\rm skin}$ in ^{48}Ca-isotope (CREX) [32], the same has been reported to be (0.121 ± 0.026) fm which is in disagreement with PREX results. CREX findings indicate the symmetry energy to be low consequently leading to more compact NSs. This indicates a need to further investigate the limitations of present dense matter models so that both of the findings can be accomodated which is beyond the scope of this work.

References

[1] N.K. Glendenning, *Compact Stars, (Springer-Verlag, New York, 2007), 2nd ed.* (1996)

[2] S. Abrahamyan, Z. Ahmed, H. Albataineh, K. Aniol, D.S. Armstrong et al. (PREX Collaboration), Phys. Rev. Lett. **108**, 112502 (2012)

[3] M. Oertel, M. Hempel, T. Klähn, S. Typel, Reviews of Modern Physics **89**, 015007 (2017), 1610.03361

[4] C. Drischler, R.J. Furnstahl, J.A. Melendez, D.R. Phillips, Phys. Rev. Lett. **125**, 202702 (2020)

[5] D. Adhikari, H. Albataineh, D. Androic et al. (PREX Collaboration), Phys. Rev. Lett. **126**, 172502 (2021)

[6] B.T. Reed, F.J. Fattoyev, C.J. Horowitz, J. Piekarewicz, Phys. Rev. Lett. **126**, 172503 (2021)

[7] Z. Arzoumanian et al. (NANOGrav), ApJ Suppl. **235**, 37 (2018), 1801.01837

[8] J. Antoniadis, P.C.C. Freire, N. Wex, T.M. Tauris, R.S. Lynch, M.H. van Kerkwijk, M. Kramer, C. Bassa, V.S. Dhillon, T. Driebe et al., Science **340**, 448 (2013), 1304.6875

[9] E. Fonseca, H.T. Cromartie, T.T. Pennucci, P.S. Ray, A.Y. Kirichenko, S.M. Ransom, P.B. Demorest, I.H. Stairs, Z. Arzoumanian, L. Guillemot et al., arXiv e-prints arXiv:2104.00880 (2021), 2104.00880

[10] M.C. Miller, F.K. Lamb, A.J. Dittmann, S. Bogdanov, Z. Arzoumanian, K.C. Gendreau, S. Guillot, A.K. Harding, W.C.G. Ho, J.M. Lattimer et al., Astro. Phys. J. Lett.**887**, L24 (2019), 1912.05705

[11] T.E. Riley, A.L. Watts, S. Bogdanov, P.S. Ray, R.M. Ludlam, S. Guillot, Z. Arzoumanian, C.L. Baker, A.V. Bilous, D. Chakrabarty et al., Astro. Phys. J. Lett.**887**, L21 (2019), 1912.05702

[12] P. Landry, R. Essick, K. Chatziioannou, Phys. Rev. D**101**, 123007 (2020), 2003.04880

[13] T.E. Riley, A.L. Watts, P.S. Ray, S. Bogdanov, S. Guillot, S.M. Morsink, A.V. Bilous, Z. Arzoumanian, D. Choudhury, J.S. Deneva et al., arXiv e-prints arXiv:2105.06980 (2021), 2105.06980

[14] M.C. Miller, F.K. Lamb, A.J. Dittmann, S. Bogdanov, Z. Arzoumanian, K.C. Gendreau, S. Guillot, W.C.G. Ho, J.M. Lattimer, M. Loewenstein et al., arXiv e-prints arXiv:2105.06979 (2021), 2105.06979

[15] B.P. Abbott, R. Abbott, T.D. Abbott, F. Acernese, K. Ackley et al. (LIGO Scientific Collaboration and Virgo Collaboration), PhRvL **119**, 161101 (2017)

[16] B.P. Abbott, R. Abbott, T.D. Abbott, F. Acernese, K. Ackley et al. (The LIGO Scientific Collaboration and the Virgo Collaboration), PhRvL **121**, 161101 (2018)

[17] B.P. Abbott, R. Abbott, T.D. Abbott et al. (LIGO Scientific Collaboration and Virgo Collaboration), Phys. Rev. X **9**, 011001 (2019)

[18] B.P. Abbott, R. Abbott, T.D. Abbott, S. Abraham, F. Acernese, K. Ackley, C. Adams, R.X. Adhikari, V.B. Adya, C. Affeldt et al., Astro. Phys. J. Lett.**892**, L3 (2020), 2001.01761

[19] R.W. Romani, D. Kandel, A.V. Filippenko, T.G. Brink, W. Zheng, arXiv e-prints arXiv:2207.05124 (2022), 2207.05124

[20] A. Drago, A. Lavagno, G. Pagliara, D. Pigato, Phys. Rev. C **90**, 065809 (2014)

[21] E. Kolomeitsev, K. Maslov, D. Voskresensky, Nuclear Physics A **961**, 106 (2017)

[22] J.J. Li, A. Sedrakian, M. Alford, Phys. Rev. D**101**, 063022 (2020)

[23] V.B. Thapa, M. Sinha, J.J. Li, A. Sedrakian, Particles **3**, 660 (2020), 2010.00981

[24] V.B. Thapa, A. Kumar, M. Sinha, Mon. Not. Roy. Astron. Soc.**507**, 2991 (2021), 2108.04318

[25] V.B. Thapa, M. Sinha, Phys. Rev. C **105**, 015802 (2022)

[26] A. Gal, E.V. Hungerford, D.J. Millener, Rev. Mod. Phys. **88**, 035004 (2016)

[27] E. Friedman, A. Gal, arXiv e-prints arXiv:2104.00421 (2021), 2104.00421

[28] J. Schaffner, C. Dover, A. Gal, C. Greiner, D. Millener, H. Stocker, Annals of Physics **235**, 35 (1994)

[29] P. Danielewicz, R. Lacey, W.G. Lynch, Science **298**, 1592 (2002), nucl-th/0208016

[30] K.D. Marquez, H. Pais, D.P. Menezes, C. Providência, arXiv e-prints arXiv:2206.02935 (2022), 2206.02935

[31] R. Gamba, M. Breschi, S. Bernuzzi, M. Agathos, A. Nagar, Phys. Rev. D**103**, 124015 (2021), 2009.08467

[32] D. Adhikari, H. Albataineh, D. Androic, K.A. Aniol, D.S. Armstrong, T. Averett et al. (CREX Collaboration), Phys. Rev. Lett. **129**, 042501 (2022)

Determination of the $^3He(\alpha,\gamma)^7Be$ and $^6Li(p,\gamma)^7Be$ astrophysical factors down to zero energy using the asymptotic normalization coefficients.

M. La Cognata[1,*], *G. G.* Kiss[2], *R.* Yarmukhamedov[3], *K. I.* Tursunmakhatov[3,4], *I.* Wiedenhöver[5], *L. T.* Baby[5], *S.* Cherubini[1,6], *A.* Cvetinović[1,7], *G.* D'Agata[1,8], *P.* Figuera[1], *G.L.* Guardo[1], *M.* Gulino[1,9], *S.* Hayakawa[10], *I.* Indelicato[1,6], *L.* Lamia[1,6,11], *M.* Lattuada[1,6], *F.* Mudò[1,6], *S.* Palmerini[12,13], *R.G.* Pizzone[1], *G.G.* Rapisarda[1,6], *S.* Romano[1,6,11], *M. L.* Sergi[1,6], *R.* Spartà[1,6], *C.* Spitaleri[1,6], *O.* Trippella[12,13], *A.* Tumino[1,9], *M.* Anastasiou[5], *S.A.* Kuvin[5], *N.* Rijal[5], *B.* Schmidt[5], *S. B.* Igamov[3], *S. B.* Sakuta[14], *Zs.* Fülöp[2], *Gy.* Gyürky[2], *T.* Szücs[2], *Z.* Halász[2], *E.* Somorjai[2], *Z.* Hons[8], *J.* Mrázek[8], *R. E.* Tribble[15], and *A.M.* Mukhamedzhanov[15]

[1]Laboratori Nazionali del Sud - INFN, Via S. Sofia 62, 95123 Catania, Italy
[2]Institute for Nuclear Research (ATOMKI), H-4001 Debrecen, POB.51, Hungary
[3]Institute of Nuclear Physics, Uzbekistan Academy of Sciences, 100214 Tashkent, Uzbekistan
[4]Physical and Mathematical Department, Gulistan State University, 120100 Gulistan, Uzbekistan
[5]Department of Physics, Florida State University, Tallahassee, Florida 32306, USA
[6]Dipartimento di Fisica e Astronomia "E. Majorana", Università di Catania, 95123 Catania, Italy
[7]Jožef Stefan Institute, Department of Low and Medium Energy Physics (F2), 1000 Ljubljana, Slovenia
[8]Nuclear Physics Institute of the Czech Academy of Sciences, 250 68 Řež, Czech Republic
[9]Facoltà di Ingegneria e Architettura, Università di Enna "Kore", 94100, Enna, Italy
[10]Center for Nuclear Study (CNS), University of Tokyo, RIKEN campus, Saitama 351-0198, Japan
[11]Centro Siciliano di Fisica Nucleare e Struttura della Materia (CSFNSM), Catania 95123, Italy
[12]Dipartimento di Fisica e Geologia, Università di Perugia, 06123 Perugia, Italy
[13]Istituto Nazionale di Fisica Nucleare, Sezione di Perugia, 06123 Perugia, Italy
[14]National Research Center "Kurchatov Institute", Moscow 123182, Russia
[15]Cyclotron Institute, Texas A&M University, College Station, Texas 77843, USA

Abstract. The $p-p$-chain reaction $^3He(\alpha,\gamma)^7Be$ can sensitively influence the prediction of the ^7Be and ^8B neutrino fluxes. Despite its importance, the knowledge of its reaction cross section at energies of the core of the Sun (15 keV - 30 keV) is limited and the accuracy far from the desired ~ 3% level. In the present paper the indirect measurement of the external capture contribution using the asymptotic normalization coefficient (ANC) technique is reported. The angular distributions of deuterons emitted in the $^6Li(^3He,d)^7Be$ α-transfer reactions were measured and the ANCs extracted from the scaling of distorted-wave Born approximation (DWBA) and coupled-channel (CC) calculations. Then, the astrophysical S-factor for the $^3He(\alpha,\gamma)^7Be$ reaction was calculated assuming $E1$ direct capture and the zero energy value turned out to be 0.534 ± 0.025 keVb. Both our experimental and theoretical approaches were benchmarked through the analysis of the $^6Li(p,\gamma)^7Be$ astrophysical factor, with interesting astrophysical applications to the understanding of the primordial lithium problem. In particular, the present work disfavors the occurrence of a claimed 200 keV resonance in the astrophysical factor.

*e-mail: lacognata@lns.infn.it

1 The astrophysical background

The $^3He(\alpha,\gamma)^7Be$ reaction is among the key ones in nuclear astrophysics. It is the first reaction of the 2nd and 3rd $p-p$ chain branch and therefore the uncertainty of its rate strongly influences the precision of the predicted flux of the ^7Be and ^8B neutrinos. While the detection of the neutrinos coming directly from the core of the Sun became more and more precise after the construction of larger and more efficient neutrino detectors, sensitive to a wider neutrino energy range, the $^3He(\alpha,\gamma)^7Be$ reaction remained critical after decades, despite the large number of experimental and theoretical studies devoted to it.

In particular, the flux of the $p-p$ neutrinos was measured with a precision of about 3.4% by the BOREXINO, SNO and Super-Kamiokande collaborations [1–3]. The precise neutrino flux measurements are used to constrain the Standard Solar Model (SSM) and provide information on the core temperature of the Sun; however, the relevant nuclear reaction cross sections are to be known with matching accuracy. At present, the uncertainties of these input parameters are far too high, typically of the order of 5-8% [4] contrary to the 3% precision required [5, 6]. Therefore, an improvement on the knowledge of the low-energy cross section of the $^3He(\alpha,\gamma)^7Be$ reaction would result in a substantial reduction of the uncertainties and might have important consequences for the SSM.

The main reason behind such uncertainties is the fact that the astrophysically relevant energy region lies between about 15 keV and 30 keV for a temperature of 15 MK, characterizing the core of the Sun, and at these temperatures the $^3He(\alpha,\gamma)^7Be$ reaction cross section is far too small to be measured directly. Theory-based extrapolations are therefore necessary to obtain the reaction rate [7–9]. Regarding the experimental methods recently adopted they can be sorted into three groups: the detection of prompt γ rays [10–13], the measurement of the ^7Be activity [14–18], and the counting of the ^7Be recoils with a recoil mass separator [19].

Regarding the theoretical description, several different models - including external capture models (e.g. [20]), potential models (e.g. [21, 22]), modified two-body potential approach [23], resonating group calculation (e.g. [24]), *ab initio* models (e.g. [25, 26]) and R-matrix theory [27, 28] - were used to describe the reaction. While the precision of the extrapolations are of the order of 6-7%, the difference between the zero-energy $^3He(\alpha,\gamma)^7Be$ astrophysical factor $S_{34}(0)$ values exceeds 10%. The predicted $S_{34}(0)$ factors are shown in fig. 1. The figure shows that the calculated $S_{34}(0)$ factors depend strongly on the model used in the extrapolations and high precision experimental data is needed to constrain the theoretical models.

2 The ANC experiment

Here we present the results of a new approach that has made it possible to deduce the $S_{34}(0)$ factor of the $^3He(\alpha,\gamma)^7Be$ reaction with no need of extrapolation, using the asymptotic normalization coefficient (ANC) technique [35]. Indeed, present-day direct measurements cannot reach down to astrophysical energies due to the vanishing cross sections, and have to rely on extrapolations, often based on theoretical models. The large scatter in the calculated $S_{34}(0)$ (see fig.1) suggests the presence of unaccounted systematic errors. Instead, the ANC indirect method makes it possible to directly access $S_{34}(0)$, leading to an improved accuracy provided that sources of systematic errors such as the model dependence and the occurrence of unwanted competitive reaction mechanisms are tested. This is done using a variety of

https://doi.org/10.1051/epjconf/202327911001

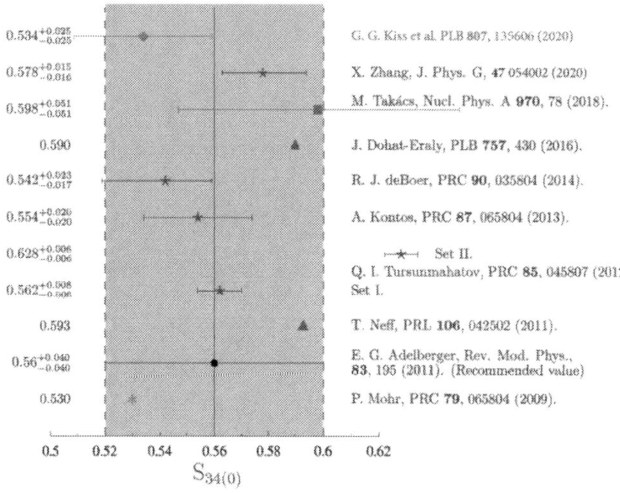

Figure 1. Summary of the most recent ^{3}He$(\alpha, \gamma)^7$Be $S_{34}(0)$ factor results (fig.1 of [32]): derived from the analysis of elastic scattering angular distributions [22] (pink star), theoretical calculations [25, 30] (dark red triangle), extrapolations of experimental data sets [13, 23, 28, 29] (blue star), prediction based on neutrino yield measurement [31] (green box) and derived using the ANC technique [32] (red diamond). The solid central line represents the recommended value of [8], with its uncertainty indicated with the shaded area. For Tursunmahatov et al. [23], the $S_{34}(0)$ value obtained by fitting [10, 12, 14, 15] is shown.

methods, e.g., testing multiple potential models and parameters, and benchmarking the deduced cross sections against known ones (see ref.[36] for a detailed discussion). The ANC approach is especially suitable since the ^{3}He$(\alpha,\gamma)^7$Be reaction at stellar energies is a pure external direct capture process [8], so it essentially proceeds through the tail of the nuclear overlap function, with no sensitivity to nuclear structure details. The shape of the overlap function in the tail region is solely determined by the Coulomb interaction and, in turn, the amplitude of the overlap function determines the rate of the capture reaction [37, 38]. Since the direct capture cross sections are proportional to the squares of the ANCs - which are found from transfer reactions - with the study of the near barrier ^{6}Li(^{3}He,d)^{7}Be α particle transfer reaction the ANCs for the ^{3}He$(\alpha,\gamma)^7$Be reaction can be obtained.

The transfer reaction was measured using the ^{3}He beams provided by the 3.5 MV singletron accelerator of the Department of Physics and Astronomy (DFA) of the University of Catania (Italy) and the FN tandem accelerator at the John D. Fox Superconducting Accelerator Laboratory at the Florida State University (FSU), Tallahassee (FL), USA. More details about the experiment and the theoretical analysis can be found in [32, 33]. To deduce the ANCs, deuteron angular distributions were measured at two energies ($E_{lab.}$ = 3 MeV and $E_{lab.}$ = 5 MeV) over a broad angular range using silicon ΔE-E telescopes mounted on a turntable. Monitor detectors were placed at fixed angles with respect to the beam axis for absolute normalization. ^{6}LiF (enriched in ^{6}Li by 95%) and pure ^{6}Li targets (enriched in ^{6}Li by 98%) were used. Thanks to the ΔE-E particle identification technique [34] and to the high-resolution achieved, clear evidence of d_0 and d_1 groups, corresponding to ^{7}Be ground and first excited states, was obtained. At the backward hemisphere the differential cross section (DCS) increases with increasing angles and this confirms the presence of a dominant

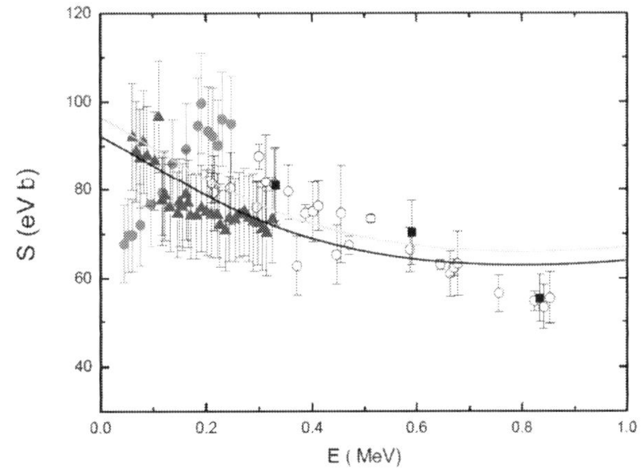

Figure 2. The experimental and calculated astrophysical S-factor for the radiative-capture ^{6}Li(p,γ)^{7}Be reaction (fig.7 of ref.[33]). The solid green line is the direct component of the astrophysical S factor, obtained using the weighted average ANC values from the near-barrier proton transfer ^{6}Li(^{3}He,d)^{7}Be reaction at $E_{^3\mathrm{He}}$=3 and 5 MeV. The black line is the S factor obtained from the ANCs deduced from the analysis of ^{6}Li(p,γ)^{7}Be directly measured reaction [44]. Blue solid triangles represent the bare-nucleus astrophysical factor from ref.[44] (including systematic error), red filled circles are the experimental astrophysical factor published in [45], empty circles are taken from [46] and black solid squares from [47].

one-step α-particle exchange mechanism. Similarly, one-step proton transfer mechanism is found to be dominant in the forwards hemisphere, with negligible interference.

The ANCs for the ^{3}He + $\alpha \rightarrow$ ^{7}Be channel was deduced in the framework of the modified Distorted Wave Born Approximation (DWBA) [39] assuming one step proton and α particle transfer [40]. By normalizing the calculated DCSs to the experimental ones for each experimental point (θ = θ^{exp}) for the backward angle regions, the ANCs for ^{3}He + $\alpha \rightarrow$ ^{7}Be$_{g.s}$ and ^{3}He + $\alpha \rightarrow$ ^{7}Be(0.429 MeV) (namely, ^{7}Be first excited state) channels were deduced. The channels coupling effects (CCE) were derived for each experimental point of θ^{exp} using the FRESCO code [41] by taking into account only one step processes with proton stripping ^{6}Li(^{3}He, d)^{7}Be and exchange mechanism with the α-particle cluster transfer ^{6}Li(^{3}He, ^{7}Be)d.

The weighed mean values of the square of the ANCs for the ^{3}He + $\alpha \rightarrow$ ^{7}Be(g.s.) and ^{3}He + $\alpha \rightarrow$ ^{7}Be(0.429 MeV) are equal to C^2= 20.84 ± 1.12 [0.82; 0.77] fm^{-1} and C^2= 12.86 ± 0.50 [0.35; 0.36] fm^{-1}, respectively, which are in an excellent agreement with those of [23] derived from the analysis of the experimental S-factor data of [10, 12, 14, 15]. The uncertainties given here includes both experimental errors in the $d\sigma^{\mathrm{exp}}/d\Omega$ (first term in square parentheses), as well as the model uncertainties (second term in square parentheses). Then, the direct capture contribution to the ^{3}He(^{4}He,γ)^{7}Be astrophysical factor at energies corresponding to the core temperature of the Sun was derived within the modified two-body potential model (MTBPM) framework [42, 43], and the resulting $S_{34}(0)$ and S_{34}(23 keV) factors were found to be $S_{34}(0)$ = 0.534 ± 0.025 [0.015; 0.019] keVb and S_{34}(23 keV) = 0.525 ± 0.022 [0.016; 0.016] keVb. The comparison with the values in the literature is performed in fig.1, and shows an improved accuracy with respect to the to-date recommended value in ref.[8] but an uncertainty still higher than the target value, calling for more work to further reduce it.

3 Independent validation of the approach: the $^6\mathrm{Li}(p,\gamma)^7\mathrm{Be}$ reaction

Since the one-step proton transfer is dominant in the forward hemisphere, the ANCs for the ^{6}Li + $p \rightarrow$ ^{7}Be channel was deduced as well, using a similar approach as discussed above. The $^6\mathrm{Li}(p,\gamma)^7\mathrm{Be}$ astrophysical factor influences a variety of astrophysical scenarios, including big bang and stellar nucleosynthesis [33]. Moreover, conflicting results of direct measurements have been published [44, 45], reporting contradictory low-energy trends. The weighted mean values of the square of the ANCs for ^{6}Li + $p \rightarrow$ ^{7}Be were found to be 4.81±0.38 fm^{-1} and 4.29±0.27 fm^{-1} for the ground and first excited states of ^{7}Be, respectively. The overall uncertainties correspond to the averaged squared errors, including both experimental errors in the $d\sigma^{\exp}/d\Omega$ and the theoretical uncertainty from the DWBA analysis. Finally, as discussed in ref.[33], the $^6\mathrm{Li}(p,\gamma)^7\mathrm{Be}$ astrophysical S-factor was calculated assuming $E1$ direct capture (DC) (green line in fig.2). At E=0, the indirect $S_{61}^{(\mathrm{DC})}(E)$ equals $65.781 \pm 5.227[3.380; 1.040; 3.859]$ eV·b and $30.675 \pm 1.957[0.464; 0.514; 1.828]$ eV·b for the ground and the first excited states of ^{7}Be, respectively, leading to a total S-factor of 96.5±5.7 eV·b. This value is in excellent agreement with the extrapolated S-factor to zero energy (S(0) = 95 ± 9 eV·b) of [44], with an uncertainty 1.6 times lower. While this result does not support the occurrence of the 200 keV resonance claimed in [45], such agreement is a validation of the method used for deducing the ^{3}He(^{4}He,γ)^{7}Be S-factor.

A further validation of the adopted theoretical approach is given by the re-analysis of the ^{6}Li(p,γ)^{7}Be directly-measured astrophysical factor [44]. The ANCs for the ^{6}Li + $p \rightarrow$ ^{7}Be(g.s.) and ^{6}Li + $p \rightarrow$ ^{7}Be(0.429 MeV) channels were derived from the experimental total astrophysical S-factor and the branching ratios of ref. [44], within the MTBPM [43]. The values of the weighted means for the ANC values for ^{7}Be ground and first excited states obtained from all the experimental data in [44] are equal to $(C^{exp}_{1\,1/2+1\,3/2})^2 = 4.345 \pm 0.576\,[0.033;\, 0.041;\, 0.574]\,\mathrm{fm}^{-1}$ and $(C^{exp}_{1\,1/2+1\,3/2})^2 = 4.571 \pm 0.595\,[0.027;\, 0.033;\, 0.594]\,\mathrm{fm}^{-1}$, respectively. These ANCs are in excellent agreement with the values of 4.81±0.38 fm^{-1} for the ground and 4.29±0.27 fm^{-1} for the first excited state of ^{7}Be extracted from the analysis of the ^{6}Li(^{3}He,d)^{7}Be transfer reaction. Furthermore, the corresponding ^{6}Li(p,γ)^{7}Be was calculated assuming $E1$ DC leading to the black line in fig.2. The excellent agreement with the astrophysical factor reported by ref.[44] is a further validation of the adopted method.

4 Acknowledgements

This work was supported by INFN (Istituto Nazionale di Fisica Nucleare), by NKFIH (NN128072, K120666), and by the ÚNKP-19-4-DE-65 New National Excellence Program of the Ministry of Human Capacities of Hungary, and supported in part by the National Science Foundation, Grant No. PHY-1712953 (USA), and by the University of Catania (Finanziamenti di linea 2 and Starting grant 2020). G. G. Kiss acknowledges the support from the János Bolyai research fellowship of the Hungarian Academy of Sciences. R. Yarmukhamedov and K.I. Tursunmakhatov acknowledge the support from the Academy of Sciences of the Republic of Uzbekistan. J. Mrázek and G. D'Agata acknowledge the support from MEYS Czech Republic under the project EF16_013/0001679. The authors acknowledge the support of prof. M. G. Grimaldi and of the technical staff of the DFA.

References

[1] The Borexino Collaboration, Nature **562** 505 (2018)

[2] Aharmim B *et al.*, Phys. Rev. C **81** 055504 (2010)
[3] Abe K *et al.* , Phys. Rev. D **83** 052010 (2011)
[4] Vinyoles N *et al.*, Astrophys. J. **852** 202 ()2017
[5] Haxton WC and Serenelli AM , Astrophys. J. **687** 678 (2008)
[6] Bahcall JN and Pinsonneault MH , Phys. Rev. Lett. **92** 121301 (2004)
[7] Adelberger EG *et al.* , Rev. Mod. Phys. **70** 1265 (1998)
[8] Adelberger EG *et al.* , Rev. Mod. Phys. **83** 195 (2011)
[9] Cyburt RH and Davids B , Phys. Rev. C **78** 064614 (2008)
[10] Bemmerer D *et al.* , Phys. Rev. Lett. **97** 122502 (2006)
[11] Brown TAD *et al.* , Phys. Rev. C **76** 055801 (2007)
[12] Confortola F *et al.* , Phys. Rev. C **75** 065803 (2007)
[13] Kontos A *et al.* , Phys. Rev. C **87** 065804 (2013)
[14] Nara Singh BS *et al.* , Phys. Rev. Lett. **93** 262503 (2004)
[15] Gyürky Gy. *et al.* , Phys. Rev. C **75** 035805 (2007)
[16] Carmona-Gallardo M *et al.* Phys. Rev. C **86** 032801(R) (2012)
[17] Bordeanu C *et al.* , Nucl. Phys. A **908** 1 (2013)
[18] Szücs T *et al.* , Phys. Rev. C **99** 055804 (2019)
[19] Di Leva A *et al.* , Phys. Rev. Lett. **102** 232502 (2009)
[20] Tombrello TA and Parker PD , Phys. Rev. **131** 2582 (1963)
[21] Mohr P *et al.* , Phys. Rev. C **48** 1420 (1993)
[22] Mohr P , Phys. Rev. C **79** 065804 (2009)
[23] Tursunmahatov QI and Yarmukhamedov R , Phys. Rev. C **85** 045807 (2012)
[24] Kajino T, Toki H and Austin SM , Astrophys. J. **319** 531 (1987)
[25] Neff T 2011, Phys. Rev. Lett. **106** 042502 (2011)
[26] Nollett KM , Phys. Rev. C **63** 054002 (2001)
[27] Descouvemont P *et al.* , At. Data Nucl. Data Tables **88** 203 (2004)
[28] deBoer RJ *et al.*, Phys. Rev. C **90** 035804 (2014)
[29] Zhang X, Nollett K and Philips DR , J. Phys. G: Nucl. Part. Phys. **47** 054002 (2020)
[30] Dohet-Eraly J *et al.* , Phys. Lett. B **757** 430 (2016)
[31] Takács MP *at al.* , Nucl. Phys. A **970** 78 (2018)
[32] Kiss GG *et al.* , Phys. Lett. B **807** 135606 (2020)
[33] Kiss GG *et al.* , Phys. Rev. C **104** 015807 (2021)
[34] Badala A *et al.* , La Rivista del Nuovo Cimento **45** 189 (2022)
[35] Mukhamedzhanov AM *et al.* , Phys. Rev. C **63** 024612 (2001)
[36] Tribble RE *et al*, Rep. Prog. Phys. **77** 106901 (2014)
[37] Xu HM *et al.* , Phys. Rev. Lett. **73** 2027 (1994)
[38] Mukhamedzhanov AM *et al.* , Phys. Rev. C **67** 065804 (2003)
[39] Mukhamedzhanov AM *et al.* , Phys. Rev. C **56** 1302 (1997)
[40] Dolinsky EI *et al.* , Nucl. Phys. **202** 97 (1973)
[41] Thompson IJ , Comput. Phys. Rep. **7** 167 (1988)
[42] Tursunmakhtov KI and Yarmukhamedov R , IJMP:Conf.Series **49** 1960017 (2019)
[43] Igamov SB and Yarmukhamedov R , Nucl. Phys. A **781** 247 (2007)
[44] Piatti D *et al.* , Phys. Rev. C **102** 052802 (2020)
[45] He JJ *et al.* , Phys. Lett. B **725** 287 (2013)
[46] Switkowski ZE *et al.* , Nucl. Phys. A **331** 50 (1979)
[47] Amar A and Burtebayev N , Journal of Nuclear Sciences **1** 15 (2014)

Towards a direct measurement of the $^{17}O(p,\gamma)^{18}F$ 65 keV resonance strength at LUNA

D. Piatti[1,2,*] and *G.F.* Ciani[3,4]for LUNA Collaboration

[1]Università degli Studi di Padova, Via Marzolo 8 35136 Padova (Italy)
[2]INFN Division of Padova, Via Marzolo 8 35136 Padova (Italy)
[3]Università degli Studi di Bari, 70125 Bari, Italy
[4]INFN, Sezione di Bari, 70125 Bari, Italy

Abstract. The $^{17}O(p,\gamma)^{18}F$ reaction plays a crucial role in the hydrogen burning phases of different stellar scenarios. At temperature of interest for AGB nucleosynthesis (20 MK < T < 80 MK) the main contribution to the astrophysical reaction rate comes from the poorly constrained 65 keV resonance. The strength of this resonance is presently determined only through indirect measurements, with a reported value of $\omega\gamma = (1.6 \pm 0.3) \times 10^{-11}$ eV. With typical experimental quantities for beam current, isotopic enrichment and detection efficiency, this strength yields to an expected count rate of less than one count per Coulomb, making the direct measurement of this resonance extremely challenging.
A new high sensitivity setup has been installed at LUNA (Laboratory for Underground Nuclear Astrophysics) of Laboratori Nazionali del Gran Sasso. The high performance LUNA 400kV accelerator underground location guarantees, indeed, a reduction of cosmic ray background by several orders of magnitude. The residual background was further reduced by a devoted shielding of lead and borated (5%) polyethylene. On the other hand, the 4π BGO detector efficiency was optimized installing aluminum target chamber and holder. With about 400 C accumulated on Ta_2O_5 targets, with nominal ^{17}O enrichment of 90%, the LUNA collaboration has performed the first direct measurement of the 65 keV resonance strength.

1 Introduction

The isotopic ratios observed in presolar grains are the fingerprint of their progenitor composition. Attributing their origin to specific type of stars, however, often proves challenging. Asymptotic Giant Branch (AGB) stars are expected to have contributed a large fraction of meteoritic stardust. Yet, models struggle to match observations. This is a long-standing puzzle, which points to serious gaps in our understanding of the lifecycle of stars and dust in our Galaxy. The oxygen isotopic ratios are strongly affected by the $^{17}O + p$ reactions, taking part to the CNO cycle active in the giant star H-burning shell [1]. A recent direct measurement of the 65 keV resonance in the $^{17}O(p,\alpha)^{14}N$ reaction ($Q = 1192$ keV) at LUNA had a paramount impact in determining the origin of some of the presolar grains [1]. On the other hand the 65 keV resonance in the (p,γ) channel is still poorly constrained. The $^{17}O(p,\gamma)^{18}F$ reaction (Q

*e-mail: denise.piatti@pd.infn.it

= 5607 keV) affects, indeed, both the ^{17}O depletion and ^{18}O production (^{18}F decays to ^{18}O with a $T_{\frac{1}{2}}$ = 109.77 m). An accurate measurement of the resonance strength can improve the reaction rate determination and will help to constrain the present AGB models.

The strength of the E_R = 65 keV resonance is presently determined only through indirect measurements. The Γ_γ and Γ_α are provided by the measurement of the ^{14}N(α,γ)^{18}F and ^{14}N(α,α)^{14}N reaction cross section respectively [2, 3]. The Γ_p is derived from the $\omega\gamma$ of the ^{17}O(p,α)^{14}N channel and it contributes the most to the final uncertainty because of the discrepant results reported in literature [4, 5]. The presently adopted resonance strength is $\omega\gamma_{(p,\gamma)}$ = (16 ± 3) peV [6]. Such a low resonance strength translates into an expected rate as low as one reaction per Coulomb, thus a direct measurement requires both a high sensitivity setup and a devoted technique to monitor and subtract potential beam-induced background (BIB). In following sections the setup, the analysis and the preliminary results are presented.

2 Experimental setup

The measurement was performed at LUNA laboratory, located in the deep underground facility of Laboratori Nazionali del Gran Sasso (LNGS) [7]. Thanks to the 1400 m overburden of rock, the muon cosmic ray background is reduced by six orders of magnitude with respect to the overground laboratories [8]. The residual background at $E_\gamma \leq$ 3 MeV is due to natural radioactivity from the laboratory and the rock while at higher energy it is mainly neutron-induced background [9]. In order to increase our sensitivity, the residual background was further reduced by a three layer shielding which was installed all around the detector and the target chamber. The shielding was made of 1 cm thick layer of borated(5%) polyethylene, 15 cm thick lead shielding and 5 cm thick borated (5%) polyethylene envelope, see Fig.1. The detected background was reduced by a factor 4.3(1) in the region 5.2 < E_γ < 6.2 MeV with respect to using an only lead shielding [10]. The LUNA 400kV accelerator [11] provided high stability and high intensity proton beam, E_p = 80 keV and I_p = 200 μA, which was delivered through a Cu pipe to the target. The copper tube was used as cold finger, to prevent carbon build-up, and as secondary electron suppressor with applied - 300 V. The Ta_2O_5 solid targets were produced by anodization of tantalum backings in 90% ^{17}O enriched water doped in ^{18}O at the level of 5% [12]. The target was water cooled to prevent target degradation, which was monitored via periodical scan of the E_R = 143 keV resonance in the ^{18}O(p,γ)^{19}F reaction [13].
In order to minimize the γ-ray absorption, both the scattering chamber and the target holder were made in aluminum, providing an increase in efficiency of more than 20% with respect to the previous stainless-steel and brass setup. The high efficiency (74% at 661 keV) Bismuth-Germanium-Oxide (BGO) detector surrounded the reaction chamber, covering a 4π solid angle. The detector is made of six optically independent crystals, coupled with a listmode DAQ which allows both a single crystal reading and the offline construction of the add-back spectrum, namely by adding coincident events in the individual crystals [14]. A 3D model of the detection setup is shown in Fig.1.

3 Analysis and preliminary results

An accurate Montecarlo simulation of the setup was crucial for the analysis of the acquired data. The simulation was developed using the Geant4 toolkit [15] and it was validated using the devoted spectra acquired with ^{60}Co and ^{137}Cs sources mounted in the same configuration as the Ta_2O_5 target. In addition a spectrum on top of the ^{14}N(p,γ)^{15}O E_R = 270 keV resonance was used to fine tune the simulation, see Fig.2 for a comparison between simulated and

 https://doi.org/10.1051/epjconf/202327911002

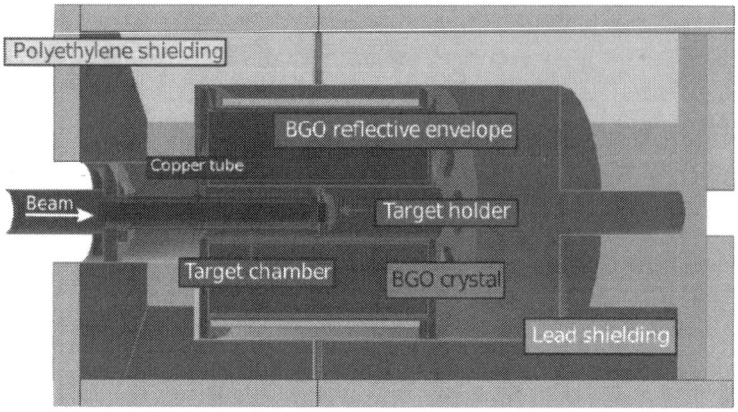

Figure 1. Section of the 3D model of the present experimental setup, which was used in the Geant4 simulation.

experimental spectrum. After proper normalisation, the average residuals between simulated and measured spectra was ≤ 3% in all three cases. This value also represents the uncertainty of detection efficiency from the simulations.

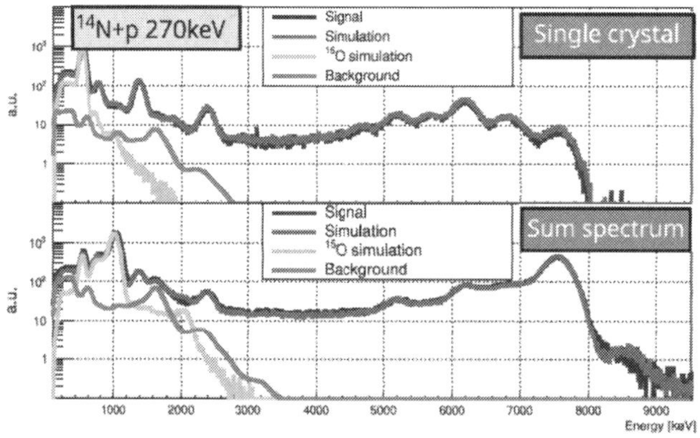

Figure 2. Comparison of simulated (red) and measured (blue) spectra of $^{14}N(p,\gamma)^{15}O$ 270 keV resonance. In green and yellow are represented environmental background and ^{15}O decay spectra respectively, included in the simulation.

During five experimental campaigns, about 400 C were accumulated on $Ta_2^{17}O_5$ targets. In order to monitor the BIB, 300 C were collected on target made as $Ta_2^{17}O_5$ targets but using Ultra Pure Water (UPW), with natural (0.04%) ^{17}O abundance. Tantalum is, indeed, a natural absorber of H and D [16] and the p+D (Q = 5493 keV) reaction produces a single γ-ray very close in energy to the $^{17}O(p,\gamma)^{18}F$ 65 keV resonance γ-ray, see Fig.3. Due to the poor BGO resolution the p+D and ^{17}O+p sum peaks overlap. Moreover the p+D cross section is much higher than the case of interest as reported by recent high precision measurement at LUNA [17, 18]. In order to subtract the p+D BIB, two approaches were applied. The first approach is based on the comparison between on resonance add-back spectra acquired with ^{17}O target and UPW targets. For the analysis a ROI $E_\gamma \in [5.35; 5.95]$ MeV was adopted and the corresponding yield was obtained for both types of spectra, namely acquired with ^{17}O

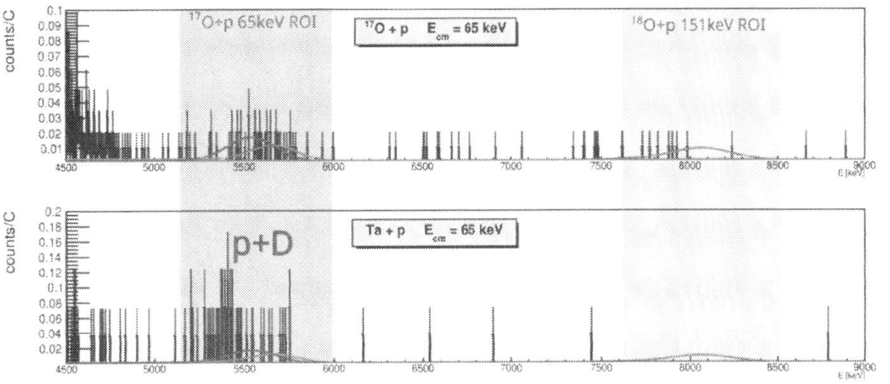

Figure 3. Spectra acquired shooting a E_{lab} = 80 keV proton beam on ^{18}O (top), ^{17}O (top) targets and naked Ta backings (bottom). The $p + d$ peak in the bottom spectra indicates a contamination in the Ta backings estimated at 3 p.p.m. level. The coloured peaks are shown to guide the eye.

targets and UPW targets. A clear excess of counts was observed in the former, Fig.4 and the net counts, calculated subtracting from the ^{17}O counts the background counts observed in the UPW spectra, were far above the critical limit. The resonance strength was derived in a monte carlo based approach [19–21]. The background and the signal, observed in the same ROI but in UPW and $^{17}\mathrm{O} + \mathrm{p}$ spectra respectively, were assumed to follow a Poisson distribution, and 10^6 samples were taken for each. The net counts were calculated at each iteration and the resultant probability density function actually shows a maximum at which the resonance strength was determined.

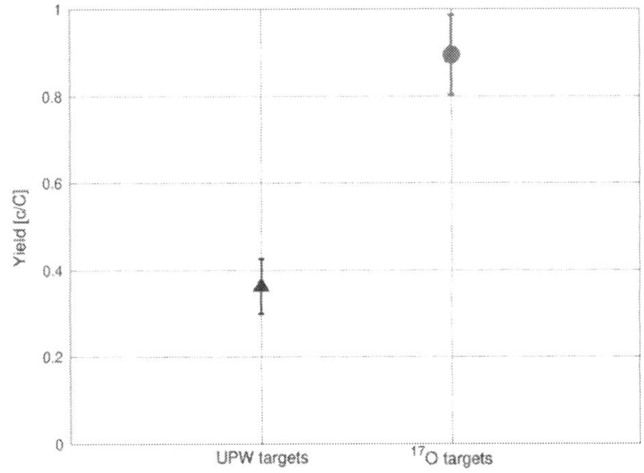

Figure 4. Yields of on resonance spectra acquired with UPW target, black, (negligible ^{17}O) and ^{17}O enriched target, red.

In addition a second technique was developed which combines our knowledge of the E = 5672 keV de-excitation branching ratios, signature of the reaction of interest only, and of the BGO detector segmentation. The technique consists of selecting multiplicity 2 and 3 transition γ-rays, contributing to the sum peak, ROI = 5350-5950 keV in the addback spectrum, and with energies matching the 5672 keV de-excitation chain. Multiplicity 1 events were rejected as due to BIB. This approach allows an almost complete background subtraction while losing only a small amount of γ-rays from the resonance, which correspond to the 5672 keV de-excitation towards the ground state (multiplicity 1 and I_γ = 6%). Residual spurious coincidences due to BIB were subtracted applying the same analysis on UPW target spectra. Both analysis found consistent preliminary $\omega\gamma$, which point to higher value than the indirect estimates.

At the time of writing this proceeding, the last acquired data is being analysed and an in-depth evaluation of the uncertainties is being performed. A technical paper on the detector, setup and analysis technique is ongoing, while the final results will be published in a devoted paper.

References

[1] M. Lugaro, A.I. Karakas, C.G. Bruno, M. Aliotta, L.R. Nittler, D. Bemmerer, A. Best, A. Boeltzig, C. Broggini, A. Caciolli et al., Nature Astronomy **1**, 0027 (2017), 1703.00276

[2] I. Berka, K.P. Jackson, C. Rolfs, A.M. Charlesworth, R.E. Azuma, Nuclear Physics A **288**, 317 (1977)

[3] H.B. Mak, G. Ewan, H. Evans, J. MacArthur, W. McLatchie, R. Azuma, Nuclear Physics A **343**, 79 (1980)

[4] C.G. Bruno, D.A. Scott, M. Aliotta, A. Formicola, A. Best, A. Boeltzig, D. Bemmerer, C. Broggini, A. Caciolli, F. Cavanna et al., Physical Review Letters **117**, 142502 (2016), 1610.00483

[5] M.L. Sergi, C. Spitaleri, M. La Cognata, L. Lamia, R.G. Pizzone, G.G. Rapisarda, X.D. Tang, B. Bucher, M. Couder, P. Davies et al., Physical Review C **91**, 065803 (2015)

[6] M.Q. Buckner, C. Iliadis, K.J. Kelly, L.N. Downen, A.E. Champagne, J.M. Cesaratto, C. Howard, R. Longland, Physical Review C **91**, 015812 (2015)

[7] F. Cavanna, P. Prati, International Journal of Modern Physics A **33**, 1843010-346 (2018)

[8] A. Best, A. Caciolli, Z. Fülöp, G. Gyürky, M. Laubenstein, E. Napolitani, V. Rigato, V. Roca, T. Szücs, European Physical Journal A **52**, 72 (2016)

[9] C. Broggini, D. Bemmerer, A. Caciolli, D. Trezzi, Progress in Particle and Nuclear Physics **98**, 55 (2018)

[10] A. Boeltzig, A. Best, F.R. Pantaleo, G. Imbriani, M. Junker, M. Aliotta, J. Balibrea-Correa, D. Bemmerer, C. Broggini, C.G. Bruno et al., Physics Letters B **795**, 122 (2019)

[11] A. Formicola, G. Imbriani, M. Junker, D. Bemmerer, R. Bonetti, C. Broggini, C. Casella, P. Corvisiero, H. Costantini, G. Gervino et al., Nuclear Instruments and Methods in Physics Research Section A: Accelerators, Spectrometers, Detectors and Associated Equipment **507**, 609 (2003)

[12] A. Caciolli, D.A. Scott, A. Di Leva, A. Formicola, M. Aliotta, M. Anders, A. Bellini, D. Bemmerer, C. Broggini, M. Campeggio et al., European Physical Journal A **48**, 144 (2012), 1210.0327

[13] F.R. Pantaleo, A. Boeltzig, A. Best, R. Perrino, M. Aliotta, J. Balibrea-Correa, F. Barile, D. Bemmerer, C. Broggini, C.G. Bruno et al. (LUNA Collaboration), Physical Review C **104**, 025802 (2021)

[14] A. Boeltzig, A. Best, G. Imbriani, M. Junker, M. Aliotta, D. Bemmerer, C. Broggini, C.G. Bruno, R. Buompane, A. Caciolli et al., Journal Physics G **45**, 025203 (2018)

[15] S. Agostinelli, J. Allison, K. Amako, J. Apostolakis, H. Araujo, P. Arce, M. Asai, D. Axen, S. Banerjee, G. Barrand et al., Nuclear Instruments and Methods in Physics

Research Section A: Accelerators, Spectrometers, Detectors and Associated Equipment **506**, 250 (2003)

[16] T. Asakawa, D. Nagano, H. Miyazawa, I. Clark, Journal of Vacuum Science and Technology B **38**, 034008 (2020)

[17] V. Mossa, K. Stöckel, F. Cavanna, F. Ferraro, M. Aliotta, F. Barile, D. Bemmerer, A. Best, A. Boeltzig, C. Broggini et al., European Physical Journal A **56**, 144 (2020), 2005.00002

[18] V. Mossa, K. Stöckel, F. Cavanna, F. Ferraro, M. Aliotta, F. Barile, D. Bemmerer, A. Best, A. Boeltzig, C. Broggini et al., Nature **587**, 210 (2020)

[19] W.A. Rolke, A.M. López, Nuclear Instruments and Methods in Physics Research A **458**, 745 (2001), hep-ph/0005187

[20] R. Depalo, F. Cavanna, M. Aliotta, M. Anders, D. Bemmerer, A. Best, A. Boeltzig, C. Broggini, C.G. Bruno, A. Caciolli et al., Physical Review C **94**, 055804 (2016), 1610.00925

[21] D. Piatti, E. Masha, M. Aliotta, J. Balibrea-Correa, F. Barile, D. Bemmerer, A. Best, A. Boeltzig, C. Broggini, C.G. Bruno et al., European Physical Journal A **58**, 194 (2022), 2209.03051

EPJ Web of Conferences **279**, 11003 (2023) https://doi.org/10.1051/epjconf/202327911003
NPA-X 2022

Measurement of the ^{86}Kr(α,n)^{89}Sr cross section at energies relevant for the weak r-process

Cameron Angus[1,2*], *Matthew* Williams[2], *Andrei* Andreyev[1], *Soumendu* Bhattacharjee[3], *Samantha* Buck[4], *Soham* Chakraborty[1,2], *Barry* Davids[2,5], *Christian* Diget[1], *Adam* Garnsworthy[2], *Chris* Griffin[2], *Asunción* Fernandez[6], *Greg* Hackman[2], *Kevan* Hudson[2,5], *Dirk* Hufschmidt[6], *Vasil* Karayonchev[2], *Yong* Kim[7], *Alison* Laird[1], *Annika* Lennarz[2], *Konstantin* Mashtakov[2,4], *Peter* Machule[2], *Connor* Natzke[8], *Kihong* Pak[7], *Athanasios* Psaltis[9], *Allison* Radich[4], *Thomas* Rauscher[10,11], *Daniel* Rhodes[2], *Thierry* Sauvage[12], *Anna* Simon[13], *Carl* Svensson[4], *Sriteja* Upadhyayula[2], *Jonathan* Williams[2], *Daniel* Yates[2,14], and *Tammy* Zidar[4]

[1]University of York, Heslington, York, YO10 5DD, United Kingdom
[2]Triumf, 4004 Wesbrook Mall, Vancouver, BC V6T 2A3, Canada
[3]CTU, 166 36 Prague 6, Czechia
[4]University of Guelph, 50 Stone Rd E, Guelph, ON N1G 2W1, Canada
[5]Simon Fraser University, 8888 University Dr, Burnaby, BC V5A 1S6, Canada
[6]Instituto de Ciencia de Materiales de Sevilla, CSIC-Univ. Seville, Avda. Américo Vespucio 49, 41092 Seville, Spain
[7]Hanyang University, 222 Wangsimni-ro, Seongdong-gu, Seoul, South Korea
[8]Colorado School of Mines, 1500 Illinois St, Golden, CO 80401, United States
[9]Technische Universität Darmstadt, Schlossgartenstrasse 2, 64289 Darmstadt, Germany
[10]Department of Physics, University of Basel, 4056 Basel, Switzerland
[11]Centre for Astrophysics Research, University of Hertfordshire, Hatfield AL10 9AB, United Kingdom
[12]CEMHTI, UPR3079 CNRS, 1D avenue de la Recherche Scientifique, 45071 Orléans, France
[13]University of Notre Dame, Notre Dame, IN 46556, United States
[14]University of British Columbia, Vancouver BC V6T 1Z4, Canada

Abstract. The r-process has been shown to be robust in reproducing the abundance distributions of heavy elements, such as europium, seen in ultra-metal poor stars. In contrast, observations of elements $26 < Z < 47$ display overabundances relative to r-process model predictions. A proposed additional source of early nucleosynthesis is the weak r-process in neutrino-driven winds of core-collapse supernovae. It has been shown that in this site (α,n) reactions are both crucial to nucleosynthesis and the main source of uncertainty in model-based abundance predictions. Aiming to improve the certainty of nucleosynthesis predictions, the cross section of the important reaction ^{86}Kr(α,n)^{89}Sr has been measured at an energy relevant to the weak r-process. This experiment was conducted in inverse kinematics at TRIUMF with the EMMA recoil mass spectrometer and the TIGRESS gamma-ray spectrometer. A novel type of solid helium target was used.

* Corresponding author: cja543@york.ac.uk

1 Background

The main r-process is a nucleosynthesis process responsible for producing around half of the elements heavier than iron. While there remains some ambiguity over the site(s) of the r-process, the observation of strontium in the first observed kilonovae (AT2017gfo) [1] indicates that r-process nucleosynthesis can at least occur in neutron star mergers. As a primary process, the r-process does not require the existence of pre-existing heavy nuclei, unlike the s-process which requires seed nuclei to be present in a star's composition in order to function. Therefore, the further back in the history of the universe, the greater the proportion of the heavier than iron elements was made via the r-process. Ultra-Metal Poor (UMP) stars formed so early in the universe that the r-process is expected to have produced most of the elements heavier than iron detected in their atmospheres [2].

This assumption is supported by studies that compare observed elemental abundances in UMP stars to predictions made based on models of r-process nucleosynthesis. Here, observations for elements $Z > 47$ agree well with model predictions; however, observations of elements $26 < Z < 47$ show a significant enrichment relative to model predictions [2]. Due to the good agreement for the heavier elements, it has been suggested that these additional lighter heavy elements, are produced by a separate, and as of yet unknown, mode of nucleosynthesis; one proposed site is the weak r-process in neutrino-driven winds of core-collapse supernovae [3].

Models predict that the potential for nucleosynthesis in the neutrino-driven winds of core-collapse supernovae is limited to nuclei $A < 130$ [4], making it a candidate for the source of the overabundant lighter heavy nuclei in UMP stars. The exact physical conditions in this site are unknown however, the value for electron fraction Y_e (the ratio of the number of protons to the total number of protons and neutrons) has been shown to have a significant impact on the predictions for nucleosynthesis. In the case that the electron fraction is slightly neutron rich (0.40—0.49), simulations have shown that (α,n) reactions are the most important for moving matter to higher atomic numbers [3]. These (α,n) reactions are also the primary source of uncertainty in models of weak r-process nucleosynthesis [3]. This uncertainty arises from significant disagreements in Hauser-Feshbach cross section predictions upon which, models of nucleosynthesis must be based as there are few cross section measurements in literature for (α,n) reactions on the lighter heavy elements within the energy range of interest ($T_9 = 2 - 5$ GK) [3].

2 Experiment

In order to reduce the uncertainty in nucleosynthesis predictions for the weak r-process, a measurement of the reaction $^{86}Kr(\alpha,n)^{89}Sr$ was conducted at the TRIUMF ISAC-II facility using the EMMA recoil mass spectrometer [5] and the TIGRESS gamma-ray spectrometer [6]. This reaction was chosen as it has been identified as affecting final elemental abundances under many astrophysical conditions [7, 8]. The reaction was studied in inverse kinematics at two beam energies, 2.6 AMeV and 3.1 AMeV: the former being a measurement within the range of interest for the weak r-process and the latter allowing for a comparison with a previous study that, while failing to produce a cross section, succeeded in measuring the gamma-ray spectrum for the reaction [9]. The recoiling ^{89}Sr nuclei were detected in the focal plane of EMMA at a position determined by the mass-to-charge ratio of each recoiling nucleus. This ratio allows for the identification of the detected nuclei which can be used to set gates on the gamma-rays detected by TIGRESS: only gamma-rays detected in coincidence with recoiling nuclei that had a measured position consistent with that of $^{89}Sr^{20+}$ were used to plot a gamma-ray spectrum. The gates used in the analysis of part one of the experiment (2.6 AMeV) are shown in Figure 1. In this experiment, the act of gating on the time-of-flight peak cut 90% of the total rate in EMMA. The real-to-random events ratio (i.e., the height of the time-of-flight peak above the random coincidence rate) is further improved by the EMMA mass-gate as the ^{89}Sr nuclei have a smaller spread on the focal plane that the random events.

The targets used in this experiment were of a novel design and consisted of helium contained within a silicon matrix; they were made using magnetron sputtering and have a density comparable to gas targets [10]. This work represents their first use in a nuclear astrophysics experiment.

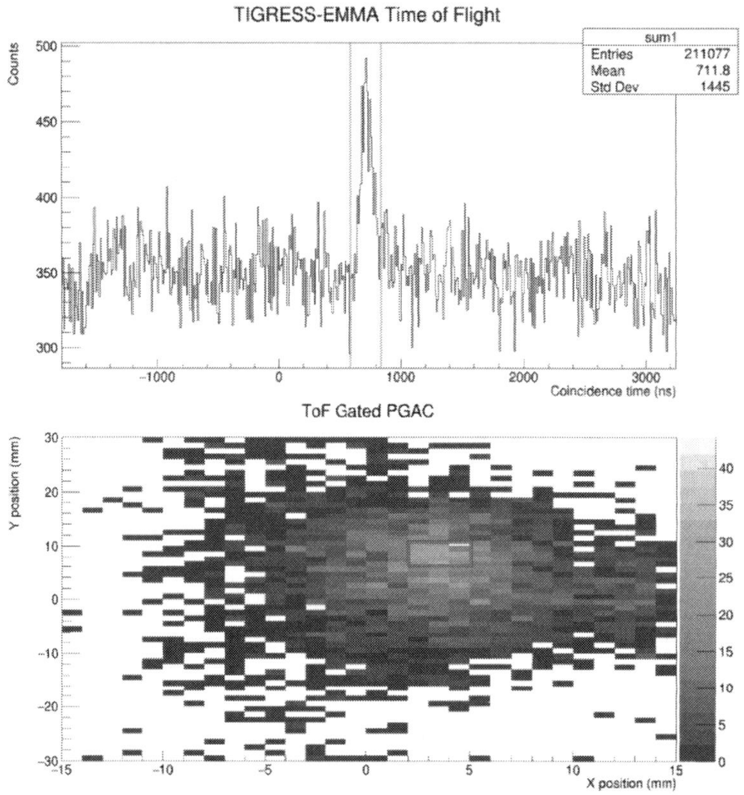

Fig. 1. Above: coincidence time gate between events in EMMA and events in TIGRESS. Below: focal plane gate set on the recoils measured by the EMMA PGAC (position-sensitive parallel grid avalanche counter).

3 Results

Analysis of the data collected in both experiments are underway and a gamma-ray spectrum from the 3.1 AMeV measurement can be seen in Figure 2. The aim is to use the measured counts of individual gamma-rays to deduce the number of times that low lying excited states in ^{89}Sr were populated and from there to infer a cross section based on theoretical predictions for the population of excited states in ^{89}Sr and their decay. Gamma-rays from the de-excitation of the second excited state ($E_x = 1473$keV) were observed at both beam energies. Gamma-rays that occur in random coincidence with events in EMMA can be observed in red on figure 2 and correspond to the background that remains after gating on the EMMA PGAC. These time random counts can be estimated by setting a second gate on the time-of-flight between EMMA and TIGRESS, this time excluding the signal peak (shown in figure 1) then normalising.

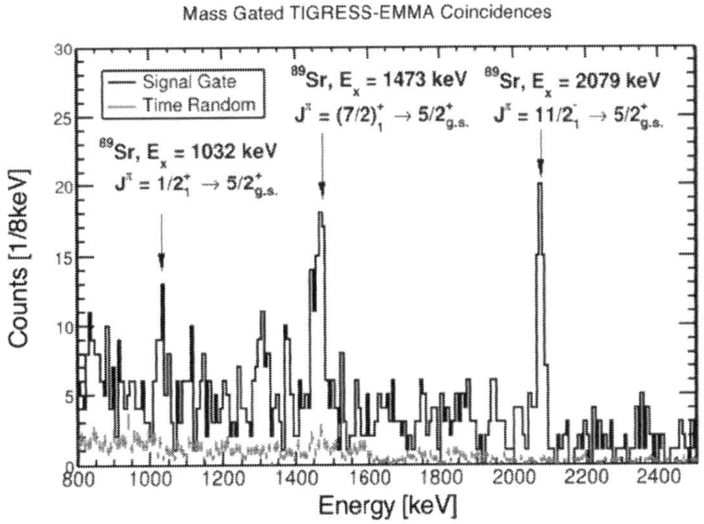

Fig. 2. Gamma spectrum resulting from the ^{86}Kr(α,n)^{89}Sr reaction with a beam of ^{86}Kr at an energy of 3.1 AMeV measured by TIGRESS after recoil mass-gating with EMMA.

4 Summary

This experiment aimed to measure the cross section of the important reaction ^{86}Kr(α,n)^{89}Sr at energies relevant to the weak r-process in the neutrino-driven winds of core-collapse supernovae in order to reduce uncertainty in model predictions for nucleosynthesis in this site. The experiment was conducted at the TRIUMF ISAC-II facility using the EMMA recoil mass spectrometer and the TIGRESS gamma-ray spectrometer along with a novel solid helium target. Analysis is ongoing.

This work was supported by the U.S. Department of Energy (DOE) under contract no. DE-FG02-93ER40789. The development of helium containing targets was supported by the Spanish and Junta de Andalucia grants no. PID2021-124439NB-I00 and P20-00239 (con-financed by EU FEDER) and the CSIC grant 202160E029. We also acknowledge the French EMIR&A network for provision of irradiation beam time and assistance in using the CEMHTI-Pelletron facility for IBA analysis. The authors would like to thank the TRIUMF beam delivery group for their efforts in providing high-quality beams. This work has been supported by the Natural Sciences and Engineering Reach Council of Canada (NSERC), the Canada Foundation for Innovation and the British Columbia Knowledge Development Fund. TRUMF receives federal funding via a contribution agreement through the National Research Council Canada.

References

1. D. Watson, C.J. Hansen, J. Selsing, A. Koch, D.B. Malesani, A.C. Andersen, J.P.U. Fynbo, A. Arcones, A. Bauswein, S. Covino, *et al.*, Nature **574(7779)**, 497-500 (2019).

2. C. Travaglio, R. Gallino, E. Arnone, J. Cowan, F. Jordan, C. Sneden, ApJ **601**, 864-884 (2004).
3. J. Bliss, A. Arcones, F. Montes, J. Pereira, J Phys G Nucl Part Phys, **44**, 054003 (2017).
4. A. Arcones, H.Th. Janka, L. Scheck, A&A **467(3)**, 1227-1248 (2007).
5. B. Davids, M. Williams, N.E. Esker, M. Alcorta, D. Connolly, D.R. Fulton, K. Hudson, N. Khan, O.S. Kirsebom, J.Lighthall, P. Machule, NIM-A **930**, 191-195 (2019).
6. G. Hackman, C.E. Svensson, Hyperfine Interact. **225(1)**, 241-251 (2014).
7. J. Bliss, A. Arcones, F. Montes, J. Pereira, PRC **101**, 055807 (2020).
8. A. Psaltis, A. Arcones, F. Montes, P. Mohr, C.J. Hansen, M. Jacobi, H. Schatz, ApJ **935**, 27 (2022).
9. E. Wallander, A. Nilsson, L.P. Ekström, G.D. Jones, F. Kearns, T.P. Morrison, H.G. Price, P.J. Twin, R. Wadsworth, N.J. Ward, Nucl. Phys. A **361(2)**, 387-398 (1981).
10. V. Godinho, F.J. Ferrer, B. Fernández, J. Caballero-Hernández, J. Gómez-Camacho, A. Fernández, ACS Omega **1**, 1229-1238 (2016).

Results of total cross-section measurements of the ^{87}Rb(p,γ)^{88}Sr reaction

Svenja Wilden[1,*], *Felix* Heim[1], *Martin* Müller[1], and *Andreas* Zilges[1]

[1]Institute for Nuclear Physics, University of Cologne, 50937 Cologne, Germany

Abstract. The existence of a set of stable proton-rich nuclei - the *p* nuclei - cannot be explained via neutron-capture reactions. Therefore, another mechanism has to exist in order to explain their origin, the most probable, especially at high masses, being photodisintegration reactions. This gives rise to the γ process. Since most photodisintegration reactions involved in the process are not experimentally accessible, reliable statistical model calculations are needed to predict cross sections and reaction rates. To improve these calculations nuclear input parameters need to be constrained and a large experimental database is needed. Via comparison of experimental data to theoretical predictions different models can be tested and constrained. In order to study the ^{87}Rb(p,γ)^{88}Sr reaction, for which previously no experimental data have been available, an in-beam experiment at at the University of Cologne's high-efficiency HPGe γ-ray spectrometer HORUS was performed. Proton beams with energies between E_p = 2000 to 5000 keV reaching deep into the Gamow window of the reaction were provided by the 10 MV FN Tandem accelerator. Cross-section values at six proton beam energies were determined. The experimental results are in good agreement with statistical model calculations. The obtained results are the first experimental cross-section values for the ^{87}Rb(p,γ)^{88}Sr reaction and help to constrain the nuclear physics input for statistical model calculations.

1 Introduction

Neutron-capture processes are responsible for the production of most elements heavier than iron. About 30 - 35 stable, proton-rich nuclei - the so called *p* nuclei [1] - are bypassed by these processes. These nuclei are predominantly produced in the γ process, which consists of different combinations of photodisintegration reactions, e.g. (γ,p) or (γ,α), on mainly unstable neutron capture seed nuclei.

For statistical model calculations, nuclear physics input parameters - e.g. particle and nucleus optical model potentials, nuclear level densities, and γ-ray strength functions - are needed and measurements in the Gamow window are of utmost importance.

Systematic studies of the N = 50 isotonic chain are of interest since the *p* nuclide ^{92}Mo shields the extinct radionuclide ^{92}Nb from being produced by the *rp*- and νp-processes [2]. ^{92}Nb can be produced by the γ process [2]. By studying the inverse radiative-capture reactions, that is (p,γ) instead of (γ,p), thermal population of excited states are accounted for.

The ^{87}Rb(p,γ)^{88}Sr reaction has not been measured before and its results hopefully helps to understand the production of ^{92}Mo and ^{92}Nb and improve statistical model calculations.

*swilden@ikp.uni-koeln.de

2 Experimental Details and Method

The $^{87}Rb(p,\gamma)^{88}Sr$ reaction has been measured in the Astrochamber installed within the HORUS γ-ray spectrometer at the 10 MV FN Tandem accelerator at the Institute for Nuclear Physics at the University of Cologne.

The in-beam technique allows to measure transitions to the ground state to determine the total cross section by detecting the γ rays stemming from a highly excited compound nucleus with an excitation energy of $E_x = E_{c.m.} + Q$ due to the impinging protons, where Q denotes the Q value and $E_{c.m.}$ denotes the center-of-mass energy. $\gamma\gamma$-coincidence measurements are sometimes necessary to find possible contaminants in the γ-ray spectra and to identify the transitions to the ground state of the reaction of interest. The prompt γ-ray decays into the ground state which were observed in the experiment are illustrated in Fig. 1 and two γ-spectra can be seen in Fig. 2.

The HORUS γ-ray spectrometer consists of up to 14 HPGe detectors where six of them are equipped with bismuth germanate (BGO) shields for Compton background suppression. These detectors cover five different angles with respect to the beam axis which allows to measure angular correlations and to perform $\gamma\gamma$-coincidence measurements. Details on the setup can be found in Ref. [4].

Proton beam energies of $E_p = 2000$ to 5000 keV were provided by the 10 MV Tandem accelerator. Irradiation times between 8 h to 20 h and beam currents of 130 to 630 nA were used.

Since ^{87}Rb has a melting point of $T_m \approx 40\,°C$, a pure ^{87}Rb target would melt during the irradiation. Therefore, Rb_2CO_3 (rubidiumcarbonate) with a melting point of $T_m \approx 840\,°C$ was used with an isotopic enrichment in ^{87}Rb of 99.2(1) %. The Rb_2CO_3 target was produced in-house by evaporating Rb_2CO_3 with a thickness of $0.65(15)\frac{\text{mg}}{\text{cm}^2}$ onto a gold backing . The thickness was determined via Rutherford backscattering spectrometry (RBS) at the RUBION facility in Bochum, Germany, after the in-beam experiment took place. This showed that the target layer and the backing layer are not clearly distinguishable after irradiation. Therefore, the determination of target thickness is ongoing and the uncertainty is not finalized. The energy loss in the target is obtained from SRIM [3] simulations and is about 25 to 46 keV depending on the proton beam's energy.

The total cross section is given by

$$\sigma_{tot} = \frac{N_R}{N_p \cdot N_t}, \tag{1}$$

where N_p and N_t denote the number of projectiles and target nuclei per area, respectively, and N_R denotes the number of proton-capture reactions. The emitted γ rays follow an angular distribution $W(\Theta)$ with respect to the beam axis. The number of γ-rays $N_\gamma(E_\gamma)$ is corrected for the dead time of the data acquisition τ as well as for the full-energy peak efficiency $\epsilon\,(E_\gamma)$ leading to the following expression:

$$W(\Theta)_\gamma = \frac{N_\gamma(E_\gamma)}{\epsilon\,(E_\gamma) \cdot \tau}. \tag{2}$$

The fitting of a sum of Legendre polynomials to the experimental data for each γ-ray transition at each beam energy leads to the angular distribution

$$W(\Theta)_\gamma = A_{0,\gamma} \cdot \left(1 + \sum_{k=2,4} \alpha_k P_k(\cos\Theta)\right). \tag{3}$$

The sum of all A_0 coefficients represent the total number of proton-capture reactions N_R.

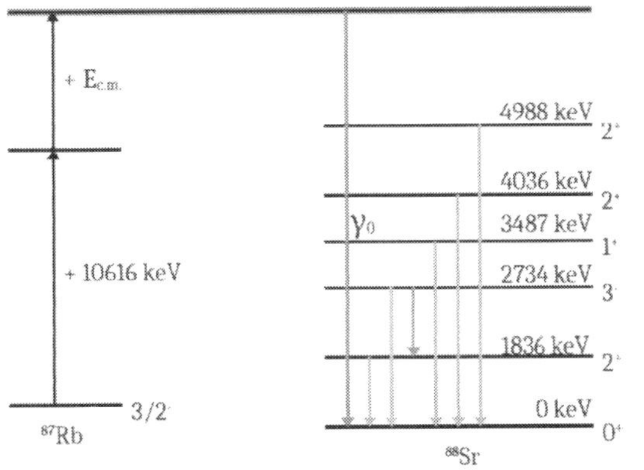

Figure 1. Schematic level scheme of the ^{87}Rb(p,γ) reaction [5]. The observed γ-ray transitions to the ground state are shown in orange. The transition in light blue was not investigated directly but indirectly via the branching ratio [5] of the transition in darker blue.

3 Results and Analysis

3.1 γ-ray Spectra

In Fig. 2 the comparison of typical γ-ray spectra at proton beam energies of E_p = 2000 keV and 5000 keV are shown. A higher beam-induced background and more transitions of the (p,n)-product ^{87}Sr are visible for a proton beam energy of E_p = 5000 keV. For E_p = 2000 keV small contributions of contaminants in the target material and the target chamber are visible due to the low beam-induced background. At the higher measured beam energies the γ-ray spectra become increasingly complicated to analyse and a precise reconstruction of peak origins is necessary.

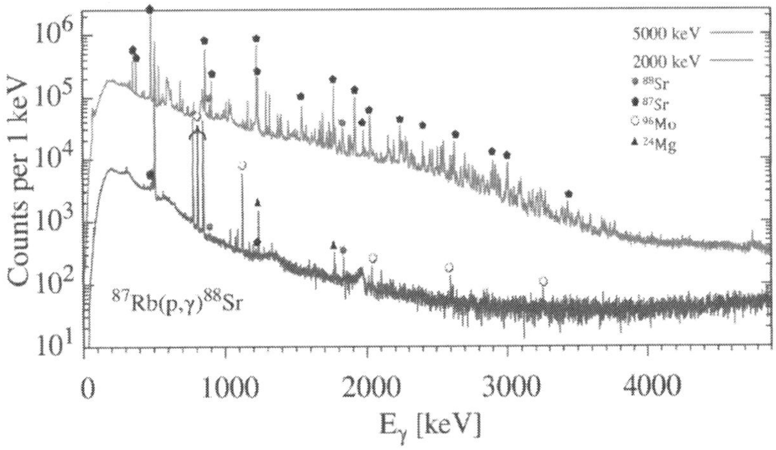

Figure 2. Comparison of γ-ray spectra obtained at beam energies of E_p = 2000 keV and 5000 keV. The different symbols represent the different origins of the peaks.

3.2 Total cross-section results

The obtained values of the total cross sections are listed in Table 1. The energies are given as center-of-mass energies corrected for the energy loss in the target. Note, that the uncertainty for the energy loss in the target obtained from SRIM [3] is much smaller than the accuracy of the accelerator. Therefore, the absolute values of this uncertainty do not differ. The uncertainties for the cross section values stem from the uncertainties of the number of projectiles ($\approx 5\,\%$), the target thickness ($\approx 25\,\%$), full-energy peak efficiency ($\approx 4\,\%$), and the statistical error after fitting the Legendre polynomials ($\approx 10\,\%$).

Table 1. Total cross-section values σ_{tot} for the ^{87}Rb(p,γ)^{88}Sr reaction depending on the center-of-mass energy corrected for the energy loss in target.

$E_{c.m.}$ [MeV]	σ [μb]
1.975(25)	0.34(10)
2.779(25)	4.2(21)
3.482(25)	17.7(83)
3.983(25)	79(28)
4.485(25)	141(56)
4.985(25)	555(192)

Figure 3 shows preliminary cross-section values obtained in this work in comparison to theoretical statistical model calculations using the TALYS 1.95 code [6]. The microscopic level density (Skyrme force) from Hilaire's combinatorial tables [7] (micr. LD5, cf. Fig. 3)

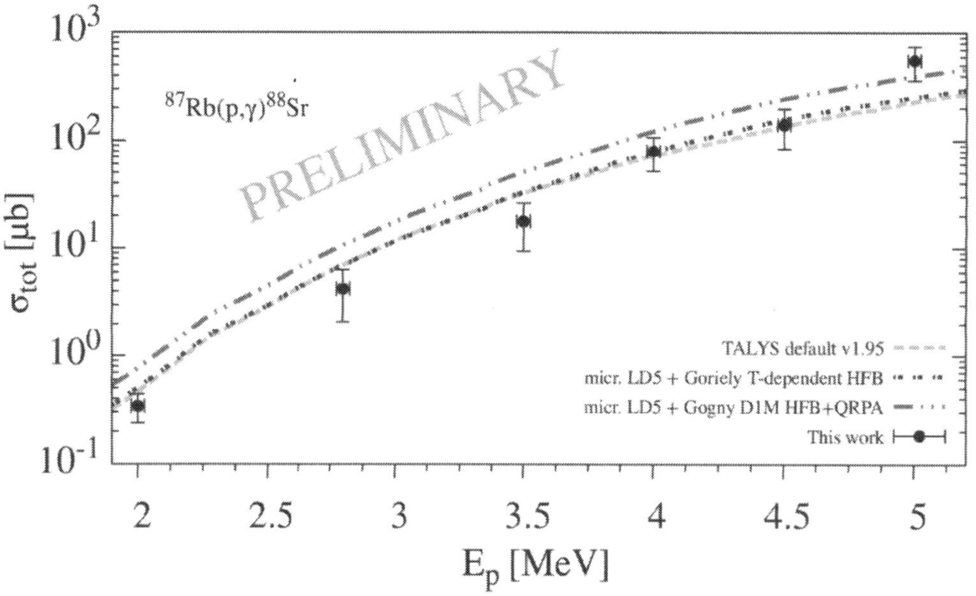

Figure 3. Experimental cross sections for the ^{87}Rb(p,γ)^{88}Sr reaction obtained in this work (black points). Statistical model calculations using the TALYS 1.95 code [6] are shown. The microscopic level density from Hilaire's combinatorial tables [7] has been used in combination with the γ-ray strength function models Goriely T-dependent HFB [10] and Gogny D1M HFB+QPRA [11].

proved to provide the best prediction of the measured cross sections [8, 9]. The experimental data show good agreement with theoretical calculations.

4 Summary and Outlook

The experimental total cross section of the ^{87}Rb(p,γ)^{88}Sr reaction at six proton beam energies between 2.0 MeV and 5.0 MeV have been studied for the first time and were compared to statistical model calculations. The total cross sections can be well-described by these calculations.

The exact determination of the target thickness is still ongoing. A more detailed analysis of the target thickness will probably affect the overall uncertainties. Further analysis of partial cross sections to obtain information on the γ-ray strength function by comparing them to statistical model calculations is still pending. It is also planned to investigate the ^{87}Rb(p,n)^{87}Sr reaction where systematic discrepancies between experimental data and statistical model calculations have been observed [12].

This work has been supported by the DFG (grant ZI510/9-2). SW is supported by the Hans-Böckler-Stiftung.

References

[1] M. Arnould and S. Goriely, Phys. Rep. **384**, 1 (2003).
[2] T. Rauscher *et al.*, Rep. Prog. Phys. **76**, 066201 (2013).
[3] J. F. Ziegler, J. P. Biersack, and M. D. Ziegler, *SRIM - The Stopping and Range of Ions in Matter* (Lulu Press Co., 2010)
[4] F. Heim *et al.*, Nucl. Instrum. Methods A **966**, 163854 (2020).
[5] Nucl. Data Sheets **115** (2014) 135.
[6] A. J. Koning *et al.*, Nucl. Data Sheets **155**, 1 (2019).
[7] S. Goriely *et al.*, Phys. Rev. Lett., **102**, 242501 (2009).
[8] F. Heim *et al.*, Phys. Rev. C, **101**, 035805 (2020).
[9] F. Heim *et al.*, Phys. Rev. C, **103**, 025805 (2021).
[10] S. Hilaire, M. Girod, S. Goriely, and A. J. Koning, Phys. Rev. C, **86**, 064317 (2021)
[11] S. Goriely *et al.*, Phys. Rev. C, **98**, 014327 (2018).
[12] E. Vagena *et al.*, Phys. Rev. C, **103**, 045806 (2021).

^{12}C+^{12}C reactions for Nuclear Astrophysics

Lizeth Morales-Gallegos[1,2,3,*], *Marialuisa* Aliotta[2,3], *Andreas* Best[1,4], *Carlo G.* Bruno[2], *Rafaelle* Buompane[3,1], *Thomas* Davinson[2], *Mario* De Cesare[1,5], *Antonino* Di Leva[1,4], *Antonio* D'Onofrio[3,1], *Jeremias* Duarte[3,1,6], *Leandro* Gasques[3,1,7], *Lucio* Gialanella[3,1], *Gianluca* Imbriani[1,4], *Giuseppe* Porzio[3], *David* Rapagnani[3,1], *Mauro* Romoli[1], and *Filippo* Terrasi[3,1]

[1]INFN, Sezione di Napoli, Napoli, Italy
[2]SUPA, School of Physics and Astronomy, University of Edinburgh, Edinburgh, UK
[3]Dipartimento di Matematica e Fisica, Università degli Studi della Campania "Luigi Vanvitelli", Caserta, Italy
[4]Dipartimento di Fisica "E. Pancini", Università degli Studi di Napoli "Federico II", Napoli, Italy
[5]Centro Italiano di Ricerche Aerospaziali, Capua, Italy
[6]Present Address: Nuclear and Particle Physics Group, Lawrence Livermore National Laboratory, Livermore CA 94550, USA
[7]Departamento de Física Nuclear, Instituto de Física da Universidade de São Paulo, São Paulo, Brazil

Abstract. ^{12}C fusion reactions are among the most important in stellar evolution since they determine the destiny of massive stars. Over the past fifty years, massive efforts have been done to measure these reactions at low energies. However, existing data present several discrepancies between sets and large uncertainties specially at the lowest energies. Factors such as beam/environmental backgrounds, extremely low cross sections and insufficient knowledge of the reaction mechanism contribute to these problems. Recently, the ERNA collaboration measured the ^{12}C+^{12}C reactions at $E_{c.m.}$ = 2.51 - 4.36 MeV with energy steps between 10 and 25 keV in the centre of mass. Representing the smallest energy steps to date. In these measurements, beam induced background was minimised and S-factors for the proton and alpha channels were calculated. Results indicate that a possible explanation for the discrepancies between data sets is the wrongly assumed constant branching ratios and isotropical angular distributions. Given the excellent performance of the detectors for low energy measurements, a collaboration with the LUNA group (LNGS) has started. Background measurements underground are being performed and results indicate it could be possible to measure the ^{12}C+^{12}C reactions directly into the Gamow Window.

1 Introduction

Carbon fusion reactions are among the most important in nuclear astrophysics because of their far-reaching impact on stellar evolution and nucleosynthesis. In particular, the ^{12}C+^{12}C reactions determine the mass threshold for carbon burning to occur, are key for supernova explosions and essential to model X-ray bursts and explosions on the surface of neutron stars.

*e-mail: moralesgallegos@na.infn.it

Several attempts have been made over the past five decades to determine the $^{12}C+^{12}C$ reactions cross-sections [1–15]. However, data still carry large uncertainties and show significant discrepancies between different data sets. Furthermore, no direct measurement has been possible at energies below $E_{c.m.}$ = 2.14 MeV and indirect measurements [16] incited an intense debate [17, 18]. For these reasons, further direct experimental investigations are required.

2 Experimental setup

Measurements of the $^{12}C+^{12}C$ reactions were performed at the 3 MV Pelletron Tandem Accelerator of the CIRCE Laboratory, Department of Mathematics and Physics of the University of Campania "Luigi Vanvitelli" in Caserta, Italy. Thick (1 mm) HOPG (Highly Ordered Pyrolitic Graphite) targets were mounted on a water-cooled target ladder surrounded by a sphere kept at 300 V for electron suppression, allowing for beam-current reading directly on target. The detection system consisted of four telescope detectors called GASTLY (GAs Silicon Two-Layer sYstem), each comprising an ionisation chamber (IC, ΔE stage) and a large area (25 cm^2) silicon strip detector (SSD, Erest stage). Further details on the full detector array and its commissioning are reported in [19]. For the present study, the silicon detector was used as a single pad. Three detectors were mounted on a vertical plane at 121° (D121) and 156° (above and below the beam axis; D156), and one on the horizontal plane at an angle 143° (D143) to the beam axis, as shown in figure 1. See [20] for a full experimental setup description.

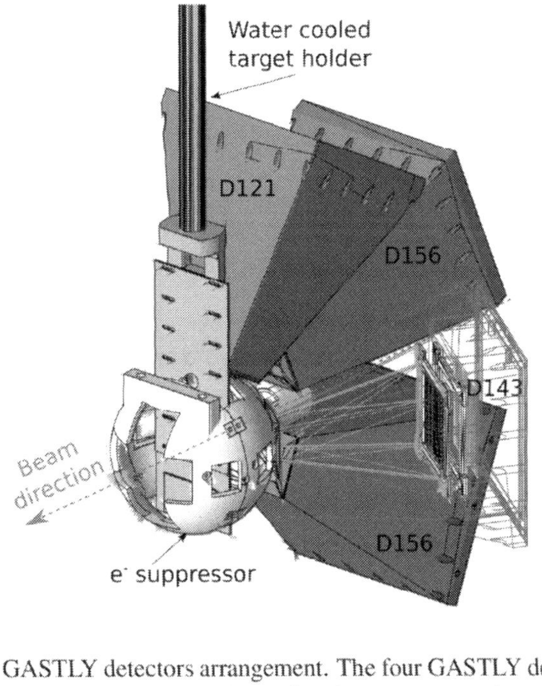

Figure 1. Sketch of the GASTLY detectors arrangement. The four GASTLY detectors are shown along with the target holder and the sphere surrounding it for electron suppression purposes. Detector D143 is shown in wire-frame to reveal its internal components. Figure taken from [20] (published under a Creative Commons Attribution 4.0 International License).

Data were taken at energy intervals of 20-50 keV in the laboratory system. Target temperature was constantly monitored with a thermocamera and maintained to at least 400°C (using intense beams) to reduce deuterium contamination on target by up to 90% its original value, as found in our previous study [21]. With these recommendations and the four GASTLY detectors, the $^{12}C+^{12}C$ reactions were measured using 35-70 mbar of CF_4 in the ionization chambers. Figure 2 shows a typical calibrated ΔE- Erest matrix for detector D121 at a pressure of 35 mbar, obtained with a $^{12}C^{+3}$ beam at E_{lab} = 8.72 MeV on the HOPG target. The two loci correspond to protons and α-particles from the $^{12}C(^{12}C,p)^{23}Na$ and $^{12}C(^{12}C,\alpha)^{20}Ne$ reactions.

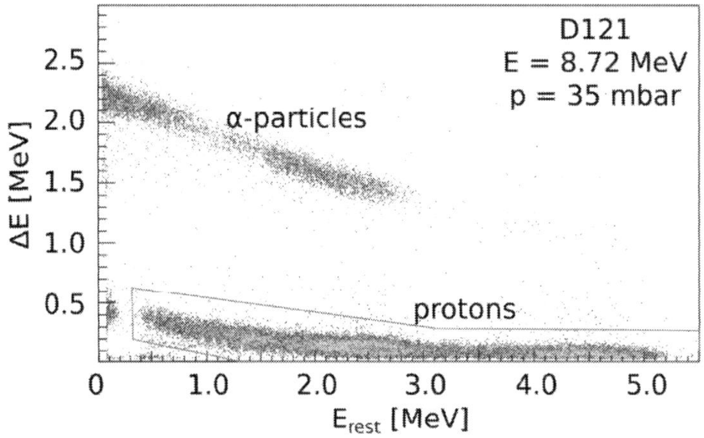

Figure 2. Typical calibrated ΔE - Erest matrix showing the α particles and protons loci. This matrix was obtained using D121 with 35 mbar of CF_4 in the IC and with a E_{lab} = 8.72 MeV $^{12}C^{+3}$ beam impinging on the HOPG target. Figure taken from [20] (published under a Creative Commons Attribution 4.0 International License).

Background runs of several days were taken in the same experimental conditions as the $^{12}C+^{12}C$ measurements and subtracted (after time normalisation) from the corresponding proton and α-particle spectra at each beam energy.

Proton and α-particle peaks from the $^{12}C+^{12}C$ reactions were identified through kinematic reconstruction and comparing with simulations. As many particle peaks overlap, the number of events within each was extracted using the maximum likelihood method from a combined fit of skewed Gaussian functions. Given that all analysed protons at the energies studied here arrive to the SSD, only its spectra were used in the proton analysis. Some deuterium-induced peaks were still visible (despite its minimization) in the proton spectra. In most cases, it was possible to disentangle this beam-induced contribution from the peaks of interest. Otherwise, the affected proton peaks were discarded from further analysis. Unlike for the proton channel, data analysis for the α-particles channel was performed on reconstructed total energy spectra, (E_{tot} = IC+SSD). Thick-target yields were calculated from the net number of events at each beam energy, then differentiated at two consecutive beam energies to finally extract the differential cross sections. Each cross section was later associated to an effective energy expressed in the centre of mass system and finally converted into S -factors. See [20] for a complete description of the data analysis.

3 Results

Differential S -factors for individual proton groups were obtained for each detector as shown in figure 3 for D156. Upper-limits are shown in the form of open symbols. Where data points are missing, this was due to either: (a) difficulties in the fitting procedure due to low statistics and/or poor kinematic discrimination between proton groups (red shaded stripped area); (b) overlap with the deuterium contaminant peak at different beam energies for different proton groups (green shaded area); or (c) low energy protons (high proton-group number) being stopped in the entrance window of the detector.

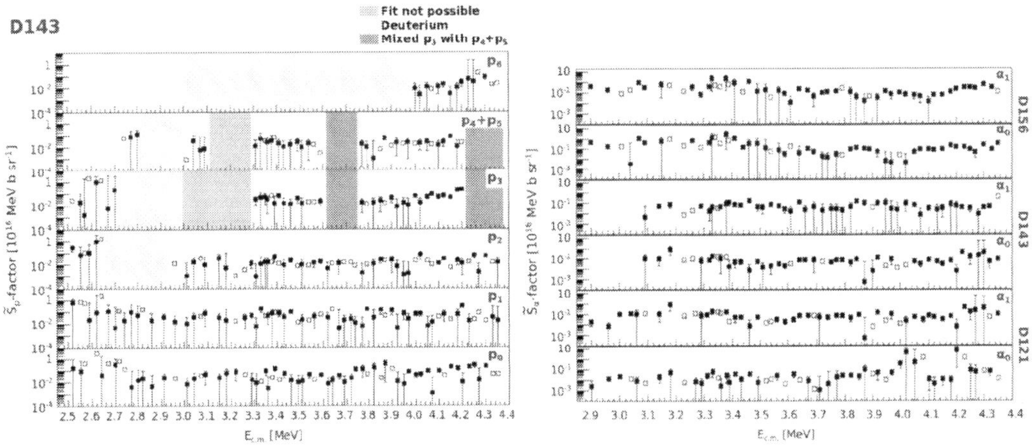

Figure 3. Differential S -factors for individual proton groups obtained with detector D143 (left) and for individual α-particles obtained at all three angles (right). Figures taken from [20] and its supplemental material (published under a Creative Commons Attribution 4.0 International License).

Similar analysis procedures were adopted for the α channel. In this case, however, the $^{12}C+^{1,2}H$ reactions do not produce α-particles within the region of interest, thus the extraction of cross sections for the $^{12}C(^{12}C,\alpha)^{20}Ne$ reaction was more straightforward. Differential S -factors are shown for individual detectors and particle group in figure 3.

Our results reveal the presence of resonance-like structures across the entire energy region explored in this work, as also reported in previous studies [7, 10, 13, 14, 22]. On the other hand, non-constant branching ratios and anisotropic angular distributions were observed for all particle groups at most energies, thus preventing the calculation of the total angular-integrated S -factors. See [20] for a complete description and figures.

4 Outlook

The $^{12}C+^{12}C$ reactions should continue to be investigated. For these reason, a collaboration with the LNGS (National Laboratories of Gran Sasso) of INFN has been established and background measurements have been performed underground using a GASTLY detector (the benefits of moving underground for in-beam measurements of nuclear astrophysical reactions involving charged particles were assessed in [23]). Preliminary results in the conditions of a real experiment indicate a very low intrinsic background affecting only the proton channel. These results already suggest the possibility to measure the $^{12}C+^{12}C$ reactions underground directly into the Gamow window. Nevertheless, in order to optimise signal to noise ratio, investigations of new materials are ongoing.

Furthermore, an in order to asses the astrophysical impact of new and existing data, an accurate measurement of the $^{12}C+^{12}C$ reactions angular distributions should be performed. Such measurements have already started at the CIRCE laboratory using the GASTLY detectors and are foreseen down to $E_{c.m.}$=2.0 MeV with fine steps

References

[1] I. Baraffe et al., *Astrophys. J.*, **615**, 378–382 (2004)
[2] D.A. Bromley et al., *Phys. Rev. Lett.*, **4**, 365 (1960)
[3] L.J. Satkowiak et al., *Phys. Rev. C*, **26**, 2027–2034 (1982)
[4] C.L. Jiang et al., *Phys. Rev. Lett.*, **110**, 072701 (2013)
[5] M.E. Bennett et al., *Proceedings of Science*, **NIC XI**, p. 182 (2010)
[6] M.G. Mazarakis, L. Stephens, *Phys. Rev. C*, **7**, 1280–1287 (1973)
[7] H.W. Becker et al., *Zeitschrift für Physik A. Atoms Nucl.*, **303**, 305–312 (1981)
[8] J. Zickefoose et al., *Phys. Rev. C*, **97**, 065806 (2018)
[9] M.D. High, B. C^ ujec, *Nucl. Phys. A*, **282**, 181 (1977)
[10] K.U. Kettner et al., *Zeitschrift für Physik A. Atoms Nucl.*, **298**, 65–75 (1980)
[11] P. Rosales, E.F. Aguilera, *Revista Mexicana de Física S4*, **49**, 88–91 (2003)
[12] L. Barrón-Palos et al., *Nuc. Phys. A*, **779**, 318-332 (2006)
[13] E. Aguilera et al., *Phys. Rev. C*, **73**, 064601 (2006)
[14] T. Spillane et al., *Phys. Rev. Lett.*, **98**, 122501 (2007)
[15] C.L. Jiang et al., *Nucl. Instrum. Methods Phys. Res. Sect. A*, **682**, 12–15 (2012)
[16] A. Tumino et al., *Nature*, **557**, 687–690 (2018)
[17] A.M. Mukhamedzhanov et al. **arXiv:1806.05921v1** (2018)
[18] A. Tumino et al. **arXiv:1807.06148** (2018)
[19] M. Romoli, L. Morales-Gallegos et al., *Eur. Phys. J. A*, **54**, 142 (2018)
[20] L. Morales-Gallegos,et al., *Eur. Phys. J. A*, **58**, 65 (2022)
[21] L. Morales-Gallegos et al., *Eur. Phys. J. A*, **54**, 132 (2018)
[22] J. Zickefoose, *12C+12C fusion: measurement and advances toward the gamow energy*, University of Connecticut, **Ph.D. Thesis** (2010)
[23] C.G. Bruno et al., *Eur. Phys. J. A*, **51**, 94 (2015)

Indirect measurements of neutron-induced reaction cross sections at storage rings

M. Sguazzin[1], B. Jurado[1*], J. Pibernat[1], J. A. Swartz[1], M. Grieser[2], J. Glorius[3], Y. A. Litvinov[3], R. Reifarth[4], K. Blaum[2], P. Alfaurt[1], P. Ascher[1], L. Audouin[5], C. Berthelot[1], B. Blank[1], B. Bruckner[4], S. Dellmann[4], I. Dillmann[6], C. Domingo-Pardo[7], M. Dupuis[8,9], P. Erbacher[4], M. Flayol[1], O. Forstner[3], D. Freire-Fernández[2,10], M. Gerbaux[1], J. Giovinazzo[1], S. Grévy[1], C. J. Griffin[6], A. Gumberidze[3], S. Heil[4], A. Heinz[11], D. Kurtulgil[4], G. Leckenby[6], S. Litvinov[3], B. Lorentz[3], V. Méot[8,9], J. Michaud[1], S. Perard[1], N. Petridis[3], U. Popp[3], D. Ramos[12], M. Roche[1], M.S. Sanjari[3], R.S. Sidhu[13], U. Spillmann[3], M. Steck[3], Th. Stöhlker[3], B. Thomas[1], L. Thulliez[14] and M. Versteegen[1]

[1]LP2I Bordeaux, CNRS/IN2P3-Université de Bordeaux, 33170 Gradignan, France
[2]Max-Planck Institut für Kernphysik, 69117 Heidelberg, Germany
[3]GSI Helmholtzzentrum für Schwerionenforschung, 64291 Darmstadt, Germany
[4] Goethe University of Frankfurt, 60438 Frankfurt, Germany
[5] Université Paris-Saclay, CNRS, IJCLab, 91405 Orsay, France
[6]TRIUMF, Vancouver, British Columbia, V6T 2A3, Canada
[7]IFIC, CSIC-Universidad de Valencia, 46980 Valencia, Spain
[8]CEA, DAM, DIF, 91297 Arpajon, France
[9]Université Paris-Saclay, CEA, LMCE, 91680 Bruyères-Le-Châtel, France
[10]Ruprecht-Karls-Universität Heidelberg, 69117 Heidelberg, Germany
[11]Chalmers University of Technology, 41296 Gothenburg, Sweden
[12]GANIL, 14000 Caen, France
[13]School of Physics and Astronomy, University of Edinburgh, UK
[14] CEA-Paris Saclay, 91191 Gif-sur-Yvette, France

Abstract. Neutron-induced reaction cross sections of unstable nuclei are essential for understanding the synthesis of heavy elements in stars. However, their measurement is very difficult due to the radioactivity of the targets involved. We propose to circumvent this problem by using for the first time the surrogate reaction method in inverse kinematics at heavy-ion storage rings. In this contribution, we describe the developments we have done to perform surrogate-reaction studies at the storage rings of GSI/FAIR. In particular, we present the first results of the proof of principle experiment, which we conducted recently at the Experimental Storage Ring (ESR).

1 Introduction

Most of the elements from iron to uranium are synthesized in stars via neutron-induced reactions during the slow (s) and rapid (r) neutron-capture processes. Many uncertainties and open questions remain regarding the r-process. For instance, it is not yet clear if the r-process abundance distribution in the solar system is the result of one or multiple scenarios. The

* Corresponding author: jurado@cenbg.in2p3.fr

measurement of neutron-induced cross sections of key neutron-rich nuclei is essential to answer this question [1]. In the neutron-star merger scenario, the fission process plays a significant role [1, 2, 3, 4], since fission sets the end of the r-process path and can significantly impact the abundances in the mass region 120 < A < 210.

A neutron-induced reaction at incident energies below a few MeV can be described as a two-step process (Fig. 1). First, the nucleus A captures the neutron forming a compound nucleus (A+1)*. The excited compound nucleus can then decay in different ways, by emitting γ rays, by emitting a neutron and by fission. Each decay mode has a given probability (P_{γ}, P_n, P_f) and the sum of the probabilities of all the open decay channels must add up to 1.

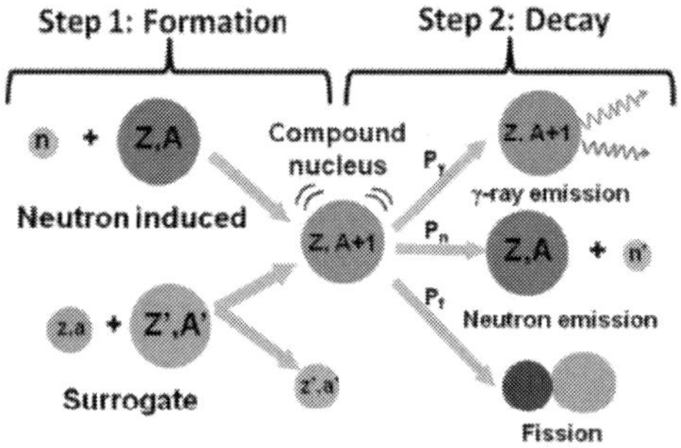

Fig. 1. Sequence of a neutron-induced reaction. First a compound nucleus is formed, which then de-excites by the emission of γ rays, a neutron or by fission. The probability P associated to each decay channel is indicated. In a surrogate reaction, the same compound nucleus as in the neutron-induced reaction is produced by a different reaction.

In traditional neutron-irradiation experiments, the direct measurement of neutron-induced cross sections of short-lived nuclei is very challenging because of the difficulties to produce and handle radioactive targets. Moreover, the associated radioactivity can generate a significant background and can damage the detectors during the measurement. Therefore, when the target nuclei are highly radioactive, experimental data are scarce and most of the neutron-induced reaction cross sections rely on theoretical model predictions. However, these predictions often have very large uncertainties due to difficulties in describing the compound-nucleus de-excitation process (step 2 in Fig. 1). Indeed, the de-excitation process is ruled by fundamental properties (level densities, fission barriers, transmission coefficients, etc.) for which the existing nuclear models give very different predictions. This leads to discrepancies between the calculated cross sections as large as two orders of magnitude or more when no experimental data are available [5].

It is possible to overcome the radioactive-target difficulties by performing the experiments in inverse kinematics with radioactive ion beams. However, this is not yet possible for neutron-induced reactions since free neutron targets are not available. An elegant alternative is to use the surrogate-reaction method in inverse kinematics. The surrogate reaction produces the compound nucleus of interest by a two-body reaction, different than the neutron-induced reaction (see Fig. 1) and the decay probabilities for γ emission, neutron emission and fission are measured as a function of the excitation energy of the compound nucleus. The measured probabilities are used to constrain the fundamental properties mentioned above and enable for much more accurate predictions of the desired neutron-induced cross sections [6, 7, 8].

Using the surrogate-reaction method in inverse kinematics enables for the formation of very short-lived nuclei and for the detection of the heavy, beam-like residues produced after the emission of γ rays and neutrons. However, the decay probabilities change very rapidly with excitation energy at the neutron separation energy and at the fission threshold. The excitation-energy resolution required to scan this rapid evolution is a few 100 keV, which is quite difficult to achieve for heavy nuclei in inverse kinematics, due to different target issues. Namely, the required large target density and thickness lead to significant energy loss and straggling effects that translate into a large uncertainty for the energy of the projectile and for the emission angle and the energy of the target-like residue. In addition, the presence of target windows and impurities induces background.

In this work, we address the discussed target issues by investigating for the first time surrogate reactions at the ESR storage ring of GSI/FAIR [9]. A key capability of storage rings is beam cooling, which allows for a significant reduction of the size and energy spread of the stored beam. If a gas target is present in the ring, the electron cooler can compensate the energy loss as well as the energy and angular straggling of the beam in the gas target. The ions pass through the target always with the same energy and the same outstanding beam quality. Hence, energy loss and straggling effects in the target are negligible, quite in contrast to single-pass experiments. Moreover, the frequent passing of the target zone (about 1 million times per second at 10 AMeV) allows gas targets with ultra-low density (10^{13} atoms/cm^2) to be used and no windows are necessary. This is a great improvement for surrogate reactions since the beam will only interact with the desired material and in a well-defined interaction zone, while the luminosity remains at a high level. Storage rings can be used to reduce the energy of the stored beam from a few 100 AMeV, the typical energy required to produce bare ions, to a few AMeV. This enables another unique feature: the production of 10 AMeV cooled beams of fully-stripped radioactive heavy ions. However, heavy-ion storage rings are operated in ultra-high vacuum (UHV) conditions (10^{-10} to 10^{-11} mbar), which poses severe constraints to in-ring detection systems. UHV-compatible silicon detectors have only started to be used since a few years for pioneering nuclear-reaction studies at the ESR, see e.g. [10].

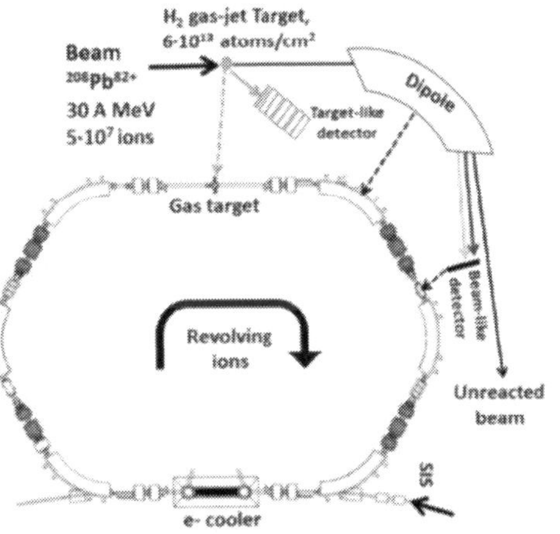

Fig. 2. The lower part shows the ESR and the upper part the setup used in our experiment. The trajectory of the scattered protons is shown in pink. The trajectories of the beam-like residues produced after γ-ray ($^{208}Pb^{82+}$) and neutron emission ($^{207}Pb^{82+}$) are shown in blue and green, respectively.

2 First proof of principle experiment

End of June 2022, we conducted our first proof-of-principle experiment at the ESR storage ring to investigate the ^{208}Pb(p,p')^{208}Pb* reaction as a surrogate for the n+^{207}Pb reaction. In this experiment, a $^{208}Pb^{82+}$ beam at 30.77 AMeV interacted with a gas-jet target of hydrogen. We had on average $5 \cdot 10^7$ $^{208}Pb^{82+}$ ions cooled and decelerated per injection and the target thickness was $6 \cdot 10^{13}$ atoms/cm^2. The ^{208}Pb projectiles were excited by inelastic scattering

 https://doi.org/10.1051/epjconf/202327911006

reactions with the target. We measured the inelastically scattered protons with a Si telescope and the beam residues produced after the de-excitation of $^{208}Pb^*$ via γ-ray and neutron emission with a position-sensitive Si strip detector placed behind the ring dipole magnet downstream from the target, see Fig. 2. This dipole acted as a recoil spectrometer separating the unreacted beam, the $^{208}Pb^{82+}$residues produced after γ-ray emission and the $^{207}Pb^{82+}$ residues produced after neutron emission.

One of the main objectives of this experiment was to evaluate the excitation energy resolution ΔE*. The left part of Fig. 3 shows the energy loss in the telescope versus the residual energy for protons detected at 64.5°. The ground state of ^{208}Pb is well separated and can be used to evaluate the excitation energy resolution. This is shown on the right part of Fig. 3, where the excitation energy spectrum of ^{208}Pb for elastically scattered protons is represented. The standard deviation of the ground-state peak is $\Delta E^* \approx 600$ keV. This resolution agrees with our expectations and is dominated by the angular uncertainty of the scattered protons caused by the large target radius of 2.5 mm. This first experiment has allowed us to validate our simulations, which predict $\Delta E^* \approx 300$ keV and 200 keV for smaller target radii of 1 or 0.5 mm, respectively. A target with a smaller radius will be available for our next experiments.

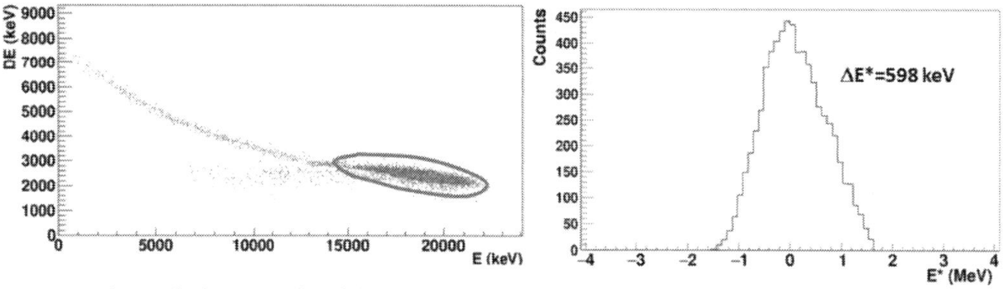

Fig. 3. (Left) Preliminary results of the energy loss versus residual energy for scattered protons detected at 64.5°. The red contour selects the elastically scattered protons. (Right) Preliminary results of the excitation energy of ^{208}Pb for the selected protons. The standard deviation of the peak is indicated.

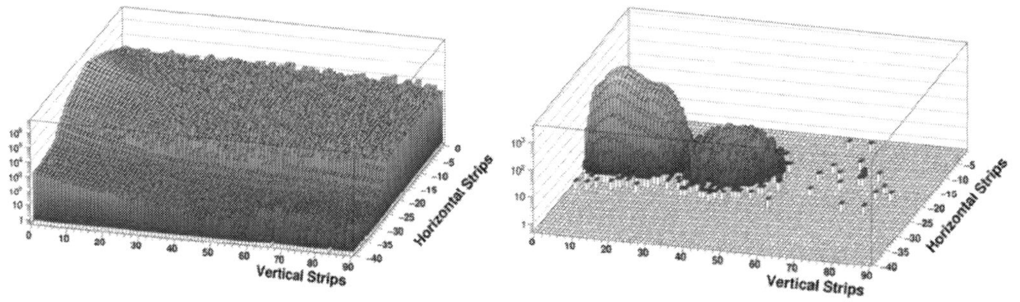

Fig. 4. Position of beam residues measured without (left) and with (right) selection of scattered protons detected in the telescope. In the right figure, $^{208}Pb^{82+}$ nuclei that de-excite by γ emission are in the left bump, while $^{207}Pb^{82+}$ nuclei produced after neutron emission are located in the right bump.

The other main objective of our first experiment was to evaluate the transmission and the separation of the heavy, beam-like residues. Fig. 4 shows the position spectra of heavy residues detected without (left) and with (right) the coincidence with the protons detected in the telescope. Due to the very intense background created by the elastic scattering of the beam, visible in the left part of Fig. 4, the heavy residues of the projectile produced after its de-excitation can only be seen in the coincidence spectrum on the right. The $^{208}Pb^{82+}$ nuclei that de-excite by γ emission are in the left bump, while $^{207}Pb^{82+}$ nuclei produced after neutron emission are in the right bump. The two bumps are very clearly separated. As expected, the beam residues are detected with very high efficiencies. We detect 100% of the residues produced after neutron emission and our preliminary analysis shows that we detect between 70 to 96% of the residues produced after γ-ray emission. These efficiencies are much larger than what can be obtained in standard surrogate-reaction experiments in direct kinematics. By counting events in each of the peaks at different E* intervals, we will be able to extract for the first time simultaneously the γ- and the neutron-emission probabilities of ^{208}Pb without the need of measuring γ rays or neutrons. To our knowledge the neutron emission probability as a function of E* has never been measured before for any nucleus.

3 Conclusion and perspectives

The surrogate-reaction method is a powerful tool to infer indirectly neutron-induced cross sections of short-lived nuclei. Storage rings offer ideal conditions to use the surrogate reaction method in inverse kinematics. We successfully conducted a first proof-of-principle experiment at the ESR storage ring using the ^{208}Pb(p,p') surrogate reaction. We were able to evaluate the excitation energy resolution and confirm the full separation and highly efficient detection of the beam-like residues produced after γ and neutron emission. In the next months, we will determine the decay probabilities for γ and neutron emission and use them to infer the neutron-induced radiative-capture and inelastic-scattering cross sections of ^{207}Pb.

In the near future, we will complete our setup with fission detectors to measure, in addition to the γ-ray and neutron emission probabilities, the fission probabilities of various U isotopes. After validation of the experimental method, we foresee to indirectly infer the neutron-induced reaction cross sections of many short-lived nuclides. First, we will use primary beams of e.g. ^{238}U and ^{208}Pb and nearby secondary beams produced by fragmentation. In the longer term, we aim to explore the region of neutron-deficient actinides and pre-actinides towards the N=126 shell closure. It will be the first time that fission probabilities are measured near a shell closure. These studies will help to provide much better theoretical predictions for the fission barriers and cross sections of neutron-rich nuclei towards the shell closure at N=184, which are essential for the r-process [2, 3] and not yet accessible to experiments.

Acknowledgements

This work is supported by the European Research Council (ERC) under the European Union's Horizon 2020 research and innovation programme (ERC-Advanced grant NECTAR, grant agreement No 884715). We thank the 80PRIME program from the CNRS for funding the PhD thesis of MS and the GSI/IN2P3 agreement 19-80. AH is grateful for funding from the Knut and Alice Wallenberg Foundation under KAW 2020.0076.

References

1. T. Kajino et al., Prog. Part. Nucl. Phys. **107**, 109 (2019).
2. S. Goriely, Eur. Phys. J. A **51**, 22 (2015).
3. N. Vassh et al., J. Phys. G: Nucl. Part. Phys. **46**, 065202 (2019).
4. Y. L. Zhu et al., Astrophys. J. **906**, 94 (2021).
5. M. Arnould et al., Phys. Reports **450**, 97 (2007).
6. J. E. Escher et al., Phys. Rev. Lett. **121**, 052501 (2018).
7. A. Ratkiewicz et al., Phys Rev Lett. **122**, 052502 (2019).
8. R. Pérez Sánchez et al., Phys. Rev. Lett. **125**, 122502 (2020).
9. B. Franzke, Nucl. Instrum. Methods B **25**, 18 (1987).

https://doi.org/10.1051/epjconf/202327911006

10. J. Glorius et al., Phys Rev Lett **122**, 092701 (2019).

New experimental study of the ^{3}He(α,γ)^{7}Be reaction around the E_{cm} = 3 MeV resonance

Ákos Tóth[1,2,*] and *Tamás* Szücs[1,**]

[1]Institute for Nuclear Research (ATOMKI), Debrecen, Hungary
[2]University of Debrecen, Doctoral School of Physics, Debrecen, Hungary

Abstract. The total cross section of the ^{3}He(α,γ)^{7}Be reaction was measured by the activation technique in the energy range of E_{cm} = 2.6 – 3.1 MeV, covering the first known resonance in the reaction. Consistent results were obtained with the only one available literature dataset in this energy range.

1 Introduction

The ^{3}He(α,γ)^{7}Be reaction was studied by many groups before the turn of the century. For a compilation, see e.g. the first *Solar fusion cross sections* review [1]. Microscopic calculations successfully described the energy dependence of the experimental datasets, thus robust extrapolation to solar and stellar energies was established. However, the absolute normalization of the theoretical calculations to the experimental data showed different scalings for datasets obtained by in-beam γ-spectroscopy and by activation method. Motivated by this ambiguity pointed out in the review, a new series of cross section data have been measured.

The newer datasets are compiled in the second *Solar fusion cross sections* review [2], where the previously inconsistent scaling was attributed to underestimated experimental uncertainties in the pre-2000 datasets. For the extrapolation to solar energies, only the well documented post-2000 works were considered, from which data only up to E_{cm} = 1.0 MeV were used. The reason behind was that the different theoretical models differ by only a few percent in this energy range, and starts to deviate above. In addition, the new experimental data based on recoil detection [3] disagreed with the only one older in-beam γ-spectroscopy measurement [4] in the slope of the S-factor[1] towards higher energies.

In the last decade there were a few works motivated to solve this tension between the higher energy datasets, and all of them [5–7] supported the newer measurement. These new data also provided a firm basis for updated theoretical calculations also toward higher energies. There is one previous study of the total radiative capture cross section with the activation technique around the ^{7}Be proton separation threshold in the energy range of E_{cm} = 4.0 – 4.4 MeV [8], which was motivated by a proposed level in ^{7}Be which appeared as a resonance in the ^{6}Li(p,γ)^{7}Be reactions [9]. The existence of the level was not confirmed either by the ^{3}He(α,γ)^{7}Be work, or by a newer ^{6}Li(p,γ)^{7}Be investigation [10].

For the precise determination of the reaction rate in different astrophysical scenarios, extrapolation of the cross section is mostly unavoidable. For example, the lowest measured

*e-mail: toth.akos@atomki.hu
**e-mail: szucs.tamas@atomki.hu
[1] $S(E)=Ee^{(2\pi\eta)}\sigma(E)$, where η is the Sommerfeld parameter

experimental cross section is at E_{cm} = 93 keV [11], while the solar Gamow peak (the energy range that gives the most contribution to the reaction rate) centers at about 23 keV. Because of the Coulomb barrier, between these two energies the cross section drops by about 7 orders of magnitude, making it impossible with the present days techniques to be measured.

Beside the direct reaction studies, indirect methods are also used to infer the solar reaction cross section. For example, with transfer reactions, the ^{3}He(α,γ)^{7}Be ANC (asymptotic normalization coefficient) can be determined [12], which directly sets the magnitude of the zero-energy S-factor extrapolation. An other possibility is to use the Standard Solar Model and the experimental solar neutrino fluxes to extract the reaction rates [13].

For the extrapolation, one of the widely used possibilities is the R-matrix theory, where multi-channel, multi-level fits can be performed, taking into account almost all the experimental information about a compound nucleus [14]. However, the fit is highly dependent on the precision and consistency of the datasets used in the analysis. For example, a new precise determination of the scattering cross section changed the extrapolation in such a way, that the previously discrepant (or off scale) low energy capture data became consistent with the fit [15]. To strengthen the predictive power of the fits, a new study is following up the one at high energy with the same setup [8]. The aim is to cover the energy range of the first resonance, where only one experimental dataset exist, and close the energy gap lacking experimental data. Even thought the new data will have minor influence on the S(0) extrapolation, it is worth while to cross validate different R-matrix calculations. A previous cross validation work [16] concentrated only on scattering data, however the treatment of the radiative capture may differ in the codes, which should also be validated.

2 Experimental aspects

For the experiment the activation technique was employed. Since the ^{7}Be reaction product is radioactive and decays by electron capture with a half-life of 53.22 days [17], it is well suited for this technique. The decay is followed by the emission of a 478-keV γ ray with 10.44 % branching [17]. This γ radiation makes it possible to measure the activity created by the irradiation via γ-ray detection.

Since the target is a noble gas, it forms no solid compound, which could be used as a reaction target. Solid target from helium can be created by implantation, which has already been used for studying the ^{3}He(α,γ)^{7}Be [18, 19]. However, the determination of the absolute amount of nuclei captured in the crystal structure of the host material, and the stability of such target under beam bombardment require extensive target characterization techniques. In most cases, helium targets in gas form are used. Either in a differential pumped windowless gas target (e. g. [3, 20]), or in gas-jet target, or enclosed in a cell (e. g. [6]). All of them may have their own difficulties, e. g. in the determination of the number of target atoms, the beam current measurement, or the overlap of the beam and target, just to mention a few. In the present work, the investigated energy range allowed to employ a gas cell closed by thin window, thus no differential pumping was necessary, and the number of projectiles was possible to be determined by measuring the electric current hitting the target. Here, the window itself may cause some problems, which has to be tackled for a precision measurement.

The continuously developed thin-windowed gas-cell setup of Atomki [6, 8, 21, 22] contains for the present study two 10 μm thick aluminium windows, which separate the ^{3}He gas volume from the beamline vacuum. The beam was produced by the Atomki cyclotron accelerator. For the targeted center-of-mass energy range, the α particles had an energy around 8 MeV. The beam lost about 1-1 MeV energy in both the entrance and exit foils, and a few keV in the gas volume. All the energy of the beam particles passing through the windows, and later the energy of the created ^{7}Be recoils passing through the gas was determined by simulations using the SRIM software package [23]. With this double-window configuration sizable portion of the total energy is dumped outside the gas cell (in a water cooled tantalum sheet), thus any temperature effect caused by the heating of the gas is reduced. The target areal density was determined from the initial temperature ($\approx$24 °C), pressure ($\approx$100 mbar) and the length of the cell ($\approx$4 cm) using the ideal gas law. The real gas behavior of helium deviates less than 0.005 % from the ideal gas low in this pressure and temperature regime [24], thus this effect is negligible compared to other uncertainties. Since the cell is closed, the number of target atoms did not change during the irradiation even if the temperature and pressure may increase as a consequence of the beam power deposition in the cell or gas desorption from the cell walls. The foils expected to be bent because of the pressure difference between its two sides. The size of this effect was tested in an off-site setup [22], and found to be in the order of 0.3 mm at the center of the foil. Additional to the foil bending effect, a slight offset between the beamline and cell axes may cause a maximum 0.1 mm effective length change. Because of the two foil configuration, a conservative 1 mm uncertainty in the cell length was adopted, which includes the possibility of a slightly higher foil deformation while it is heated by the beam. There was a maximum 0.2 mbar and 2 °C fluctuation observed in the initial pressure and temperature, respectively, thus these values were used as a conservative estimate in the calculation of the number of target atoms. In total, 2.5 % uncertainty is assigned to the areal number density of the target.

The whole cell was attached to an electrically isolated activation chamber serving as a Faraday-cup. At the entrance of the chamber, a beam defining aperture ensured that only those α particles enter the chamber, which also hit the gas volume. Electrons that may emerge from the entrance aperture, or would escape from the chamber, are suppressed by employing -300 V on an aperture right behind the beam defining one. Since the gas volume was inside the Faraday-cup, charge exchange inside the gas did not affect the current measurement. An overall current reading accuracy of 3 % is taken into account as a systematic uncertainty.

The ^{7}Be formed in the reaction had a forward momentum with an energy of 4.15 – 3.83 MeV and a maximum angle of about 1°. Recoils may lose maximum about 0.4 MeV energy in the gas volume, when they arrive to the exit foil serving as a catcher. Since the implantation occur with a minimum of 3.4 MeV, negligible backscattering is expected. The implantation depth of ^{7}Be with these energies is about 5 μm, thus sputtering of the already implanted activity by the unreacted α beam is also negligible.

The catchers were extracted and replaced after each irradiation, and their activity was determined using a lead shielded high purity germanium (HPGe) detector. The detection efficiency of the HPGe was determined by using multi-line γ-ray standards at far geometry (detector-source distance of 27 cm). From the calibration, the detection efficiency at 478-keV was interpolated. The actual, low activity samples created by the ^{3}He(α,γ)^{7}Be reaction were measured in close geometry (detector-source distance of 1 cm). To determine the ratio of the far and close geometry efficiencies, a stronger ^{7}Be source created via the ^{7}Li(p,n)^{7}Be reaction was used. With this approach the true coincidence summing effect relevant for multi-line calibration sources in close geometry is avoided. The final uncertainty assigned to the γ detection efficiency is 3 % incorporating the uncertainty of the activity of the γ-ray standards, and that of the interpolation and the scaling.

For the center-of-mass energy determination, the thickness of each entrance foil was precisely measured after the irradiation in a setup consisting of a triple nuclide alpha source and a particle detector. From the energy loss of the α particles passing through the foils, their thickness was determined. For the conversion between the energy loss and thickness, the SRIM tables [23] were used. Each foil was measured at several points. No significant thickness variation was observed between the region where the beam hit the foils and those which where unexposed to the beam or to the tension created by the pressure difference.

3 Preliminary results

The work is in progress, however, a few preliminary data points are shown in Fig. 1. The newly obtained results are inline with the literature data. Additional to the statistical uncertainty plotted in the figure, a systematic uncertainty of about 5 % has to be considered, which affects the absolute scale of the whole dataset. This systematic uncertainty accounts for the uncertainty of the γ branching, γ detection efficiency (3 %), current reading (3 %), cell-length (2.4 %); common for each data point.

The horizontal error bar shows the absolute energy uncertainty. It stems from the uncertainty of the entrance foil thickness, and the uncertainty of the stopping power of the incoming α particles in aluminium, and the energy uncertainty of the initial beam.

The dataset will be extended towards higher energies, to close the gap between E_{cm} = 3.1 – 4.0 MeV, where experimental data are lacking.

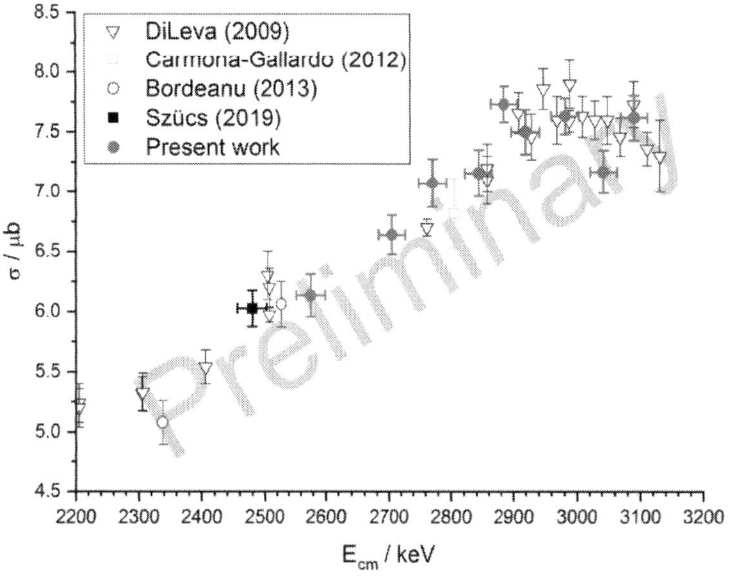

Figure 1. The new preliminary data is compared with literature datasets in a similar energy range [3, 5, 6, 8]. Only the statistical uncertainties are plotted on each datapoint. The horizontal error bars show the beam energy uncertainty in the gas.

Acknowledgment

This work was supported by NKFIH Grant No. FK134845, and New National Excellence Programs of the Ministry of Human Capacities of Hungary under nr. ÚNKP-22-3-II-DE-31 and ÚNKP-22-5-DE-428. T.S. acknowledges support from the János Bolyai research fellowship of the Hungarian Academy of Sciences.

References

[1] E.G. Adelberger, S.M. Austin, J.N. Bahcall, A.B. Balantekin, G. Bogaert, L.S. Brown, L. Buchmann, F.E. Cecil, A.E. Champagne, L. de Braeckeleer et al., Rev. Mod. Phys. **70**, 1265 (1998)

[2] E.G. Adelberger, A. García, R.G.H. Robertson, K.A. Snover, A.B. Balantekin, K. Heeger, M.J. Ramsey-Musolf, D. Bemmerer, A. Junghans, C.A. Bertulani et al., Rev. Mod. Phys. **83**, 195 (2011)

[3] A. Di Leva, L. Gialanella, R. Kunz, D. Rogalla, D. Schürmann, F. Strieder, M. De Cesare, N. De Cesare, A. D'Onofrio, Z. Fülöp et al., Phys. Rev. Lett. **102**, 232502 (2009), erratum: Phys. Rev. Lett. 103, 159903 (2009)

[4] P.D. Parker, R.W. Kavanagh, Phys. Rev. **131**, 2578 (1963)
[5] M. Carmona-Gallardo, B.S.N. Singh, M.J.G. Borge, J.A. Briz, M. Cubero, B.R. Fulton, H. Fynbo, N. Gordillo, M. Hass, G. Haquin et al., Phys. Rev. C **86**, 032801 (2012)
[6] C. Bordeanu, G. Gyürky, Z. Halász, T. Szücs, G. Kiss, Z. Elekes, J. Farkas, Z. Fülöp, E. Somorjai, Nucl. Phys. A **908**, 1 (2013)
[7] A. Kontos, E. Uberseder, R. deBoer, J. Görres, C. Akers, A. Best, M. Couder, M. Wiescher, Phys. Rev. C **87**, 065804 (2013)
[8] T. Szücs, G.G. Kiss, G. Gyürky, Z. Halász, T.N. Szegedi, Z. Fülöp, Phys. Rev. C **99**, 055804 (2019), erratum: Phys. Rev. C. 105, 069901(E) (2022)
[9] J. He, S. Chen, C. Rolfs, S. Xu, J. Hu, X. Ma, M. Wiescher, R. deBoer, T. Kajino, M. Kusakabe et al., Phys. Lett. B **725**, 287 (2013)
[10] D. Piatti, T. Chillery, R. Depalo, M. Aliotta, D. Bemmerer, A. Best, A. Boeltzig, C. Broggini, C.G. Bruno, A. Caciolli et al. (LUNA Collaboration), Phys. Rev. C **102**, 052802 (2020)
[11] D. Bemmerer, F. Confortola, H. Costantini, A. Formicola, G. Gyürky, R. Bonetti, C. Broggini, P. Corvisiero, Z. Elekes, Z. Fülöp et al. (LUNA Collaboration), Phys. Rev. Lett. **97**, 122502 (2006)
[12] G. Kiss, M.L. Cognata, C. Spitaleri, R. Yarmukhamedov, I. Wiedenhöver, L. Baby, S. Cherubini, A. Cvetinović, G. D'Agata, P. Figuera et al., Phys. Lett. B **807**, 135606 (2020)
[13] M.P. Takács, D. Bemmerer, T. Szücs, K. Zuber, Phys. Rev. D **91**, 123526 (2015)
[14] R.E. Azuma, E. Uberseder, E.C. Simpson, C.R. Brune, H. Costantini, R.J. deBoer, J. Görres, M. Heil, P.J. LeBlanc, C. Ugalde et al., Phys. Rev. C **81**, 045805 (2010)
[15] D. Odell, C.R. Brune, D.R. Phillips, R.J. deBoer, S.N. Paneru, *Performing bayesian analyses with AZURE2 using BRICK: an application to the ^{7}Be system* (2021), `arXiv:2112.12838`
[16] I.J. Thompson, R.J. deBoer, P. Dimitriou, S. Kunieda, M.T. Pigni, G. Arbanas, H. Leeb, T. Srdinko, G. Hale, P. Tamagno et al., Eur. Phys. J. A **55**, 92 (2019)
[17] D. Tilley, C. Cheves, J. Godwin, G. Hale, H. Hofmann, J. Kelley, C. Sheu, H. Weller, Nucl. Phys. A **708**, 3 (2002)
[18] T. Alexander, G. Ball, W. Lennard, H. Geissel, H.B. Mak, Nucl. Phys. A **427**, 526 (1984)
[19] S. Turkat, K. Zuber, S. Akhmadaliev, D. Bemmerer, M. Grieger, S. Hammer, T. Hensel, L. Hübinger, F. Ludwig, S. Reinicke et al., *Measurement of the $^{3}He(\alpha,\gamma)^{3}Be$ γ-ray angular distribution*, in *Solar Neutrinos*, edited by M. Meyer, K. Zuber (WORLD SCIENTIFIC, 2019), p. 513
[20] H. Costantini, D. Bemmerer, F. Confortola, A. Formicola, G. Gyürky, P. Bezzon, R. Bonetti, C. Broggini, P. Corvisiero, Z. Elekes et al. (LUNA Collaboration), Nucl. Phys. A **814**, 144 (2008)
[21] C. Bordeanu, G. Gyürky, Z. Elekes, J. Farkas, Z. Fülöp, Z. Halász, G. Kiss, E. Somorjai, T. Szücs, Nucl. Instrum. Meth. A **693**, 220 (2012)
[22] Z. Halász, E. Somorjai, G. Gyürky, Z. Elekes, Z. Fülöp, T. Szücs, G.G. Kiss, N.T. Szegedi, T. Rauscher, J. Görres et al., Phys. Rev. C **94**, 045801 (2016)
[23] J.F. Ziegler, M. Ziegler, J. Biersack, Nucl. Instrum. Meth. B **268**, 1818 (2010), http://www.srim.org
[24] G. Imbert, P. Dejours, G. Hildwein, Undersea Biomed Res **9**, 305 (1982)

Experimental work on Nuclear Astrophysics at JRC GELINA facility

Carlos Paradela[1*], *Stefan* Kopecky[1], and *Peter* Schillebeeckx[1]

[1]European Commission Joint Research Center, 2440 Geel, Belgium

Abstract. JRC Geel operates a neutron time-of-flight facility based on an electron accelerator, GELINA. Experimental setups to determine total and reaction cross sections (capture, elastic and inelastic scattering, fission and charged particle reactions) are available at measurement stations different flight-path length. While most of the experimental work is focussed on nuclear energy applications, regularly cross section of interest for astrophysical problems are measured. Examples of those are capture experiments of nuclei relevant for the s-process such as ^{89}Y or the ^{16}O (n,α) reaction which is the inverse of the $^{13}C(\alpha,n)$ reaction, an important source of neutrons feeding the s-process.

1 Introduction

Joint Research Centre (JRC) is the science and knowledge service of the European Commission, providing advice and support to European policies. JRC-Geel hosts the GELINA facility [1], an intense pulsed white neutron source driven by a linear electron accelerator which is dedicated to high-resolution time-of-flight measurements. GELINA combines four specially designed and distinct units: a high-power pulsed linear electron accelerator, a post-accelerating beam compression magnet, a mercury-cooled uranium target, and measurement stations at flight path lengths ranging from 10 m to 400 m. Thanks to the multiple neutron beam lines, as many as 10 experiments can be carried out simultaneously.

Research at GELINA is focused on the following priority topics:

- Neutron cross sections measurements for energy and non-energy nuclear applications
- Measurements of nuclear data standards
- Integral experiments for the validation of nuclear data libraries and testing of nuclear transport codes
- Better understanding of the nuclear fission process
- Development of advanced detection methods and scientific concepts in nuclear technologies
- Basic physics such as nuclear reaction theory or nuclear astrophysics

Since the early 90's, an extensive investigation on neutron cross sections relevant for nuclear astrophysics was performed at IRMM (former JRC-Geel), led by Franco Corvi in the field of neutron capture cross sections and by Cyriel Wagemans with neutron-induced charge particle reactions. Their work can be summarized by the contributions to Nuclei in Cosmos

* Corresponding author: carlos.paradela-dobarro@ec.europa.eu

proceedings, the reference conference in the field. An overview of these contributions is detailed in Table 1.

Table 1. Nuclear astrophysics contributions to Nuclei in Cosmos conferences with experimental data obtained at the GELINA facility.

(n,γ)	(n,α),(n,p)	Conference	Year
^{138}Ba	^{32,33}S, ^{14}N	NIC-II	1992
^{208}Pb	^{35,36}Cl, ^{41}Ca	NIC-III	1994
^{136}Ba		NIC-IV	1996
^{84}Kr, ^{207}Pb, ^{209}Bi	^{37}Ar	NIC-V	1998
^{99}Tc	^{37}Ar, ^{39}Ar	NIC-VI	2000
	^{26}Al, ^{36}Cl	NIC-VIII	2004
	^{41}Ca	NIC-IX	2006

2 Recent contributions to nuclear astrophysics

During last decade most of the contributions relevant for nuclear astrophysics have been carried out in collaboration with external partners. The majority of the projects were carried out as part of the EUFRAT open access program or collaboration agreements with EU Member State organisations. Several nuclides of interest for nuclear astrophysics have been measured at the GELINA capture and transmission measurement stations, e.g. ^{139}La, ^{142}Ce [2], ^{154,155,156,157,158}Gd [3,4], ^{92}Zr [5], ^{88}Sr and ^{89}Y. With respect to neutron-induced charge particle reactions, two different external users have carried experiments at GELINA with their own detection setups in order to study ^{26}Al(n,α) [6] and ^{16}O(n,α) cross sections. Here, we will provide more details for two examples of measurements.

2.1 Capture cross sections in the vicinity of N=50: the ^{89}Y(n,γ) reaction

The slow neutron capture processes (s-process) is responsible for the abundances of a large fraction of the nuclides of masses between 50 and 210 and, in particular, for the presence of two main peaks in the elemental abundance distribution. The first s-process peak is due to ^{88}Sr, ^{89}Y and ^{90}Zr having a magic number of neutrons equal to 50. The cross section of these nuclides strongly influences the production of all heavier elements, at least up to the second s-process peak corresponding to the next bottleneck at Ba, La and Ce with a neutron magic number of 82. In particular, for the yttrium case its stellar abundance can be derived from high-resolution spectra thanks to the large number of optical lines, being extensively used to constrain stellar models. An accurate determination of the ^{89}Y capture cross section is a primordial input for these models.

At GELINA, transmission measurements on yttrium samples of different thicknesses between 1 and 8 mm and neutron capture measurement with a disk of 80 mm diameter and 2 mm thickness were carried out by researchers from INFN Bari. Capture measurements with a Y disk of 30 mm diameter and 1 mm thickness were also carried out at the n_TOF facility.

The transmission measurements were performed at the 50 m transmission station of FP4. A ^{10}B overlap filter was used to reduce the contribution of slow neutrons coming from previous accelerator bursts and Na, Co and W black resonance filters were used to determine the background contribution. The neutron beam passing through the sample and filters was detected by a cylindrical NE912 Li-glass scintillator of 6.35 mm thickness, placed at 47 m from the neutron target.

Fig. 1. Neutron capture setup based on C_6D_6 detectors located at GELINA FP14-60m measurement station

The capture measurements at n_TOF and GELINA were performed by applying the total energy detection principle using C_6D_6-based liquid scintillators which are place at 125° with respect to the beam direction. The GELINA setup, which is presented in Figure 1, was located at the 60 m station of FP14. The energy distribution of the incident neutron beam was measured with a ^{10}B ionisation chamber placed 80 cm in front of the sample.

Resonance parameters were derived from a combined analysis of capture yields and transmission data, following the prescription given in Ref [7]. The yields were analysed by means of the R-matrix codes SAMMY [8] and REFIT [9] in the Reich-Moore approximation. The Doppler broadening, neutron self-shielding and multiple interactions and the facility TOF response function, were taken into account. The whole set of measurements enabled the identification of 112 resonances belonging to the ^{89}Y for neutron incident energies up to 95 keV. Thanks to the combination of experimental data obtained in different experimental conditions bias effect due to the TOF response function, the normalization of the capture data and the background contributions are largely reduced.

2.2 Neutron-induced charge particle reactions: $^{16}O(n,\alpha)^{13}C$

The $^{16}O(n,\alpha)^{13}C$ reaction plays an important role in nuclear technology. It is the inverse reaction of the main source of neutrons for the slow neutron capture process in the nucleosynthesis. The measurements at GELINA were led by the HZDR group through the EUFRAT open access program. They provided the Frisch Grid Ionization Chamber (FGIC) for measuring the reaction in ^{16}O, while the well characterized ^{235}U chamber H19 [10] from PTB Braunschweig, was used for an absolute determination of the incident neutron spectrum using the ^{235}U standard. The DFGIC was operated with a gas mixture of 95% Kr and 5% CO_2 at 2 bar. The oxygen of the CO_2 between the cathode and the grid was the effective target nuclei.

The signals from the H19 and the FGIC were digitized and processed off-line to determine the time-of-flight and the deposited energy. For the FGIC, the drift time of the particle was used to only select events with full energy deposition between the cathode and the grid and accurately provide the number of oxygen nuclei acting as a target.

Fig. 2. Detection setup installed at GELINA FP16-60m to measure the $^{16}O(n,\alpha)$ cross section including the Frisch grid ionization chamber from HZDR (DFGIC) and the ^{235}U loaded fission chamber from PTB (H19)

The data analysis allowed to discriminate $^{16}O(n,\alpha_0)$ events from scattering or other reactions in the gas and electrodes. The e method to determine the absolute cross section was validated with the determination of the elastic scattering cross section on C nuclei. The results obtained for neutron energies between 3 and 9 MeV suggests a bias of -9% for the cross section recommended in ENDF/B-VIII.0 and of +26% for the one in JEFF-3.3. For the $^{13}C(\alpha,n)$reaction, a correction factor is proposed which is very close to one derived from the thick target yields obtained by West and Sherwood [11] between 3.5 and 5 MeV.

Table 2. Correction factors for $^{13}C(\alpha,n)$ experimental data proposed in this work compared with those proposed in literature

	This work	West and Sherwood [11]	Pigni and Croft [12]	Ciani et al. [13]
Bair and Hass [14]	0.87	0.85	0.85	
Harissopoulos [15]	1.30	1.27	1.15	1.37

Shekharan [16]	1.37	1.35		

3 Summary and perspectives

In this work, we have provided an overview of different experimental work carried out at JRC GELINA facility on neutron-induced reactions of relevance for nuclear astrophysics, detailing two examples on neutron capture and neutron-induced charge particle reactions. This experimental program performed in collaboration with external partners is pursued with capture measurements on ^{90,91}Zr and ^{94,95,96}Mo isotopes. The EUFRAT open access program provides the framework to researchers from EU Member States to propose further research on nuclear astrophysics using the neutron facilities at JRC-Geel.

This work was supported by the EUFRAT open access programme of the Joint Research Centre (JRC).

References

1. W. Mondelaers and P. Schillebeeckx, Notiziario Neutroni e Luce di Sincrotrone **11**, 19 (2006)
2. D. H. Moon, C. Paradela, G. Alaerts, V. Chavan, K. Guber, J. Heyse, S. W. Hong, S. Kopecky, P. Schillebeeckx and R. Wynants, Tech. Rep. INDC(EUR)-0038 (2020)
3. A. Mazzone et al. Phys. Lett. B **804**, 1354 (2020)
4. Y. Kye et al. Eur. Phys. J. A, **56**, 30 (2020)
5. G. Tagliente et al., Phys. Rev. C **105**, 025805 (2022)
6. C. Lederer et al., Phys. Rev. C **104**, L032803 (2021)
7. P. Schillebeeckx et al., Nuclear Data Sheets **113**, 3054 (2012)
8. N. M. Larson, *Updated users' guide for SAMMY : Multilevel R-matrix fits to neutron data using Bayes' equations*, Tech. Rep. ORNL/TM-9179/R8 Oak Ridge National Laboratory (2008)
9. M. C. Moxon, J. B. Brisland, *GEEL REFIT, A least squares fitting program for resonance analysis of neutron transmission and capture data computer code*, Tech. Rep. AEA-InTec-0630, AEA Technology (2008)
10. D. B. Gayther, Metrologia **27**, 221 (1990)
11. D. West and A. Sherwood, Annals of Nucl. Energy **9**, 551 (1982)
12. M. T. Pigni and S. Croft, Phys. Rev. C **102,** 04618 (2020)
13. G. Ciani et al., Rhys. Rev. Lett. **127**, 152701 (2021)
14. J. K. Bair, Nucl. Sci. Eng. **51**, 83 (1973)
15. S. Harissopoulos, H. W. Becker, J. W. Hammer, A. Lagoyannis, C. Rolfs and F. Strieder, Phys. Rev. C **72**, 062801 (2005)
16. K. K. Shekharan, A. S. Divatia, M. K. Mehta, S. S. Kerakatte, and K. B. Nambiar, Phys. Rev. C **156**, 1187 (1967)

Neutron-induced cross section measurements

C. Massimi*[1,2], O. Aberle[3], V. Alcayne[4], S. Altieri[5], S. Amaducci[6], J. Andrzejewski[7], V. Babiano-Suarez[8], M. Bacak[3], J. Balibrea[8], C. Beltrami[5], S. Bennett[9], A. P. Bernardes[3], E. Berthoumieux[10], M. Boromiza[11], D. Bosnar[12], M. Caamaño[13], F. Calviño[14], M. Calviani[3], D. Cano-Ott[4], A. Casanovas[14], D. M. Castelluccio[1,28], F. Cerutti[3], G. Cescutti[15,16], S. Chasapoglou[17], E. Chiaveri[3,9], P. Colombetti[18], N. Colonna[19], P. Console Camprini[1,28], G. Cortés[14], M. A. Cortés-Giraldo[20], L. Cosentino[6], S. Cristallo[21,22], S. Dellmann[23], M. Di Castro[3], S. Di Maria[24], M. Diakaki[17], M. Dietz[25], C. Domingo-Pardo[8], R. Dressler[26], E. Dupont[10], I. Durán[13], Z. Eleme[27], S. Fargier[3], B. Fernández[20], B. Fernández-Domínguez[13], P. Finocchiaro[6], S. Fiore[28,43], V. Furman[29], F. García-Infantes[30], A. Gawlik-Ramięga[7], G. Gervino[18], S. Gilardoni[3], E. González-Romero[4], C. Guerrero[20], F. Gunsing[10], C. Gustavino[43], J. Heyse[32], W. Hillman[9], D. G. Jenkins[33], E. Jericha[34], A. Junghans[35], Y. Kadi[3], K. Kaperoni[17], G. Kaur[10], A. Kimura[36], I. Knapová[37], M. Kokkoris[17], Y. Kopatch[29], M. Krtička[37], N. Kyritsis[17], I. Ladarescu[8], C. Lederer-Woods[38], J. Lerendegui-Marco[8], G. Lerner[3], A. Manna[1,2], T. Martínez[4], A. Masi[3], P. Mastinu[39], M. Mastromarco[19,40], E. A. Maugeri[26], A. Mazzone[19,41], E. Mendoza[4], A. Mengoni[28,1], V. Michalopoulou[17], P. M. Milazzo[15], R. Mucciola[21,44], F. Murtas[†31], E. Musacchio-Gonzalez[39], A. Musumarra[45,46], A. Negret[11], A. Pérez de Rada[4], P. Pérez-Maroto[20], N. Patronis[27], J. A. Pavón-Rodríguez[20], M. G. Pellegriti[45], J. Perkowski[7], C. Petrone[11], E. Pirovano[25], J. Plaza[4], S. Pomp[42], I. Porras[30], J. Praena[30], J. M. Quesada[20], R. Reifarth[23], D. Rochman[26], Y. Romanets[24], C. Rubbia[3], A. Sánchez[4], M. Sabaté-Gilarte[3], P. Schillebeeckx[32], D. Schumann[26], A. Sekhar[9], A. G. Smith[9], N. V. Sosnin[38], M. E. Stamati[27], A. Sturniolo[18], G. Tagliente[19], D. Tarrío[42], P. Torres-Sánchez[30], S. Urlass[35,3], E. Vagena[27], S. Valenta[37], V. Variale[19], P. Vaz[24], G. Vecchio[6], D. Vescovi[23], V. Vlachoudis[3], R. Vlastou[17], A. Wallner[35], P. J. Woods[38], T. Wright[9], R. Zarrella[1,2], P. Žugec[12], and the n_TOF Collaboration

[1]Istituto Nazionale di Fisica Nucleare, Sezione di Bologna, Italy
[2]Dipartimento di Fisica e Astronomia, Università di Bologna, Italy
[3]European Organization for Nuclear Research (CERN), Switzerland
[4]Centro de Investigaciones Energéticas Medioambientales y Tecnológicas (CIEMAT), Spain
[5]Istituto Nazionale di Fisica Nucleare, Sezione di Pavia & University of Pavia, Italy
[6]INFN Laboratori Nazionali del Sud, Catania, Italy
[7]University of Lodz, Poland
[8]Instituto de Física Corpuscular, CSIC - Universidad de Valencia, Spain
[9]University of Manchester, United Kingdom
[10]CEA Irfu, Université Paris-Saclay, F-91191 Gif-sur-Yvette, France
[11]Horia Hulubei National Institute of Physics and Nuclear Engineering, Romania
[12]Department of Physics, Faculty of Science, University of Zagreb, Zagreb, Croatia
[13]University of Santiago de Compostela, Spain
[14]Universitat Politècnica de Catalunya, Spain
[15]Istituto Nazionale di Fisica Nucleare, Sezione di Trieste, Italy

*e-mail: massimi@bo.infn.it (corresponding author)

[16]Osservatorio Astronomico di Trieste, Italy
[17]National Technical University of Athens, Greece
[18]Istituto Nazionale di Fisica Nucleare, Sezione di Torino & University of Torino, Italy
[19]Istituto Nazionale di Fisica Nucleare, Sezione di Bari, Italy
[20]Universidad de Sevilla, Spain
[21]Istituto Nazionale di Fisica Nucleare, Sezione di Perugia, Italy
[22]Istituto Nazionale di Astrofisica - Osservatorio Astronomico di Teramo, Italy
[23]Goethe University Frankfurt, Germany
[24]Instituto Superior Técnico, Lisbon, Portugal
[25]Physikalisch-Technische Bundesanstalt (PTB), Bundesallee 100, 38116 Braunschweig, Germany
[26]Paul Scherrer Institut (PSI), Villigen, Switzerland
[27]University of Ioannina, Greece
[28]Agenzia nazionale per le nuove tecnologie (ENEA), Italy
[29]Joint Institute for Nuclear Research (JINR), Dubna, Russia
[30]University of Granada, Spain
[31]Laboratori Nazionali di Frascati, Italy
[32]European Commission, Joint Research Centre (JRC), Geel, Belgium
[33]University of York, United Kingdom
[34]TU Wien, Atominstitut, Stadionallee 2, 1020 Wien, Austria
[35]Helmholtz-Zentrum Dresden-Rossendorf, Germany
[36]Japan Atomic Energy Agency (JAEA), Tokai-Mura, Japan
[37]Charles University, Prague, Czech Republic
[38]School of Physics and Astronomy, University of Edinburgh, United Kingdom
[39]INFN Laboratori Nazionali di Legnaro, Italy
[40]Dipartimento Interateneo di Fisica, Università degli Studi di Bari, Italy
[41]Consiglio Nazionale delle Ricerche, Bari, Italy
[42]Uppsala University, Sweden
[43]Istituto Nazionale di Fisica Nucleare, Sezione di Roma1, Roma, Italy
[44]Dipartimento di Fisica e Geologia, Università di Perugia, Italy
[45]Istituto Nazionale di Fisica Nucleare, Sezione di Catania, Italy
[46]Department of Physics and Astronomy, University of Catania, Italy
[†] deceased

Abstract. Neutron-induced cross sections represent the main nuclear input to models of stellar and Big-Bang nucleosynthesis. While (n,γ) reactions are relevant for the formation of elements heavier than iron, (n,p) and (n,α) reactions can play an important role in specific cases. The time-of-flight method is routinely used at n_TOF to experimentally determine the cross section data. In addition, recent upgrades of the facility will allow the use of activation techniques as well, possibly opening the way to a systematic study of neutron interaction with radioactive isotopes. In the last 20 years n_TOF has provided a large amount of experimental data for Nuclear Astrophysics. Our plan is to carry on challenging measurements and produce nuclear data in the next decades as well.

1 Introduction

The prominent role of neutron-induced reactions in the nucleosynthesis of isotopes with A>60 is well-established. In particular, the slow neutron capture process (the *s* process, see, e.g., [1] and [2]) is responsible for the formation of approximately half of the elemental abundances heavier than iron. It consists of a sequence of neutron radiative captures and β-decays. Starting from a seed distribution around iron, the *s* process builds up elements up to Bismuth. Overall, β-decay rates are faster than neutron capture rates, which means that reactions proceed along the valley of β stability. The s process is thought to take place in 2 different scenarios:

- the main component takes place during the asymptotic giant branch (AGB) phase [3–6] of low mass stars ($1.5 \leq M/M_\odot < 4.0$) and, to a lower extent, intermediate mass stars ($4.0 \leq M/M_\odot < 7.0$). Here, a sequence of short convective He-burning runaways (thermal pulses, TPs) and long periods of quiescent H-burning (interpulses) take place.
- the weak component, activated during the core He-burning and C-shell burning of massive stars ($M/M_\odot > 10.0$) largely contributes to abundances between Fe and Zr (e.g. [7]).

Together with β-decay rates, neutron capture cross sections are the basic nuclear physics input to the slow neutron capture process. However, to a minor extent (n,p) and (n,α) reactions on a few light elements can play some role as neutron poisons.

Experimentally, neutron-induced cross section data are determined in two ways:

- time-of-flight (TOF); energy-dependent cross sections $\sigma(E_n)$ are deduced by inferring the neutron energy from the time of flight of the neutron over a certain flight path length. $\sigma(E_n)$ is then averaged over the thermal energy distribution of neutrons inside stars, so to be used in stellar models.
- activation technique; the sample of interest is first irradiated with a continuous neutron beam, and subsequently the activity of the resulting product nucleus is counted. Clearly, this method can be applied only with radioactive products, i.e. provided that neutron captures result in unstable isotopes.

In both cases, the astrophysical reaction rate is extracted, i.e. the number density of interacting particles times the reaction rate per particle pair $\langle\sigma v\rangle$. Since the interacting particles are in thermodynamic equilibrium in a stellar plasma, the relative velocity v can be described by a Maxwell-Boltzmann distribution. Consequently, the reaction rate per particle pair can be expressed in terms of the Maxwellian-averaged cross section (MACS):

$$\text{MACS} = \frac{\langle\sigma v\rangle}{v_T} = \frac{2}{\sqrt{\pi}(kT)^2}\int_0^\infty E\sigma(E)e^{-\frac{E}{kT}dE}, \tag{1}$$

If a cross section can be determined with the TOF method, it is possible to calculate MACSs at all relevant stellar temperatures, i.e from $kT = 8$ keV to $kT = 90$ keV. On the other hand, in case of rare and/or short-lived isotopes it is unlikely to collect few milligrams of mass (i.e., the minimum quantity required for a TOF measurement) as the activity of the sample pose severe constraints. In these cases, an alternative to TOF measurements is the activation method.

2 The n_TOF Facility at CERN

n_TOF is a white spallation neutron source based on the CERN proton synchrotron (PS) accelerator. In particular, neutrons are produced by 20 GeV/c protons impinging onto a massive Pb block, surrounded by a water layer acting as moderator of the initially fast neutron spectrum. The neutron-producing target [8] was conceived to produce a wide neutron energy spectrum at experimental areas [9, 10], ranging from meV to GeV. Two beam lines, EAR1 located at 185 m and EAR2 at 19 m from the neutron-producing target are equipped with detection systems for TOF measurements. In addition, an irradiation station named NEAR was recently built at 3 m from the spallation target. The high peak current of the PS accelerator, with nearly 10^{13} protons per pulse, results in 5×10^5 and 2×10^7 neutrons per pulse delivered to EAR1 and EAR2. In addition to the high instantaneous neutron flux, another important feature is the high resolution in neutron energy (reconstructed from the time of flight). The 185 m long flight path, in combination with the narrow proton-bunch width of 6 ns FWHM, result in an energy resolution $\Delta E/E \approx 10^{-4}$ in the eV region and $\Delta E/E \approx 10^{-3}$ in the keV region.

The n_TOF project has benefited from the close collaboration with other neutron facilities, in particular GELINA [11], at the European Commission Joint Research Center in Belgium and SARAF [12] in Israel. In fact, several cooperative projects were carried out in order to improve the quality of the cross-section data (see for example, the case of ^{197}Au [13–16] or ^{171}Tm [17].)

3 n_TOF: 20 years of nuclear data

In addition to some relevant cases not directly related to the *s*-process - as for example ^{7}Be(n,α) and ^{7}Be(n,p) [18, 19] of interest to Big Bang Nucleosynthesis, or ^{26}Al(n,p) and ^{26}Al(n,α) [20, 21] essential for understanding the observation of the ^{26}Al γ-ray emitter in our Galaxy - the n_TOF collaboration has investigated different aspects of the *s*-process: (i) the main neutron sources in Red Giants stars (e.g. [22–25]); (ii) The seeds nuclei for the *s*-process (e.g.^{58}Ni [26] and ^{62}Ni [27, 28]); (iii) end-point and isotopes with a closed shell configuration (^{204,206,207}Pb [29–31] and ^{209}Bi [32]); (iv) isotopes whose production is entirely ascribed to the *s* process (e.g. ^{58}Ni [26], ^{70}Ge [33], ^{154}Gd [34] and ^{186}Os [35, 36]); unstable isotopes where β decay competes with neutron capture (e.g.^{63}Ni [37] and ^{171}Tm [17]).

These examples represent a non-exhaustive overview of the n_TOF contribution to nucleosynthesis studies (for more details see [38]). In the near future, n_TOF will perform challenging cross-section measurements exploiting the improved characteristics of the renovated experimental areas EAR1 and EAR2, as well as the new irradiation station NEAR. The recent upgrade of the spallation target, has improved the energy resolution at EAR2 while increasing the instantaneous neutron flux. Therefore (n,γ) cross sections on radiositopes with very short half-life could become feasible either via TOF or activation. In addition, the possibility to produce rare isotopes at other facilities such as ISOLDE and PSI could lead to fruitful collaborations between these different facilities in the future.

References

[1] E.M. Burbidge, G.R. Burbidge, W.A. Fowler, F. Hoyle, Reviews of Modern Physics **29**, 547 (1957)
[2] F. Käppeler, R. Gallino, S. Bisterzo, W. Aoki, Rev. Mod. Phys. **83**, 157 (2011)
[3] M. Busso, R. Gallino, G.J. Wasserburg, Annual Review of Astronomy and Astrophysics **37**, 239 (1999)
[4] F. Herwig, Annual Review of Astronomy and Astrophysics **43**, 435 (2005)
[5] O. Straniero, R. Gallino, S. Cristallo, Nuclear Physics A **777**, 311 (2006), astro-ph/0501405
[6] A.I. Karakas, J.C. Lattanzio, Publications of the Astronomical Society of Australia **31**, e030 (2014)
[7] M. Pignatari, R. Gallino, M. Heil, M. Wiescher, F. Käppeler, F. Herwig, S. Bisterzo, The Astrophysical Journal **710**, 1557 (2010)
[8] R. Esposito, M. Calviani, O. Aberle, M. Barbagallo, D. Cano-Ott, T. Coiffet, N. Colonna, C. Domingo-Pardo, F. Dragoni, R. Franqueira Ximenes et al., Physical Review Accelerators and Beams **24**, 093001 (2021), 2106.11242

[9] M. Barbagallo, C. Guerrero, A. Tsinganis, D. Tarrío, S. Altstadt, S. Andriamonje, J. Andrzejewski, L. Audouin, V. Bécares, F. Bečvář et al., European Physical Journal A **49**, 156 (2013)

[10] M. Sabaté-Gilarte, M. Barbagallo, N. Colonna, F. Gunsing, P. Žugec, V. Vlachoudis, Y.H. Chen, A. Stamatopoulos, J. Lerendegui-Marco, M.A. Cortés-Giraldo et al., European Physical Journal A **53**, 210 (2017)

[11] W. Mondelears, P. Schillebeeckx, *GELINA, a Neutron Time-of-Flight Facility for High-Resolution Neutron Data Measurements. Research Infrastructures II (2)*, https://publications.jrc.ec.europa.eu/repository/handle/JRC35644 (2006)

[12] M. Paul, M. Tessler, M. Friedman, S. Halfon, T. Palchan, L. Weissman, A. Arenshtam, D. Berkovits, Y. Eisen, I. Eliahu et al., European Physical Journal A **55**, 44 (2019)

[13] C. Massimi, C. Domingo-Pardo, G. Vannini, L. Audouin, C. Guerrero, U. Abbondanno, G. Aerts, H. Álvarez, F. Álvarez-Velarde, S. Andriamonje et al., Physical Review C **81**, 044616 (2010)

[14] I. Sirakov, B. Becker, R. Capote, E. Dupont, S. Kopecky, C. Massimi, P. Schillebeeckx, European Physical Journal A **49**, 144 (2013)

[15] C. Lederer, N. Colonna, C. Domingo-Pardo, F. Gunsing, F. Käppeler, C. Massimi, A. Mengoni, A. Wallner, U. Abbondanno, G. Aerts et al., Physical Review C **83**, 034608 (2011)

[16] C. Massimi, B. Becker, E. Dupont, S. Kopecky, C. Lampoudis, R. Massarczyk, M. Moxon, V. Pronyaev, P. Schillebeeckx, I. Sirakov et al., European Physical Journal A **50**, 124 (2014)

[17] C. Guerrero, J. Lerendegui-Marco, M. Paul, M. Tessler, S. Heinitz, C. Domingo-Pardo, S. Cristallo, R. Dressler, S. Halfon, N. Kivel et al., Physical Review Letters **125**, 142701 (2020)

[18] M. Barbagallo, A. Musumarra, L. Cosentino, E. Maugeri, S. Heinitz, A. Mengoni, R. Dressler, D. Schumann, F. Käppeler, N. Colonna et al. (n_TOF Collaboration), Phys. Rev. Lett. **117**, 152701 (2016)

[19] L. Damone, M. Barbagallo, M. Mastromarco, A. Mengoni, L. Cosentino, E. Maugeri, S. Heinitz, D. Schumann, R. Dressler, F. Käppeler et al. (The n_TOF Collaboration [www.cern.ch/ntof]), Phys. Rev. Lett. **121**, 042701 (2018)

[20] C. Lederer-Woods, P.J. Woods, T. Davinson, D. Kahl, S.J. Lonsdale, O. Aberle, S. Amaducci, J. Andrzejewski, L. Audouin, M. Bacak et al. (n_TOF Collaboration), Phys. Rev. C **104**, L022803 (2021)

[21] C. Lederer-Woods, P.J. Woods, T. Davinson, A. Estrade, J. Heyse, D. Kahl, S.J. Lonsdale, C. Paradela, P. Schillebeeckx, O. Aberle et al. (n_TOF Collaboration), Phys. Rev. C **104**, L032803 (2021)

[22] S. Cristallo, M. La Cognata, C. Massimi, A. Best, S. Palmerini, O. Straniero, O. Trippella, M. Busso, G.F. Ciani, F. Mingrone et al., The Astrophysical Journal **859**, 105 (2018), 1804.10751

[23] P. Adsley, U. Battino, A. Best, A. Caciolli, A. Guglielmetti, G. Imbriani, H. Jayatissa, M. La Cognata, L. Lamia, E. Masha et al., Physical Review C **103**, 015805 (2021), 2005.14482

[24] C. Massimi, P. Koehler, S. Bisterzo, N. Colonna, R. Gallino, F. Gunsing, F. Käppeler, G. Lorusso, A. Mengoni, M. Pignatari et al., Physical Review C **85**, 044615 (2012)

[25] C. Massimi, S. Altstadt, J. Andrzejewski, L. Audouin, M. Barbagallo, V. Bécares, F. Bečvář, F. Belloni, E. Berthoumieux, J. Billowes et al., Physics Letters B **768**, 1 (2017)

[26] P. Žugec, M. Barbagallo, N. Colonna, D. Bosnar, S. Altstadt, J. Andrzejewski, L. Audouin, V. Bécares, F. Bečvář, F. Belloni et al., Physical Review C **89**, 014605 (2014)

[27] C. Lederer, C. Massimi, E. Berthoumieux, N. Colonna, R. Dressler, C. Guerrero, F. Gunsing, F. Käppeler, N. Kivel, M. Pignatari et al., Physical Review C **89**, 025810 (2014), `1403.4778`

[28] C. Lederer, C. Massimi, E. Berthoumieux, N. Colonna, R. Dressler, C. Guerrero, F. Gunsing, F. Käppeler, N. Kivel, M. Pignatari et al., Physical Review C **92**, 019903 (2015)

[29] C. Domingo-Pardo, U. Abbondanno, G. Aerts, H. Álvarez-Pol, F. Alvarez-Velarde, S. Andriamonje, J. Andrzejewski, P. Assimakopoulos, L. Audouin, G. Badurek et al., Physical Review C **75**, 015806 (2007), `nucl-ex/0610033`

[30] C. Domingo-Pardo, U. Abbondanno, G. Aerts, H. Álvarez, F. Alvarez-Velarde, S. Andriamonje, J. Andrzejewski, P. Assimakopoulos, L. Audouin, G. Badurek et al., Physical Review C **76**, 045805 (2007), `0707.3679`

[31] C. Domingo-Pardo, U. Abbondanno, G. Aerts, H. Álvarez-Pol, F. Alvarez-Velarde, S. Andriamonje, J. Andrzejewski, P. Assimakopoulos, L. Audouin, G. Badurek et al., Physical Review C **74**, 055802 (2006), `nucl-ex/0610039`

[32] C. Domingo-Pardo, U. Abbondanno, G. Aerts, H. Álvarez-Pol, F. Alvarez-Velarde, S. Andriamonje, J. Andrzejewski, P. Assimakopoulos, L. Audouin, G. Badurek et al., Physical Review C **74**, 025807 (2006), `nucl-ex/0610040`

[33] A. Gawlik, C. Lederer-Woods, J. Andrzejewski, U. Battino, P. Ferreira, F. Gunsing, S. Heinitz, M. Krtička, C. Massimi, F. Mingrone et al., Physical Review C **100**, 045804 (2019), `1910.09360`

[34] A. Mazzone, S. Cristallo, O. Aberle, G. Alaerts, V. Alcayne, S. Amaducci, J. Andrzejewski, L. Audouin, V. Babiano-Suarez, M. Bacak et al., Physics Letters B **804**, 135405 (2020), `2002.02322`

[35] M. Mosconi, K. Fujii, A. Mengoni, C. Domingo-Pardo, F. Käppeler, U. Abbondanno, G. Aerts, H. Álvarez-Pol, F. Alvarez-Velarde, S. Andriamonje et al., Physical Review C **82**, 015802 (2010)

[36] K. Fujii, M. Mosconi, A. Mengoni, C. Domingo-Pardo, F. Käppeler, U. Abbondanno, G. Aerts, H. Álvarez-Pol, F. Alvarez-Velarde, S. Andriamonje et al., Physical Review C **82**, 015804 (2010)

[37] C. Lederer, C. Massimi, S. Altstadt, J. Andrzejewski, L. Audouin, M. Barbagallo, V. Bécares, F. Bečvář, F. Belloni, E. Berthoumieux et al., Physical Review Letters **110**, 022501 (2013), `1304.3310`

[38] C. Massimi, S. Cristallo, C. Domingo-Pardo, C. Lederer-Woods, Universe **8** (2022)

Indirect measurement of the $(n,\gamma)^{127}$Sb cross section

Francesco Pogliano[1,*], *Ann-Cecilie* Larsen[1,**], *Frank Leonel* Bello Garrote[1], *Marianne* Møller Bjørøen[1], *Thomas* Kvalheim Eriksen[1], *Dorthea* Gjestvang[1], *Andreas* Görgen[1], *Magne* Guttormsen[1], *Kevin Ching Wei* Li[1], *Maria* Markova[1], *Eric Francis* Matthews[2], *Wanja* Paulsen[1], *Line* Gaard Pedersen[1], *Sunniva* Siem[1], *Tellef* Storebakken[1], *Tamás* Gabor Tornyi[1], and *Julian* Ersland Vevik[1]

[1]Department of Physics, University of Oslo, N-0316 Oslo, Norway
[2]Department of Nuclear Engineering, University of California, Berkeley, California 94720 U.S.A.

Abstract. Sensitivity studies of the *i* process have identified the region around ^{135}I as a bottleneck for the neutron capture flow. Nuclear properties such as the Maxwellian-averaged cross section (MACS) are key to constrain the uncertainties in the final abundance patterns. With the Oslo method, we are able to indirectly measure such properties for the nuclei involved in this process. From the ^{124}Sn$(\alpha, p\gamma)^{127}$Sb reaction data we extract the nuclear level density and γ-ray strength function for ^{127}Sb. The level density at higher excitation energies is compatible with the constant-temperature model, while the γ-ray strength function presents features like an upbend and a pygmy-like structure below S_n. From these two quantities we can calculate the MACS for the ^{126}Sb$(n,\gamma)^{127}$Sb reaction using the Hauser-Feshbach formalism, and constrain its unceraintities from the theoretical ones. Libraries such as JINA REACLIB, TENDL and BRUSLIB agree well with the experimental results, while ENDF/B-VIII.0 predicts a higher rate.

1 Introduction

Burbidge et al. [1] in 1957 in their seminal paper identified the *s* and the *r* processes (standing for the slow and rapid neutron-capture processes, respectively) as the main mechanisms behind the creation of the elements heavier than iron in our universe. The two nucleosynthesis processes produce different elemental abundance patterns, and while our Sun's abundance presents characteristics from both processes, in many stars only the pattern from one of these may be observable [2]. This is the case for carbon-enhanced, metal-poor stars (CEMPs), which may be enriched in *r* process elements [3], *s* process elements or both [4]. This last case is challenging to explain, as the two nucleosynthesis processes are thought to happen in different astrophysical sites, and CEMPs are thought to have formed before the interstellar medium from them could mix. A possible solution to this problem is the introduction of the *i* process. The *i* process stands for intermediate neutron-capture process, has neutron densities between those of the *s* and the *r* process, and was first proposed in 1977 by Cowan and Rose [5]. Sensitivity simulations of this process are able to reproduce the abundances for

*e-mail: francesco.pogliano@fys.uio.no
**e-mail: a.c.larsen@fys.uio.no

these CEMPs-s/r (CEMPs presenting elements characteristics to both the s and r process), and identify the region around ^{135}I as a bottleneck [6]. However, one of the main sources of uncertainty lies in the correct estimation of nuclear properties such as the neutron-capture rates of the nuclei involved in the i process. By carrying out experimental studies in the ^{135}I region we are able to constrain uncertainties for the neutron-capture rates. This also provides useful experimental measurements of nuclear properties so that better and more precise theoretical models may be developed, these being very important for both i and r process simulations. This region of the nuclear chart involves many unstable, neutron-rich nuclei, for whose it is very difficult to carry direct measurements of the (n, γ) reaction cross section. For this reason an indirect approach is used. With the Oslo method we are able to extract two statistical properties of the nucleus, the nuclear level density (NLD) and the γ-ray strength function (GSF). The NLD and GSF are statistical properties of the nucleus, the first one being the "continuous" equivalent of counting energy levels, and the second of the reduced transition probabilities. These are the ingredients needed in the Hauser-Feshbach formalism (together with the optic model potential) to calculate the neutron-capture rate (see [7] and references therein.) Here we present the results from the ^{124}Sn$(\alpha, p\gamma)^{127}$Sb experiment. By analyzing the data with the Oslo method we obtain the NLD and the GSF of ^{127}Sb, part of the ^{135}I region. The $(n, \gamma)^{127}$Sb reaction rate is then calculated, presented and compared to known libraries.

2 The extraction of the NLD and GSF

The ^{124}Sn$(\alpha, p\gamma)^{127}$Sb reaction experiment was carried out at the Oslo Cyclotron Laboratory for a period of five days. a 24 MeV α-beam was impinged on a self-supporting ^{124}Sn target. In order to use the Oslo method, we are interested in particle-γ coincidences. For this reason the reaction data was collected using a silicon particle detector (SiRi [8],) and a $LaBr_3$ γ-ray detector (OSCAR, [9]). The particle detector consists of a thin front detector (ΔE) and a thicker back detector (E), and this allows us to separate the different reaction channels and only select the (α, p) data. From the energy of the beam and that of the ejected proton, we are able to calculate the excitation energy the resulting ^{127}Sb is left in, and associate it to the coinciding γ-rays from its de-excitation to the ground state. These different γ-ray spectra for each excitation energy can be plotted together in a matrix, called raw coincidence matrix. The Oslo method is a procedure for which we can obtain the NLD and the GSF of a nucleus starting from a raw coincidence matrix [10–12]. First the matrix has to be unfolded to get rid of the detector response. Secondly, The first-generation matrix is obtained with a subtraction technique, where we only select the first γ-rays from each de-excitation. The NLD and the GSF are used to describe the nucleus in the quasicontinuum excitation energy region, where the level density becomes so large ($\gtrsim$ 50 per MeV) that it is more useful to describe nuclear properties statistically. These are then extracted from the first-generation matrix from this region using a χ^2 minimization algorithm explained in Ref. [12] and finally normalized following the procedure described in the main article about this experiment (Ref. [13]). The extracted NLD and GSF are shown in Figure 1. Here we see how the NLD follows an exponential curve compatible with the constant-temperature model. The GSF shows features such as the upbend at low energies, the pygmy resonance and possibly a small structure at $\approx$ 3 MeV. For each graph, all the available TALYS 1.95 [15, 16] models were plotted as comparison, showing a general disagreement to the experimental data.

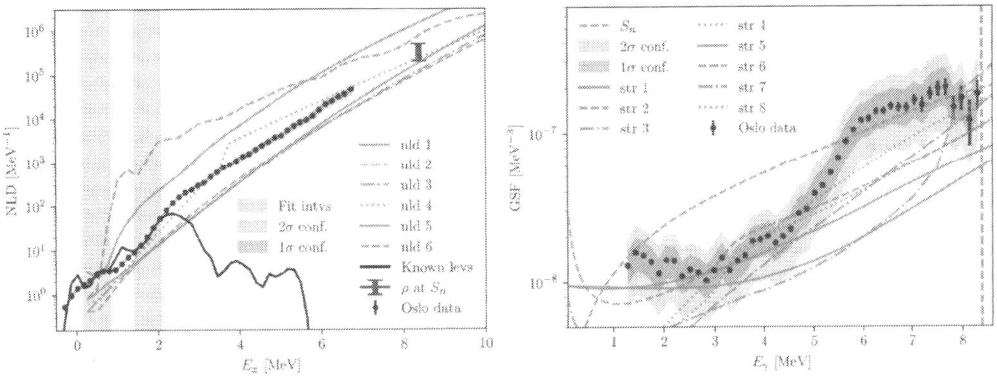

Figure 1. The NLD (left) and the GSF (right) for ^{127}Sb. `nld` and `str` indicate the model number obtained from TALYS 1.95 [15, 16]. The red bars in the left plot indicate the low excitation energy regions the NLD was normalized to (see Ref. [13] for details.)

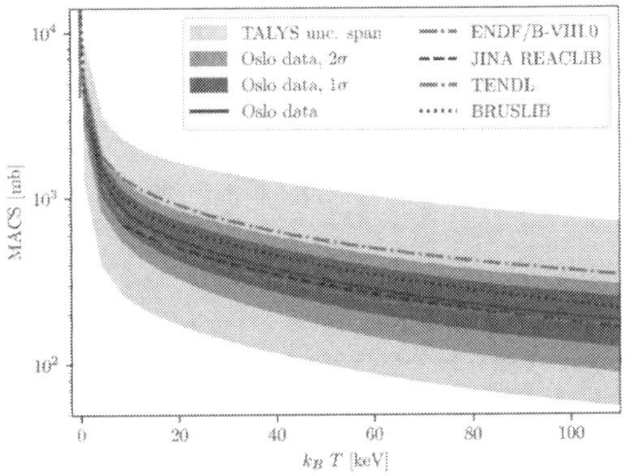

Figure 2. The constrained MACS uncertainty for the $(n,\gamma)^{127}$Sb reaction, compared to the uncertainty range from combining all the NLD and GSF theoretical models from TALYS.

3 The MACS

The Maxwellian-averaged cross section (MACS) is a quantity easily translatable to the neutron-capture rate [14] and is an important nuclear input parameter for nucleosynthesis simulations. Theoretical MACSs can be calculated with TALYS in the Hauser-Feshbach framework by choosing a NLD, a GSF and a optical model potential. While the latter does not influence the MACS too much, combining all the models gives an uncertainty span of more than one order of magnitude (see Figure 2). The experiment data constrains this uncertainty and is compared to common libraries such as JINA REACLIB [17], BRUSLIB [18], TENDL [16] and ENDF/B-VIII.0 [19]. While the first three give predictions within 1σ uncertainty from the experimental result, the latter predicts values outside the 2σ uncertainty region.

References

[1] M. Burbidge et al., Rev. Mod. Phys. **29**, 547 (1957)
[2] M. Arnould et al., Physics Reports **450**, Pages 97-213 (2007)
[3] T. Hansen et al., ApJ **807**, 173 (2015)
[4] C. Sneden et al., Ann. Rev. of A&A **46:1**, 241-288 (2008)

[5] J. J. Cowan et al., ApJ **212**, 149-158 (1977)
[6] M. Hampel et al., ApJ **831**, 171 (2016)
[7] A.C. Larsen et al., PPNP **107**, Pages 69-108 (2019)
[8] M. Guttormsen et al., Nucl. Instrum. Methods Phys. Res. Sect. A **648**, 168 (2011)
[9] F. Zeiser et al., Nucl. Instrum. Methods Phys. Res. Sect. A **985**, 164678 (2020)
[10] M. Guttormsen et al., Nucl. Instrum. Methods Phys. Res. Sect. A **374**, 371 (1996)
[11] M. Guttormsen et al., Nucl. Instrum. Methods Phys. Res. Sect. A **255**, 518 (1987)
[12] A. Schiller et al., Nucl. Instrum. Methods Phys. Res. Sect. A **447**, 498 (2000)
[13] F. Pogliano et al. Phys. Rev. C **106**, 015804 (2022)
[14] C. Iliadis, N*uclear Physics of Stars* (Wiley-VCH, Weinheim, 2015) 147-150
[15] A. Koning et al., User manual, Tech. Rep., (2017)
[16] A. Koning et al., Nucl. Data Sheets **155**, 1 (2019)
[17] R. H. Cyburt et al., ApJ Suppl. Series **189**, 240 (2010)
[18] Y. Xu et al., Astron. Astrophys. **549**, A106 (2013)
[19] D. Brown et al., Nucl. Data Sheets **148**, 1 (2018)

EPJ Web of Conferences 279, 11011 (2023) https://doi.org/10.1051/epjconf/202327911011
NPA-X 2022

The ASTRAL database for neutron-capture nucleosynthesis studies

Diego Vescovi[1,*], *René* Reifarth[1], *Enis* Lorenz[1], and *Andreas* Elbe[1]

[1]Goethe University Frankfurt, Max-von-Laue-Strasse 1, Frankfurt am Main 60438, Germany

Abstract. Present nuclear reaction network computations for astrophysical simulations involve many different types of rates, including neutron-capture reactions of interest for the modeling of heavy-element nucleosynthesis. While for many of them we still have to rely on theoretical calculations, an increasing number of experimentally-determined cross sections have now become available. In this contribution, we present "ASTrophysical Rate and rAw data Library" (ASTRAL), a new online database for neutron-capture cross sections based on experimental results, mainly obtained through activation and time-of-flight measurements. For the evaluation process, cross sections were recalculated starting from raw data and by considering recent changes in physical properties of the involved isotopes (e.g., half-life and γ-ray intensities). We show the current status of the database, the techniques adopted to derive the recommended Maxwellian-averaged cross sections, and future developments.

1 Introduction

The understanding of nucleosynthesis in stars has become a forefront field of experimental and theoretical research owing to recent developments in observational techniques, astrophysical models, and laboratory experiments. The vast majority of nuclei heavier than iron are synthesized through the slow process [1] and the rapid process [2]. Experimental determination of the (n, γ) cross sections at stellar thermal energies is of utmost importance in studying these processes. In light of this, it is of paramount importance to have a database of neutron-capture cross sections to perform accurate astrophysical calculations of heavy element nucleosynthesis. To date, there is no recent database of this kind, the last such work being more than 10 years old and no longer maintained [3]. We aim to fill this gap by creating a new, updated, and homogeneous database of Maxwellian-averaged cross sections (MACS) to be used for astrophysical computations, named ASTRAL - "ASTrophysical Rate and rAw data Library".

2 The ASTRAL database

In contrast to previous databases, ASTRAL does contain not only actual cross sections but also experimentally-determined raw data that are directly used to evaluate the MACS from 1 to 500 keV. To cover the whole energy range, energy-differential cross sections are needed

*e-mail: vescovi@iap.uni-frankfurt.de

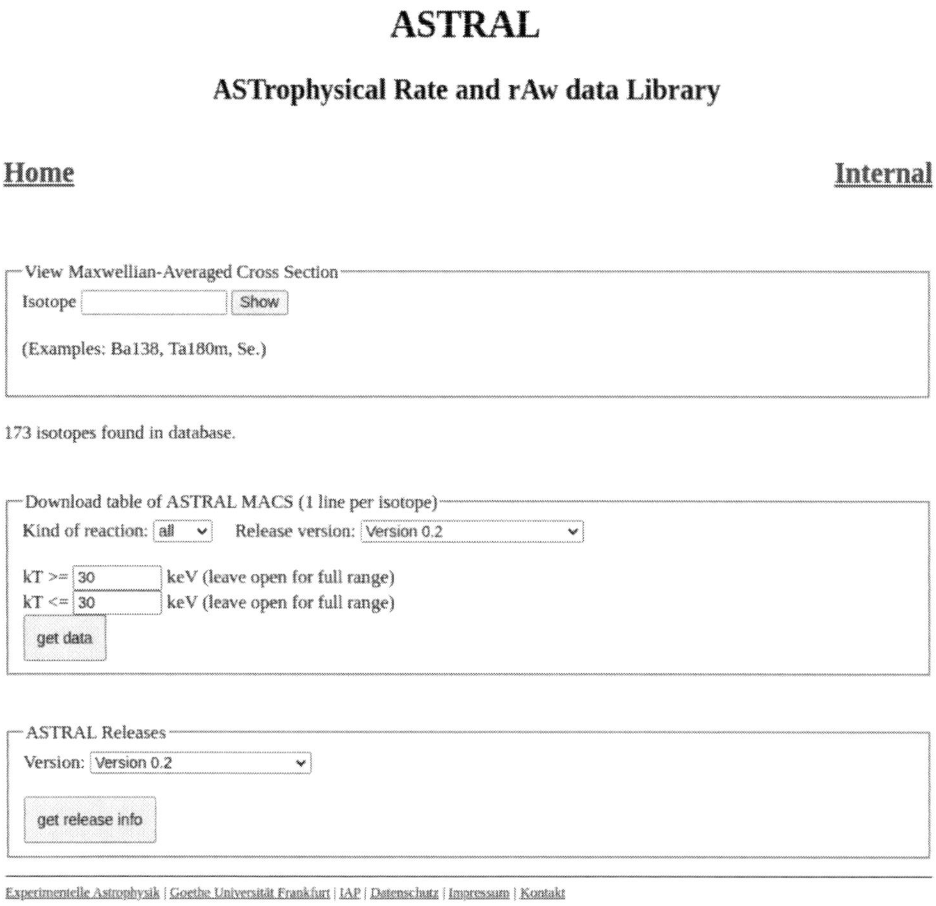

Figure 1. Web appearance of the ASTRAL database [4].

for neutron energy up to a few MeV. When experimental data in this range is missing, evaluated cross sections from latest data libraries (ENDF, JEFF, JENDL, TENDL) are taken into account. The measured data can be cross section ratios between the isotope under study and the reference cross section or resonance parameters, to be used for computing the differential cross section. The MACS are evaluated upon various experimental measurements of (n,γ) cross sections, based on activation and time-of-flight (TOF) techniques. ASTRAL is an online database [4] (see Fig. 1). The first release (v0.1) of ASTRAL occurred in 2018, in which about 70 isotope MACS, based on TOF measurements, were re-evaluated by taking into account the new recommended value for gold differential cross section [5]. In 2022, the ASTRAL v0.2 has been released [6]. It contains an updated version of the 70 isotopes already evaluated and, in addition, about 50 new isotopes whose experimental cross section data were based on activation measurements.

3 ASTRAL status: activations

The current version of the ASTRAL database (v0.2) includes raw data for around 200 activation measurements, for isotopes from ^{9}Be to ^{209}Bi. For each of them, the needed information to determine the MACS are stored, in particular, the measured ratio (with errors) of

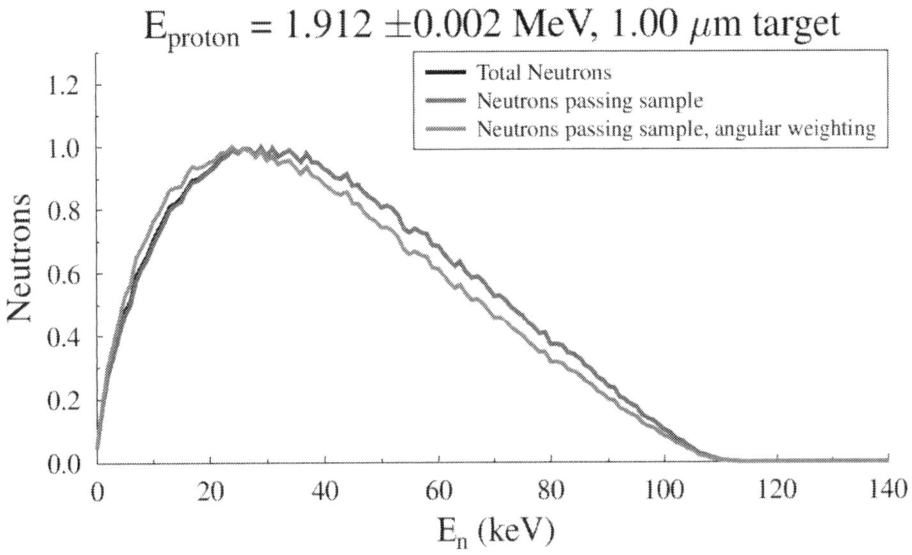

Figure 2. Comparison of the number of angle-integrated neutrons per linear energy bin for a simulation that contain weighting and no weighting for a sample with a radius of 3 mm. The radius of the Li-spot is 3 mm. The simulated spectrum is normalized to a common maximum of 1.

the neutron-capture cross sections relative to gold. In addition, the beam proton-energy distribution, the properties of the neutron sources and target samples, and the geometry of the activation setup are collected as well. These quantities are required to properly simulate the experimental neutron spectrum with a Monte Carlo approach. For this purpose, we make use of the PINO code [7]. An example of the reconstructed neutron spectrum with PINO can be found in Fig. 2. Moreover, cross sections are corrected by considering new values for the adopted gamma-ray intensities of the activation products from the NuDat database[1]. These corrections have to be taken into account since they affect the cross-section value by more than 1%. For some unstable targets, a further corrective factor is applied as well to account for the effect of improved half-life (e.g., ^{60}Fe).

The MACS evaluation is performed in a new manner. In activation measurements, neutron energies are unknown and only spectrum-averaged cross sections (SACS) can be determined. The SACS represent an average over the respective neutron spectrum, i.e.

$$\text{SACS} = \frac{\int \sigma(E_n)\Phi(E_n)\,dE_n}{\int \Phi(E_n)\,dE_n}, \tag{1}$$

where the integral runs from the lower to the upper limit of the experimental neutron spectrum $\Phi(E_n)$. Should $\Phi(E_n)$ be a perfect Maxwellian distribution, the SACS would equal the MACS up to a factor $2/\sqrt{\pi}$. However, due to energy cutoffs in the neutron spectra and variations between experimental and Maxwell-Boltzmann distribution, corrections have to be considered to compute MACS at different thermal energies. This can be accomplished by the use of the experimentally-determined SACS to normalize the energy-dependent cross sections from the evaluated data libraries by means of a suited normalization factor (NF).

[1]https://www.nndc.bnl.gov/nudat3

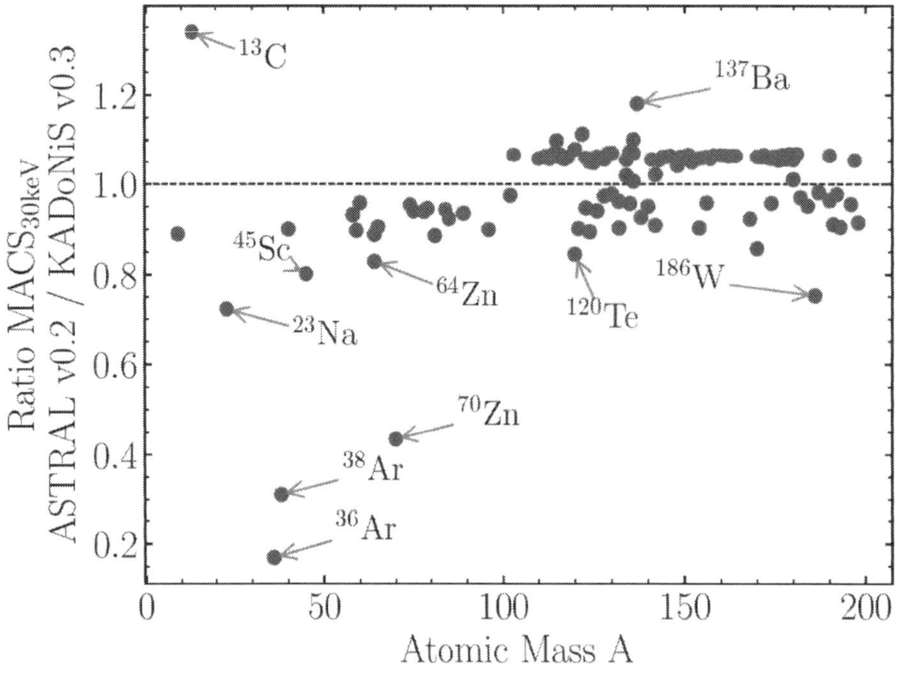

Figure 3. Comparison of MACS at kT=30 keV from the present ASTRAL v0.2 and the KADoNiS v0.3 compilation [3].

In general, the normalization factor should be temperature-dependent and if several activation measurements with spectra corresponding to different thermal energies are available, they have to be properly considered. For this reason, in ASTRAL, activation measurements at different energy regimes have been accounted for by computing proper weighting factors, based on the overlap between the experimental neutron spectra, simulated with a Monte Carlo approach with the PINO code, and the Maxwellian energy distribution. As a result, the relative weights of each activation vary depending on the temperature. This method assures that the experimental spectrum with the highest weight is the one closest to the stellar energy distribution. The normalization factor was calculated based on the weights to scale the evaluated cross sections [6]. The corresponding MACS for each thermal energy kT was then determined using this energy-dependent evaluated cross section as

$$\mathrm{MACS} = \frac{2}{\sqrt{\pi}} \frac{NF}{(kT)^2} \int_0^\infty \sigma_{\mathrm{lib}}(E_n)\, E_n \exp\left(-\frac{E_n}{kT}\right) dE_n. \tag{2}$$

In this way, about 50 isotopes, whose SACS are determined by activation measurements, were evaluated [6].

4 ASTRAL status: TOF

In ASTRAL v0.2, we made use of the latest evaluated data to recompute MACS of around 70 isotopes already present in ASTRAL v0.1. For them, TOF data are available across a narrow neutron energy range, while MACS computation requires data virtually ranging from zero to infinity. Therefore, the use of cross sections from evaluated data library, properly normalized to the measured data from TOF experiments, is needed in the energy region not covered by experimental data [5]. While ASTRAL v0.1 was largely based on ENDF-B/VII.1 [8], in version 0.2 we mostly adopted ENDF-B/VIII [9] and, in some cases, JENDL-4.0 [10] or JEFF-3.3 [11].

5 Conclusions and perspectives

Figure 3 gives an overview of the changes between the present version of ASTRAL (v0.2) and the KADoNiS v0.3 compilation [3]. The isotopes with the largest changes are labeled. Whereas most of these changes are due to the revised reference gold cross section [5], some of these nuclei were measured for the first time and thus replace theoretical estimates (e.g., ^{36}Ar and ^{38}Ar). Other important variations are also due to the adoption of new data libraries (e.g., ^{70}Zn) or revised gamma-intensities (e.g., ^{186}W). Another database version is foreseen to be released by the end of 2022, containing more MACS based on activation and TOF measurements.

DV acknowledges financial support from the German-Israeli Foundation (GIF No. I-1500-303.7/2019).

References

[1] J.J. Cowan, C. Sneden, J.E. Lawler, A. Aprahamian, M. Wiescher, K. Langanke, G. Martínez-Pinedo, F.K. Thielemann, Reviews of Modern Physics **93**, 015002 (2021), 1901.01410

[2] F. Käppeler, R. Gallino, S. Bisterzo, W. Aoki, Reviews of Modern Physics **83**, 157 (2011), 1012.5218

[3] I. Dillmann, R. Plag, F. Käppeler, T. Rauscher, *KADoNiS v0.3 - the third update of the Karlsruhe Astrophysical Database of Nucleosynthesis in Stars*, in *EFNUDAT Fast Neutrons - Proceedings of the Scientific Workshop on Neutron Measurements, Theory and Applications - Nuclear Data for Sustainable Nuclear Energy - 28-30 April 2009, Geel, Belgium*, edited by F. Hambsch (2010), pp. 55–58

[4] R. Reifarth, D. Vescovi, *Astral - astrophysical rate and raw data library*, https://exp-astro.de/astral/ (2022)

[5] R. Reifarth, P. Erbacher, S. Fiebiger, K. Göbel, T. Heftrich, M. Heil, F. Käppeler, N. Klapper, D. Kurtulgil, C. Langer et al., European Physical Journal Plus **133**, 424 (2018), 1902.09838

[6] R. Reifarth, Tech. rep., IAEA (2021), https://www-nds.iaea.org/media/docs/tr/Reifarth_Summary_Report_TAL2021.pdf

[7] R. Reifarth, M. Heil, F. Käppeler, R. Plag, Nuclear Instruments and Methods in Physics Research A **608**, 139 (2009), 1310.1637

[8] M.B. Chadwick, M. Herman, P. Obložinský, M.E. Dunn, Y. Danon, A.C. Kahler, D.L. Smith, B. Pritychenko, G. Arbanas, R. Arcilla et al., Nuclear Data Sheets **112**, 2887 (2011)

[9] D.A. Brown, M.B. Chadwick, R. Capote, A.C. Kahler, A. Trkov, M.W. Herman, A.A. Sonzogni, Y. Danon, A.D. Carlson, M. Dunn et al., Nuclear Data Sheets **148**, 1 (2018)

[10] K. Shibata, O. Iwamoto, T. Nakagawa, N. Iwamoto, A. Ichihara, S. Kunieda, S. Chiba, K. Furutaka, N. Otuka, T. Ohsawa et al., Journal of Nuclear Science and Technology **48**, 1 (2011)

[11] A.J.M. Plompen, O. Cabellos, C. De Saint Jean, M. Fleming, A. Algora, M. Angelone, P. Archier, E. Bauge, O. Bersillon, A. Blokhin et al., European Physical Journal A **56**, 181 (2020)

EPJ Web of Conferences **279**, 11012 (2023) https://doi.org/10.1051/epjconf/202327911012
NPA-X 2022

LUNA measurement found no evidence of a low-energy resonance in ^{6}Li(p, γ)^{7}Be reaction

D. Piatti[1,2,*] for LUNA Collaboration

[1]Università degli Studi di Padova, Via Marzolo 8 35136 Padova (Italy)
[2]INFN Division of Padova, Via Marzolo 8 35136 Padova (Italy)

Abstract. The ^{6}Li(p, γ)^{7}Be reaction is mainly at work in three nucleosynthesis scenarios: Big Bang Nucleosynthesis, ^{6}Li depletion in pre-main and in main sequence stars and cosmic ray interaction with interstellar matter. The ^{6}Li(p, γ)^{7}Be S-factor trend was poorly constrained at astrophysical energies because of conflicting experimental results reported in literature. A recent direct measurement, indeed, found a resonance-like structure at $E_{c.m.}$ = 195 keV, corresponding to an excited state at $E_x \sim 5800$ keV in ^{7}Be which, however, has not been confirmed by either other direct measurements or predicted by theoretical calculations.
In order to clarify the existence of this resonance, a new experiment was performed at the Laboratory for Underground Nuclear Astrophysics (LUNA), located deep underground in Gran Sasso Laboratory. Thanks to the extremely low background environment, the ^{6}Li(p, γ)^{7}Be cross section was measured in the center-of-mass energy range E = 60-350 keV with unprecedented sensitivity. No evidence for the alleged resonance was found. LUNA results was confirmed by latest published indirect determination of ^{6}Li(p, γ)^{7}Be S-factor and it is supported by a recent theoretical study.

1 Introduction

Lithium abundance involves mainly three nucleosynthesis scenarios. The Galactic chemical evolution models predict, indeed, that most of the solar lithium was provided by low-mass stars [1] while the rest was produced by Big Bang Nucleosynthesis (BBN) [2, 3] or by Galactic cosmic rays interacting with interstellar matter.

The ^{6}Li/^{7}Li isotopic ratio has been proposed as a tool to constrain non-standard lithium production mechanisms [4] and pollution of stellar atmospheres [5] in the context of the cosmological lithium problem. Recent (re-)observations of metal poor stars either severely reduced or provided only upper limits for the lithium isotopic ratio [6–8], suggesting that ^{6}Li depletion must occur in halo stars, which in turn call into question the ^{7}Li abundance observed in these stars corresponds to the primordial value [9].

The ^{6}Li(p,γ)^{7}Be reaction (Q value = 5606.85(7) keV) has a crucial role in determining the stellar ^{6}Li/^{7}Li ratio. The ^{6}Li(p,γ)^{7}Be reaction not only deplete ^{6}Li but it also convert some of it to ^{7}Li, through ^{7}Be radioactive decay.

*e-mail: denise.piatti@pd.infn.it

The slope of the astrophysical S-factor is poorly constrained at low energies given the inconsistent results reported in literature [10, 11]. Moreover, a new resonance at $E_{c.m.}$ = 195 keV, corresponding to an excited level at $E_x \approx 5800$ keV with $J^\pi = (1/2^+, 3/2^+)$ and $\Gamma_p \approx$ 50 keV, was claimed by [12]. In a recent comprehensive study of the ^{3}He(^{4}He,γ)^{7}Be reaction (Q value = 1587.14(7)) no evidence for such a resonance was found at $E_{c.m.}$ = 4210 keV [13].

None of the theoretical calculations of the ^{6}Li(p,γ)^{7}Be S-factor can reproduce the newly-reported resonance [14, 15, and references therein], unless this is added *ad-hoc* to reproduce the experimental data [16].

2 Experimental Setup

A new experiment [17] was performed at the Laboratory for Underground Nuclear Astrophysics (LUNA), at Laboratori Nazionali del Gran Sasso (Italy) [18]. The LUNA deep-underground location guarantees the reduction of environmental background by several orders of magnitude with respect to overground laboratories, enabling high-sensitivity measurements to be performed.

A schematic view of the experimental setup is shown in Fig.1. The high-intensity pro-

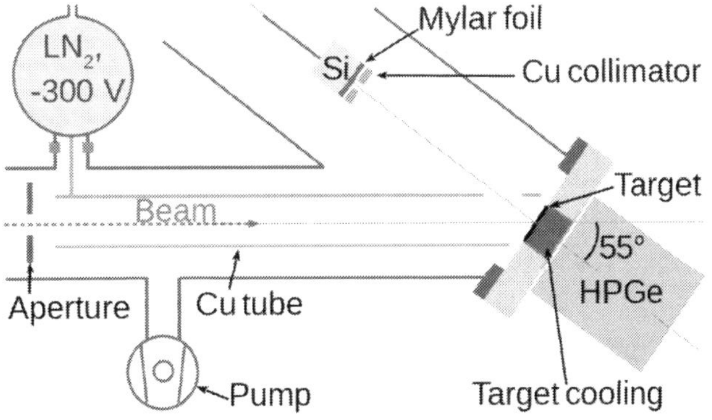

Figure 1. Sketch of the experimental setup used for the measurement of the ^{6}Li(p,γ)^{7}Be cross section at LUNA [17].

ton beam was provided by LUNA-400 accelerator [19] and it was collimated and delivered through a copper pipe to the target, mounted at 55° with respect to the beam direction. The Cu tube was used both as a cold trap, to improve the scattering chamber vacuum and prevent carbon build-up on target, and for secondary electron suppression. The evaporated targets were made from ^{6}Li$_2$WO$_4$ or ^{6}Li$_2$O powder, with thicknesses 100 – 200 μg/cm^2 and 20 μg/cm^2 respectively. The ^{6}Li isotopic enrichment level was 95% for all targets, which were water cooled to limit target degradation during irradiation [17].

To detect ^{6}Li(p,γ)^{7}Be reaction γ-rays a High-Purity Germanium (HPGe) detector was positioned in close geometry to the target and at 55° with respect to the beam direction. In addition a Silicon (Si) detector was installed at 125° from the beam direction to detect the α and ^{3}He particles from the ^{6}Li(p,α)^{3}He reaction concurrently with the gamma rays from the ^{6}Li(p,γ)^{7}Be reaction. Efficiencies for both detectors were obtained using GEANT simulations, fine tuned through the comparison with experimental results for γ and α standards as well as for known resonances of ^{14}N(p,γ)^{15}O and ^{18}O(p,α)^{15}N reactions [17]. The total uncertainty is 4% and 8% for the HPGe and Si detector efficiency respectively.

3 Results and Discussion

A measurement of the ^{6}Li(p,γ)^{7}Be and ^{6}Li(p,α)^{3}He excitation functions was performed for each target in the whole dynamic range of the LUNA-400 accelerator in order to make consistency checks and verify results are unaffected by systematic effects [17].

The ^{6}Li(p,γ)^{7}Be experimental yield was calculated as the sum of the contributions from the direct capture to the ground state (γ_0) and to the 429 keV excited state of ^{7}Be (γ_1).

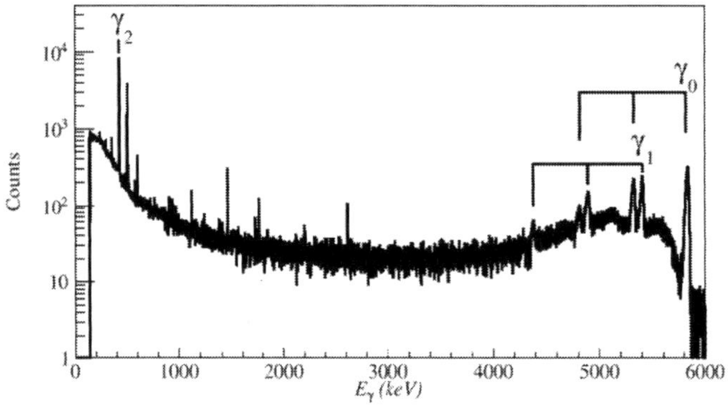

Figure 2. Experimental γ spectrum acquired at E_p = 265 keV. The ^{6}Li(p,γ)^{7}Be proceeds through direct capture (DC) to either the ground state of ^{7}Be, γ_0, or its first excited state, γ_1, with subsequent emission of a 429 keV secondary gamma ray, γ_2.

For the calculation of the ^{6}Li(p,γ)^{7}Be reaction S-factor, we adopted a relative approach [17]: the (p,γ) yield was normalized at each energy to the (p,α) yield. This ratio can be expressed in terms of the ratio between (p,γ) and (p,α) S-factors. We adopted for the ^{6}Li(p,α)^{3}He reaction the S-factor parametrization reported in [20]. For the (p,α) channel, the angular distribution coefficients A_k and related uncertainties were taken from [21, and references therein]. For the (p,γ) channel we adopted the theoretical angular distribution described in [14]. The measured S-factor was corrected for electron screening using the approximation in [22] and assuming a screening potential U_e = 273 eV [20].

The present S-factor has a monotonic dependence on the energy and show no evidence of the resonance reported in [12], see Fig.3. The measurement covered the center-of-mass energy range 60 – 350 keV and the reported statistical and systematic uncertainty were $\leq$2% and 12% respectively. Combining current data and the high energy results reported in [23] an R-matrix fit was performed providing an extrapolated S-factor to zero energy $S(0)$ = 95 $\pm$ 5 eV b. The R-matrix fit was used to calculate a new ^{6}Li(p,γ)^{7}Be reaction rate, which is 9% lower than NACRE [24] and 33% higher than reported in NACREII [16] at temperatures relevant for ^{6}Li depletion in pre-main sequence stars. Moreover the reaction rate uncertainty has been significantly reduced [17], see Fig.4.

The result of a subsequent indirect study confirms LUNA extrapolation down to low energies for the ^{6}Li(p,γ)^{7}Be S-factor, reporting an $S(0)$ = 92 $\pm$ 12 eV b [25]. A recent theoretical study found a consistent trend for the ^{6}Li(p,γ)^{7}Be S-factor predicting a S-factor to zero energy of 98.3 eV b [26]

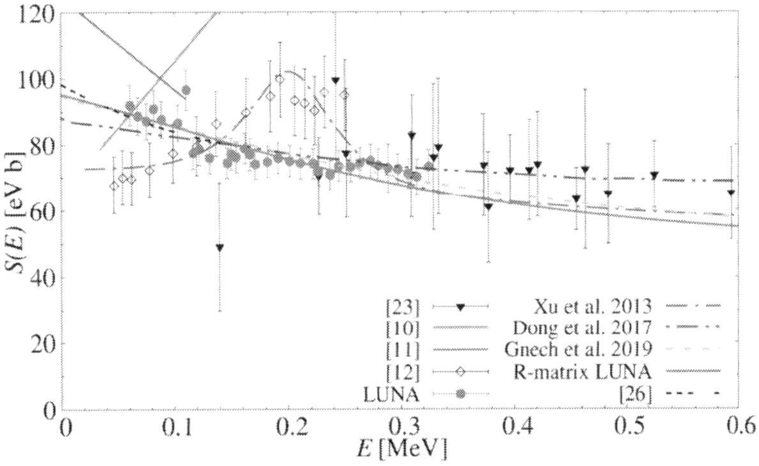

Figure 3. Astrophysical S-factor for the ^{6}Li(p,γ)^{7}Be reaction as obtained by LUNA in red [17]. Previous experimental data and theoretical evaluations are also shown for comparison. The solid red line represents an R-matrix fit of LUNA data and data from [23].

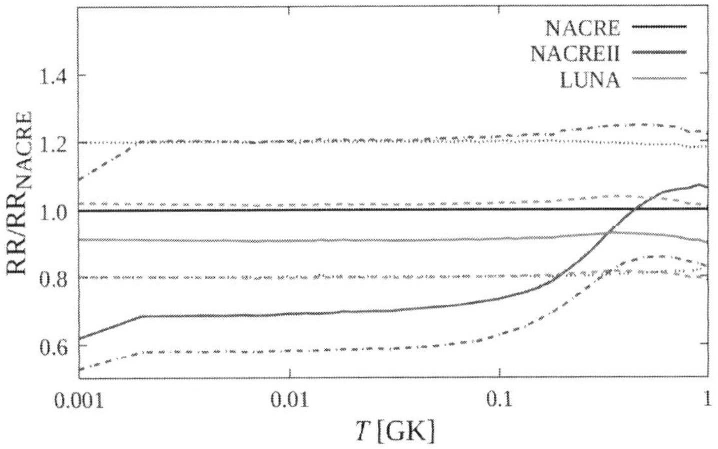

Figure 4. Reaction rate for the ^{6}Li(p,γ)^{7}Be reaction, normalized to the NACRE rate [24]. The NACRE II rate [16] is also shown for comparison. Dashed lines represent the uncertainty on the NACRE rate (black), on NACREII rate (blue) and on LUNA rate (red).

References

[1] N. Prantzos *et al.*, Astronomy and Astrophysics **542**, A67 (2012)
[2] C. Pitrou *et al.*, Physics Reports **754**, 1 (2018)
[3] B. D. Fields *et al.*,Journal of Cosmology and Astroparticle Physics **2020**, 010 (2020)
[4] J. C. Howk *et al.*, Nature **489**, 121 (2012)
[5] G. Harutyunyan *et al.*, Astronomy and Astrophysics **618**, A16 (2018)
[6] A.E.G. Perez *et al.*, Astronomy and Astrophysics **504**, 213 (2009)
[7] K. Lind *et al.*, Astronomy and Astrophysics **544**, A96 (2013)
[8] E.X. Wang *et al.*, Mon. Not. Roy. Astron. Soc.**509**, 1521-1535 (2022)

[9] B. Fields and K. Olive, arXiv e-prints, 2204.03167 (2022)
[10] F. Cecil *et al.*, Nuclear Physics A **539**, 75 (1992)
[11] R. M. Prior *et al.*, Physical Review C **70**, 055801 (2004)
[12] J. He *et al.*, Physics Letters B **725**, 287 (2013)
[13] T. Szücs *et al.*, Physical Review C **99**, 055804 (2019)
[14] A. Gnech and L. E. Marcucci, Nuclear Physics A **987**, 1 (2019)
[15] G. X. Dong *et al.*, Journal Physics G Nuclear and Particle Physics **44**, 045201 (2017)
[16] J. Huang *et al.*, Atomic Data and Nuclear Data Table **96**, 824 (2010)
[17] D. Piatti *et al.*, Physical Review C **102**, (2020)
[18] F. Cavanna and P. Prati, International Journal of Modern Physics A **33**, 1843010 (2018)
[19] A. Formicola *et al.*, Nuclear Instruments and Methods in Physics Research **507**, 609 (2003)
[20] J. Cruz *et al.*, Physics Letters B **624**, 181 (2005)
[21] C. Brune *et al.*, Nuclear Instruments and Methods in Physics Research A **389**, 421 (1997)
[22] H. J. Assenbaum, K. Langanke and C. Rolfs, Zeitschrift für Physik A Atomic Nuclei **327**, 461 (1987)
[23] Z. Switkowski *et al.*, Nuclear Physics A **331**, 50 (1979)
[24] C. Angulo *et al.*, Nuclear Physics A **656**, 3 (1999)
[25] G.G. Kiss *et al.*, Physical Review C **104**, 015807 (2021)
[26] S.B. Dubovichenko *et al.*, Physical Review C **105**, 065806 (2022)

Fusion of ^{12}C + ^{24}Mg at far sub-barrier energies

Alberto M. Stefanini[1,*], Giovanna Montagnoli[2], Mirco Del Fabbro[3], Giulia Colucci[4], Rosanna Depalo[5], Alain Goasduff[1], and Irene Zanon[1]

[1]INFN - Laboratori Nazionali di Legnaro, I-35020 Legnaro (Padova), Italy
[2]Department of Physics and Astronomy, Univ. of Padua and INFN, Padova, Italy
[3]University of Ferrara and INFN, Padova, Italy
[4]Heavy Ion Laboratory, University of Warsaw, Warsaw, Poland
[5]Department of Physics, University of Milan and INFN, Milan, Italy

Abstract. Recent experiments on ^{12}C + ^{24}Mg and preliminary data on ^{12}C + ^{26}Mg show that we observe the hindrance phenomenon in these two systems that are close to the lighter ones relevant for astrophysics. The cross section at the hindrance threshold has a remarkably high value for ^{12}C + ^{24}Mg, while it is closer to an empirical systematics for ^{12}C + ^{26}Mg. The lowest-energy fusion cross sections of ^{12}C + ^{24}Mg are consistent with simple one-dimensional barrier penetration calculations, i.e. the coupling strengths seem to be strongly damped far below the barrier. Measurements at slightly lower energies would be essential to discriminate between different models and to allow reliably extrapolating to the lighter systems producing energy and elemental synthesis in stellar environments.

1 The phenomenon of fusion hindrance

Heavy-ion fusion reactions are essential to investigate the fundamental problem of quantum tunnelling of many-body systems in the presence of intrinsic degrees of freedom and allow to extend our knowledge of the nuclear chart. In addition, the fusion of light systems serves as a base for the understanding of the astrophysical reaction networks responsible for the energy production and nucleosynthesis in stellar environments [1–3]. It is well known that the nuclear structure of the colliding nuclei produces large enhancements of the fusion cross section around and below the Coulomb barrier, due to couplings with the low-energy collective modes of the colliding nuclei, producing a splitting of the original barrier into a multitude of barrier, called "barrier distribution". However, at very low energies, below the lowest barrier, the fusion dynamics is made more complex by the appearance of the hindrance phenomenon discovered by Jiang *et al.* in ^{60}Ni + ^{89}Y [4], as a fast decrease of the excitation function below the result of standard CC calculations. Hindrance is experimentally identified by the observation of a maximum of the astrophysical S factor vs energy, and it was soon recognised as a general effect in heavy-ion fusion at deep sub-barrier energies.

Fusion hindrance was first recognized in a number of medium-heavy systems and has created a link between astrophysics and heavy-ion fusion. One of the first, and more clear, examples of hindrance [5] is shown in Fig. 1 for the system ^{64}Ni + ^{64}Ni (left panel). The

*e-mail: alberto.stefanini@lnl.infn.it

 https://doi.org/10.1051/epjconf/202327911013

difference between the measurements and the black curve (simple tunnelling through the potential barrier) displays the fusion enhancement, while the difference between the red curve (standard CC calculations with a Woods-Saxon potential) and the data shows the fusion hindrance at extreme sub-barrier energies. The S factor develops a maximum at the energy where the logarithmic slope $L(E)$, larger than the CC results, reaches the value $L_{CS} = \pi\eta/E$, as reported in the figure.

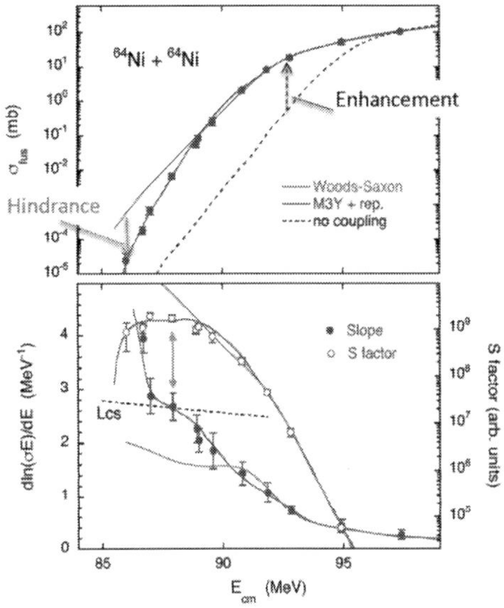

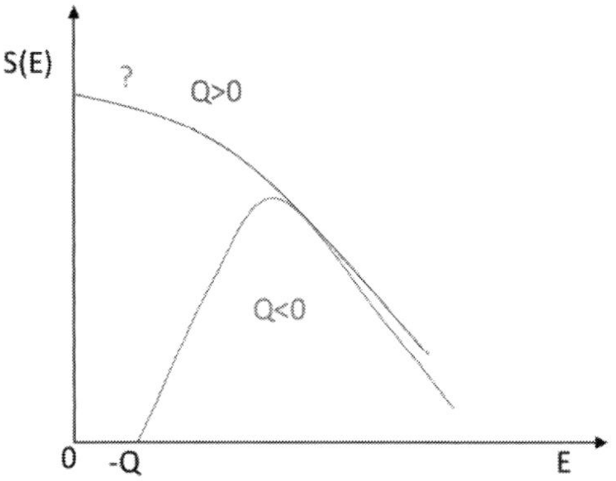

	E_{min}	$e^{2\pi\eta}$	S(E)
Q<0	-Q	finite	0
Q>0	0	→∞	finite ?

Figure 1. (left) Fusion excitation function, logarithmic slope and S factor for ^{64}Ni + ^{64}Ni. (right) Schematic representation of the S factor trend for Q <0 and Q >0.

The physical mechanism underlying the hindrance effect is still a matter of debate in the nuclear community, however, it has been recently proposed that the Pauli exclusion principle affects fusion of atomic nuclei [6]. When that principle is considered in the ion-ion interaction, the Coulomb barrier turns out to be higher and thicker, thus hindering sub-barrier fusion.

2 Fusion of light nuclei

Fusion of light heavy-ions is characterised by a positive Q-value. This brings a substantial difference from the heavier systems where Q is negative and the cross section goes to zero at positive energy. This implies that S must show a maximum at some higher energy, but, conversely, if Q >0 the cross section stays in principle finite down to E= 0, so that S may not show any maximum. The situation is schematically shown in Fig. 1 (right panel).

It has been observed that for light systems $L(E)$ and L_{CS} tend to be two nearly parallel curves and the crossing point (if existing) is rather undetermined. The S factor maximum becomes broader and establishing the presence of hindrance requires precise and challenging measurements.

Concerning the very important case of ^{12}C + ^{12}C, recent data are shown in Fig. 2 (left) [7–9]. We also recall the results obtained using the Trojan Horse Method by Tumino et al. [10]. The stellar evolution of massive stars, after the phase of helium burning, is determined by the reactions ^{12}C + ^{12}C, ^{16}O, whose measured low-energy S factors suffer from large uncertain-

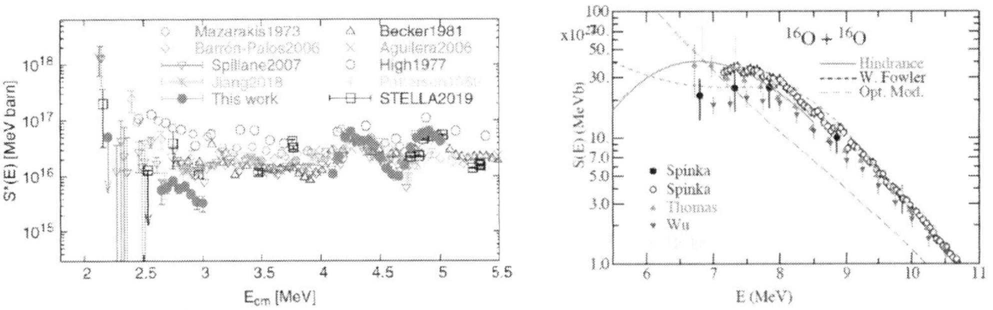

Figure 2. S factor measured by Tan (red symbols [7]) and by other groups for $^{12}C + ^{12}C$ (left) and for $^{16}O + ^{16}O$ (right) [3], compared to several calculations.

ties and from the presence of resonances. It is not yet clear whether hindrance is effective in such cases.

We point out that in two recent experiments for $^{12}C + ^{13}C$ [11, 12], both done with the thick target technique, one does not observe resonances and no S factor maximum shows up. For the case of $^{16}O + ^{16}O$, evidence of hindrance is found in most measurements (see right panel of Fig. 2) [13], even if the various sets of data show disagreements in the absolute scale.

3 The case of $^{12}C + ^{24}Mg$

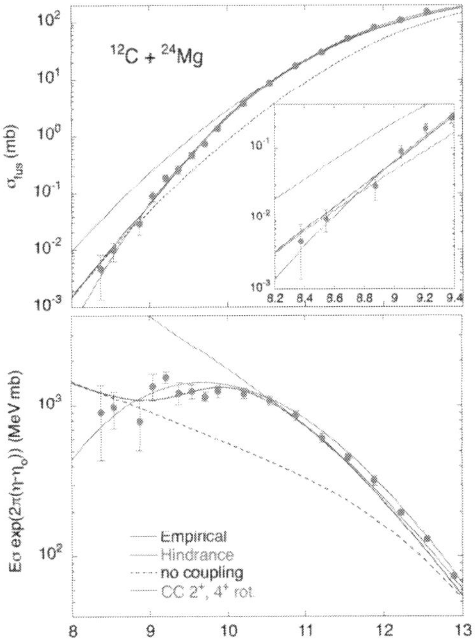

Figure 3. *Measured excitation function (top) and S factor (bottom) of $^{12}C + ^{24}Mg$ compared to calculations.*

The existence and features of hindrance in the systems of astrophysical interest are not at all established, so that the study of slightly heavier cases is appealing, with the aim of allowing a reliable extrapolation towards the lighter systems. $^{12}C + ^{24}Mg$ is a medium-light system with positive fusion Q-value = +16.3 MeV, very close to the "astrophysical" cases. We measured its excitation function, by detecting the fusion evaporation residues with the set-up PISOLO [14, 15] at the Tandem XTU of INFN – Laboratori Nazionali di Legnaro (LNL).

In Fig. 3 the data are compared to no-coupling (dashed line) and CC calculations [16] where the ground state rotational band of ^{24}Mg has been taken into account (green line). One sees that the standard CC calculation largely over-predicts the cross section and the S-factor at low energies. The blue line is the result of an "empirical" calculation in the spirit of the adiabatic model [17]. The red line is instead obtained using the hindrance parametrisation proposed by Jiang et al. [18].

One notices that fusion hindrance shows up already at ~0.75 mb, which is a very large value compared to many other systems presenting hindrance. Moreover, the data are well fitted using both the empirical calculation and the hindrance parametrisation, and it appears that tunnelling a one-dimensional barrier reproduces the lowest energy points. Finally, it is clear that discriminating between the two models requires lower energy measurements and smaller experimental errors.

4 Recent results on ^{12}C + ^{26}Mg

Very recently, the fusion of ^{12}C + ^{26}Mg has been measured from above to far below the barrier at LNL. The main motivation was to investigate whether the anomalously high threshold energy of hindrance, observed with ^{24}Mg, shows up also in this system. Here we show the preliminary results of the experiment. Fig. 4 (left panel) reports the excitation function of ^{12}C + ^{26}Mg (the lowest data point is an upper limit), compared to CC calculations using a Woods-Saxon potential whose parameters have been only slightly modified with respect to ^{12}C + ^{24}Mg, to reproduce the data near the barrier.

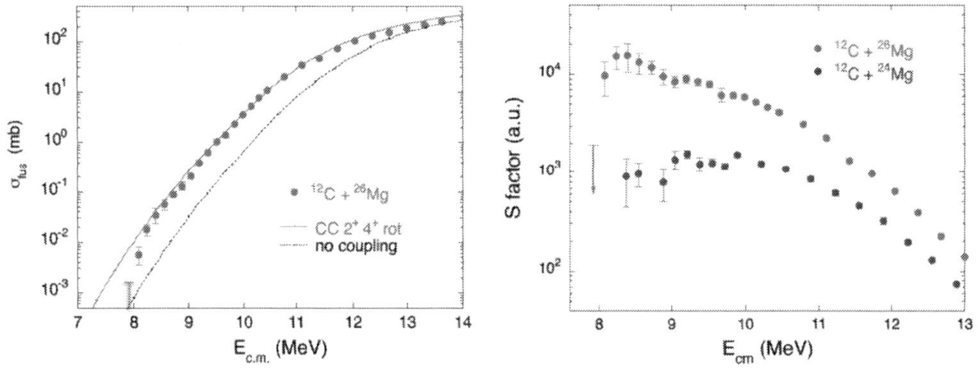

Figure 4. Fusion excitation function measured for ^{12}C + ^{26}Mg (preliminary data) compared to CC calculations (left) and S factor of the two systems (right panel).

The CC calculation gives a very good account of the data in a large energy range, however, we observe that it starts over-predicting the experimental cross sections at the level of $\approx$0.03 mb.

This means that the hindrance threshold for this system is significantly lower than for ^{12}C + ^{24}Mg. This difference is confirmed in the comparison of the two S factors, reported in the right panel of Fig. 4, from which we have also an indication that the S factor maximum of ^{12}C + ^{26}Mg is significantly narrower.

In Fig. 5 we have inserted the threshold energies for hindrance of ^{12}C + ^{24}Mg, ^{26}Mg in the phenomenological systematics of Jiang [18], where each system is characterized by the parameter $\zeta = Z_1 Z_2 \mu^{1/2}$. Please note that the points for the C+C, C+O

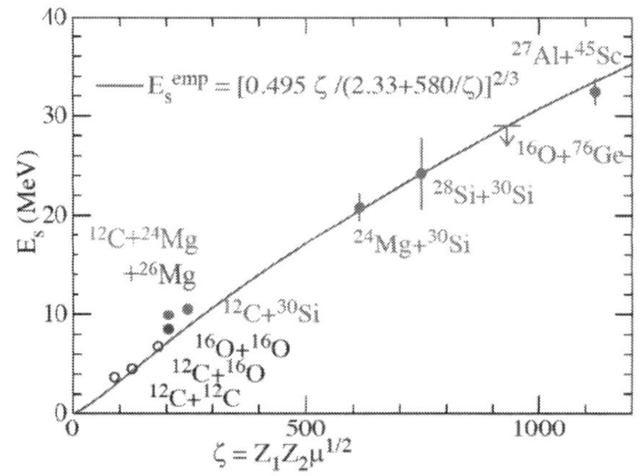

Figure 5. *Threshold energies for hindrance in light and medium-light systems.*

and O+O systems (open symbols) are only obtained by extrapolating from higher energies. All the plotted systems have positive fusion Q values. The case of ^{12}C + ^{30}Si is also reported, which was investigated by our group a few years ago [19]. The line is the phenomenological formula proposed in Ref. [18], as shown in the figure.

The two systems discussed in this contribution have ζ parameters very close to the lighter cases important for stellar evolution. The difference between these two cases can clearly be seen in this representation, and it is relevant for the wished possibility of extrapolating to the lighter systems on a reliable basis. Indeed, the different behaviour of ^{12}C + ^{24}Mg and ^{12}C + ^{26}Mg "complicates" the situation, and throws doubt on that procedure should be performed.

The final analysis of the ^{12}C + ^{26}Mg data will be ready soon. In any case, we can state that consequences for the dynamics of stellar evolution arising from the phenomenon of fusion hindrance have to be clarified by further experimental and theoretical work.

Acknowledgments

We acknowledge the very professional work of the XTU Tandem staff, and of M. Loriggiola for preparing targets of excellent quality. The research leading to these results has received funding from the European Union's Horizon 2020 research and innovation programme under grant agreement No 654002. R.D. acknowledges funding from the Italian Ministry of Education, University and Research (MIUR) through the "Dipartimenti di Eccellenza" project Science of the Universe.

References

[1] B. B. Back, H. Esbensen, C. L. Jiang, and K. E. Rehm, Rev. Mod. Phys. **86**, 317 (2014)
[2] G. Montagnoli and A.M. Stefanini, Eur. Phys.J.A **53**, 169 (2017)
[3] C.L. Jiang, B. B. Back, K. E. Rehm, K.Hagino, G. Montagnoli and A.M. Stefanini, Eur. Phys. J. A **57**, 235 (2021)
[4] C.L. Jiang et al., Phys. Rev. Lett. **89**, 052701 (2002)
[5] C.L.Jiang et al., Phys. Rev. Lett. **93**, 012701 (2004)
[6] C. Simenel, A.S. Umar, K. Godbey, M. Dasgupta, and D.J. Hinde, Phys. Rev. C **95**, 031601(R) (2017)
[7] W.P. Tan et al., Phys. Rev. Lett. **124**, 192702 (2020)
[8] C.L. Jiang et al., Phys. Rev. C **97**, 012801(R) (2018)
[9] G. Fruet et al., Phys. Rev. Lett. **124**, 192701 (2020)
[10] A. Tumino et al., Nature **557**, 687 (2018)
[11] M. Notani et al., Phys. Rev. C **85**, 014607 (2012)
[12] N.T. Zhang et al., Phys. Lett. B, **801**, 135170 (2019)
[13] C.L. Jiang, K. E. Rehm, B. B. Back, and R. V. F. Janssens, Phys. Rev. C **75**, 015803 (2007)
[14] G. Montagnoli et al., Phys. Rev. C **101**, 044608 (2020)
[15] G. Montagnoli et al., J. Phys. G **49**, 095101 (2022)
[16] K. Hagino, N. Rowley, and A.T. Kruppa, Comput. Phys. Comm. **123**, 143 (1999)
[17] T. Ichikawa, Phys. Rev. C**92**, 064604 (2015)
[18] C. L. Jiang, K.E. Rehm, B.B. Back and R.V.F. Janssens, Phys. Rev. C **79**, 044601 (2009)
[19] G. Montagnoli et al., Phys. Rev. C **97**, 024610 (2018)

Experimental study of alpha-induced nuclear reactions on Tellurium isotopes for the astrophysical γ-process

Zs. Mátyus[1,2*], *Gy. Gyürky*[1]

[1]Institute for Nuclear Research (ATOMKI), Bem tér 18/c, H-4001 Debrecen, Hungary
[2]University of Debrecen Doctoral School of Physics, Egyetem tér 1, H-4032 Debrecen, Hungary

Abstract. Heavier p isotopes are believed to be produced through the γ-process. The path of γ-process can be modelled with network calculations. These calculations include about 20000 reactions on more than 2000, mostly unstable nuclei, many of these reactions involves alpha particles. Theoretical cross sections of reactions involving alpha particles are found to be very sensitive to the alpha-nucleus optical potential, which is not known well enough at low, astrophysical energies. This potential can be studied experimentally by measuring the cross section of (α,n) reactions close above the reaction threshold. The aim of the present work is thus to measure the cross section of the ^{122}Te(α,n)^{125}Xe, ^{124}Te(α,n)^{127}Xe and ^{130}Te(α,n)^{133}Xe reactions for which no data exist in literature. The half-life of the produced Xe isotopes allows us to use the activation technique. Tellurium targets with natural isotopic composition will be used in order to measure the three reactions in parallel. The experiments are in progress using the cyclotron accelerator of ATOMKI. In this paper details of the experimental technique and the preliminary results of the first test measurements are presented.

1. Introduction

The astrophysical γ-process [1] is the main part of the more general p-process, which is responsible for the production of the heavy, proton rich isotopes. These isotopes are inaccessible for the s- and r-processes. Modelling the path of the γ-process requires to take into account a complicated reaction network. These networks include more than 2000 nuclei, most of them being radioactive. The crucial nuclear reaction cross sections in these networks are mostly unknown at astrophysical relevant energies. In the networks the γ-induced reactions play the main role, but other reactions, such as α-captures, are also included. Moreover, among the γ-induced reactions the (γ,α) reactions become especially important in the region of the more and more proton rich isotopes.

Due to the lack of experimental data and their size, the networks utilize reaction rates based on theoretical cross sections. These reaction rates are very sensitive to various

* Corresponding author: matyus.zsolt@atomki.hu

uncertainties. Therefore, the theoretical calculations have relatively poor predictive power, experimental data are thus needed to make the theoretical calculations more reliable.

Theoretical calculations describe the projectile-nucleus interaction with the so-called alpha-nucleus optical potential (α-OMP) [2]. The theory is very sensitive to the parameters of the α-OMP. The (α,n) reactions give a chance to study the α-OMP experimentally.

2. Preparation and characterization of the Te targets

The aim of the present work is to measure (α,n) cross sections on Te isotopes with the activation method which proved very powerful in cross section measurements for the γ-process [3]. In the case of the ^{120}Te(α,n) reaction there are data in the literature [4]. Therefore, in the present work three other Te isotopes are investigated (^{122}Te, ^{124}Te, ^{130}Te) where the activation technique is applicable for measuring the (α,n) cross section thanks to the lifetime of the created Xenon isotopes. The abundance of these isotopes is high enough to avoid isotopically enriched targets. Therefore, the targets were made from natural isotopic composition metallic Tellurium with vacuum evaporation on 10μm thick Aluminum foils.

The first information about the target thickness, which is an important quantity for the cross-section determination, was obtained by weighing. We measured the mass of the target backing foils before and after the evaporation to a precision of 1 μg. The difference of the two masses gives the mass of the Te layer.

For the further measurement of the target thickness, we used Rutherford Backscattering Spectroscopy (RBS) with proton and alpha beams. Proton RBS is less suited for the target thickness measurement than alpha-RBS, but it was the only option before the first test experiments. Nevertheless, the result of the proton RBS provided enough target thickness information for the evaluation of the first cross section data. A typical proton-RBS spectrum can be seen in Fig.1, where the region of the Te peak and Al plateau are clearly visible and from the width of the Te peak the target thickness can be derived.

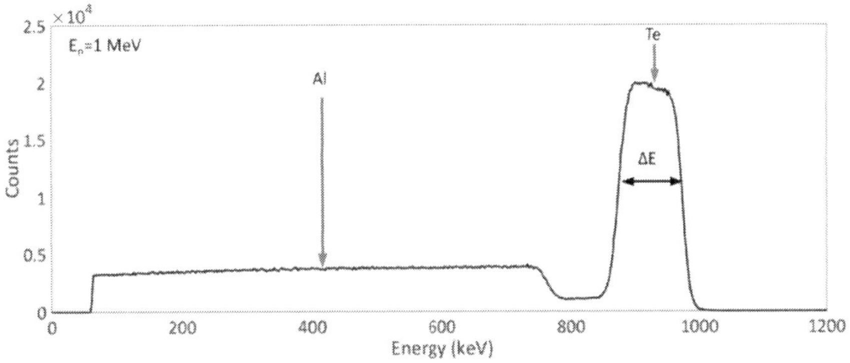

Fig.1. A typical spectrum of the proton RBS.

The target thickness obtained from weighing and from the proton-RBS have typically 7% difference. This may be caused by tellurium oxide formation during or after the irradiation. The target composition is therefore to be further investigated by e.g. alpha-RBS.

Xenon is a noble gas, so we need to be sure about the stability of the created Xenon in the backing foil. For this purpose, we are planning to do lifetime calculations based on our experiments and compare the results with the precise half-life data from literature. Earlier experiments showed that Xenon seems stable inside the target foil [5].

3. The irradiation and the γ-detection

The alpha beam was provided by the cyclotron accelerator of ATOMKI. The first two measurements were carried out at energies of 12 MeV (20-hour long irradiation) and 17 MeV (3-hour long irradiation) with 1 μA intensity. After the irradiation, we used a HPGe detector to count the gamma rays coming from the produced Xe isotopes. A typical γ-spectrum can be seen in Figure 2. The detector was calibrated with three calibration sources in three different geometries. Measurements in far geometry and their comparison with close geometry ones allows to eliminate the summing-in effect typical in close geometries.

The $^{130}Te(\alpha,n)^{133}Xe$ reaction leads to two distinct final states because of the isomeric state of ^{133}Xe. The calculations to determine the ground state cross section are still in progress.

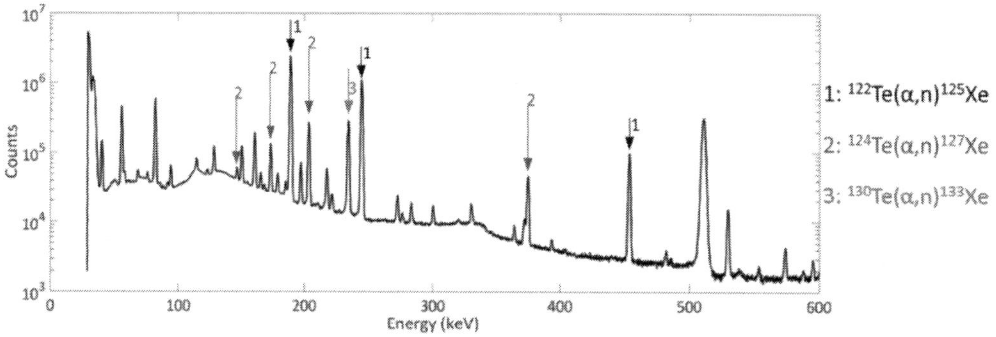

Fig. 2. The γ-spectrum after the 17MeV irradiation. The relevant γ-peaks are labelled. In the case of $^{130}Te(\alpha,n)^{133}Xe$ only the isomeric state γ-peak is labelled.

We made a TALYS [6] calculation to compare our preliminary results. Figure 3 shows the preliminary results of our work in the case of the $^{124}Te(\alpha,n)^{127}Xe$ reaction. The error bar of the measured points is smaller than the size of the symbols. In the case of 17 MeV data point, the $^{125}Te(\alpha,2n)^{127}Xe$ reaction gives a contribution to the measured yield, explaining the rather high value compared to TALYS. This will have to be taken into account in the analysis. Due mainly to the not final target thickness determination, however, the results may still change and can be considered as preliminary.

4. Conclusion and outlook

The experimental procedure seems feasible. Further measurements are in progress. The cross section of the three reactions will be measured at different beam energies from the thresholds up to about 18 MeV. The results of the measurements will be compared with calculations using different alpha nucleus optical potentials to choose the best models and improve their parameters.

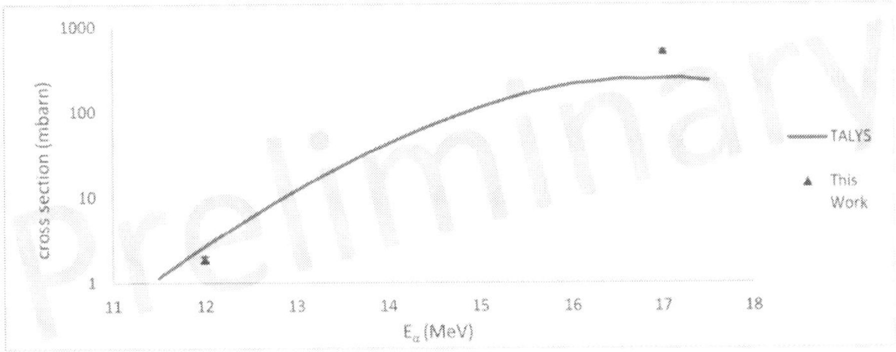

Fig. 3. The preliminary results of this work in the case of $^{124}Te(\alpha,n)^{127}Xe$.

5. Acknowledgments

This work was supported by the National Research, Development and Innovation Office NKFIH (contract numbers K134197).

References

[1]: T. Rauscher et. al. Rep. Prog. Phys. **76** 66201 (2013)

[2]: P. Mohr At. Data Nucl. Data Tables **142** 101453 (2021)

[3]: Gy. Gyürky et. al. Eur. Phys. J. A **55**, 41 (2019)

[4]: A. Palumbo et. al. Phys. Rev. C **85**, 028801 (2012)

[5]: G.G. Kiss et. al. Nuc. Phys. A **986**, 213-222 (2019)

[6]: A.J. Koning et. al. Computer code TALYS http://www.talys.eu/

EPJ Web of Conferences **279**, 11015 (2023) https://doi.org/10.1051/epjconf/202327911015
NPA-X 2022

Study of the $^{12}C + ^{16}O$ fusion reaction in carbon burning via the Trojan Horse Method

A. A. Oliva[1,2,*], A. Tumino[2,3], N. Soic[4], P. M. Prajapati[2,5], L. Acosta[6], R. Alba[2], F. Barba[7], S. Cherubini[1,2], G. D'Agata[1,2], D. Dell'Aquila[4], A. Di Pietro[2], J.P. Fernandez[8], P. Figuera[2], D. Galaviz Redondo[7], L. Guardo[1,2], M. Gulino[2,3], F. Hammache[9], D. Jelavic Malenica[4], A.I. Kiliç[10], M. La Cognata[2], M. La Commara[11,12], L. Lamia[1,2,13], D. Lattuada[2,3], C. Maiolino[2], G. Manicò[1,2], M. Mazzocco[14,15], M. Milin[4,16], Ma Nanru[17], A. Nurmukhanbetova[18], D. Nurkic[4,16], S. Palmerini[19,20], T. Parascandolo[11], D. Pierroutsakou[11], R.G. Pizzone[2], R. Popocovski[4], G.G. Rapisarda[1,2], S. Romano[1,2,13], D. Santonocito[2], M.L. Sergi[1,2], A. Shotter[21], R. Spartà[1,2], A. Spiridon[21], L. Trache[22], N. Vukman[4], and H. Yamaguchi[17]

[1]Dipartimento di Fisica e Astronomia "E. Majorana", Univ. di Catania, Italy
[2]INFN-LNS, Laboratori Nazionali del Sud, Catania, Italy
[3]Facoltà di Ingegneria e Architettura, Università degli Studi di Enna "Kore", Enna, Italy
[4]Rudjer Boskovic Institute, Zagreb, Croatia
[5]Manipal Centre for Natural Sciences, Manipal, India
[6]Instituto de Física, Universidad Nacional Autónoma de México, Mexico City, Mexico
[7]Departamento de Física, Faculdade de Ciências da Universidade de Lisboa, Lisboa, Portugal
[8]University of Seville, Seville, Spain
[9]IPN-Orsay, IN2P3-CNRS, Universitè Paris XI, Orsay, France
[10]Nuclear Physics Institute of ASCR, Rez near Prague, Czech Republic
[11]Dipartimento di Fisica, Università degli Studi di Napoli "Federico II", Napoli, Italy
[12]INFN-Sezione di Napoli, Napoli, Italy
[13]Centro Siciliano Fisica Nucleare e Struttura della Materia, Catania, Italy
[14]Dipartimento di Fisica, Università degli Studi di Padova, Padova
[15]INFN-Laboratori Nazionali di Legnaro, Legnaro, Italy
[16]Physics Department, Faculty of Science, University of Zagreb, Zagreb, Croatia
[17]Center for Nuclear Study, The University of Tokyo, Tokyo, Japan
[18]Nazarbayev University, Astana, Kazakhstan
[19]Dipartimento di Fisica e Geologia, Università di Perugia, Italy.
[20]INFN-Sezione di Perugia, Perugia, Italy.
[21]School of Physics - The University of Edinburgh, Edinburgh, Scotland
[22]IFIN-HH, Bucharest-Magurele, Romania

Abstract. $^{12}C + ^{12}C$ is the main reaction during core and shell carbon burning in massive stars, however, at temperatures higher than $10^9 K$ when most of the carbon is depleted and its abundance is lower than ^{16}O, the $^{12}C + ^{16}O$ fusion can also become relevant. Moreover, $^{12}C + ^{16}O$ reaction can ignite also in the scenario of explosive carbon burning. The astrophysical energy region of interest thus ranges from 3 to 7.2 MeV in the center-of-mass frame. There are various measurements of the cross-section available in the literature, however, they all stop around 4 MeV, making extrapolation necessary at lower energies.

*e-mail: aoliva@lns.infn.it

To try to solve this uncertainty and corroborate direct measurement the Trojan Horse Method was applied to three-body processes $^{16}O(^{14}N, \alpha^{24}Mg)^{2}H$ and $^{16}O(^{14}N, p^{27}Al)^{2}H$ to study the $^{16}O(^{12}C, \alpha)^{24}Mg$ and $^{16}O(^{12}C, p)^{27}Al$ reactions.

1 Astrophysical Scenario

Carbon burning is a fundamental process for the advanced life of a massive stars since it determines their properties and evolution as well as their contribution to the chemical evolution of the universe. Both during the core and shell carbon burning, the $^{12}C + ^{12}C$ fusion is the main reaction and the first to ignite since it has the lowest Coulomb barrier among all the possible reactions [1, 2]. In truth, the ashes of helium burning are composed also of ^{16}O nuclei, thus, when most of the carbon nuclei in the core are depleted, the abundance of ^{16}O is actually significantly higher [3, 4] than that of ^{12}C. Considering also that at this point of the evolution, the environmental temperature is higher than the one it had when the carbon burning started, other reactions, even with a higher Coulomb barrier, can now occur. In this context, the $^{12}C + ^{16}O$ reaction can indeed play a significant role at temperatures higher than 10^9 K [5]. Carbon burning is also a key process in the evolution of Type Ia supernovae and once again the $^{12}C + ^{12}C$ fusion is thought to be the main energy source. As shown by recent studies [6], the $^{12}C + ^{16}O$ fusion can also play a significant role in this scenario, at temperatures around $3.6 \cdot 10^9$ K. Therefore, to assess the impact of the $^{12}C + ^{16}O$ fusion in the whole context of carbon burning, a precise study of its reaction rate in the energy range of astrophysical interest, namely between 3 and 7.2 MeV in the center of mass frame, is required.

2 Status of the $^{12}C + ^{16}O$

In the literature, various measurements of the cross-section and rate for the $^{12}C + ^{16}O$ fusion can be found, however, they all stop at around 4 MeV, which imposes the use of extrapolation at lower energies [7]. Unfortunately, this procedure is strongly dependent on the model used, on the presence of nuclear effects not included in the model (clustering, molecular-like behaviour, etc.) and on the stellar environment itself [8]. Therefore, a new measurement is needed to better explore the low-energy region. During both hydrostatic and explosive carbon burning, the $^{12}C + ^{16}O$ fusion process proceeds mainly through $^{12}C(^{16}O, \alpha)^{24}Mg$ (Q = 6.77 MeV) and $^{12}C(^{16}O, p)^{27}Al$ (Q = 5.17 MeV) reaction channels and partially through the $^{12}C(^{16}O, n)^{27}Si$, less favored due to its negative Q-value. Other reactions such as $^{12}C(^{16}O, 2\alpha)^{20}Ne$ are hindered by the presence of the Coulombian barrier in the exit channel [7].

3 THM Approach

3.1 The Trojan Horse Method

The Trojan Horse Method (THM) is based on the theoretical formalism of Quasi-Free (QF) break-up reactions: it studies the QF contribution of an appropriate three-body reaction $a + A \rightarrow b + B + s$ measured at energies above the Coulomb barrier to evaluate the cross-section of the two-body reaction of interest $x + A \rightarrow b + B$ at sub-Coulomb energies. The so-called Trojan Horse (TH) nucleus a is assumed to possess a clustered state with a $x + s$ configuration, where the x cluster is called the participant while s is the spectator of the interaction because, under QF kinematic conditions, the momentum transfer to the cluster s is negligible. These conditions correspond to the region of the phase space where the intercluster $x+s$ momentum distribution has its maximum value or, equally, where the distance between the two clusters inside the TH nucleus is maximized. The main advantage of the method is that the two-body cross-section is not affected by any barrier effect, either Coulombian or centrifugal, or by electron screening effects. Moreover, taking advantage of the intercluster motion, with just one fixed beam energy it is possible to actually study an energy range and even cover the subthreshold region. A detailed explanation of the THM formalism can be found in Ref. [9] and references therein.

3.2 The Experiment

The THM has been used to evaluate the $^{12}C(^{16}O,\alpha)^{24}Mg$ and $^{12}C(^{16}O,p)^{27}Al$ reactions at astrophysical energies by studying the $^{16}O(^{14}N,\alpha^{24}Mg)^{2}H$ and $^{16}O(^{14}N,p^{27}Al)^{2}H$ three-body reactions. The chosen TH nucleus was the ^{14}N since it has a well-known ^{12}C + d cluster structure [10, 11] and it has already successfully used in past THM measurements [11]. The reactions were investigated at the Laboratori Nazionali del Sud (LNS-INFN) in Catania using a 33.7 MeV ^{14}N beam, accelerated by the Van de Graaff tandem, impinging on a 458 μg/cm^2 WO_3 target. The experimental setup was similar to previous THM experiments [11, 12] and consisted of an array of four telescopes, each of which was made up of three silicon detectors, was used to correctly identify the particle of interest in the exit channel. A 35 μm Position Sensitive Detector (PSD) and a 100 μm one were used, respectively, for the first and second stage of each telescope while for the third one a 1500 μm pad detector was used. The experimental setup covered an angular range from 7° to 30° and from 45° to 68° in the laboratory frame, precisely chosen to cover the so-called QF angles, where the contribution of the mechanism is maximized. The setup was symmetrical with respect to the beam axis to increase the statistical count.

3.3 Data Analysis

The energy and angle calibration of the detectors was performed with dedicated experimental runs using an eight-peak alpha source, elastic scattering of ^{14}N and $^{1,2}H$, and the $^{2}H(^{14}N,\alpha)^{12}C$) reaction. Since the PSDs do not have any segmentation, the angle calibration was performed by covering the detector with a grid with equally spaced slits each of which placed at a known angle. To select the reaction channel of interests, detected particles must be identified first: by means of the ΔE-E technique, using the first two stages of each telescope, it is possible to select only the events in which a deuteron and either an alpha particle or a proton were detected in coincidence. The remaining ^{24}Mg or ^{27}Al nuclei in the exit channel were not detected; therefore, the experimental Q-value was reconstructed assuming the presence of either one in the final state and using the measured energies and the known beam energy. This assumption was then corroborated by comparing the reconstructed experimental Q-value with the theoretical one. In Fig. 1 the presence of events for both the α_0 and α_1 channels for the $^{16}O(^{14}N,\alpha^{24}Mg)^{2}H$ reaction ca be seen. In Fig. 1, to better see the relevant events, a minimum threshold was set in the color axis. However, the presence of background events can still be seen: this is related to an incomplete separation between the deuterons and protons loci in the central ΔE-E telescopes. Further analysis of the data, also including the two peripheral telescopes is thus needed.

4 Conclusions

In this contribution were presented the preliminary results of the data analysis which clearly show the proper execution of the experiment and assess the presence of the

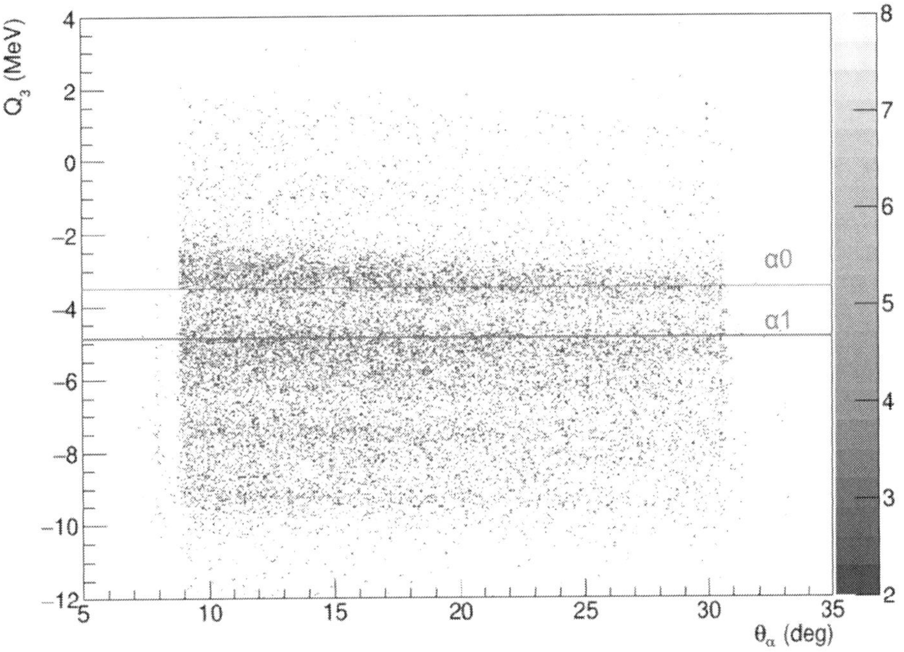

Figure 1. Experimental Q-value vs detection angle of the alpha particle for the $^{16}O(^{14}N, \alpha^{24}Mg)^{2}H$ reaction channel. Colored dots represent the experimental data, according to the scale in the right side figure, while red lines represents the theoretical value for both the ground state (α_0) and the first excited state ($\alpha 1$).

$^{16}O(^{14}N, \alpha^{24}Mg)^{2}H$ reaction channels of interest. Future developments of the data analysis foresee the accurate evaluation of the momentum distribution of the spectator P_s, in order to assess the presence of the QF contribution, which will be then selected to finalize the analysis. The same procedure will be applied to the other reaction channel, the $^{16}O(^{14}N, p^{27}Al)^{2}H$ and, finally, the THM analysis will be then completed with the evaluation of the $^{12}C +^{16} O$ fusion cross-section and reaction rate.

References

[1] Pignatari M. et al., Astrophys. J. **762**, 31 (2013)
[2] Chieffi A. et al., Astrophys. J. **916-2**, 79 (2021)
[3] Weaver T. A. et al., Phys. Rep.**227**, 65 (1993)
[4] Buchmann L., Astrophys. J. **468**, L127 (1996)
[5] Woosley S.E. et al., Phys. Rev. Lett. **27**, 213 (1971)
[6] Martínez-Rodríguez H. et al., Astrophys. J. **843**, 35 (2017)
[7] Fang X. et al., Phys. Rev. C **96-4**, 045804 (2017)
[8] Jiang C. L. et al., Phys. Rev. C **75**, 015803 (2007)
[9] Tumino A. et al., Annu. Rev. Nucl. Part. S. **71-1**, 345–76 (2021)
[10] Zurmühle R. W. et al., Phys. Rev. C **49**, 2549 (1994)
[11] Tumino A. et al., Nature **557-7707**, 687 (2018)
[12] Palmerini S. et al., Eur. Phys. J. Plus **136**, 898 (2021)

A new ^{12}C+^{12}C reaction rate: Impact on stellar evolution

E. Monpribat[1,*], *A.* Choplin[2], *S.* Martinet[2], *S.* Courtin[1,3], *M.* Heine[1], *P.* Adsley[4,5], *D.* Curien[1], *T.* Dumont[1], *S.* Ekström[6], *D.G.* Jenkins[3,7], *M.* Moukaddam[1], *J.* Nippert[1], *S.* Tsiatsiou[6], and *G.* Meynet[6]

[1]Université de Strasbourg, CNRS, IPHC UMR 7178, F-67000 Strasbourg, France
[2]Institut d'Astronomie et d'Astrophysique, Université Libre de Bruxelles, CP 226, 1050, Brussels, Belgium
[3]University of Strasbourg Institute of Advanced Studies (USIAS), Strasbourg, France
[4]Cyclotron Institute, Texas A&M University, College Station, Texas 77843, USA
[5]Department of Physics & Astronomy, Texas A&M University, College Station, Texas 77843, USA
[6]Department of astronomy, University of Geneva, chemin Pegasi 51, 1290 Versoix, Switzerland
[7]Department of Physics, University of York, Heslington, York YO10 5DD, United Kingdom

Abstract. Among the reactions driving stellar evolution during carbon burning, ^{12}C + ^{12}C fusion provides the key ingredients. This system reveals many resonances, but also regions with suppressed fusion cross-sections. The reaction was recently measured by the STELLA collaboration utilizing the gamma-particle coincidence technique for precise cross-section measurements reaching down to the Gamow window of massive stars. From the experimental data, reaction rates were determined by approximating a hindrance parametrization and by adding on top a resonance at the lowest measured energy. The impact of these reaction rates on the evolution of massive stars was explored with models of 12 and 25 $M_\odot$ using the stellar evolution code GENEC, and a detailed study of the resulting nucleosynthesis with a 1454 elements network was performed. The sensitivity of the STELLA experimental cross-sections on the temperature range for C-burning for the stellar models studied were presented. The final abundances and their impacts on stellar evolution were discussed.

1 Introduction

The origin of the chemical elements, the nucleosynthesis, has always been an important and fundamental subject in physics. By changing the internal composition of stars, nuclear physics plays a key role in this process. That is deeply linked to stellar evolution, to the different types of stars, and their combustion phases. Among these, the carbon-burning phase is the most interesting one for the STELLA collaboration. Indeed, it is a key phase of the stellar evolution–that only massive stars enter and it has an important role in type 1a supernova–during which the ^{12}C + ^{12}C fusion reaction takes place. The latter set the conditions of subsequent important nuclear mechanisms, like fusion of heavy-ions or the neutron seed generation.

During the past decades, numerous experiments have been performed, aiming at the direct measurement of the carbon fusion cross section at sub-barrier energies. These efforts revealed

*e-mail: emma.monpribat@iphc.cnrs.fr

the presence of resonances, possibly due to molecular ^{12}C configurations of the ^{24}Mg nucleus [1]. The most recent results suggest that fusion hindrance, a behaviour observed in a large number of heavier systems [2], is present in the $^{12}C+^{12}C$ fusion reaction [3–5].

However, at astrophysical energy, *i.e.* at deep sub-barrier energy, in the Gamow Window, experimental results have large uncertainties, and the different theoretical models, like fusion hindrance and the CF88 model – the one currently used in astrophysics [6]– diverge with orders of magnitudes between them. The lack of precise data is caused by tremendous experimental challenges. Indeed, the small cross section associated to this system, around picobarn, requires extended beamtime of months and at sufficiently high intensity. Furthermore, spectra are dominated by background that comes from contaminant reactions, natural radioactivity, and cosmic rays. As a solution, the coincidence method for the $^{12}C+^{12}C$ fusion reaction has been set-up for reliable measurements deep sub-barrier energies [7].

2 Direct measurements with STELLA UK-FATIMA

By using the coincidence method, the STELLA station is a straight answer to the experimental challenges of the $^{12}C+^{12}C$ cross section in direct measurements [8]. This method is based on the simultaneous detection of the emitted light particles, here an alpha or a proton, and the associated gamma rays of de-excitation of the daughter nucleus.

The $C^{2+,3+}$ beam required by the ^{12}C + ^{12}C cross section measurement is delivered by the Andromède facility at IJCLab, in Orsay (France) with an intensity up to 6 pμA. In order to allow the heat dissipation and prevent target deterioration, a system of rotating targets has been developed. The large thin targets are made of carbon graphite, with a diameter of about 5 cm and a thickness of 20 to 70 μg/cm^2, and can spin with up to 1000 rpm. Carbon buildup on the target is prevented by the high vacuum (10^{-8} mbar) of the reaction chamber. The detection of light particles is done by two annular charged-particle silicon detectors, with high granularity and nanoseconds timing (trigger). The gamma rays are detected by the $LaBr_3(Ce)$ scintillators from the UK-FATIMA (FAst TIMing Array) collaboration.

3 Reaction rates and STELLA sensitivity

In [9], two different fusion scenarios for the excitation functions of the $^{12}C+^{12}C$ fusion reaction have been studied, based on direct measurement performed with the STELLA experiment [4] and following a χ^2 analysis of the data: the first one employs the so-called fusion hindrance (Hin model) described in [7], and the other is composed of fusion hindrance with a resonance at E_{rel} = 2.14 MeV proposed by [10] (HinRes model). The parameters of the fusion hindrance were kept free during the fits of the excitation functions, made simultaneously in both exit channels, *i.e.* for alpha and proton emission. The obtained fitting parameters are consistent with the ones suggested by [3] from a systematic phenomenological study. We observe in particular a good compatibility of the HinRes model and the STELLA measurements.

The reaction rates have been determined for both fusion scenarios. They are generally lower than the one currently used during the last decades in astrophysical simulations, determined in [6] (CF88 model). The resonance has a strong impact on the reaction rate, where it significantly increases the latter to a level comparable with the CF88 model rate for a temperature around T = 0.85 GK (see Fig. 2 in [9]).

The STELLA sensitivity corresponds to the temperature range where the reaction rates are determined based only on the interpolation of the excitation function proposed by the STELLA collaboration [9]. The temperature range in which the reaction rate is probed by the STELLA experiment was determined by interpreting the relative energy, at which the $^{12}C+^{12}C$ cross sections were measured, as a temperature. The lowest energy measured is $E_{\rm rel} = 2.03$ MeV. By using the Gamow window definition and with the approximation proposed in [11], the STELLA sensitivity reached a minimal temperature $T = 0.6$ GK at 1σ uncertainty width. In this temperature range, the relative uncertainties are around 15% and 30% for the Hin and HinRes models, respectively.

4 Impact on stellar evolution

In order to study the impact of the presented reaction rates, hydrodynamics and nucleosynthesis simulations were carried out using the Geneva stellar evolution code GENEC [12] for two stellar models with $12\,M_\odot$ and $25\,M_\odot$ at solar initial metallicity and without rotation. The evolution was followed until the end of the carbon burning phase, where the fusion temperature for the Hin model is 10% higher than for HinRes model, which reduces the carbon burning lifetime by a factor of two. We explain this by the substantially lower rate of the former model as compared to the latter with lower heat output, thus succumbing the gravitation pressure of the star resulting in contraction and higher temperature for carbon burning. The counter intuitive finding is that lower reaction rates result in higher effective temperatures [13] from permanent readjustment of the star.

Another finding is clearly depicted in the Kippenhahn diagram at the end of core carbon burning of the $25\,M_\odot$ star (see Fig. 4 in [9]) where both reaction rate models evolve in the same way, with the exception that the convective pocket of Hin model extends much further as compared to HinRes model. This is due to the higher temperature of the former with a stronger temperature gradient. Finally, the cooling of the star by neutrino emission (see Fig. 7 in [9]) is affected by the variation in carbon burning lifetimes which likely results in different dynamics during the core collapse and has an impact on the remnant nature.

The $25\,M_\odot$ model was further investigated with the complementary One-Layer code [14] to generate a single layer model of a star, but employing an extended nuclear reaction network, that takes into account more than one thousand isotopes and reactions in between. The temperature trajectory was adapted to the one of the CF88 model. In Fig. 9 in [9], minor variations of the final abundances obtained with CF88, Hin and HinRes models are visible for sodium, aluminium and phosphorus, and some heavier elements. These variations may eventually have an impact on the stellar evolution in the following phases, which would require further investigation to clarify.

We note that in the GENEC simulations, the results of CF88 and HinRes models are very similar as can be expected from comparable reaction rates at the temperatures during carbon burning (see Fig. 2 in [9]). In contrary, the abundances from the One-Layer code of CF88 and Hin models are closer, presenting a somewhat swapped situation. As the CF88 and HinRes rates are comparable with identical temperature trajectories during this run, the reasoning might be the branching with α- and proton-emission, that was adapted to the experimental findings given in Tab. 1 in [9] in the One-Layer code, but could only be accounted for indirectly in the GENEC package [15]. Indeed, the branching during carbon burning from the One-Layer simulation given in Fig. 1, indicates a situation where the CF88 and Hin models are closer to each other than HinRes model (branching of 0.65/0.35 [16]). The latter is given by the ratio of strengths of the resonance at $E = 2.14$ MeV in the carbon-carbon system where α emission is dominating. However, the results need to be taken with care as the temperature trajectories need to be adapted to the actual hydrodynamics constraints in CF88, Hin and HinRes models separately and different paths of nucleosynthesis might open with more robust assumptions. In conclusion, such a finding can demonstrate the sensitivity to resonances in

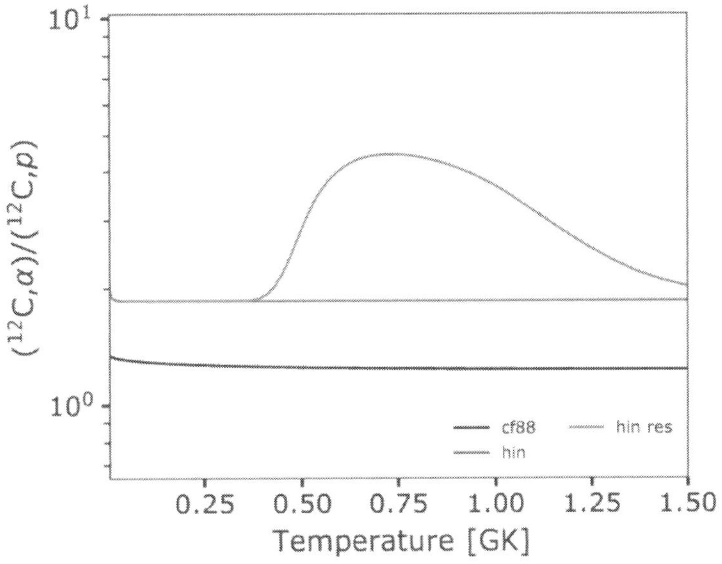

Figure 1. Ratio of the reaction channels in the valid temperature range for carbon burning for CF88, the Hin and HinRes models in the One-Layer code.

the branching of reaction rates for key reactions during nucleosynthesis calculations where straight factorizing of entire energy regions might yield only approximate results.

Acknowledgments

STELLA was funded by the University of Strasbourg IdEX program and CNRS Strasbourg. JN acknowledges support from the Interdisciplinary Thematic Institute QMat, as part of the ITI 2021 2028 program of the University of Strasbourg, CNRS and Inserm, which was supported by IdEx Unistra (ANR 10 IDEX 0002), and by SFRI STRAT'US project (ANR 20 SFRI 0012) and EUR QMAT ANR-17-EURE-0024 under the framework of the French Investments for the Future Program. SM, GM, and SE acknowledge the STAREX grant from the ERC Horizon 2020 research and innovation programme (grant agreement No 833925), and the COST Action ChETEC (CA 16117) supported by COST (European Cooperation in Science and Technology). This work was supported by the Fonds de la Recherche Scientifique-FNRS under Grant No IISN 4.4502.19.

References

[1] D. Jenkins and S. Courtin, Journal Physics G **42**, 034010 (2015)
[2] C.L. Jiang *et al.*, Physical Review Letters **89**, 052701 (2002)
[3] C.L. Jiang *et al.*, Physical Review C **97**, 012801(R) (2018)
[4] G. Fruet *et al.*, Physical Review Letters **124**, 192701 (2020)
[5] W.P. Tan *et al.*, Physical Review Letters **124**, 192702 (2020)
[6] G.R. Caughlan, W.A. Fowler, Atomic Data and Nuclear Data Tables **40**, 283 (1988)
[7] C.L. Jiang *et al.*, Nuclear Instruments and Methods A **682**, 12 (2012)
[8] M. Heine *et al.*, Nuclear Instruments and Methods A **903**, 1 (2018)

[9] E. Monpribat *et al.*, A&A **660**, A47 (2022)
[10] T. Spillane *et al.*, Physical Review Letters **98**, 122501 (2007)
[11] C. Illiadis, *Nuclear Physics of Stars*, 2nd edn. (Wiley-VCH, 2015), ISBN 978-3-527-33648-7
[12] P. Eggenberger *et al.*, Astrophysics and Space Science **316**, 43 (2008)
[13] M. Pignatari *et al.*, Astrophysical Journal **762**, 31 (2013)
[14] A. Choplin *et al.*, A&A **593**, A36 (2016)
[15] R. Hirschi, *Massive rotating stars : the road to supernova explosion* (Université de Genève. Thèse, 2004)
[16] M.E. Bennett *et al.*, Monthly Notices of the Royal Astronomical Society **420**, 3047 (2012)

Onset of pressure gradient on collective flow through balance and transition geometry

Sangeeta[1,*] and *Varinderjit* Kaur[2,**]

[1]Department of Applied Sciences, Chandigarh Engineering College, Landran (Punjab) - 140307, India
[2]Department of Physics, Khalsa College, Patiala (Punjab)-147001, India

Abstract. The impression of mean field and spectator matter has been studied through the collision geometry dependence of reduced flow and elliptical flow in heavy-ion collisions (HICs) at intermediate energies using Isospin-dependent Quantum Molecular Dynamics (IQMD) model for the reaction of $^{50}_{20}$Ca+$^{50}_{20}$Ca and $^{197}_{79}$Au+$^{197}_{79}$Au at incident energy between 50 MeV/nucleon and 450 MeV/nucleon. The collision geometry at which the reduced flow and elliptical flow changes its sign is termed as balance geometry ($\hat{b}_{bal}$) and transition geometry ($\hat{b}_{trans}$) of flow, respectively. Our study reveals that ($\hat{b}_{bal}$) and ($\hat{b}_{trans}$) increases with increase in incident energy. In addition to this, the effect of nuclear charge radius has also been studied by using the isospin-independent as well dependent nuclear charge radii parameterizations.

1 Introduction

The collective flow is a phenomena which is related to the amount of momentum attained by the emitted particle and its direction of emission from the fire ball [1].The strength of compression extensively depends upon various physical quantities like incident energy (E), impact parameter (b), composite mass of colliding nuclei (A$_{tot}$), mass asymmetry (η) etc. The compression in nuclear matter generates pressure gradient and due to this pressure gradient, there exist azimuthal distribution of fragments. The collective flow is a measure of the transverse motion imparted to particles and fragments during the collision of two nuclei and is defined as [2]:

$$\frac{dN}{d\phi} \propto 1 + 2\sum_{n=1}^{\infty} v_n Cos(n\phi). \tag{1}$$

The flow parameter v_n is called n^{th} harmonic function or Fourier coefficient of Fourier expansion where, n = 1, 2, 3....... and so on. The first two harmonics, v_1 and v_2 are the directed flow and elliptical flow respectively. The directed flow is defined as:

$$v_1 = < Cos\phi > = \langle \frac{p_x}{p_t} \rangle. \tag{2}$$

The nucleonic flow basically arises from the most compressed zone of the system, i.e. participant zone, which measure the total momentum transfer in the collision process. The

*e-mail: sangeeta.ar003@gmail.com
**e-mail: drvarinderjit@gmail.com

slope parameter around mid-rapidity obtained by plotting v_1 as a function of rapidity, measures the strength of directed flow called *reduced flow* and represented as: $\frac{\partial v_1}{\partial Y^{red}}$ [3]. The second harmonic, v_2 can be written as [4, 5]:

$$v_2 = < Cos2\phi > = \langle \frac{p_x^2 - p_y^2}{p_t^2} \rangle. \tag{3}$$

mean transverse momentum (p_t) given by, $p_t = \sqrt{p_x^2 + p_y^2}$ where, p_x and p_y are the x and y components of momentum. The elliptical flow represents the ellipse like distribution of emitted particles in the reaction plane[6–8]. In the recent communications, the impact parameter dependance of flow has been depicted to quantify the effect of spectator matter in the reaction dynamics at intermediate energy regime [9]. The study shows that, at a fixed incident energy of 150 MeV/nucleon, the impact parameter dependence of transverse in-plane flow ($< p_x >_{dir}$) crosses zero value at specific collision geometry termed as geometry of vanishing flow (GVF). Since then, study has been embellished to explore the role of different NEOS and mass-asymmetry of the reaction by keeping the total mass fixed [10] as well as with fixed target nuclei [11]. In spite these studies, the information is silent about the excitation energy dependence of GVF. In addition to that, very less literature exist on the impact parameter dependence of collective flow at mid-rapidity region. In this manuscript, we intend to study the collision geometry dependence of reduced flow and elliptical flow for a wide range of incident energy. It is to be explore through analysis that, whether reduced and elliptical flow also vanishes at particular collision geometry for a fixed incident energy?. If yes, then what range of incident energy contribute in this scenario.

In addition to geometry dependence of v_1 and v_2, the structural effects through nuclear charge radii parameterizations have also been studied by taking two different radii parameterizations i.e. the liquid drop model (LDM) R_{LDM} [12, 13] (isospin-independent) and a recently parameterized form of the nuclear charge radius proposed by Royer and Rousseau [14–16] R_{RR} (isospin-dependent). The study made on various phenomena like multi-fragmentation[17], nuclear stoping [18–20] and excitation function of collective flow [21] occurring at intermediate energies reveals that, the collective flow is the most affected phenomena. Since, change in radius at initial stage of experiment influence the pressure developed during the collision mechanism. Therefore, it will be interesting to observe the role of change in nuclear charge radius on the collision geometry dependence of v_1 and v_2 at various incident energies. Hence R_{LDM} and R_{RR} parameterizations of nuclear charge radii have been used in this manuscript to study the impact of nuclear charge radius on ($\hat{b}_{bal}$) and ($\hat{b}_{trans}$) at intermediate energies.

2 Result and discussion

The simulations have been carried out for the reactions of ${}^{50}_{20}$Ca+${}^{50}_{20}$Ca and ${}^{197}_{79}$Au+${}^{197}_{79}$Au at incident energy between 50 MeV/nucleon and 450 MeV/nucleon within the framework of IQMD model [22, 23]. The reactions are simulated at seven scaled impact parameter ($\hat{b}$) bins of [a] 0-0.28, [b] 0.28-0.39, [c] 0.39-0.48, [d] 0.48-0.56, [e] 0.56-0.62, [f] 0.62-0.76, [g] 0.76-0.88 covering the wide range of collision geometry. However, while obtaining the required figures, the magnitude of reduced flow and elliptical flow are plotted in the center of each impact parameter bin. The outcome of reactions is analyzed at saturation time of 200 fm/c by using minimum spanning tree (MST) method [24, 25].

Fig. 1 display the scaled impact parameter dependence of reduced flow for the reactions of ${}^{50}_{20}$Ca+${}^{50}_{20}$Ca (upper panels) and ${}^{197}_{79}$Au+${}^{197}_{79}$Au (lower panels) at different incident energies

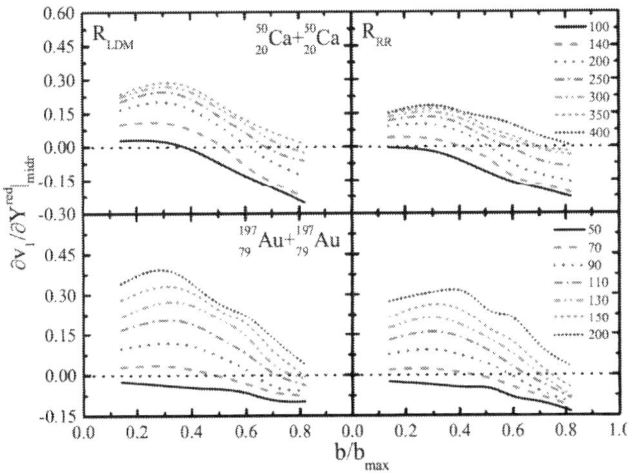

Figure 1. Scaled impact parameter dependence of reduced flow for the reactions of ${}^{50}_{20}$Ca+${}^{50}_{20}$Ca (upper panel) and ${}^{197}_{79}$Au+${}^{197}_{79}$Au (lower panel) at different incident energies for two nuclear charge radii parameterizations: R_{LDM} (left panels) and R_{RR} (right panels). Different lines in figure represents different incident energies in upper and lower panels.

(represented by different line in the figure) for particles having Z=1 with R_{LDM} (left panels) and R_{RR} (right panels) nuclear charge radii parameterizations. One can observe that, for both the reactions, the magnitude of reduced flow decreases with increase in collision geometry and at some point it passes through a zero value. This is because, as the spectator matter increases, the inward flow of particles decreases and the flow of particles shift to outward flow. Moreover, with increase in impact parameter, the strength of repulsive nucleon-nucleon collision reduces as the participant matter decreases. Therefore, the attractive mean filed dominates the reaction dynamics resulting in negative values of reduced flow. The collision geometry at which the flow is zero or one can say that both the attractive mean field and repulsive nucleon-nucleon collision counter balance each other and the reduced flow vanishes, is labelled as *"balance geometry"* and represented by $\hat{b}_{bal}$. One can observe that the reduced flow passes through zero value at particular incident energy range only. At low energy (for example below 100 MeV/nucleon in the case of ${}^{50}_{20}$Ca+${}^{50}_{20}$Ca and below 70 MeV/nucleon in the case of ${}^{197}_{79}$Au+${}^{197}_{79}$Au reaction), the value of reduced flow remains negative through out the collision geometry. This is because the attractive mean field increases with increase in the spectator matter. Therefore, the strength of the attractive force further increase and positive value of reduced flow can not be achieved. On the other hand, at high incident energies (for example, beyond 350 MeV/nucleon and 200 MeV/nucleon for the case of ${}^{50}_{20}$Ca+${}^{50}_{20}$Ca and ${}^{197}_{79}$Au+${}^{197}_{79}$Au respectively), the strength of repulsive nucleon-nucleon collision is much higher that it can not be overcome by decreasing the participant matter. Similar observation has been made in Fig. 2 for the collision geometry dependence of elliptical flow in the mid-rapidity region for the reaction of ${}^{50}_{20}$Ca+${}^{50}_{20}$Ca (upper panels) and ${}^{197}_{79}$Au+${}^{197}_{79}$Au (lower panels) at different incident energies (represented by different line in the figure) with R_{LDM} (left panels) and R_{RR} (right panels) nuclear charge radii parameterizations. One can observe that, the elliptical flow increases with increase in impact parameter. The negative value of v_2 corresponds to the strength of compressed nuclear matter and positive value is due to shadowing of the spectator matter. For fixed incident energy, the negative value of elliptical flow passes the zero value to become positive with as the collision geometry increases. The scaled impact parameter at which the transition of elliptical flow from out-of-plane to in-plane emission occurs is labelled as *"transition geometry"* and represented by $\hat{b}_{trans}$. This occurs because the

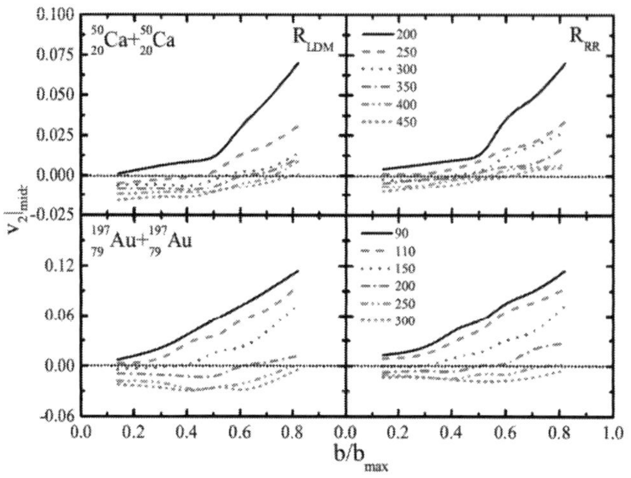

Figure 2. Same as in Fig. 1, but for elliptical flow v_2.

impression of spectator matter movement increases with increase in collision geometry. At relatively lower incident energies (i.e. below 200 MeV/nucleon and 110 MeV/nucleon for $^{50}_{20}$Ca+$^{50}_{20}$Ca and $^{197}_{79}$Au+$^{197}_{79}$Au respectively) the value of elliptical flow is already positive and can not show transition effect with increase in $\hat{b}$. Moreover, at higher incident energies (like beyond 450 and 300 MeV/nucleon in the case of $^{50}_{20}$Ca+$^{50}_{20}$Ca and $^{197}_{79}$Au+$^{197}_{79}$Au respectively), the strength of compressed nuclear matter can not be overcome by increasing the spectator matter.

The above stated observations made on impact parameter dependence of v_1 and v_2 is same for both the nuclear charge radii parameterizations. It is to be noted that, the calculated radius of $^{50}_{20}$Ca and $^{197}_{79}$Au increases when we switch from R_{LDM} to R_{RR}. With increase in nuclear radius the values of reduced flow (or elliptical) flow sharply turns from positive to negative (or negative to positive). Therefore larger radius leads to $\hat{b}_{bal}$ (or $\hat{b}_{trans}$) at lower impact parameter because the percentage of spectator matter in larger radius (R_{RR}) is more as compared to small radius (R_{RR}).

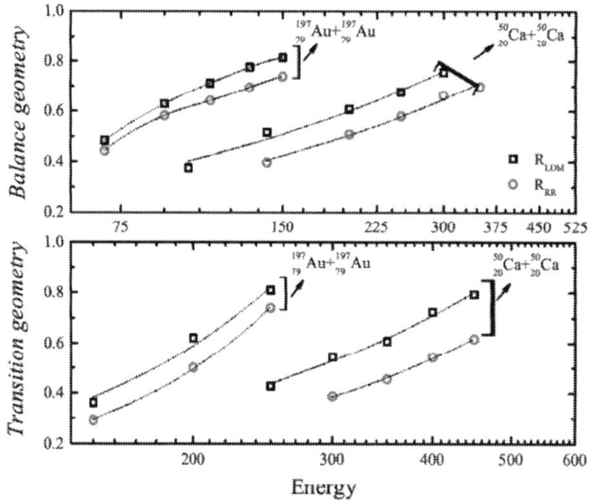

Figure 3. Incident energy dependence of *"balance geometry"* ($\hat{b}_{bal}$) (upper panel) and *"transition geometry"* ($\hat{b}_{trans}$) (lower panel) for the reactions of $^{50}_{20}$Ca+$^{50}_{20}$Ca and $^{197}_{79}$Au+$^{197}_{79}$Au.

To explore the effect of change in radius results more deeply, the values of $\hat{b}_{bal}$ and $\hat{b}_{trans}$ obtained from Fig. 1 and 2 are displayed in upper and lower panels of Fig. 3 respectively for the reactions of $^{50}_{20}$Ca+$^{50}_{20}$Ca and $^{197}_{79}$Au+$^{197}_{79}$Au. The open squares and circles in this figure represents the calculations with R_{LDM} and R_{RR} respectively. One can see that the values of $\hat{b}_{bal}$ and $\hat{b}_{trans}$ increases with increase in incident energy for both reactions and radii parameterizations. The magnitudes of both balance geometry and transition geometry are lower for larger radius. This is because, with increase in radius the size of spectator matter also increases (since scaled impact parameter is taken here which corresponds to percentage of interacting matter), which consequently increase the role of mean field. The range of incident energy which contributes to the observation of balance and transition geometries is comparatively lower for heavier colliding nuclei and higher for lighter colliding nuclei. Because at fixed incident energy, the impression of spectator matter is larger in the heavier colliding nuclei compared to lighter systems. The figure also reveals that the role of nuclear charge radii parametrization is more for lighter colliding systems. Also, the influence of increase in nuclear radius is more in $\hat{b}_{trans}$ as compared to $\hat{b}_{bal}$, which proves elliptical flow to be more sensitive towards the choice of nuclear charge radii parametrization than directed or reduced flow.

3 Summary and Conclusions

We presented a systematic study of the influence of nuclear charge radii parameterizations on the reduced flow and elliptical flow throughout the scaled impact parameter range for the reactions $^{50}_{20}$Ca+$^{50}_{20}$Ca and $^{197}_{79}$Au+$^{197}_{79}$Au at incident energy between 50 MeV/nucleon and 450 MeV/nucleon within the framework of IQMD model. The study concludes that the values of $\hat{b}_{bal}$ and $\hat{b}_{trans}$ increases with increase in incident energy. It has also been observed that with increase in radius (while switching from R_{LDM} and R_{RR} radii parameterization) $\hat{b}_{bal}$ and $\hat{b}_{trans}$ occurs at relatively lower impact parameter. The effect of nuclear charge radius in more on elliptical flow as compared to reduced flow. Therefore, elliptical flow is more sensitive towards the impression of pressure gradient and nuclear charge radius.

References

[1] S. A. Voloshin, Nucl. Phys. A **715**, 379c-388c (2003)
[2] S. A. Voloshin, Phys. Rev. C **55**, R1630 (1997)
[3] P. Russoto *et al.*, Eur. Phys. J. A **50**, 38 (2014)
[4] H. Sorge, Phys. Rev Lett., **78**, 2309-2312 (1997)
[5] A. Andronic, et al., Nucl. Phys. A **679**, 765-792 (2001)
[6] A. Bonasera and L. P. Csernai, Phys. Rev. Lett. **59**, 630 (1987)
[7] C. A. Ogilvie *et al.*, Phys. Rev. C **40**, 2592 (1989)
[8] J. Y. Ollitrault, Phys. Rev. D **46**, 229 (1992)
[9] R. Chugh and A. D. Sood, Pram. J Phys. **77**, 289 (2011)
[10] S. Goyal, Nucl. Phys. A **856**, 154 (2011)
[11] Lovejot and S. Gautam, Phys. of Part. and Nucl. Lett. **11**, 232 (2014)
[12] A. Bohr and B. R. Mottelson, Nuclear Structure, Vol. 1, *Nuclear Structure* (World Scientific, Singapore, 1998)
[13] P. G. Reinhard, M. Bender, W. Nazarewicz and T. Vertse, Phys. Rev. C **73**, 014309 (2006)
[14] G. Royer, Nucl. Phys. A **807**, 105 (2008)

[15] G. Royer and R. Rousseau, Eur. Phys. J. A **42**, 541 (2009)
[16] V. Y. Denisov, Phys. lett. B **526**, 315 (2002)
[17] Sangeeta, A. Jain and S. Kumar, Nucl. Phys. A **927**, 220-231 (2014)
[18] Sangeeta, Acta Phys. Pol. B **47**, 991-996 (2016)
[19] Sangeeta and V. Kaur, Acta Phys. Pol. B **48**, 623-627 (2017)
[20] Sangeeta and V. Kaur, Ind. J. Sci. and Tech. **10**, 31 (2017)
[21] Sangeeta and V. Kaur, Nucl. Phys. A **966**, 20-33 (2017)
[22] C. Hartnack, J. Aichelin, H. Stöcker, and W. Greiner, Mod. Phys. Lett. A **9**, 1151 (1994); *ibid*, Phys. Lett. B **336**, 131 (1994)
[23] C. Hartnack *et al.*, Phys. Rep. **510**, 119 (2012)
[24] J. Aichelin, Phys. Reports **202**, 233-360 (1991)
[25] Y. K. Vermani and R. K. Puri, Eur. Phys. Lett. **85**, 62001 (2009)

EPJ Web of Conferences **279**, 11018 (2023) https://doi.org/10.1051/epjconf/202327911018
NPA-X 2022

Proton capture on stored radioactive ^{118}Te ions

S. F. Dellmann[1,*], *J.* Glorius[2], *Y. A.* Litvinov[2], *R.* Reifarth[1], *K.* Al-Khasawneh[1], *M.* Aliotta[3], *L.* Bott[1], *B.* Brückner[1], *C.* Bruno[3], *R. J.* Chen[2,4], *T.* Davinson[3], *T.* Dickel[2,5], *I* Dillmann[6], *D.* Dmytriev[2], *P.* Erbacher[1], *D.* Freire-Fernández[7,8], *O.* Forstner[2,9], *H.* Geissel[5,2], *K.* Göbel[2], *C. J.* Griffin[6], *R.* Grisenti[2], *A.* Gumberidze[2], *E.* Haettner[2], *S.* Hagmann[2], *M.* Heil[2], *R.* Heß[2], *P.-M.* Hillenbrand[2], *R.* Joseph[2], *B.* Jurado[10], *C.* Kozhuharov[2], *I.* Kulikov[2], *B.* Löher[2], *C.* Langer[2,11], *G.* Leckenby[6], *C.* Lederer-Woods[3], *M.* Lestinsky[2], *S.* Litvinov[2], *B. A.* Lorenz[2], *E.* Lorenz[12], *J.* Marsh[3], *E.* Menz[2], *T.* Morgenroth[2], *N.* Petridis[2], *J.* Pibernat[10], *U.* Popp[2], *A.* Psaltis[12], *S.* Sanjari[2,11], *C.* Scheidenberger[2,5], *M.* Sguazzin[10], *R. S.* Sidhu[2,3,8], *U.* Spillmann[2], *M.* Steck[2], *T.* Stöhlker[2,9], *A.* Surzhykov[13,14], *J. A.* Swartz[10], *H.* Törnqvist[2,12], *L.* Varga[2,3], *D.* Vescovi[1], *H.* Weick[2], *M.* Weigand[1], *P.* Woods[3], *Y.* Xing[2], and *T.* Yamaguchi[15]

[1]Goethe University, Frankfurt, Germany
[2]GSI Helmholtzzentrum für Schwerionenforschung Darmstadt, Germany
[3]School of Physics and Astronomy, University of Edinburgh, United Kingdom
[4]CAS Key Laboratory of High Precision Nuclear Spectroscopy, Lanzhou, China
[5]Justus-Liebig-University Gießen, Germany
[6]TRIUMF, Canada
[7]Ruprecht-Karls-University Heidelberg, Germany
[8]Max-Planck-Institut für Kernphysik, Heidelberg, Germany
[9]Helmholtz-Institut Jena, Germany
[10]LP2I Bordeaux, CNRS/IN2P3-Université de Bordeaux, Gradignan, France
[11]Aachen University of Applied Sciences, Aachen, Germany
[12]TU Darmstadt, Germany
[13]Physikalisch Technische Bundesanstalt Braunschweig, Germany
[14]Braunschweig University of Technology, Germany
[15]Saitama University, Saitama, Japan

Abstract. Experimental determination of the cross sections of proton capture on radioactive nuclei is extremely difficult. Therefore, it is of substantial interest for the understanding of the production of the p-nuclei. For the first time, a direct measurement of proton-capture cross sections on stored, radioactive ions became possible in an energy range of interest for nuclear astrophysics. The experiment was performed at the Experimental Storage Ring (ESR) at GSI by making use of a sensitive method to measure (p,γ) and (p,n) reactions in inverse kinematics. These reaction channels are of high relevance for the nucleosynthesis processes in supernovae, which are among the most violent explosions in the universe and are not yet well understood. The cross section of the ^{118}Te(p,γ) reaction has been measured at energies of 6 MeV/u and 7 MeV/u. The heavy ions interacted with a hydrogen gas jet target. The radiative recombination process of the fully stripped ^{118}Te ions and electrons from the hydrogen target was used as a luminosity monitor. An overview of the experimental method and preliminary results from the ongoing analysis will be presented.

*e-mail: dellmann@iap.uni-frankfurt.de

1 Astrophysical motivation

Most of the elements heavier than iron are synthesized by neutron-induced reactions. However, there are around 35 stable neutron-deficient nuclei which cannot be made by either the s process or the r process. The understanding of the production of these so called p-nuclei is a current challenge in nuclear astrophysics. They have the lowest relative abundance and their production cannot be described by neutron capture processes. A production process that is currently considered to be likely to create the majority of p-nuclei is the γ process [1, 2]. By photodisintegration of pre-existing seed nuclei and subsequent β decays, p-nuclei may be produced. Simulations of the γ process involve several thousand nuclei that are connected by tens of thousands of reactions. It requires correspondingly large reaction networks to model the abundance distributions that follow from these scenarios [3]. Given the low cross sections at energies of interest for nuclear astrophysics, limited experimental data for proton capture reactions are available. Additionally, the majority of the involved nuclei are unstable, since the γ-process nucleosynthesis path moves far from the valley of stability. Experimentally there is nearly no data available, so the calculations are affected by large uncertainties [1, 2, 4, 5]. However, it is of utmost importance that these calculations are based on experimentally determined data. The measurement of a (p,γ) cross section provides, besides the determination of a (p, γ) reaction under stellar conditions, the determination of the inverse (γ,p) reaction rates, that are important for the γ process models. ^{118}Te is part of the γ process network and was chosen for these first radioactive beam studies for feasibility reasons as well as for the ability to achieve high intensities. With this experiment, for the first time a proton capture was successfully performed on stored radioactive ions in an energy region relevant for p-nucleosynthesis.

2 Experimental realization at the Experimental Storage Ring (ESR) at GSI

This experiment has become possible at GSI, by combining two unique facilities, the Fragment Seperator (FRS) [6] and the Experimental Storage Ring (ESR) [7]. In a classical nuclear reaction measurement a light projectile impinges on a heavy target, but if the target is radioactive and short-lived a sample can not be prepared. The desired nuclei must be produced online during the experiment. The study is then done in inverse kinematics. Due to the use of a radioactive isotope, with a half-life of 6 days in the case of ^{118}Te, a measurement in inverse kinematics is necessary. Due to the expected small cross sections, especially in the relevant energy range, very small counting rates are expected. Heavy ion storage rings like the ESR can compensate for low reaction rates and enable for the use of thin targets accompanied by low energy losses of the projectiles. Due to multiple passages of the projectile through the target, luminosities comparable to single pass experiments with thick targets can be achieved. By impinging stable ^{124}Xe ions on a Be production target a variety of fragments were produced, among them the radioactive isotope ^{118}Te. By magnetic separation combined with energy losses in dedicated energy degraders it was possible to filter out a clean beam of ^{118}Te in the FRS. Afterwards, ^{118}Te was stored in the ESR at 400 MeV/u. By accumulating up to 20 injections from the FRS, high beam intensities could be achieved. Inside the ESR, the fully-ionized ions were cooled and decelerated down to 6 and 7 MeV/u providing about 10^6 ions for the measurement [8]. By collisions of the radioactive beam and a thin hydrogen gas jet [9] target, proton capture reactions could take place. A proof-of-principle experiment had been performed in 2009 [10] for ^{96}Ru(p,γ) and an improved setup was introduced in 2016 to validate the method with ^{124}Xe(p,γ) close to the astrophysical Gamow window [11].

To determine the cross section in a conventional way, the number of ions inside the ring as well as the target density would have to be known exactly. The determination of these would only be possible with large uncertainties. It is therefore of significant advantage to measure the cross section relative to a well understood simultaneously occurring process. The cross section for the proton capture process is determined relative to that of the radiative electron capture (REC) of electrons from the hydrogen atoms into the bare Te ions. To measure the emitted photons X-ray detectors were used in this experiment [12]. The cross section is derived by:

$$\sigma_{(p,\gamma)} = \frac{N_{(p,\gamma)}}{N_{K-REC}} \epsilon_{K-REC,intr} \int_{\Delta\Omega} \frac{d\sigma_{K-REC}}{d\Omega} d\Omega \quad . \tag{1}$$

Here, $N_{(p,\gamma)}$ describes the number of protons captured, N_{K-REC} the number of emitted photons after electrons capture in the K-shell and $\epsilon_{K-REC,intr}$ the efficiency of the X-ray detectors used. The angle dependent cross section $\int_{\Delta\Omega} \frac{d\sigma_{K-REC}}{d\Omega} d\Omega$ of the radiative electron capture into the K-shell can be predicted very accurately by theory [13]. Figure 1 shows an X-ray spectrum that has been taken during the experiment with a High Purity Germanium (HPGe) detector installed at an angle of 90°. Highlighted in red is the corresponding peak for photons emitted after an electron is captured directly into the K-shell of the ^{118}Te projectiles.

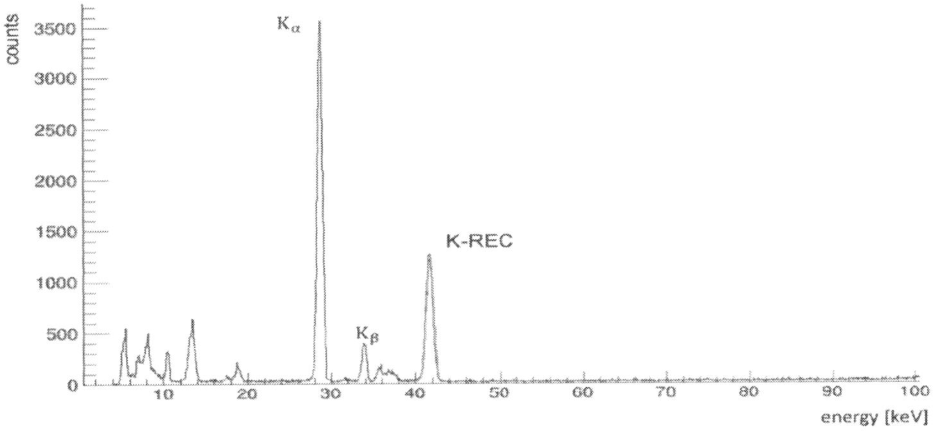

Figure 1. X-ray spectrum recorded by a HPGe detector positioned at an angle of 90° at the target region. The peak corresponding to the photons emitted after an electron is captured into the K-shell of the ^{118}Te projectiles is marked in red.

In order to determine the number of reaction products, they must be detected separately from the stored beam and the background. A dipole magnet behind the target section provides magnetic separation between the emitted reaction products and the stored beam. The bending radius ρ of an ion within a homogeneous magnetic field B can be derived from the equivalence of acting Lorentz and centripetal force:

$$B\rho = p/q \tag{2}$$

After a proton capture, the momentum p remains almost unchanged whereas the charge q increases by 1e. This leads to a decreased bending radius. The reaction products thus shift to the inner side of the ring. By placing a detector at the inner side of the ring the reaction products can be observed. For this measurement, a Double Sided Silicon Strip Detector (DSSSD) was used, consisting of 16 x 16 strips in x and y directions, respectively. This allows

for an exact determination of the number of the incident reaction products. Figure 2 shows the ion hit distributions that have been measured with the DSSSD during the experiment. The left plot shows data that has been taken with ^{118}Te at 7 MeV/u for 48 h and the right plot for 52 h with an energy of 6 MeV/u.
Due to the small nuclear cross sections, proton capture at the target occurs only rarely. In contrast, many ions undergo scattering off the target atoms. The dominant process is Rutherford scattering, which results in a large background contribution [14]. Due to the divergence of the cross sections of Rutherford scattering and proton-capture towards lower energies, the peak-to-background ratio becomes worse at low energies. This relation could be improved drastically by the usage of a scraper based ERASE technique [15], that physically blocks the disturbing Rutherford distribution, while leaving the proton-capture products untouched. This has made it possible to reach energies of 6 MeV/u for this experiment.

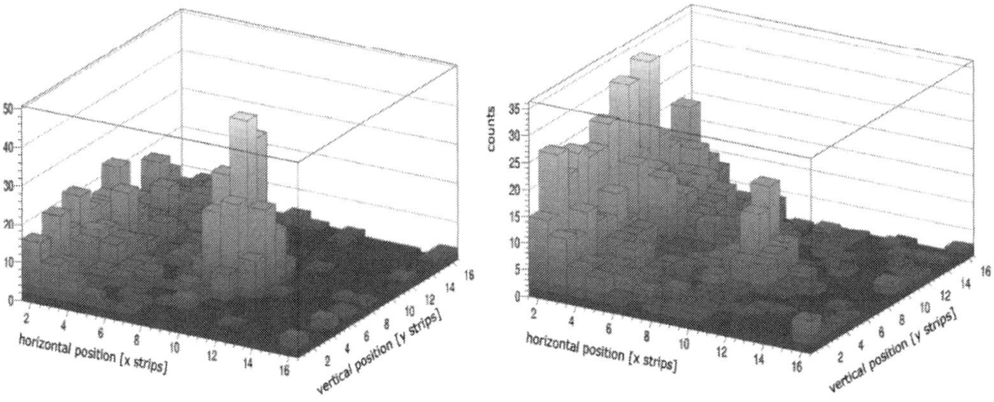

Figure 2. Hit distribution taken with the DSSSD. The left plot shows data for ^{118}Te at 7 MeV/u that has been taken for 48 h and the right plot shows data of ^{118}Te at 6 MeV/u taken for 52 h.

This enables the determination of the number of events by integrating over the peak. To subtract the background in the region of the peak, currently a special fit is used. The fit of the entire spectrum includes a 2D Gaussian fit for the peak and an additional 2D Gaussian fit for the background. In order to achieve a more precise description of the data simulations are in progress. For this purpose, a Monte Carlo simulation code will be used, namely MOCADI, a code developed at GSI [16].

3 Preliminary results

The conducted analysis provides the first preliminary constraints from experimentally determined cross sections. The preliminary cross sections have expected uncertainties of less than 5 %. Figure 3 shows the experimentally determined cross sections from this work and for an energy of 10.05 MeV [15]. The center of mass energy is plotted on the abscissa. Additionally, the theoretical predictions by NON-SMOKER [17] are shown in blue, as well as the predictions by the TALYS code [18] shown in green. The results seem to deviate from the theory for lower energies, whereas for higher energies the values agree within the uncertainties, with the predictions by NON-SMOKER. Since these are still preliminary results, this may be subject to further adjustments.
Further uncertainty calculations as well as simulations for a more precise determination of the background in the region of the peak are the main outstanding tasks.

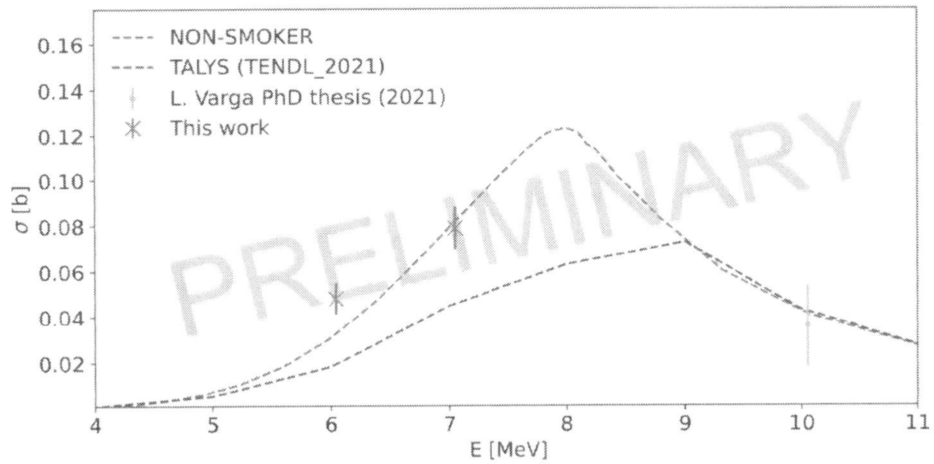

Figure 3. Preliminary cross sections obtained in this work in comparison with theoretical predictions.

References

[1] M. Pignatari, K. Göbel, R. Reifarth, C. Travaglio, International Journal of Modern Physics E **25**, 1630003 (2016)
[2] C. Travaglio, T. Rauscher, A. Heger, M. Pignatari, C. West, ApJ **854**, 18 (2018)
[3] H. Schatz, K.E. Rehm, Nuclear Physics A **777**, 601 (2006)
[4] M. Arnould, S. Goriely, Physics Reports **384**, 1 (2003)
[5] K. Göbel, J. Glorius, A. Koloczek, M. Pignatari, R. Reifarth, R. Schach, K. Sonnabend, EPJ Web of Conferences **93**, 03006 (2015)
[6] H. Geissel, P. Armbruster, K.H. Behr, A. Brünle, K. Burkard, M. Chen, H. Folger et al., Nuclear Instruments and Methods in Physics Research B **70**, 286 (1992)
[7] B. Franzke, Nucl. Instr. Meth. B **24/25**, 18 (1987)
[8] M. Steck, Y.A. Litvinov, Progress in Particle and Nuclear Physics **115**, 103811 (2020)
[9] M. Kühnel, N. Petridis, D. Winters, U. Popp, R. Dörner, T. Stöhlker, R. Grisenti, Nucl. Inst. Meth. A **602**, 311 (2009)
[10] B. Mei, T. Aumann, S. Bishop, K. Blaum, K. Boretzky, F. Bosch, C. Brandau et al., Phys. Rev. C **92**, 035803 (2015)
[11] J. Glorius, C. Langer, Z. Slavkovská, L. Bott, C. Brandau, B. Brückner, K. Blaum et al., Phys. Rev. Lett. **122**, 092701 (2019)
[12] J. Eichler, T. Stöhlker, Phys. Rep. **439**, 1 (2007)
[13] A.N. Artemyev, A. Surzhykov, S. Fritzsche, B. Najjari, A.B. Voitkiv, Phys. Rev. A **82**, 022716 (2010)
[14] Y.M. Xing, J. Glorius, L. Varga, L. Bott, C. Brandau, B. Brückner, R.J. Chen et al., Nuclear Instruments and Methods in Physics Research A **982**, 164367 (2020)
[15] L. Varga, Ph.D. thesis, University of Heidelberg (2020)
[16] N. Iwasa, H. Geissel, G. Münzenberg, C. Scheidenberger, T. Schwab, H. Wollnik, Nucl. Inst. Meth. B **126**, 284 (1997)
[17] T. Rauscher, F.K. Thielemann, Atomic Data and Nuclear Data Tables **79**, 47 (2001)
[18] A.J. Koning, S. Hilaire, M.C. Duijvestijn, *TALYS-1.0*, in *ND2007* (EDP Sciences, 2007)

Studies of low-energy K$^-$-nucleus/nuclei interactions with light nuclei by AMADEUS

Magdalena Skurzok[1,2,3,*], *Massimiliano* Bazzi[3], *Mario* Bragadireanu[4], *Damir* Bosnar[5], *Michael* Cargnelli[6], *Alberto* Clozza[3], *Luca* De Paolis[3], *Raffaele* Del Grande[7,3], *Laura* Fabbietti[7,9], *Carlo* Guaraldo[3], *Mihai* Iliescu[3], *Paolo* Levi Sandri[3], *Johann* Marton[6], *Marco* Miliucci[3], *Paweł* Moskal[1,2], *Kristian* Piscicchia[8,3], *Angels* Ramos[10], *Alessandro* Scordo[3], *Michał* Silarski[1,2], *Diana Laura* Sirghi[3,5], *Florin* Sirghi[3,5], *Antonio* Spallone[3], *Oton* Vazquez Doce[3], *Sławomir* Wycech[11], *Johann* Zmeskal[6], and *Catalina* Curceanu[3]

[1]Institute of Physics, Jagiellonian University, ul. Łojasiewicza 11, 30-348 Cracow, Poland
[2]Center for Theranostics, Jagiellonian University, Krakow, Poland
[3]INFN, Laboratori Nazionali di Frascati, Via Enrico Fermi 54, 00044 Frascati RM, Italy
[4]Horia Hulubei National Institute of Physics and Nuclear Engineering, Str. Reactorului 30, P.O. BOX MG-6, Bucharest - Magurele, Romania
[5]Department of Physics, Faculty of Science, University of Zagreb, Horvatovac 102a, HR-10000 Zagreb, Croatia
[6]Stefan-Meyer-Institut für subatomare Physik, Kegelgasse 27, 1030 Vienna, Austria
[7]Physik Department E62, Technische Universität München, James-Franck-Str. 1, 85748 Garching, Germany
[8]Centro Ricerche Enrico Fermi - Museo Storico della Fisica e Centro Studi e Ricerche "Enrico Fermi", Piazza del Viminale 1, 00184 Roma, Italy
[9]Excellence Cluster "Origin and Structure of the Universe", Boltzmann str. 2, D-85748 Garching, Germany
[10]Departament de Fisica Quantica i Astrofisica and Institut de Ciencies del Cosmos, Universitat de Barcelona, Marti i Franques 1, Martí i Franquès 1, 08028 Barcelona, Spain
[11]Department of Theoretical Physics, National Centre for Nuclear Research, ul. Pasteura 7, 02-093 Warsaw, Poland

Abstract. The AMADEUS Collaboration aims to provide unique experimental constraints to the antikaon-nucleon strong interaction in the regime of non-perturbative QCD. The K$^-$ nuclear captures, both at-rest and in-flight, are studied using the monochromatic low-momentum kaon beam (p_K ~120 MeV/c) produced at the DAΦNE collider, interacting with the KLOE detector materials. The studies are performed by reconstructing the hyperon-pion and hyperon-nucleon final states. In this work a brief description of AMADEUS results for $\Lambda\pi^-$ and Λp final states is presented.

1 Introduction

The AMADEUS Collaboration [1–4] investigates the low-energy K$^-$ induced reactions in light nuclear targets corresponding to components of the KLOE detector [5], in order to

*e-mail: magdalena.skurzok@uj.edu.pl

provide experimental constraints to the non-perturbative QCD in the strangeness sector. It is performed by exploiting the low momentum ($p_K \sim 127$ MeV/c), almost monochromatic, charged kaons produced in the decay of ϕ mesons at-rest at the DAΦNE accelerator [6].

Models of low-energy strong interaction, in the strangeness sector, face difficulties mainly related to the appearance of the broad Λ(1405) (I=0) and Σ(1385) (I=1) resonances just below the $\bar{K}$N threshold. To deal with this problem, chiral unitary models [7–13] and phenomenological potential models [14–19] were developed, leading however to contrasting predictions for the Λ(1405) parameters and related kaonic nuclear bound states.

The nature of Λ(1405) resonance remains still an open issue since experiments result in observation of different masses and widths of this resonance, depending on the production channel, as well as the observed decay mode [20]. In order to understand the line-shape of the Λ(1405) in kaon induced reactions the background due to the non-resonant $K^-N \to Y\pi$ reactions has to be taken into account. The position of the Λ(1405) reflects the strength of the $\bar{K}$N interaction, thus influencing the possible formation of more exotic kaonic bound states (for a review see [21]). The phenomenological models [14–16, 18, 19] interpret the I=0 Λ(1405) as a pure $\bar{K}$N bound state, leading to the prediction of deeply bound kaonic nuclear states. According to Chiral models [8–12], however, the Λ(1405) emerges as a superposition of two states and, as a consequence the K^-N interaction is much less attractive, which implies the prediction of only slightly bound kaonic nuclear states.

The activities of the AMADEUS Collaboration are centered on the experimental investigations of the low-energy charged kaon-nucleon/nuclei interaction, aiming to unveil the controversial nature of the Λ(1405) state, and deepen our understanding of the K^- single- and multi-nucleon interaction processes, and bound states formation.

This article reports on the studies of K^- single-nucleon absorption in ^{4}He leading to the first measurement of the non-resonant contribution in $K^-N \to Y\pi$ reaction [24] below the $\bar{K}$N threshold which is essential for studies of the Λ(1405) resonance properties. Moreover, the investigation of the K^- interactions in ^{12}C nuclei resulting in the first complete characterization of the K^- two-, three- and four-nucleon absorptions (2NA, 3NA and 4NA) in the Λp and Σ^0p final states is presented [25, 26].

2 The measurement of the $K^-n \to \Lambda\pi^-$ non-resonant transition amplitude below the $\bar{K}$N threshold

The investigation of the Λ(1405) properties, produced through the K^-p mechanism in light nuclear targets, requires two biases to be taken into account.

The first bias is the energy threshold imposed by the absorbing nucleon binding energy (for K^- capture at rest on ^{4}He the $\Sigma\pi$ invariant mass threshold is about 1412 MeV, while for ^{12}C it is about 1416 MeV). The access to $\bar{K}N$ sub-threshold region corresponding to the Λ(1405) high-mass predicted pole (about 1420 MeV), is possible by exploiting K^-N absorption in-flight [27]. For a mean kaon momentum of 100 MeV/c, the $\Sigma\pi$ invariant mass threshold is shifted upwards by about 10 MeV.

The second bias is related to the non-resonant $K^-N \to Y\pi$ reaction, which was experimentally investigated for the first time in the $K^-n \to \Lambda\pi^-$ process, considering K^-n single nucleon absorptions in ^{4}He [24]. In this work the modulus of non-resonant transition amplitude $|T_{K^-n \to \Lambda\pi^-}|$ at (33±6) MeV below the $\bar{K}N$ threshold was measured by fitting simultaneously the experimentally extracted $\Lambda\pi^-$ invariant mass, momentum and angular distributions with dedicated Monte Carlo simulations. In the fit all the contributing reactions were taken into account: non-resonant processes, resonant processes and the primary production of a Σ followed by the $\Sigma N \to \Lambda N'$ conversion process. The simulations for non-resonant/resonant processes were based on the results of [28]. The non-resonant transition amplitude modulus is equal to $|T_{K^-n\to\Lambda\pi^-}|$=0.334 ± 0.018 (*stat.*) $^{+0.034}_{-0.058}$ (*syst.*) fm and is shown in Fig. 1 together with the theoretical predictions (see Ref: [29] (P), [30] (KM), [31] (M1,M2), [32] (B2,B4), [33] (BCN)) rescaled for the $K^-n \to \Sigma\pi$ transition probabilities. This measurement can be used to test and constrain the S-wave $K^-n \to \Lambda\pi^-$ transition amplitude calculations.

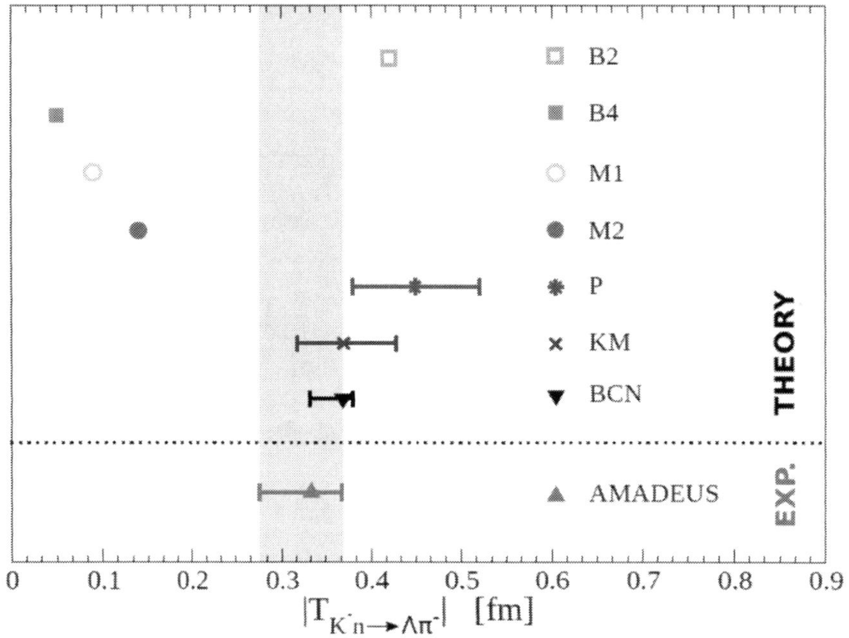

Figure 1. Modulus of the measured non resonant $K^-n \to \Lambda\pi^-$ transition amplitude compared with theoretical calculations, see details in the text. Figure is adapted from Ref. [34].

3 K^- multi-nucleon absorption cross sections and branching ratios in Λp and $\Sigma^0 p$ final states

Recent $\Lambda(\Sigma^0)$p final states investigation in K^- capture processes on ^{12}C nuclei [25, 26] by the AMADEUS collaboration results in a complete characterization of the K^- two-, three- and four-nucleon absorption (2NA, 3NA and 4NA) processes. The studies were performed [26], based on the phenomenological model for the K^- captures at-rest and in-flight on light nuclei described in Refs. [28, 40]. The K^- 2NA, 3NA and 4NA branching ratios (BRs) and cross sections for low-momentum kaons in Λp and Σ^0p channels were determined by performing a simultaneous fit of the measured Λp invariant mass, Λp angular correlation, Λ and proton momentum distribution with the corresponding simulated distributions for all the processes contributing to the measured Λp final state, including the primary Σ^0 productions followed by $\Sigma^0 \to \Lambda\gamma$ decay. For the 2NA the contributions of both elastic final state interactions (FSI) of the primary produced hyperon and nucleon with the residual nucleus were taken into account, as well as the conversion of primary produced sigma particles ($\Sigma N \to \Lambda N'$); this allows to disentangle the quasi-free (QF) Λp and Σ^0p productions. The obtained BRs and cross sections are summarized in Table 1.

process	BR (%)	cross section (mb)	@	p_K (MeV/c)
2NA-QF Λp	0.25 ± 0.02 (stat.) $^{+0.01}_{-0.02}$(syst.)	2.8 ± 0.3 (stat.) $^{+0.1}_{-0.2}$ (syst.)	@	128 ± 29
2NA-FSI Λp	6.2 ± 1.4(stat.) $^{+0.5}_{-0.6}$(syst.)	69 ± 15 (stat.) ± 6 (syst.)	@	128 ± 29
2NA-QF Σ^0p	0.35 ± 0.09(stat.) $^{+0.13}_{-0.06}$(syst.)	3.9 ± 1.0 (stat.) $^{+1.4}_{-0.7}$ (syst.)	@	128 ± 29
2NA-FSI Σ^0p	7.2 ± 2.2(stat.) $^{+4.2}_{-5.4}$(syst.)	80 ± 25 (stat.) $^{+46}_{-60}$ (syst.)	@	128 ± 29
2NA-CONV Σ / Λ	2.1 ± 1.2(stat.) $^{+0.9}_{-0.5}$(syst.)	-		
3NA Λpn	1.4 ± 0.2(stat.) $^{+0.1}_{-0.2}$(syst.)	15 ± 2 (stat.) ± 2 (syst.)	@	117 ± 23
3NA Σ^0pn	3.7 ± 0.4(stat.) $^{+0.2}_{-0.4}$(syst.)	41 ± 4 (stat.) $^{+2}_{-5}$ (syst.)	@	117 ± 23
4NA Λpnn	0.13 ± 0.09(stat.) $^{+0.08}_{-0.07}$(syst.)	-		
Global $\Lambda(\Sigma^0)$p	21 ± 3(stat.) $^{+5}_{-6}$(syst.)	-		

Table 1. Branching ratios (for the K^- absorbed at-rest) and cross sections (for the K^- absorbed in-flight) of the K^- multi-nucleon absorption processes. The K^- momentum is evaluated in the centre of mass reference frame of the absorbing nucleons, thus it differs for the 2NA and 3NA processes. The statistical and systematic errors are also given. The Table is adapted from Ref. [26].

The sum of the K^- 2NA BRs is found to be (16.1 ± 2.9(stat.)$^{+1.0}_{-0.9}$(syst.)) % and is the main 2NA contribution to the global branching ratio shown in Table 1. The missing contribution to the K^- 2NA from other few processes without Λp pair in the final state of the interaction is also estimated combining both the measured branching ratios and the available theoretical information, and amounts to (5.5 ± 0.1(stat.)$^{+1.0}_{-0.9}$(syst.)) % (for more details see [35]). Including the missing component, the total BR of the K^- 2NA in Carbon is found to be (21.6 ± 2.9(stat.)$^{+4.4}_{-5.6}$(syst.)) %.

The BR of the Λp QF production in K^- 2NA interaction is found to be smaller than the Σ^0p QF production, the ratio of the BRs in Table 1 is: $R = \frac{BR(K^-(pp)\rightarrow\Lambda p)}{BR(K^-(pp)\rightarrow\Sigma^0 p)} = 0.7 \pm 0.2(stat.)^{+0.2}_{-0.3}(syst.)$. The ratio of the corresponding phase spaces is instead $R' = 1.22$. This result was interpreted in Ref. [41] and was found to be in agreement with the theoretical calculations (BCN and P models) when the in medium effect due to the Pauli blocking is considered.

The possible contribution of K^-pp bound system in Λp spectra was also investigated. It was found that the signal of the K^-pp bound state formation in K^- induced reactions on carbon completely overlaps with the K^- 2NA-QF process. The two components can be disentangled only for narrow states ($\Gamma < 15$ MeV/c^2) which are excluded by the theoretical calculations. In the further step, in order to compare the spectra with the corresponding FINUDA measurement, back-to-back Λp events were selected ($\cos\theta_{\Lambda p} < -0.8$). As in the previous case, the obtained spectra can be completely described in terms of K^- multi-NA processes.

The presented results demonstrate that the DAΦNE collider is a unique facility for the study of low-energy kaon physics.

Acknowledgements

We acknowledge the KLOE/KLOE-2 Collaboration for their support and for having provided us with the data and the tools to perform the analysis presented in this paper. Part of this work was supported by the EU STRONG-2020 project (grant agreement No 824093); Minstero degli Affari Esteri e della Cooperazione Internazionale, EXOTICA project, PO21MO03; the Polish National Science Center through grant No. UMO-2016/21/D/ST2/01155. The project is co-financed by the Polish National Agency for Academic Exchange, grant no

PPN/BIT/2021/1/00037. The authors acknowledge support from the SciMat and qLife Priority Research Areas budget under the program Excellence Initiative-Research University at the Jagiellonian University.

References

[1] R. Del Grande et al., J. Phys.: Conf. Ser. **1643** (2020) 012081.
[2] M. Skurzok et al., Springer Proc. Phys. **238** (2020) 937.
[3] M. Skurzok et al., JPS Conf. Proc. **26** (2019) 023011.
[4] C. Curceanu et al., Acta. Phys. Pol. B **46** (2015) 203.
[5] F. Bossi et al., Riv. Nuovo Cim. **31** (2008) 531.
[6] A. Gallo et al., Conf. Proc. C **060626** (2006) 604.
[7] A. Dote, T. Hyodo, W. Weise, Phys. Rev. C **79** (2009) 014003.
[8] A. Dote, T. Inoue, T. Myo, Phys. Lett. B **784** (2018) 405.
[9] N. Barnea, A. Gal, E. Z. Liverts, Phys. Lett. B **712** (2012) 132.
[10] Y. Ikeda, H. Kamano and T. Sato, Prog. Theor. Phys. **124** (2010) 533.
[11] P. Bicudo, Phys. Rev. D **76** (2007) 031502.
[12] M. Bayar and E. Oset, Nucl. Phys. A **914** (2013) 349.
[13] T. Sekihara, E. Oset, A. Ramos, PTEP **2016** (2016) 12.
[14] Y. Akaishi, T. Yamazaki, Phys. Rev. C **65** (2002) 044005.
[15] Y. Ikeda and T. Sato, Phys. Rev. C **76** (2007) 035203.
[16] S. Wycech, A. M. Green, Phys. Rev. C **79** (2009) 014001.
[17] N. V. Shevchenko, A. Gal, J. Mares, Phys. Rev. Lett. **98** (2007) 082301.
[18] J. Revai and N. V. Shevchenko, Phys. Rev. C **90** (2014) 034004.
[19] S. Maeda, Y. Akaishi, T. Yamazaki, Proc. Jpn. Acad. B **89** (2013) 418.
[20] T. Hyodo and D. Jido, Prog. Part. Nucl. Phys. **67** (2012) 55.
[21] T. Hyodo and W. Weise, arXiv:2202.06181 (2022).
[22] S. Wycech, Nucl. Phys. A **450** (1986) 399c.
[23] Y. Akaishi, T. Yamazaki, Phys. Lett. B **535** (2002) 70.
[24] K. Piscicchia et al., Phys. Lett. B **782** (2018) 339.
[25] O. Vazques Doce et al., Phys. Lett. B **758**, 134 (2016).
[26] R. Del Grande et al., Eur. Phys. J. C **79** (2019) 190.
[27] K. Piscicchia et al., EPJ Web Conf. **166** (2018) 00020.
[28] K. Piscicchia, S. Wycech and C. Curceanu, Nucl. Phys. A **954** (2016) 75.
[29] A. Cieplý and J. Smejkal, Nucl. Phys. A **881** (2012) 115.
[30] Y. Ikeda, T. Hyodo and W. Weise, Nucl. Phys. A **881** (2012) 98.
[31] Z. H. Guo and J. A. Oller, Phys. Rev. C **87** (2013) 035202.
[32] M. Mai and U. G. Meissner, Eur. Phys. J. A **51** (2015) 30 .
[33] A. Feijoo, V. Magas and A. Ramos, Phys. Rev. C **99**, no.3 (2019) 035211.
[34] K. Piscicchia et al., EPJ Web Conf. **262** (2022) 01006.
[35] R. Del Grande, K. Piscicchia et al., Phys. Scripta **95** (8) (2020) 084012.
[36] P. A. Zyla et al. (Particle Data Group), Prog. Theor. Exp. Phys. **2020** (2020) 083C01.
[37] K. Piscicchia et al., Acta. Phys. Polon. B Proc. Suppl. **11** (2018) 609.
[38] K. Piscicchia, PhD thesis (2013), http://www.infn.it/thesis/thesis_dettaglio.php?tid=7097.
[39] A. Feijoo, V. K. Magas and A. Ramos, AIP Conf. Proc. **2130** no. 1 (2019) 040013.
[40] R. Del Grande, K. Piscicchia, S. Wycech, Acta Phys. Pol. B **48** (2017) 1881.
[41] J. Hrtankova and A. Ramos, Phys. Rev. C **101** (2020) 035204.

The study of the $^{20}Ne(p,\gamma)^{21}Na$ reaction at LUNA

Sandra Zavatarelli[1*] and *Eliana* Masha[2] on behalf of the LUNA Collaboration

[1]INFN Sezione di Genova, 16146 Via Dodecaneso 33, Genoa, Italy
[2]Helmholtz-Zentrum Dresden-Rossendorf, 01328 Bautzner Landstraße 400, Dresden Germany

Abstract. The NeNa and MgAl cycles have been the subject of much experimental activity during the last decade because of their relevance to the synthesis of Ne, Na, and Mg isotopes during the H burning in several astrophysical scenarios. Key reactions in these cycles are also believed to be the main agents of the observed anti-correlations in O-Na and Al-Mg abundances exhibited by the stars of Galactic globular clusters. The $^{20}Ne(p,\gamma)^{21}Na$ is the first reaction and the bottleneck of the NeNa cycle: having the slowest reaction rate, it controls the speed of the entire cycle. In order to better constrain the overall astrophysical reaction rate of this important reaction, the LUNA collaboration has started a new experimental effort to study the 366 keV resonance and to improve the knowledge of the cross section at proton energies below 400 keV. This contribution describes the experimental setup and preliminary results.

1 Introduction

Proton captures on ^{20}Ne nuclei could proceed effectively if the temperature reaches T = 0.05 GK [1] (Table 1). They may occur in different stellar scenarios such as red giant stars, asymptotic giant branch stars, novae, and massive stars.

The LUNA collaboration has recently measured the strength and properties of low energies states populated by the $^{22}Ne(p,\gamma)^{23}Na$ and $^{23}Na(p,\gamma)^{24}Mg$ reactions [2-6]. As results of these improved studies, existing uncertainties on the $^{20}Ne(p,\gamma)^{21}Na$ are now the ones that affect the estimated elements production in the NeNa cycle [7, 8] such as ^{22}Na, a stellar γ-ray signature.

In the temperature range between T = 0.1 – 1.0 GK (Fig.1) the rate of the $^{20}Ne(p,\gamma)^{21}Na$ reaction (Q-value = 2431.68 keV) is dominated by the narrow resonance at E_R = 366 keV, corresponding to an excited level at Ex = 2797.5 keV in ^{21}Na and by the direct capture component with contributions from different cascade branching [9]. At lower temperatures the high energy tail of a sub-threshold state at E_R = –6.7 keV, corresponding to the E_x = 2424.9 keV excited level in ^{21}Na, contributes significantly.

The strength of the resonance at 366 keV was measured with a 20% uncertainty by Rolfs and co-workers in 1975 at the California State University Van der Graaf facility [10]: the quoted value is $\omega\gamma$=0.11±0.02 meV. A second and more recent study (2019) at the Triangle University Laboratory for Nuclear Astrophysics, is reported but still unpublished, in

* Corresponding author: zavatare@ge.infn.it

Cooper's PhD Thesis [11]: the deduced weighted average for the resonance strength, $\omega\gamma$ = (0.0722 ± 0.0068) meV, is 35% lower than the Rolfs *et al* value.
The direct capture process was measured by Rolfs *et al* and by Lyons *et al* (Notre Dame University, 2018) [9] in the energy range between 0.4 and 2 MeV: Fig. 2 summarises the results available in literature for strongest transition, the DC→ 2425 keV.
In general, only poorly precision data exists at $E_{c.m.}$ < 400 keV, making relevant further experimental investigations.

Table 1. Energy and temperature ranges for the $^{20}Ne(p,\gamma)^{21}Ne$ reaction in the relevant astrophysical sites.

Astrophysical site	Initial mass ($M_{\odot}$)	T [MK]	E_G (keV)
RGB, Low mass AGB	1 - 3	20 - 60	40-85
Interm. and massive AGB	3 - 8	20 - 120	40-135
Novae		200 - 500	190-350

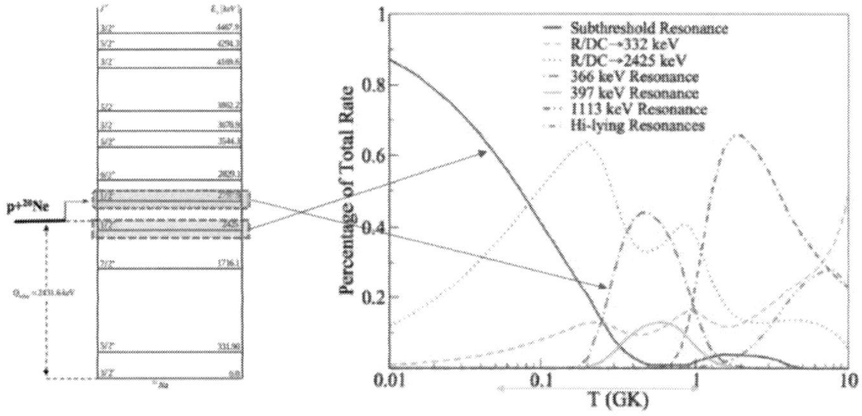

Fig. 1. Dominant contribution to the stellar rates as a function of the temperature [9].

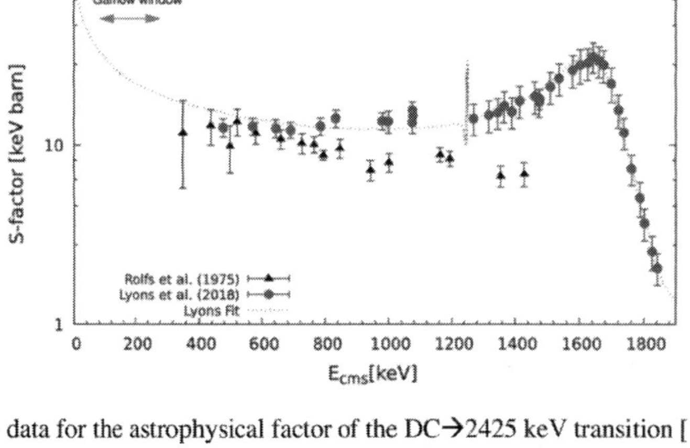

Fig. 2. Literature data for the astrophysical factor of the DC→2425 keV transition [10-11].

2 Experimental setup

The low background environment of the Gran Sasso National Laboratory, about one million times lower than that on Earth's surface, together with the high intensity beams of protons and alpha particles of the LUNA-400 accelerator, guarantees enhanced sensitivity to the cross section studies of stellar fusion processes. Details about the LUNA-400kV accelerator facility can be found in [12].
In the case of the reaction under study, because of the relatively low Q-value (2431.68 keV), most of the expected gamma-ray lines are below 3 MeV, i.e. in a region of the spectrum dominated by ambient radioactivity [13, 14]. At the beam energies achievable at LUNA-400 (E_p = 50 – 400 keV) we expect to observe mainly the transitions to the 2424.9 keV and the 331.9 keV excited states in ^{21}Na, while transitions to the ground state should be much weaker. In order to better reject environmental background, HPGe detectors are the best solution thanks to their very good resolution. In any case, a fully shielded setup should be used to reduce the ambient background. Thus, the same setup already used to successfully study the ^{22}Ne(p,γ)^{23}Na reaction has been adopted (Fig.3) [15, 16]. It consists of two HPGe: an Ortec low background detector with 90% relative efficiency (named GeDD) and a Canberra low background detector with 130% relative efficiency (named GePD).
Both detectors are shielded by a few centimetres of high-conductivity oxygen-free copper and by about 25 cm of low radioactivity lead. An anti-radon box, made of Plexiglas and flooded with nitrogen gas, was added to remove any residual radon contamination. Thanks to the overall shielding, the environmental background is reduced by about three orders of magnitude [16, 17].

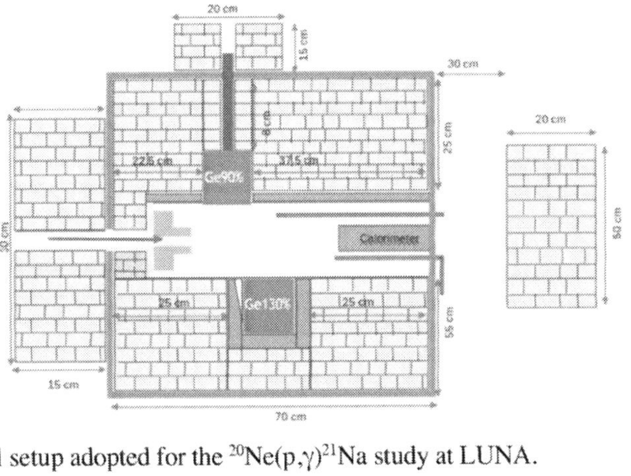

Fig. 3. Experimental setup adopted for the ^{20}Ne(p,γ)^{21}Na study at LUNA.

Natural composition neon gas (90.48% ^{20}Ne) is used inside a windowless gas target system [18], with a typical working pressure of 2 mbar. The beam intensity (I_{beam} ~ 300μA) is measured following a calorimetric approach in order to avoid possible uncertainties due to charge state fluctuations when the beam crosses the gas target [18, 19].
The adopted setup has allowed not only to study the 366 keV resonance but also to cover the proton energy range down to 250 keV, where the contribution due to the DC process and sub-threshold resonance become relevant.

3 First results

For the study of the 366 keV resonance a complete scan has been performed by changing the beam energy and therefore populating the resonance in different positions along the target [17]. The yield measured by the two detectors is shown in Fig. 4. The resonance energy and the branching ratios for the transitions to the 2424.9 keV and the 331.9 keV excited states in ^{21}Na, as well to the ground state have been determined. Once corrected for the detection efficiency, each position of the resonance scan returns a very well consistent value for the strength for both detectors (Fig. 5). The average value confirms the result obtained by Rolfs and co-workers: after a final refinement of the analysis, a 5% precision will be achieved.

In March 2022 we have started a first campaign to measure the direct capture process for energies $E_p = 260$ - 380 keV taking long runs at p = 2 mbar. At $E_p = 400$ keV the pressure was decreased to 0.5 mbar in order to not excite the 366 keV resonance. We observed both the primary and secondary gamma line for the strongest transition, the DC→ 2425 keV, getting consistent result for the astrophysical factor. In July 2022 we concluded the measurements by increasing the statistics at two energies ($E_p = 260$ and 310 keV) where the peaks from weaker transition were visible, with the aim to derive the branching ratios for all the direct capture components and find the total S-factor. New results will come soon.

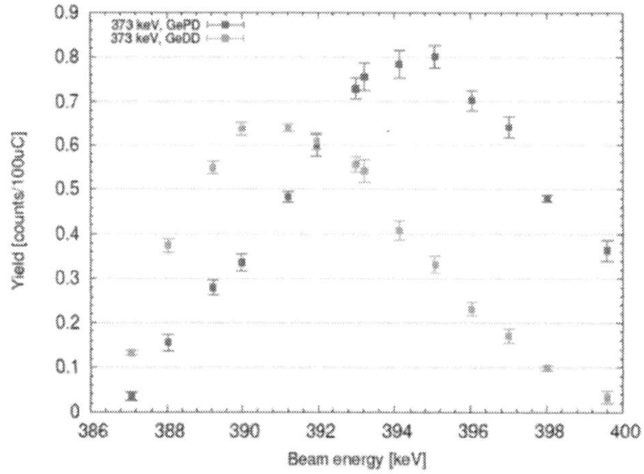

Fig. 4 Gamma yield measured by the two detectors for the DC-> 2425 keV transition (E_γ=373 keV).

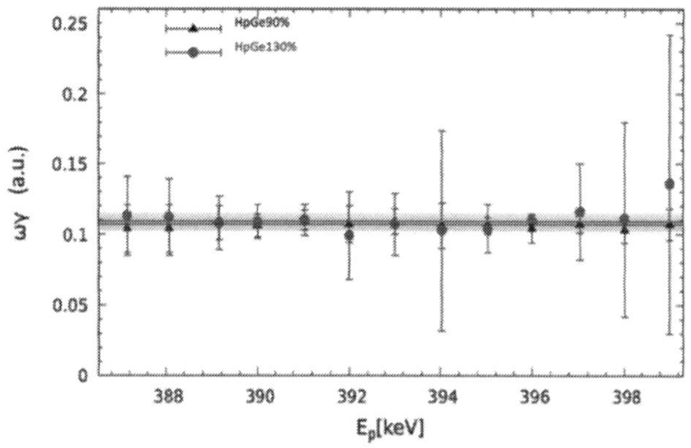

Fig. 5. The 366 keV resonance strength evaluate for each position of the scan and for the two detectors.

References

1. J. B. Marion and W. A. Fowler, ApJ **125**, 221 (1957)
2. F. Ferraro, M. P. Takacs, D. Piatti, F. Cavanna, R. Depalo, M. Aliotta, D. Bemmerer, A. Best, A. Boeltzig, C. Broggini, et al. (LUNA Collaboration), Phys. Rev. Lett. **121**, 172701 (2018)
3. A. Boeltzig, C. G. Bruno, F. Cavanna, S. Cristallo, T. Davinson, R. Depalo, R. J. deBoer, A. Di Leva, F. Ferraro, G. Imbriani, et al., Eur. Phys. J. A **52** (2016), ISSN 1434-6001, 1434-601X
4. F. Cavanna, R. Depalo, M. Aliotta, M. Anders, D. Bemmerer, A. Best, A. Boeltzig, C. Broggini, C. G. Bruno, A. Caciolli, et al., Phys. Rev. Lett. **115**, 252501 (2015), 1511.05329.
5. R. Depalo, F. Cavanna, M. Aliotta, M. Anders, D. Bemmerer, A. Best, A. Boeltzig, C. Broggini, C. G. Bruno, A. Caciolli, et al. (LUNA Collaboration), Phys. Rev. C **94**, 055804 (2016).
6. F. Cavanna, R. Depalo, M. Aliotta, M. Anders, D. Bemmerer, A. Best, A. Boeltzig, C. Broggini, C. G. Bruno, A. Caciolli, et al. (The LUNA Collaboration), Phys. Rev. Lett. **120**, 239901 (2018).
7. A. Slemer, P. Marigo, D. Piatti, M. Aliotta, D. Bemmerer, A. Best, A. Boeltzig, A. Bressan, C. Broggini, C. G. Bruno, et al., Monthly Notices of the Royal Astronomical Society **465**, 4817 (2017).
8. C. Iliadis, J. M. D'Auria, S. Starrfield, W. J. Thompson, and M. Wiescher, Astrophys. J. Suppl. Ser. **134**, 151 (2001).
9. S. Lyons, J. Görres, R. J. deBoer, E. Stech, Y. Chen, G. Gilardy, Q. Liu, A. M. Long, M. Moran, D. Robertson, et al., Phys. Rev. C **97**, 065802 (2018)
10. C. Rolfs, W. S. Rodney, M. H. Shapiro, and H. Winkler, Nucl. Phys. A 241, 460 (1975).
11. A. A. Cooper, A Low-Energy Study of the ^{20}Ne(p, γ)^{21}Na Reaction with High-Current Proton Beams at LENA, Ph.D. thesis, University of North Carolina, Chapel Hill, Department of Physics and Astronomy (2019).
12. A. Formicola, G. Imbriani, M. Junker, D. Bemmerer, R. Bonetti, C. Broggini, C. Casella, P. Corvisiero, H. Costantini, G. Gervino, C. Gustavino, A. Lemut, P. Prati, V. Roca, C. Rolfs, M. Romano, D. Schürmann, F. Strieder, F. Terrasi, H.P. Trautvetter and S. Zavatarelli, Nucl. Instr. Meth. Phys. Res. A **507**, 609–616 (2003)
13. Broggini C, Bemmerer D, Caciolli A and Trezzi D. Progress in Particle and Nuclear Physics **98**, 55-94 (2018)
14. F. Ferraro F, G. F. Ciani, A. Boeltzig, F. Cavanna and S. Zavatarelli, Frontiers in Astronomy and Space Sciences **7** (2021) ISSN 2296-987X
15. A. Caciolli, L. Agostino, D. Bemmerer, R. Bonetti, C. Broggini, F. Confortola, P. Corvisiero, H. Costantini, Z. Elekes, A. Formicola, et al., Eur. Phys. J. A **39**, 179 (2009)
16. F. Cavanna, R. Depalo, M.-L. Menzel, M. Aliotta, M. Anders, D. Bemmerer, C. Broggini, C. G. Bruno, A. Caciolli, P. Corvisiero, et al., Eur. Phys. J. A **50**, 179 (2014), 1411.2888.
17. E. Masha, Astrophysical nuclear reaction on Neon isotopes at LUNA, Ph.D. thesis, Università degli studi di Milano (2022)
18. F. Ferraro, M. P. Takacs, D. Piatti, V. Mossa, M. Aliotta, D. Bemmerer, A. Best, A. Boeltzig, C. Broggini, C. G. Bruno, et al., The European Physical Journal A **54**, 44 (2018), ISSN 1434-601X.

19. V. Mossa et al, Eur. Phys. J. A **56**, 144 (2020)

The evaluation of the fission mode and fragment yields of neutron-rich nuclei by the dynamical model

Shoya Tanaka[1,*], *Nobuya* Nishimura[1,2,**], *Itoshi* Nishimura[3,***], and *Yoshihiro* Aritomo[3,****]

[1]RIKEN Nishina Center for Accelerator-Based Science, RIKEN, Saitama 351-0198, Japan
[2]Astrophysical Big Bang Laboratory, CPR, RIKEN, Saitama 351-0198, Japan
[3]Graduate School of Science and Engineering Research, Kindai University, Higashiosaka, Osaka, 577-8502, Japan

Abstract. Nucleosynthesis by the rapid neutron-capture process (r-process) produces elements heavier than iron via neutron-rich nuclides, observed in the solar system and stars with various metallicities. In the r-process, fission plays a fundamental role by recycling the matter during neutron irradiation and by shaping the final r-abundance distribution. Nevertheless, due to the difficulty of experimental approaches, most of the fission data available for r-process calculations are based on theoretical predictions with phenomenological models. In this study, we focused on the transition of fission mode from asymmetric to symmetric in neutron-rich isotopes, which has been suggested in recent experiments for fermium isotopes. We investigated the fission of neutron-rich nuclei by a theoretical calculation based on the dynamical model and employed Langevin equations.

1 Introduction

Nucleosynthesis by the rapid-neutron-capture process, r-process, represents a significant contribution to the abundance beyond the iron-group peak, and the main process for the heaviest elements (thorium, uranium). Although several astrophysical scenarios have been proposed, the mechanism of the r-process is not completely understood. One of the main reasons is large uncertainty in the nuclear-physics properties of very neutron-rich nuclei, e.g., neutron capture rates and several decay half-lives. To determine nucleosynthesis flows on the r-process "path", the β-decay and the neutron capture (strongly depending on theoretical mass prediction) are significant. Nuclear fission is a key ingredient of the termination of and the final abundance distribution if r-processing is strong enough to reach actinoids. Nuclear fission, therefore, plays an essential role under certain r-process conditions , in particular, in very neutron-rich environments, e.g., neutron star mergers. The nucleosynthesis path goes into the very neutron-rich trans-uranium region. The effects of fission are significant to shape the r-process abundances due to fission recycling, of which fission products become seed nuclei ($A < 200$) for the next r-processing during a single nucleosynthesis process. Besides the abundance prediction, fission is also a key role as the heating source of kilonovae at late

*e-mail: shoya.tanaka@riken.jp
**e-mail: nobuya.nishimura@riken.jp
***e-mail: 2233340472d@kindai.ac.jp
****e-mail: aritomo@ele.kindai.ac.jp

times (~ 10 days to months) electromagnetic transients of neutron star mergers. A sign of fission heating may have been observed in the light curve of the kilonova associated with the gravitational wave, GW170817. The precise understanding of fission becomes much crucial in the era of gravitational astronomy.

In this study, we calculate the fission-fragment mass distributions (FFMDs) of very neutron-rich nuclei. Fission product distributions are important for the r-process, but experimental data are not available. We adopt the Langevin method [1], widely adopted in the study of low-energy fission.

2 Model

We use the fluctuation-dissipation model and employ Langevin equations to calculate the evolution of nuclear shape with time [2, 3]. The nuclear shape is defined by the two-center parametrization [4, 5], which has three deformation parameters, z, δ and α to serve as collective coordinates, abbreviated as $q=\{z,\delta,\alpha\}$. The symbol z is the distance between two potential centers, the symbol δ denotes the deformation of the fragments, and $\alpha = (A_1 - A_2)/(A_1 + A_2)$ is the mass asymmetry of the two fragments [2], where A_1 and A_2 denote the mass numbers of heavy and light fragments.

For a given value of the temperature of a system T, the potential energy is defined as a sum of the liquid-drop (LD) part and a microscopic (SH) part:

$$\begin{aligned}
V(q,T) &= V_{\rm LD}(q) + V_{\rm SH}(q,T), \\
V_{\rm LD}(q) &= E_{\rm S}(q) + E_{\rm C}(q), \\
V_{\rm SH}(q,T) &= [\Delta E_{\rm shell}(q) + \Delta E_{\rm pair}(q)]\Phi(T), \\
\Phi(T) &= \exp\left(-\frac{aT^2}{E_d}\right).
\end{aligned} \tag{1}$$

Here, the potential energy $V_{\rm LD}$ is calculated with the finite-range liquid drop model [6], given as a sum of the surface energy $E_{\rm S}$ and the Coulomb energy $E_{\rm C}$. The microscopic energy $V_{\rm SH}$ at $T = 0$ is calculated as the sum of the shell correction energy $\Delta E_{\rm shell}$, evaluated by the Strutinski method [7, 8], and the pairing correlation correction energy $\Delta E_{\rm pair}$ [9]. The shell correction energy has a temperature dependence expressed by a factor $\Phi(T)$ in which the shell damping energy E_d is chosen as 20 MeV [10] and a is the level density parameter [3, 11]. To define the potential of the two-center shell model a neck parameter of $\epsilon = 0.35$ $(0 \le \epsilon \le 1)$ has been routinely used. However, this value is not appropriate for heavier actinide nuclides as pointed out in We adopt the optimal ϵ values following the empirical relation

$$\epsilon(A_c) = 0.01007A_c - 1.94, \tag{2}$$

where A_c is the mass of the fissioning nucleus [12].

The multidimensional Langevin equations [2] are given as

$$\begin{aligned}
\frac{dq_i}{dt} &= (m^{-1})_{ij}p_j, \\
\frac{dp_i}{dt} &= -\frac{\partial V}{\partial q_i} - \frac{1}{2}\frac{\partial}{\partial q_i}(m^{-1})_{jk}p_jp_k \\
&\quad - \gamma_{ij}(m^{-1})_{jk}p_k + g_{ij}R_j(t),
\end{aligned} \tag{3}$$

where $q_i = \{z, \delta, \alpha\}$ and $p_i = m_{ij} dq_i/dt$ is a momentum conjugate to coordinate q_i. In the Langevin equation, m_{ij} and γ_{ij} are the shape-dependent collective inertia and the friction tensors, respectively. The wall-and-window one-body dissipation [13–15] is adopted for the friction tensor. The normalized random force $R_i(t)$ is assumed to be that of white noise, *i.e.*,

$$\langle R_i(t) \rangle = 0, \quad \langle R_i(t_i) R_j(t_2) \rangle = 2\delta_{ij}\delta(t_1 - t_2). \tag{4}$$

The strength of the random force g_{ij} is related to the friction tensor γ_{ij} by the classical Einstain relation,

$$\sum_k g_{ik} g_{jk} = \gamma_{ij} T. \tag{5}$$

We calculate the charge distribution based on the assumption of unchanged charge distribution (UCD). The charge distribution (charge density) remains unchanged during the whole fission process, i.e., the charge density of the compound nucleus is maintained.

3 Results and Discussion

3.1 A dramatic change of the fission mode at Fermium isotopes

The fission of fermium nuclides $^{254-257}$Fm at low excitation energy was studied using the dynamical model as shown in Fig. 1. We adopt lower excitation energy, E* = 7 MeV, corresponding to low energy environments where the r-process occurs. The mass distributions of fission fragments show a dramatic change from an asymmetric shape for the lighter fermium isotopes to sharp symmetric fission for the heavier isotopes. This trend has been already observed experimentally [16]. The sudden change of the FFMD is strongly regulated by the structure of the second fission barrier and the dynamical motion of the nucleus in the second minimum [12].

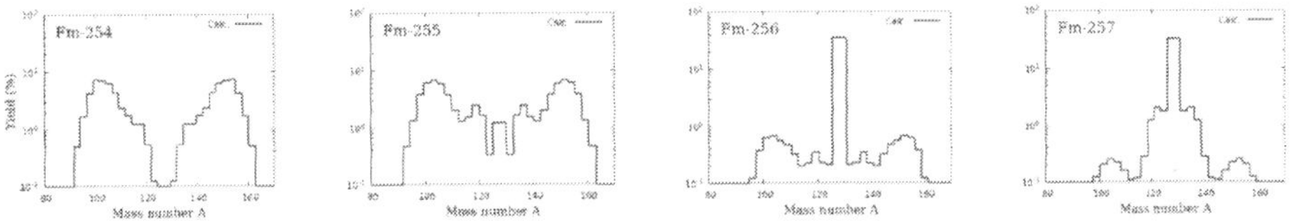

Figure 1. Calculation results of a fission fragment mass distribution for $^{254-257}$Fm (E*=7 MeV).

3.2 Fission of neutron-rich nuclei at Uranium isotopes

We perform a series of fission calculations for U (Z = 92) isotopes with 10–20 more neutron-rich from the β-stability line. The calculated FFMDs for ^{250}U to ^{256}U are shown in Fig. 2. A similar trend which is the dramatic change of FFMDs for fermium isotopes was also observed in uranium isotopes in this calculation. This trend did not appear in the results of the GEF model code [17].

3.3 Fission fragment charge distribution by the UCD assumption

For the application to the r-process simulations, we need fission properties for very neutron-rich nuclei far from the β-stability line. However, experimental data in the such region are

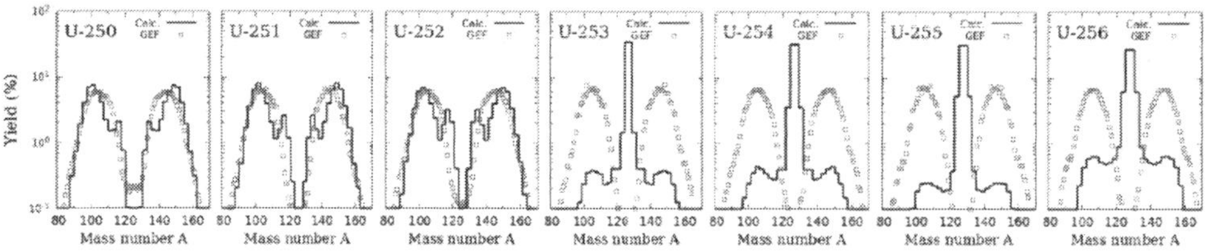

Figure 2. The calculation results of fission fragment mass distributions (FFMDs) for ^{250}U to ^{256}U with the excitation energy of E*=7 MeV compared with the data from GEF code [17].

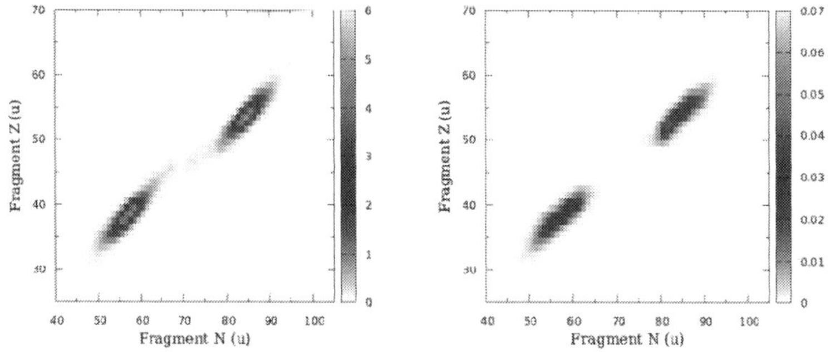

(a) Calculation result   (b) Experimental data

Figure 3. The calculation result of fission fragment distribution on the N-Z plane for ^{236}U ($E^* = 10$ MeV) is plotted. It is compared with the experimental data of ^{235}U neutron-induced fission ($E_k = 500$ KeV) from JENDL-4.0 [18].

not available yet. Therefore, we first carry out fission calculations for U isotopes near the β-stability line. Fig. 3 shows the results of the mass and charge distribution of fission fragments for ^{236}U. The fission calculation is performed with the excitation energy, $E^* = 10$ MeV. The distributions are compared with experimental data in JENDL-4.0 [18]. The calculated charge distribution with the current simple treatment, based on the UCD, shows a good agreement with experimental data. The charge distribution of fission products is a fundamental quantity for the determination of the production rate of individual isotopes. Thus, it is essential nuclear data to compare experiments with r-process calculations.

4 Summary

In the presented work, we have performed fission calculations of neutron-rich nuclei using a dynamical model for application to r-process simulations. We found that the fission fragment distribution changes from the two peak feature (asymmetric fission) to the one-peak (symmetric fission) with increasing the neutron number. We also found that the calculated fission mass distributions for uranium, of which the charge (Z) distribution is obtained by the UCD (unchanged charge density assumption), can well reproduce experimental data in JENDL (thermal neutron induced fission of ^{235}U) [18]. Our results, including further improvements, are expected to contribute to the understanding of the r-process.

This study is financially supported by JSPS KAKENHI (19H00693, 20K04003, 20H05648, 21H01087, 22K20373). Parts of computations are performed on computer facilities at CfCA in NAOJ and at YITP in Kyoto University.

References

[1] S. Tanaka, and Y. Aritomo, Y. Miyamoto, K. Hirose, K. Nishio, Phys. Rev. C **100**, 064605 (2019).

[2] Y. Aritomo and M. Ohta, Nucl. Phys. A **744**, 3 (2004).

[3] Y. Aritomo, S. Chiba, and F. Ivanyuk, Phys. Rev. C **90**, 054609 (2014).

[4] J. Maruhn and W. Greiner, Z. Phys. **251**, 431 (1972).

[5] K. Sato, A. Iwamoto, K. Harada, S. Yamaji, and S. Yoshida, Z. Phys. A **288**, 383 (1978).

[6] H. J. Krappe, J. R. Nix, and A. J. Sierk, Phys. Rev. C **20**, 992 (1979).

[7] V. M. Strutinsky, Nucl. Phys. A **95**, 420 (1967).

[8] V. M. Strutinsky, Nucl. Phys. A **122**, 1 (1968).

[9] S. G. Nilsson, C. F. Tsang, A. Sobiczewski et al., Nucl. Phys. A **131**, 1 (1969).

[10] A. V. Ignatyuk, G. N. Smirenkin, and A. S. Tishin, Sov. J. Nucl. Phys. **21**, 255 (1975).

[11] J. Toke and W. J. Swiatecki, Nucl. Phys. A **372**, 141 (1981).

[12] Y. Miyamoto, Y. Aritomo, S. Tanaka, K. Hirose, and K. Nishio, Phys. Rev. C **99**, 051601(R) (2019).

[13] J. Blocki, Y. Boneh, J. R. Nix, J. Randrup, M. Robel, A. J. Sierk, and W. J. Swiatecki, Ann. Phys. **113**, 330 (1978).

[14] J. R. Nix and A. J. Sierk, Nucl. Phys. A **428**, 161c (1984).

[15] H. Feldmeier, Rep. Prog. Phys. **50**, 915 (1987).

[16] D. C. Hoffman, J. B. Wilhelmy, J. Weber, W. R. Daniels, E. K. Hulet, R. W. Lougheed et al., Phys. Rev. C **21**, 972 (1980).

[17] K.-H. Schmidt, B. Jurado, C. Amouroux, and C. Schmitt, Nucl. Data Sheets **131**, 107 (2016).

[18] K. Shibata, O. Iwamoto, T. Nakagawa, N. Iwamoto, A. Ichihara, S. Kunieda, S. Chiba, J. Katakura, and N. Otuka: Journal of Korean Physical Society **59** (2011) 1046.

EPJ Web of Conferences **279**, 12001 (2023) https://doi.org/10.1051/epjconf/202327912001
NPA-X 2022

Nuclear level densities and γ-ray strength functions of ^{120,124}Sn for astrophysical purposes

Maria Markova[1,*], *Ann-Cecilie* Larsen[1], and *Frank Leonel* Bello Garrote[1]

[1]Department of Physics, University of Oslo, N-0316 Oslo, Norway

Abstract. The nuclear level densities (NLDs) and γ-ray strength functions (GSFs) of ^{120,124}Sn were extracted with the Oslo method, compared and combined with other experimental data to be used as inputs in the TALYS reaction code to constrain the Maxwellian-averaged cross sections (MACS) for the ^{119,123}Sn$(n,\gamma)^{120,124}$Sn reactions. For ^{120}Sn, the MACS are in relatively good agreement with the JINA REACLIB, BRUSLIB libraries, while no clear agreement was found for ^{124}Sn. In addition, the pygmy dipole resonance (PDR) was found to contribute with up to $\approx 3-8\%$ of the total MACS in these nuclei.

1 Introduction

Information on the neutron-capture cross sections and rates for nuclei in the vicinity of neutron rich stable and unstable Sn isotopes provides a deeper insight into the astrophysical s- and r-processes and abundances of elements produced in these processes in this mass region. In cases of unstable nuclei, any direct measurements of neutron-capture cross sections are greatly hindered and one has to exploit the potential of indirect techniques. The nuclear reaction codes, *e.g.* TALYS [1, 2], incorporate the Hauser-Feshbach framework [3], combining the statistical properties of nuclei, such as the nuclear level density (NLD) and γ-ray strength function (GSF), with other nuclear inputs to calculate the astrophysical values of interest. The NLD is given by the number of nuclear levels per excitation energy unit, while the GSF characterizes the reduced average γ-decay probability.

In cases of unstable neutron rich nuclei no experimental data on NLDs and GSFs are available, and constraining the neutron-capture cross sections and rates is limited by the set theoretical models, including microscopic calculations, available in TALYS as predefined calculation options. While the spread of theoretical NLDs can be considered somewhat small and the region of the giant dipole resonance is well-described by the available GSFs, the disagreement in low-lying dipole strengths below the neutron threshold is quite critical. Experimental data on the low-lying part of the $E1$ response, preferably obtained together with the NLDs, are highly desired in order to specify the constants on the neutron-capture cross sections. Gaining more knowledge on systematics of NLDs and GSFs in different isotopic chains will further facilitate estimating these cross sections indirectly for unstable, very neutron rich isotopes.

The Oslo method [5] is an experimental technique providing a simultaneous measurement of the GSF and NLDs in light charged particle induced reactions. In this work, we present

*e-mail: maria.markova@fys.uio.no

the NLDs (Sec. 2.1) and GSFs (Sec. 2.2) of ^{120,124}Sn used to experimentally constrain the Maxwellian-averaged cross sections (MACS) for the ^{119,123}Sn$(n,\gamma)^{120,124}$Sn reactions, presented in Section 3. Combining the Oslo $E1$ strength with the $E1$ and $M1$ GSFs deduced from spectra of relativistic Coulomb excitation in forward-angle inelastic proton scattering [4] provides an opportunity to study the contribution of the low-lying $E1$ strength, often referred to as the pygmy dipole resonance (PDR), to the total MACS. This study was performed for ^{120,124}Sn and will be discussed in Section 4.

2 The Oslo method

Both ^{120}Sn and ^{124}Sn nuclei were studied at the Oslo Cyclotron Laboratory (OCL) in the $(p,p'\gamma)$ reaction with 16 MeV protons. The setup at the OCL is presented by the γ-detector array OSCAR and the particle telescope SiRi (for more details see. [6]). OSCAR comprises of 30 large-volume LaBr$_3$(Ce) scintillator detectors fixed in a truncated icosahedron frame, surrounding the target chamber. The full-energy peak efficiency and energy resolution of OSCAR have been measured to be $\approx$ 20% and $\approx$ 2.7%, respectively, at $E_\gamma \approx$ 662 keV. The particle stripped telescope exploiting the $E-\Delta E$ technique is mounted in the backward position, cowering from 126° to 140°. Such form of the setup allows to collect p-γ coincidence events and, thus, the γ-spectra at each excitation energy of the nucleus below the neutron separation energy. These spectra are unfolded to correct for the detector response [6]. At the final stage of the coincidence event analysis, the first generation photons for each cascade at each excitation energy E_x are singled out to form a so-called primary matrix $P(E_x,E_\gamma)$. This matrix is the main input of the Oslo method providing a decomposition of the primary matrix into the transmission coefficient $\mathcal{T}(E_\gamma)$ and the NLD $\rho(E_x-E_\gamma)$ (for more details see [5, 6]):

$$P(E_x,E_\gamma) \propto \mathcal{T}\cdot\rho(E_x-E_\gamma), \tag{1}$$

where $\mathcal{T}(E_\gamma)$ is independent of the initial and final excitation energies according to the Brink-Axel hypothesis [5] and proportional to the GSF $f(E_\gamma)$ as $\mathcal{T}(E_\gamma)=2\pi E_\gamma^{2L+1}f(E_\gamma)$. In the Oslo analysis it is assumed that the extracted GSF is predominantly of the dipole $L=1$ nature.

This decomposition provides only the functional forms of the NLD and GSF and some additional experimental data are required to constrain the absolute values of these functions. The slope and the absolute value of the NLD are fixed with the discrete low-lying states and the value of the NLD at the neutron separation energy, obtained from the average s-wave neutron resonance spacing available from neutron resonance experiments. The GSF shares the same slope with the NLD and its absolute value is fixed with the total average radiative width obtained in the same neutron resonance experiments. All the details of the normalization procedure and parameters used are provided in [6] and the corresponding Supplementary Material.

2.1 Nuclear level densities of ^{120,124}Sn

The NLD of ^{120}Sn extracted with the Oslo method is shown in Fig. 1(a) together with the theoretical NLD provided as available options in TALYS. The total NLD of ^{124}Sn is very similar in slope and absolute value within the experimental error bands to ^{120}Sn and the previously published NLDs for other even-even ^{116,118,122}Sn isotopes (see [7] and references therein for further details). Indeed, no significant structural changes from ^{120}Sn to ^{124}Sn are expected to cause a noticeable difference in NLDs. Some minor differences in the slopes of NLDs for ^{120,124}Sn and the older data ^{116,118,122}Sn stem primarily from differences in the normalization procedure.

The NLDs for ^{120,124}Sn agree well with the low-lying discrete states [8] up to ≈ 3 MeV excitation energy, where the level scheme can be considered complete. Up to the neutron separation energy the NLDs demonstrate a rather smooth constant temperature behavior.

In order to crosscheck the normalization of the NLD in ^{124}Sn we compare a reduced Oslo NLD for $J = 1^-$ states with the corresponding NLD extracted with the fluctuation analysis of high resolution inelastic proton scattering spectra above 6 MeV [4]. The experimental data agree quite well within the error bars, however, demonstrate some disagreement with the TALYS input NLDs. As shown in Fig. 1(a) this difference might reach up to approximately a factor of 3. However, in general, the TALYS input NLDs follow the same trend as the experimental NLD and they are not expected to be the main source of TALYS uncertainty span for the MACS.

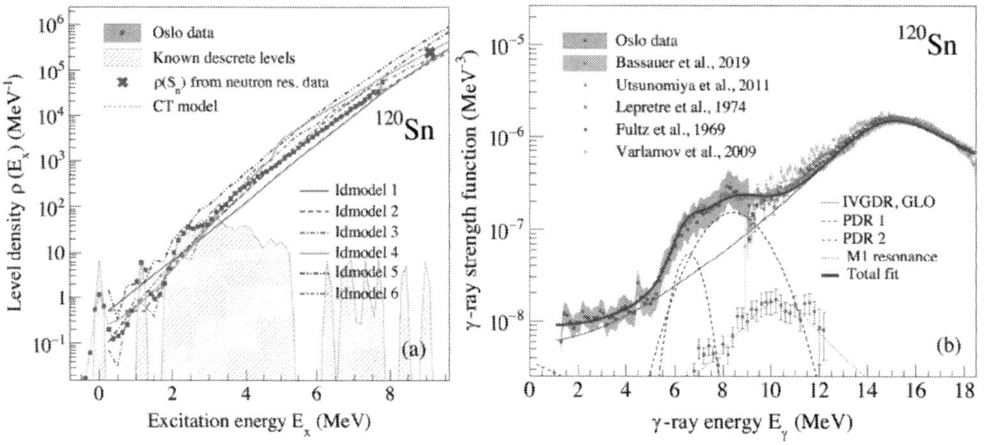

Figure 1. (a) The experimental NLD of ^{120}Sn shown with the discrete states from [8] and the theoretical NLD models available in TALYS. (b) The experimental GSF of ^{120}Sn shown together with the $E1$ and $M1$ strengths from Ref. [4] and (γ, n) experiments from Refs. [9–12]

2.2 γ-ray strength functions of ^{120,124}Sn

The GSFs of ^{120,124}Sn extracted with the Oslo method reach up to the neutron separation energy only and, therefore, need to be combined with the data above the neutron separation energy to present the total $E1 + M1$ response of the nucleus used for astrophysical calculations. The GSFs deduced from forward-angle inelastic proton scattering (p, p') at relativistic beam energies [4] have an overlap with the Oslo data between 6 MeV and the neutron separation energy. Combined with the Oslo data, it covers the whole γ-energy region of interest, including the low-lying part and the giant dipole resonance (GDR). Both data sets agree well within the error bands and reveal a double-peak structure below the neutron separation energy, often referred to as the PDR. The inelastic relativistic scattering experiment also provides the data on the $M1$ part of the response in both nuclei, which allows to perform the decomposition of the total $E1 + M1$ part of the response into the $M1$ part and the $E1$ part, comprising of the GDR and the PDR as a two-peak structure. The single and the generalized Lorentzian functions were used to parametrize the $M1$ and the GDR parts, respectively, while the PDR was best reproduced by two Gaussian peaks. Such a decomposition of different parts of the nuclear response for ^{120}Sn is shown in Fig. 1(b).

3 The MACS of ^{119,123}Sn(n, γ)^{120,124}Sn reactions

Both the experimental NLD and the GSF for each isotope were futher used as the inputs to calculate the MACS of ^{119,123}Sn$(n,\gamma)^{120,124}$Sn reactions with the TALYS code. The experimental MACS for both cases are shown in Fig. 2 together with the span of MACS obtained with all available calculation options in TALYS and the data from several commonly used libraries (see Fig. 2 caption for Refs.).

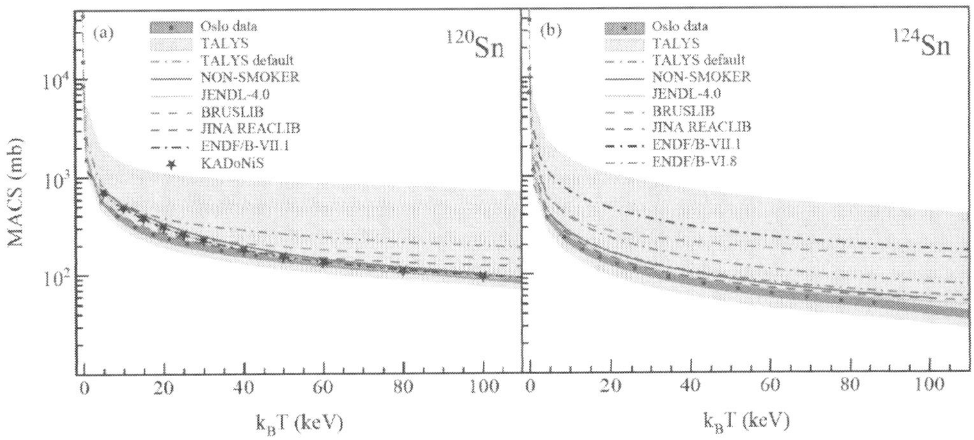

Figure 2. MACS of ^{119,123}Sn$(n,\gamma)^{120,124}$Sn reactions obtained with the Oslo method input NLDs and GSFs compared to the span of TALYS predictions (light blue band), NON-SMOKER [13], JENDL-4.0 [14], BRUSLIB [15], JINA REACLIB [16], ENDF/B-VII.1 [17], ENDF/B-VI.8 [18], KADoNiS [19] libraries.

For both ^{120}Sn and ^{124}Sn the experimental cross sections lie closer to the bottom part of the TALYS predictions. As the spread of the theoretical NLDs available in TALYS can be somewhat considered small, the main contribution to the TALYS uncertainty band is due to the spread of avalable TALYS GSFs, which poorly reproduce the experimental result. For the case of ^{120}Sn, the experimental MACS are in a fair agreement with the KADoNiS data point at 30 keV energy. For the energies below 30 keV, the cross section agrees quite well with the JINA-REACLIB and BRUSLIB data, while for higher energies it becomes closer to the ENDF/B-VII.1 and JENDL-4.0 cross sections. For ^{124}Sn the data from different libraries disagree with each other and the experimental cross section. JINA-REACLIB and JENDL-4.0 data agree the best with the experimental error band along the whole shown energy range. The spread of predictions from libraries is significantly larger than for ^{120}Sn, which might be expected for the radiative neutron capture reaction on an unstable ^{123}Sn target.

4 The role of the pygmy dipole resonance

As the combination of the Oslo and the (p,p') data allows for the decomposition of the total nuclear response into the $E1$ and $M1$ components, it becomes possible to study the astrophysical role of the low-lying dipole strength, namely the PDR. The MACS were calculated with the total $E1$ response, comprising of the GDR and two peaks of the PDR, and with the $E1$ response presented by GDR only. These results are shown together with the TALYS uncertainty span in Fig. 3. The fraction of the energy weighted sum rule for dipole transitions

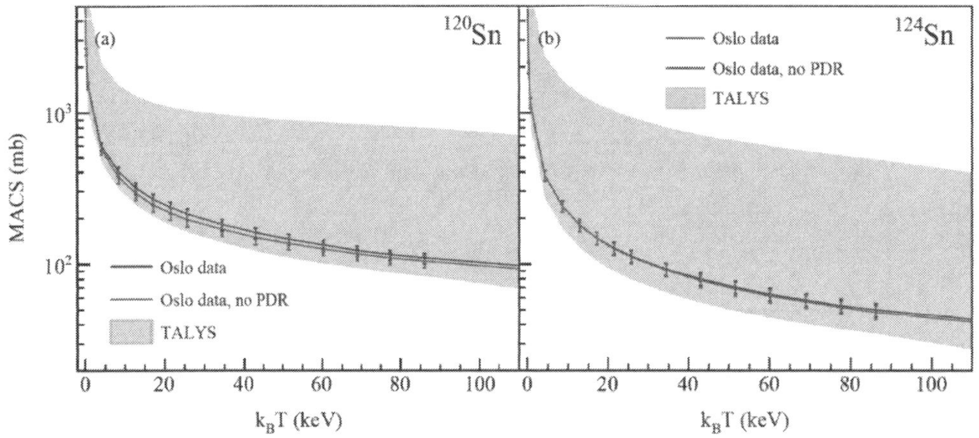

Figure 3. The MACS of ^{119,123}Sn$(n, \gamma)^{120,124}$Sn reactions obtained with and without the PDR part of the total $E1$ response compared with the TALYS span of cross sections.

exhausted by the PDR is ≈3% in ^{120}Sn and ≈2% in ^{124}Sn. For the largest PDR in ^{120}Sn, its contribution to the total MACS reaches up to ≈8%, while for the smallest PDR in ^{124}Sn it reaches up to ≈3%. Even though these contributions for stable ^{120,124}Sn are quite moderate, for the neutron rich experimentally studied ^{130}Sn [20] with the PDR of ≈ 7% we might expect this contribution to be larger that ≈ 10% of the total MACS.

5 Conclusions

The experimental NLDs and GSFs of ^{120,124}Sn were extracted with the Oslo method and combined with the relativistic inelastic proton scattering data to experimentally constrain the radisative neutron capture ^{119,123}Sn$(n, \gamma)^{120,124}$Sn MACS with the reaction code TALYS. The spread of available theoretical $E1$ GSFs in TALYS contributes to a large span of TALYS calculations. The experimental constrains narrow down the spread of MACS significantly as compared to the theoretical models. The MACS calculated for ^{120}Sn were found to be in good agreement with JINA REACLIB, BRUSLIB libraries, while no clear agreement is seen in ^{124}Sn. The pygmy dipole resonances in ^{120,124}Sn were found to contribute with up to ≈ 3 – 8% of the total MACS. For the heavier neutron rich Sn isotopes this contribution might be expected to be quite significant, and, therefore, to be of importance for reproducing solar system abundances of elements produced in the r-process.

References

[1] A. Koning, S. Goriely, and S. Hilaire, TALYS-1.9, A nuclear reaction program, user manual, Tech. Rep. (2017).
[2] A. Koning, *et al.*, Nuclear Data Sheets 155, 1 (2019).
[3] W. Hauser and H. Feshbach, Phys. Rev. 87, 366 (1952).
[4] S. Bassauer, *et al.*, Phys. Rev. C 102, 034327 (2020).
[5] A. C. Larsen, *et al.*, Phys. Rev. C 83, 034315 (2011).
[6] M. Markova, *et al.*, Phys. Rev. Lett. 127, 182501 (2021).
[7] M. Markova, *et al.*, to be published (2022).

[8] Data taken from the ENSDF database of the NNDC online data service, https://www.nndc.bnl.gov/ensdf/. Last accessed: January 19, 2022.
[9] H. Utsunomiya, S. Goriely, M. Kamata, *et. al.*, Phys. Rev. C 84, 055805 (2011).
[10] V. V. Varlamov, *et al.*, Bull. Rus. Acad. Sci. 74, 833 (2010).
[11] A.Leprêtre, *et.al.*, Nucl. Phys. A 219, 39 (1974).
[12] S. C. Fultz *et.al.*, Phys. Rev. 186, 1255 (1969).
[13] T. Rauscher and F. K. Thielemann, At. Data Nucl. Data Tables **75**, 1 (2000).
[14] K. Shibata, *et al.*, J. Nucl. Sci. Technol. **48(1)**, 1-30 (2011)
[15] Y. Xu, *et al.* , A&A **549**, A106 (2013).
[16] R. H. Cyburt, *et al.*, Astrophys. J., Suppl. Ser. **189**, 240 (2010).
[17] M. B. Chadwick, *et al.*, Nucl. Data Sheets **112**, 2887-2996 (2011).
[18] H. D. Lemmel, *et al.*, IAEA-NDS-100, Rev. **11** 1-32 (2001).
[19] Data extracted using the KADoNiS On-Line Data Service, http://www.kadonis.org, accessed 2022-08-05.
[20] P. Adrich, *et. al*, Phys. Rev. Lett. 95, 132501 (2005).

 https://doi.org/10.1051/epjconf/202327912002

β-decay spectroscopy of neutron-deficient nuclei

Sonja E. A. Orrigo[1,*], *Berta* Rubio[1], and *William* Gelletly[2]

[1]Instituto de Física Corpuscular, CSIC-Universidad de Valencia, E-46071 Valencia, Spain
[2]Department of Physics, University of Surrey, Guildford GU2 7XH, Surrey, UK

Abstract. A systematic study of the β-decay of neutron-deficient nuclei has been carried out and has provided spectroscopic information of importance for both nuclear structure and nuclear astrophysics. Following an overview of the most relevant achievements, we focus on the latest results on the β decay of ^{60}Ge and ^{62}Ge. We also summarise our results on the mass excesses in comparison with systematics and a recent measurement. Finally, we present updated half-life trends for T_z = -1/2, -1 and -2 neutron-deficient nuclides.

1 Introduction

Decay spectroscopy experiments with implanted radioactive ion beams (RIBs) are a powerful tool for the study of the structure of exotic nuclei. They provide rich spectroscopic information [1–6]: firstly the half-life of the β-decaying nucleus, secondly the branching ratios for β-delayed γ or particle emissions (where the particle can be a proton, an α particle, one or two neutrons, etc., depending on whether the unstable nucleus is neutron-deficient or neutron rich), thirdly information on the energy levels populated in the daughter nucleus together with their β-feedings and, last but not least, the absolute values of the Fermi B(F) and Gamow-Teller B(GT) transition strengths. In some cases one is also able to determine the mass of the daughter nucleus and estimate the mass of the parent. Furthermore, the structural information on exotic nuclei is relevant for nuclear astrophysics since many of these nuclei are involved in one of the processes of nucleosynthesis. Focusing on the neutron-deficient side of the nuclear chart around the $N = Z$ line, many heavy proton-rich elements are produced in the rapid proton-capture process (*rp-process*) happening in explosive stellar environments, passing through neutron-deficient nuclei in the *fp*-shell and above.

We have carried out a systematic study of neutron-deficient nuclei in β-decay spectroscopy experiments with implanted RIBs at GANIL and RIKEN obtaining remarkable results [1–6]. Among the results from the GANIL experiments we emphasise the discovery of a new exotic decay mode in the *fp*-shell in ^{56}Zn (β-delayed γ-proton decay [1]) and the first observation of the 2^+ isomer in ^{52}Co [3]. For some of the nuclides under study [1, 2, 5] it was possible to enrich the β-decay data by comparison with complementary (^{3}He,t) charge exchange (CE) reactions performed on the mirror stable target at RCNP Osaka [7–9]. CE reactions are the mirror strong interaction process of β decay and provide information on relative B(GT) values without energy restrictions [7, 10], allowing us to investigate fundamental questions such as isospin symmetry in mirror nuclei.

*e-mail: Sonja.Orrigo@ific.uv.es

The unprecedented statistics available at the Radioactive Isotope Beam Factory (RIBF) of the RIKEN Nishina Center allowed us to extend these studies to higher masses and more extreme nuclear conditions. In Sect. 2 we focus on the most recent results on the β decay of ^{60}Ge and ^{62}Ge obtained from the RIKEN experiment [6]. Sect. 3 summarises our results on the mass excesses and compares them with the mass evaluation systematics and a recent measurement [11]. In Sect. 4 we present half-life trends for T_z = -1/2, -1 and -2 nuclei, where we include all the newly measured values.

2 β decay of ^{60}Ge and ^{62}Ge

^{60}Ge and ^{62}Ge were produced at RIBF by fragmentation of a ^{78}Kr primary beam, accelerated to 345 MeV/nucleon with an intensity up to 250 pnA, on a Be target. The BigRIPS separator was used to select and identify the fragments with the $B\rho - \Delta E - ToF$ method [12]. The isotopes were implanted in the WAS3ABi setup, using three double-sided Si strip detectors 60 mm $\times$ 40 mm $\times$ 1 mm. WAS3ABi was surrounded by the EURICA array (12 clusters containing 7 high-purity Ge crystals each) employed for γ detection. The identification matrix is shown in figure 1, where the positions of ^{60}Ge and ^{62}Ge are indicated. 1.5$\times$10^4 ^{60}Ge and 2.1$\times$10^6 ^{62}Ge ions were recorded [6].

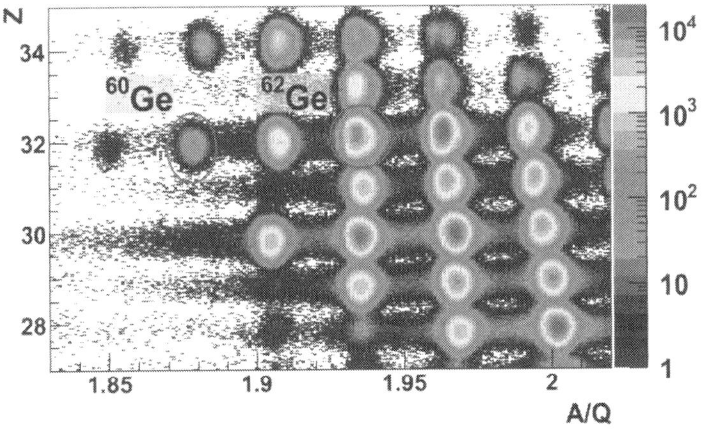

Figure 1. Identification plot for the ^{60}Ge and ^{62}Ge implants.

^{60}Ge is a semi-magic, N = 28 isotone with T_z = -2 whose decay was almost unknown before the present experiment, as were the energy levels of its daughter nucleus ^{60}Ga. The latter lies right at the proton drip-line, thus its structural properties are of relevance for the rp-process [11]. Here a half-life value of $T_{1/2}$ = 25.0(3) ms has been measured for ^{60}Ge as well as the first experimental information on both the β-delayed proton and γ emissions [6]. 7.6$\times$10^5 implants of the ^{60}Ga ions were also observed and we obtained a half-life of 69.4(2) ms for ^{60}Ga. The precision of both the ^{60}Ge and ^{60}Ga half-lives has been improved in comparison with values in the literature. In the β decay of ^{60}Ge we have also observed competition between the γ de-excitation and the (isospin-forbidden) proton emission from the T = 2 isobaric analogue state (IAS) populated in the daughter nucleus. This exotic feature was already seen in all the lighter T_z = -2 systems previously studied [1, 2, 5].

Little was known about the β decay of the T_z = -1, ^{62}Ge nucleus. Furthermore, ^{62}Ge is of particular interest because in previous studies of T_z = -1 nuclei a suppression of isoscalar γ transitions between $J^\pi = 1^+$, T = 0 states (Warburton and Weneser *quasi-rule* [13, 14]) has been observed [15]. A half-life value of $T_{1/2}$ = 73.5(1) ms has been extracted for ^{62}Ge

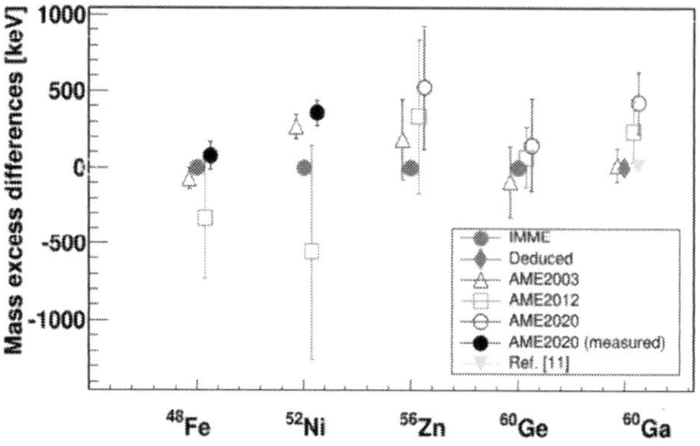

Figure 2. Mass excesses of ^{48}Fe, ^{52}Ni, ^{56}Zn, ^{60}Ge and ^{60}Ga. The difference is shown between the values we obtain (red filled circles and diamond) [2, 6] and values from the 2003, 2012 and 2020 AMEs [20–22]. Open symbols represent values from systematics, while filled symbols are experimental values. The cyan filled triangle is the measurement from Ref. [11]. The data points belonging to each nucleus are slightly displaced to show the error bars better.

[6]. This new value agrees with and improves the precision of our previous measurement ($T_{1/2}$ = 76(6) ms, obtained with 6.1×10^3 implants of ^{62}Ge [4]). New information on the β-delayed γ emission has been obtained, indicating the persistence of the *quasi-rule* [13, 14] in ^{62}Ge. Finally, we do not find evidence of enhanced low-lying Gamow-Teller strength in ^{62}Ga due to isoscalar proton-neutron pairing, confirming the findings of a previous measurement [16].

3 Mass excess of neutron-deficient nuclei

Knowing the mass excesses of the neutron-deficient nucleus undergoing β decay and its daughter nucleus is key to determining important quantities such as the Q_β value of the decay which, in turn, enters in the determination of B(F) and B(GT). Moreover, the information on the mass excesses of the β- and β-proton daughter nuclei allows the calculation of the proton separation energy S_p. Since there is a lack of mass measurements in this mass region, one has to rely often on the atomic mass evaluation (AME) systematics. However, if the mass excesses of at least three members of an isospin multiplet are known, another option is to determine the mass excess of the remaining member of the multiplet from the Isobaric Multiplet Mass Equation (IMME) [17–19]:

$$\Delta m(\alpha, T, T_z) = a + b\, T_z + c\, T_z^2 \,. \tag{1}$$

where T_z is the third component of the isospin T and α stands for all the other quantum numbers. We used our β-decay data to determine the ground-state (g.s.) mass excesses of the T_z = -2 nuclei ^{48}Fe, ^{52}Ni and ^{56}Zn in Ref. [2] and ^{60}Ge in Ref. [6] from the IMME, knowing four members of each quintuplet. We have also deduced the g.s. mass excess of ^{60}Ga and its S_p [6]. In figure 2 our deduced mass excesses (red filled circles and diamond) are compared to the values obtained from the 2003 [20], 2012 [21] and 2020 [22] AME systematics. In the figure open (filled) symbols represent the AME systematics (measured or deduced) values.

Since our first study of ^{56}Zn [1] we have observed that for proton-rich nuclei in this region of the mass chart the AMEs subsequent to that in 2003 give a worse agreement for the mass excess in comparison with our IMME values. In figure 2 one can see that the values from the 2003 AME lie closer to our estimates than the values from the 2012 AME and a very similar behaviour happens for the 2016 AME [23] (not shown in the figure). Other authors then reported similar issues [11, 24]. The 2020 AME also has a similar behaviour for the unmeasured nuclei, but includes new measured values for ^{48}Fe and ^{52}Ni (black filled circles). The AME systematic evaluation does not include isobaric multiplets. The reasoning is that the IAS might be mixed and consequently its energy can deviate from the isobaric multiplet formula. The ^{56}Zn IAS is indeed mixed [1]. We think that, with caution, this knowledge can help to calculate extrapolated values. The future AME would benefit from new mass measurements in this region.

Figure 2 also shows a recent measurement of the mass excess in ^{60}Ga (cyan filled triangle) [11], in excellent agreement with our indirect determination based on the β-decay data [6]. Paul *et al.* have also determined S_p= 78(30) keV in ^{60}Ga, in agreement with our value of 90(15) keV [6] (the 2020 AME systematic value is -340(200) keV [22]) and combined the two experimental values to give $S_p(^{60}\text{Ga})$ = 88(18) keV, establishing the proton-bound nature of ^{60}Ga. Then, since ^{59}Ga was not observed in fragmentation reactions at NSCL, there is strong evidence that ^{60}Ga is the last proton-bound gallium isotope [11]. Our recent experiment at RIKEN confirms that ^{59}Ga is not observed, providing further evidence that ^{60}Ga marks the location of the proton drip line.

4 Half-life trends

In figure 7 from Ref. [4] we have analysed the trend observed in the half-lives as a function of atomic number for the T_z = -1/2 and -1 nuclei. Here, in figure 3, we present a new version of that figure, where we have updated the half-life values and included the trend for T_z = -2 nuclei. In particular, we have included/updated the half-lives of the following nuclides: ^{56}Zn [1]; ^{48}Fe and ^{52}Ni [2]; ^{52}Co [3]; ^{60}Ge, ^{60}Ga, ^{62}Ge and ^{59}Zn [6] and ^{44}Cr [25].

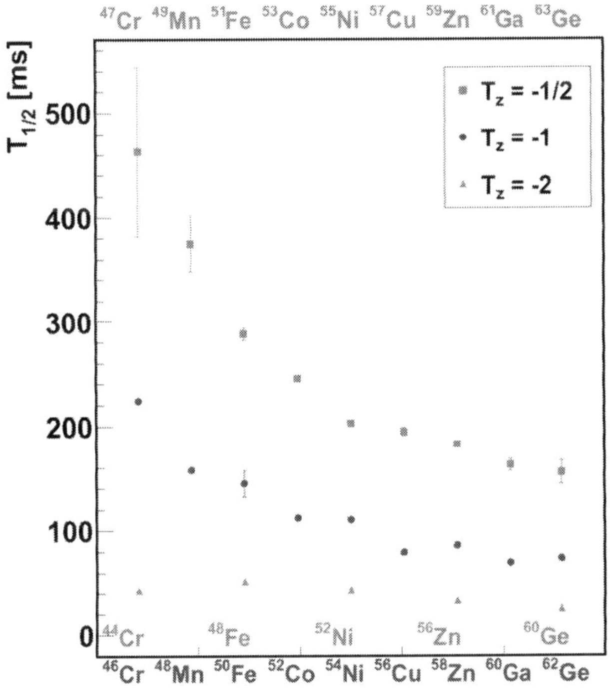

Figure 3. $T_{1/2}$ values measured for the T_z = -1/2 (red squares), -1 (blue dots) and -2 (green triangles) nuclei as a function of atomic number.

As discussed in Ref. [4], the systematic decrease of the $T_{1/2}$ values with the mass, observed in the T_z = -1/2 nuclei, reflects the increase in the Q_β value. A similar decreasing pattern is also found in the half-lives of the T_z = -1 nuclei and, on top of this behaviour, a typical odd-odd and even-even effect is observed. This is due to the fact that in the odd-odd nuclei there exist other excited states below the IAS that receive a significant amount of β feeding, which makes their half-lives slightly shorter in comparison with their even-even neighbours [4]. In the T_z = -2 nuclei, which are all even-even, a similar decreasing trend is observed, with the half-life of the most exotic ^{60}Ge being only 25.0(3) ms.

5 Conclusions

A panoramic view of the most relevant achievements from our β-decay spectroscopy experiments, with a focus on the most recent results, has been presented. The half-life trends as a function of the mass number and a comparison of our deduced mass excesses with different AME systematics have also been discussed. Altogether, it is a rich example of the valuable spectroscopic information that can be obtained from this kind of experiment.

Acknowledgments. This work was supported by the Generalitat Valenciana Grant No. PROMETEO/2019/007, Spanish Grants No. PID2019-104714GB-C21, FPA2017-83946-C2-1-P and FPA2014-52823-C2-1-P (MCIN,MINECO/AEI/FEDER), Centro de Excelencia Severo Ochoa del IFIC SEV-2014-0398; *Junta para la Ampliación de Estudios* Programme (CSIC JAE-Doc) co-financed by FSE.

References

[1] S. E. A. Orrigo *et al.*, Phys. Rev. Lett. **112**, 222501 (2014)
[2] S. E. A. Orrigo *et al.*, Phys. Rev. C **93**, 044336 (2016)
[3] S. E. A. Orrigo *et al.*, Phys. Rev. C **94**, 044315 (2016)
[4] L. Kucuk, S. E. A. Orrigo *et al.*, Eur. Phys. J. A **53**, 134 (2017)
[5] S. E. A. Orrigo *et al.*, J. Phys.: Conf. Ser. **1056**, 012041 (2018)
[6] S. E. A. Orrigo *et al.*, Phys. Rev. C **103**, 014324 (2021)
[7] Y. Fujita, B. Rubio, W. Gelletly, Prog. Part. Nucl. Phys. **66**, 549 (2011)
[8] H. Fujita *et al.*, Phys. Rev. C **88**, 054329 (2013)
[9] E. Ganioğlu *et al.*, Phys. Rev. C **93**, 064326 (2016)
[10] Y. Fujita *et al.* , Phys. Rev. Lett. **95**, 212501 (2005)
[11] S. F. Paul *et al.*, Phys. Rev. C **104**, 065803 (2021)
[12] N. Fukuda *et al.*, Nucl. Instrum. Methods B **317**, 323 (2013)
[13] G. Morpurgo, Phys. Rev. **110**, 721 (1958)
[14] D. H. Wilkinson, *Isospin in Nuclear Physics* (1969)
[15] F. Molina *et al.*, Phys. Rev. C **91**, 014301 (2015)
[16] E. Grodner *et al.*, Phys. Rev. Lett. **113**, 092501 (2014)
[17] E. Wigner, *Proc. Robert A. Welch Found. Conf.* (Robert A. Welch Foundation, Houston, Texas, 1957) 67
[18] W. Benenson, E. Kashy, Rev. Mod. Phys. **51**, 527 (1979)
[19] M. MacCormick, G. Audi, Nucl. Phys. A **925**, 61 (2014)
[20] G. Audi *et al.*, Nucl. Phys. A **729**, 3 (2003)
[21] G. Audi *et al.*, Chin. Phys. C **36**, 1157 (2012)
[22] M. Wang *et al.*, Chin. Phys. C **45**, 030003 (2021)
[23] M. Wang *et al.*, Chin. Phys. C **41**, 030003 (2017)
[24] M. Del Santo *et al.*, Phys. Lett. B **738**, 453 (2014)
[25] C. Dossat *et al.*, Nucl. Phys. A **792**, 18 (2007)

EPJ Web of Conferences **279**, 12003 (2023) https://doi.org/10.1051/epjconf/202327912003
NPA-X 2022

Study of decay properties for Ba to Nd nuclei relevant for the formation of the r-process rare-earth peak (A∼160)

M. Pallàs[1*], A. Tarifeño-Saldivia[2,1†], G. G. Kiss[3], J. L. Tain[2], A. Tolosa-Delgado[5,2], A. Vitéz-Sveiczer[3,4], F. Calviño[1], J. Agramunt[2], P. Aguilera[6], A. Algora[2], J. M. Allmond[7], H. Baba[8], N. T. Brewer[9,7], R. Caballero-Folch[10], P. J. Coleman-Smith[11], G. Cortes[1], T. Davinson[12], I. Dillmann[10,13], C. Domingo-Pardo[2], A. Estrade[14], N. Fukuda[8], S. Go[15,16], C. J. Griffin[10], R. K. Grzywacz[9,7], O. Hall[12], L. J. Harkness-Brennan[17], T. Isobe[8], D. Kahl[12], T. T. King[9], A. Korgul[18], S. Kovács[4], S. Kubono[8], M. Labiche[11], J. Liu[19], M. Madurga[9], K. Miernik[18], F. Molina[6], N. Mont-Geli[1], A. I. Morales[2], E. Nácher[2], A. Navarro[1], N. Nepal[8], S. Nishimura[8], M. Piersa-Silkowska[18], V. Phong[8], B. C. Rasco[9,7], J. Romero-Barrientos[6], B. Rubio[2], K. P. Rykaczewski[7], Y. Saito[10,20], H. Sakurai[8], Y. Shimizu[8], M. Singh[9], T. Sumikama[8], H. Suzuki[8], T. N. Szegedi[3], H. Takeda[8], K. Wang[14], M. Wolińska-Cichocka[21], P. J. Woods[12], and R. Yokoyama[16] for the BRIKEN collaboration[22]

[1]Institut de Tècniques Energètiques (INTE), Universitat Politècnica de Catalunya (UPC), 08028 Barcelona, Spain.
[2]Instituto de Física Corpuscular (IFIC), CSIC-UV, E-46980 Paterna, Spain.
[3]Institute for Nuclear Research (ATOMKI), 4026 Debrecen, Bem tér 18/c, Hungary.
[4]University of Debrecen, 4032 Debrecen, Egyetem tér 1, Hungary.
[5]Department of Physics, University of Jyväskylä, Finland.
[6]Centro de Investigación en Física Nuclear y Espectroscopía de Neutrones (CEFNEN). Comisión Chilena de Energía Nuclear, Nueva Bilbao 12501, Las Condes, Santiago-Chile.
[7]Physics Division, Oak Ridge National Laboratory, Oak Ridge, TN 37830, USA.
[8]RIKEN Nishina Center, 2-1 Hirosawa, Wako, Saitama 351-0198, Japan.
[9]Department of Physics and Astronomy, University of Tennessee, Knoxville, Tennessee 37996, USA.
[10]TRIUMF, 4004 Wesbrook Mall, Vancouver, BC V6T 2A3, Canada.
[11]STFC Daresbury Laboratory, Daresbury, Warrington WA4 4AD, United Kingdom.
[12]School of Physics and Astronomy, The University of Edinburgh, Edinburgh EH9 3FD, United Kingdom.
[13]Department of Physics and Astronomy, University of Victoria, Victoria, BC V8P 5C2, Canada.
[14]Central Michigan University, Mt. Pleasant, MI 48859, USA.
[15]Department of Physics, Kyushu University, 744 Motooka, Fukuoka 819-0395, Japan.
[16]Center for Nuclear Study, The University of Tokyo, 2-1 Hirosawa, Wako, Saitama 351-0106, Japan.
[17]Department of Physics, University of Liverpool, Liverpool L69 7ZE, United Kingdom.
[18]Faculty of Physics, University of Warsaw, 02-093 Warsaw, Poland.
[19]Department of Physics, the University of Hong Kong, Pokfulam Road, Hong Kong.
[20]Department of Physics and Astronomy, The University of British Columbia, Vancouver BC V6T 1Z1, Canada.
[21]Heavy Ion Laboratory, University of Warsaw, Pasteura 5A, 02-093 Warsaw, Poland.

[22]www.wiki.ed.ac.uk/display/BRIKEN/Home

†atarisal@ific.uv.es
*max.pallas@upc.edu

Abstract. At the RIKEN Nishina Center, exotic neutron-rich isotopes of Ba, La, Ce, Pr, and Nd were measured. This work reports their half-lives ($T_{1/2}$) and β-delayed neutron-emission probabilities (P_{xn}). The setup consisted of the BigRIPS in-flight separator for particle identification, the Advanced Implantation Detector Array (AIDA) for ions and β detection, and the BRIKEN neutron counter for neutron detection. Using this arrangement, 4 new $T_{1/2}$ and 14 new P_{1n} were measured. Furthermore, 38 $T_{1/2}$ and 2 P_{1n} values were remeasured, decreasing the uncertainties for some of them. In addition to improving predictions of nucleosynthesis models that describe the production of the rare-earth peak at A~160 via the r-process, these additional experimental data should help to constrain theoretical models for calculations of $T_{1/2}$ and P_{xn} in this region.

1 Introduction

Nearly half of the nuclei heavier than iron are produced via rapid neutron capture (the r-process) in explosive stellar scenarios. Above the mass number $A = 100$,, there are two main peaks in the r-process solar-system abundances, located at $A \sim 130$ and $A \sim 195$. Located between them, the Rare-Earth Peak (REP) is a tiny but definite peak at mass $A \sim 160$ that results from the freeze-out during the last stages of neutron exposure. Determining the late-time conditions at the r-process site may thus be possible using the formation of the REP as a special probe. According to theoretical models and sensitivity studies, the half-lives ($T_{1/2}$) and β-delayed neutron emission probabilities (P_{xn}) of very neutron-rich nuclei, in the mass region $A \sim 160$ for $55 \leq Z \leq 64$, are critical for the formation of the REP [1, 2].
From 2016 to 2021, the BRIKEN project operated at the Radioactive Isotope Beam Factory (RIBF) in the RIKEN Nishina Center. This collaboration has carried out a thorough measurement program of the β-decay properties of the most neutron-rich nuclei experimentally available. The BRIKEN REP proposal focused on measurements from Ba to Gd ($Z = 56 - 64$). The entire REP proposal comprised three experimental campaigns. Recently, a first publication containing the experimental findings for Pm to Gd ($Z = 61 - 64$) species has been released [3]. The main goal of this study is the data analysis for Ba to Nd ($Z = 56 - 60$) species. Figure 1 shows the overlap among the targeted area in our study and the high astrophysical impact region proposed by Mumpower et al. [2]. The experimental setup, the status of the data analysis in this region, and future perspectives are discussed in this work.

2 Experimental setup

The exotic neutron-rich isotopes were produced at the RIKEN Nishina Center using a ^{238}U primary beam, bombarding a ^{9}Be target. The products from the fragmentation and fission reaction were selected using the BigRIPS in-flight separator and directed to the F11 focal plane through the Zero-Degree Spectrometer (ZDS) [5]. Each ion reaching the experimental area was identified by measuring its atomic charge (Z), and its mass-to-charge ratio (A/Q). Identified ions were implanted in the Advanced Implatation Detector Array (AIDA) [6]. AIDA comprises a stack of six silicon Double-Sided Strip Detectors (DSSD). Each DSSD has a thickness of 1 mm, and 128 strips per side.

 https://doi.org/10.1051/epjconf/202327912003

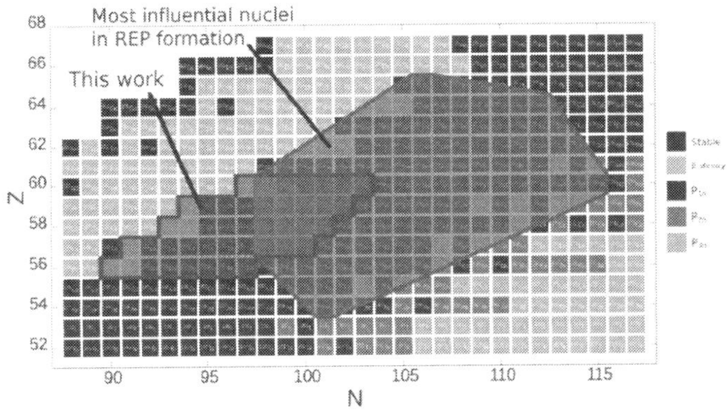

Figure 1. (Color online) This scheme shows a section of the table of isotopes. The green region indicates the most influential nuclei in rare-earth peak formation [2]. The purple area shows the isotopes that were measured in this experiment. The figure also depicts the main decay channel predicted by theoretical models for each isotope [4].

The BRIKEN neutron counter [7] was placed surrounding AIDA in order to detect the β-delayed neutrons. It comprised an array of 140 ^{3}He-filled proportional tubes embedded in a 90 cm x 90 cm x 75 cm High-Density PolyEthylene (HDPE) moderator. Given the large cross section for thermal neutrons in the capture reaction n_{th} + ^{3}He$\rightarrow$ ^{3}H+p, neutron moderation is necessary to achieve high overall detection efficiencies. The HDPE block has a central hole where AIDA is inserted. The block also has two additional holes at the lateral sides, where two CLARION-type clover detectors are placed to offer γ detection capabilities [7]. The neutron detection efficiency of the BRIKEN detector has been calculated by Monte Carlo simulations and validated experimentally using a ^{252}Cf neutron source [8]. The BRIKEN counter offers a nominal value for the efficiency of 68.6% up to 1 MeV. At higher energies, the efficiency decreases smoothly. Concerning the data acquisition, the raw data is collected by three independent data acquisition systems (DAQs) from BigRIPS, AIDA, and BRIKEN detectors. To create a common timestamp and enable the reconstruction of the identified events, a synchronization signal has been fed into each DAQ. The primary objective of this experiment was to measure the $T_{1/2}$ and P_{xn} linked to the implanted ions' decay. To achieve this, for each identified isotope, we built correlation histograms between implant, beta, and neutron events. Then, we fitted the histograms using the solution to the Bateman equations [9] for all the decay descendants. The fit procedure was run on all histograms associated with the isotope of interest at the same time to achieve self-consistency in the fit parameters [10].

3 Preliminary results

The top panel of Figure 2 presents the particle identification (PID) of the ions implanted on AIDA. In the lower part, there is the projection of the PID matrix onto the A/Q axis for the Pr isotopes. The A/Q resolution obtained is sufficient to ensure a good separation of fully stripped and H-like ions in the less exotic region ($A < 159$). Figure 3 presents the preliminary findings of the measured $T_{1/2}$ in this work. This study yielded 4 new half-lives: ^{157}La, ^{159}Ce, ^{161}Pr, and ^{163}Nd. For these new half-lives, the determination of the precise values and uncertainties is still ongoing. By improving particle identification, we expect to reduce the charge-state contamination. Additionally, a total of 38 $T_{1/2}$ have been remeasured with an improved accuracy in several cases. The P_{1n} values were measured for the first time for a vast majority of the isotopes. In this work, 14 new P_{1n} values ($^{151-152}$Ba, $^{151-153}$La, $^{154-156}$Ce, $^{155-158}$Pr, and $^{160-160}$Nd) are preliminary reported. Two P_{1n} values (^{149}La and ^{150}La) have also been remeasured with an increased accuracy. The P_{2n} values in this region are 0, because the Q_{b2n} values are less than zero [12]. Figure 3 compares our $T_{1/2}$ preliminary findings with assessed nuclear data from the ENSDF and some theoretical models: Relativistic Hartree-Bogoliubov plus proton-neutron Relativistic Quasi-Particle Random-Phase

https://doi.org/10.1051/epjconf/202327912003

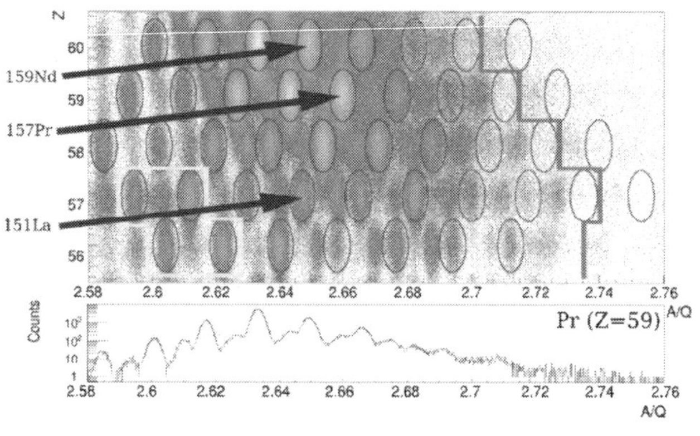

Figure 2. (Color online) Identification plot of the REP-BRIKEN 2018 experiment. Each red circumference corresponds to an identified isotope. The gray line represents the upper bound of the previously measured $T_{1/2}$ [11]. Yellow box highlights previously measured P_{1n} values. The bottom panel depicts the projection of the PID matrix on the A/Q axis for the Pr (Z = 59) isotopes.

Approximation (RHB+pn-RQRPA, 2016) [13], Finite-Range Droplet-Model mass formula (2012) plus Quasiparticle-Random-Phase Approximation (FRDM+QRPA, 2018) [4], and the proton-neutron Finite Amplitude Method (pnFAM, 2020) [14]. Compared with the evaluated nuclear data, our results are consistent within a 3σ error band. Furthermore, in Figure 4, our preliminary results for the P_{1n} values are compared with evaluated nuclear data from the ENSDF and some theoretical predictions: RHB+pn-RQRPA (2016) [13], FRDM+QRPA (2018) [4], and the proton-neutron Relativistic Quasiparticle Random-Phase Approximation plus Hauser-Feshbach statistical model (pn-RQRPA + HFM, 2021) [15]. Regarding the comparison with theoretical models, our data supports the overall trend for all P_{1n} predictions. In particular, the RHB+pn-RQRPA model gives estimates for Ba, Ce, Pr, and Nd (Z = 56, 58, 59, and 60) that are in close agreement with the experiment, but it significantly overestimates La (Z = 57) nuclei. In a comparison of the three models under consideration, pn-RQRPA + HFM best replicates the experimental data and provides good agreement for most of the isotopes.

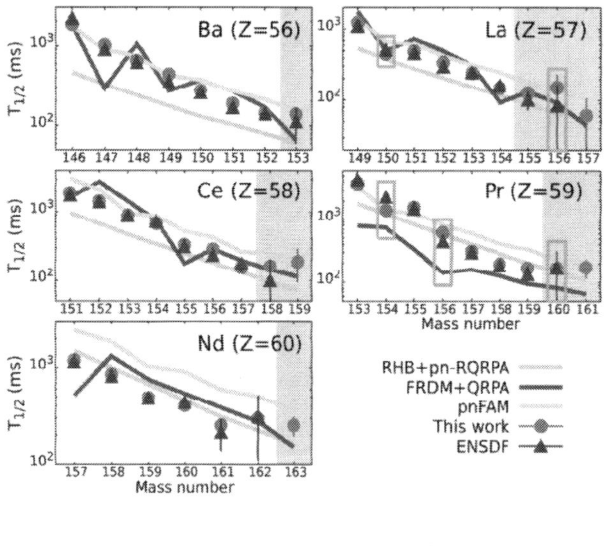

Figure 3. (Color online) Preliminary half-lives derived in this work (red circles) and previous measurements (blue triangles). These results are compared to three theoretical models: RHB+pn-RQRPA [13] (green), FRDM+QRPA [4] (purple), and pnFAM [14] (light-blue). The grey region highlights expected improvements in the determination of $T_{1/2}$ by the end of the data analysis. Potential isomer contributions are indicated by orange boxes [16].

4 Remarks and future work

In this work, preliminary results from measurements of exotic isotopes in the REP region at the RIKEN Nishina Center are presented. Four new $T_{1/2}$ and 14 new P_{1n} values are reported for the first time. 38 $T_{1/2}$ and 2 P_{1n} values were remeasured, improving the data accuracy for many of them. As we covered above, there is still some work ongoing. By reducing

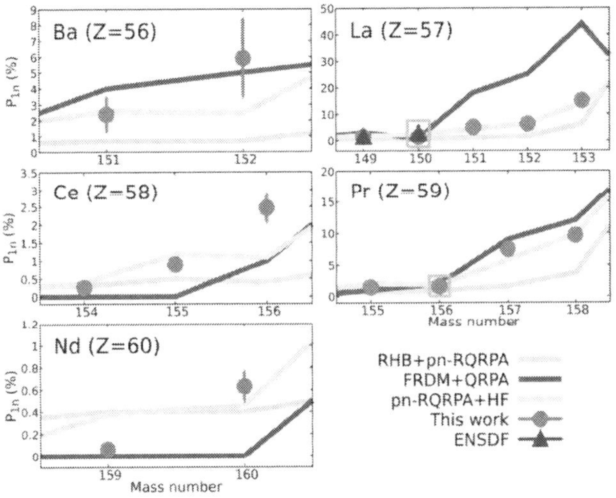

Figure 4. (Color online) Preliminary results for P_{1n} from this work (red circles) and previous measurements (blue triangles). These experimental results are compared to the following theoretical models: RHB+pn-RQRPA [13] (green), FRDM+QRPA [4] (purple), and pn-RQRPA + HFM [15] (light-blue). Orange boxes indicate potential isomer contribution [16].

the contamination from charge states, the data reported for the most exotic region should be improved. Aside from that, the impact of potential isomers on the studied isotopes must be assessed. Lastly, the evaluation of the astrophysical impact of the new nuclear data on the description of the r-process in the REP region will be developed in the following months.

5 Acknowledgments

This work has been supported by the Spanish Ministerio de Economía y Competitividad under Grants nos. FPA2014-52823-C2-1-P, FPA2014-52823-C2-2-P, FPA2017-83946-C2-1-P, FPA2017-83946-C2-2-P and grants from Ministerio de Ciencia e Innovacion nos PID2019-104714GB-C21, PID2019-104714GB-C22. It was also supported by NKFIH (NN128072) and by the ÚNKP-22-2 and ÚNKP-22-3 New National Excellence Program of the Ministry for Culture and Innovation from the source of the National Research, Development and Innovation Fund (Project numbers: ÚNKP-22-2-I-DE-75; ÚNKP-22-3-II- DE-132).

References

[1] A. Arcones and G. Martinez Pinedo, *Phys. Rev. C* **83**, 045809 (2011).
[2] M. R. Mumpower, et al., *Phys. Rev. C* **85**, 045801 (2012).
[3] G. G. Kiss, et al., *The Astrophysical Journal* **936 2**, 107 (2022).
[4] P. Möller, et al., *At. Data Nucl. Data* **125** 1 – 192 (2019).
[5] T. Kubo, et al., *Prog. Theor. Exp. Phys.* 03C002 (2012).
[6] C. J. Griffin, et al., *Jap. Phys. Soc. Conf. Proc.* **14**, 020622 (2017).
[7] A. Tarifeño-Saldivia, et al., *Journal of Instrumentation* **12(04)**:P04006 (2017).
[8] M. Pallàs, et al., ArXiv:2204.13379 (2022).
[9] H. Bateman, et al., *Proc. Cambridge Philos* **10 5**, 423 – 427 (1910).
[10] A. Tolosa-Delgado, et al., *NIM A* **925**, 133 – 147 (2019).
[11] J. Wu et al., *Phys. Rev. C* **101**, 042801(R) (2020).
[12] NuDat 3. https://www.nndc.bnl.gov/nudat3/xundl/>
[13] T. Marketin, et al., *Phys. Rev. C* **93 2** 025805 (2016).
[14] E. M. Ney, et al., *Phys. Rev. C* **102 3** 034326 (2020).
[15] F. Minato, et al., *Phys. Rev. C* **104 4** 044321 (2021).
[16] Experimental Unevaluated Nuclear Data List. <https://www.nndc.bnl.gov/ensdf/xundl/>

Beta decay along the N=Z line and its relevance in rp-process and X-Ray bursts

E. Nácher[1*], *S.* Parra[1], *J.A.* Briz[2], *P.* Aguilera[3], *J.* Agramunt[1], *A.* Algora[1], *T.* Berry[4], *M.J.G.* Borge[5], *M.* Carmona[5], *L.M.* Fraile[2], *E.* Ganioglu[6], *W.* Gelletly[4], *V.* Guadilla[1], *A.* Illana[5], *R.* Lică[7], *I.* Marroquín[5], *F.* Molina[3], *A.I.* Morales[1], *N.* Orce[8], *S.E.J.* Orrigo[1], *A.* Perea[5], *J.* Romero[3], *B.* Rubio[1], *C.* Sotty[7], *J.L.* Taín[1], *O.* Tengblad[5], *A.* Tolosa[1]
for the IS570 collaboration.

[1]Instituto de Física Corpuscular, CSIC - Univ. de Valencia, E-46980 Paterna, Valencia, Spain
[2]Fac. de CC Físicas, Universidad Complutense de Madrid - CEI Moncloa, 28040 Madrid, Spain
[3]Comisión Chilena de Energía Nuclear, P.O. Box 188-D, Santiago, Chile.
[4]Dep. of Physics, University of Surrey, Guildford, GU2 7XH, United Kingdom
[5]Instituto de Estructura de la Materia, CSIC, E-28006 Madrid, Spain
[6]Dep. of Physics, Istanbul University, Istanbul, 34134, Turkey
[7]Horia Hulubei Nat. Inst. for Physics and Nuclear Engineering, Bucharest 077125, Romania
[8]Dep. of Physics & Astronomy, Univ. of the Western Cape, P/B X17 Bellville 7535, South Africa

Abstract. Nucleosynthesis in Type I X-ray bursts (XRB) proceeds eventually through the rp-process near the proton drip-line. Several N=Z nuclei act as waiting points in the reaction network chain. Astrophysical calculations of XRB light curves depend upon the theoretical modelling of the beta decays of interest, with the N=Z waiting points ^{64}Ge, ^{68}Se, ^{72}Kr, ^{76}Sr, and their second-neighbours N=Z+2 being key nuclei in this context. We have carried out different experimental campaigns at ISOLDE (CERN) to determine the B(GT) distributions, in the decay of several N=Z, N=Z+2 and their daughters, of particular relevance in rp-process calculations. To this aim the Total Absorption Spectroscopy technique is applied. Here we present results on the beta decay of ^{64}Ga and the status of the analysis of ^{64}Ge. Our results provide benchmarks for testing and constraining models under terrestrial conditions that can be used later for predictions in stellar environments.

1 Introduction

Nucleosynthesis in explosive hydrogen burning at high temperatures (T > 10^8 K) is characterized mainly by the rapid proton capture (rp-) process [1]. Discussions of the possible scenarios for such extreme conditions can be found in Refs. [2] and [3], where Type I X-ray bursts (XRBs) are suggested as possible sites for the rp-process. These explosions are produced in binary systems in which a neutron star accretes hydrogen-rich material from a low-mass companion star, typically a Main Sequence or a Red-Giant star. Thermonuclear ignition takes place in semi-degenerate conditions, when the temperature and density in the

* Corresponding author: enrique.nacher@csic.es

accreted envelope become high enough to allow for a breakout from the hot CNO cycle. Nucleosynthesis eventually proceeds near the proton drip-line via the rp-process [4]. Type I XRBs are characterized by $T_{peak} = 1 - 3$ GK and $\rho = 10^6 - 10^7$ g cm^{-3}. To date, 118 XRB sources exhibiting these characteristics have been discovered, the last one reported in June-2022 [5].

Discussions of the main features and observations of XRBs can be found in Refs. [4][6][7]. In these works, different models of Type I XRBs are presented, with a focus on the nuclear physics processes involved along the rp-process path, and the sensitivity of the luminosity curves to changes in particular reaction cross sections and/or weak-decay rates (see for instance Fig. 1). It turns out that the beta decay of the waiting points and second-neighbours ^{64,66}Ge, ^{68,70}Se, ^{72,74}Kr and ^{76,78}Sr are particularly relevant for the energy generation, reaction flow, and final composition of the ashes from the burst [4]. Some of these decays, namely Kr and Sr, have been studied by our collaboration in the past at ISOLDE [8-11]. Lately, we have concentrated our efforts in the lighter systems, more specifically Ge and Se isotopes and their decay daughters, pointed out as key decays in most of the XRB models described e.g., in [4].

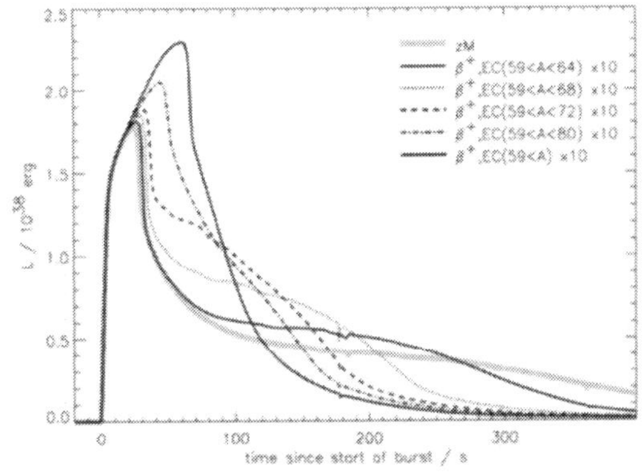

Fig. 1. Taken from [6]: sensitivity of the light curve of the first pulse in model zM to variations along the waiting points in the vicinity of A = 60, 64, and 68. The nominal light curve is shown along with the result when all weak rates above A = 59 are multiplied by 10. Also shown are the results of progressively adding in accelerations to flows in the mass ranges A = 60–63, 64–67, 68–71, and 72–79

Accurate measurements in terrestrial conditions of the aforementioned waiting points, daughters and neighbours are therefore essential to validate and constrain the theoretical models used for the computation of stellar decay rates because, as pointed out in [7], many decay rates computed in stellar conditions do not converge to their laboratory values for terrestrial conditions, thus putting in question the model used for these calculations. In this context, it was first pointed out by Sarriguren [12][13], pursued later by others [14][15], that the process of electron capture from the continuum (cEC), as well as the decay from thermally populated states in the WP nuclei and neighbours, actually play an important role in the weak-decay rates of nuclei close to the proton drip-line in XRB calculations. Quoting Ref. [12]: *"Although these decay properties (B(GT) distributions and half-lives) may be different at high ρ and T existing in rp-process scenarios, success in their description under terrestrial conditions is a requirement for a reliable calculation of the weak decay rates in more general conditions"*. In all the theoretical works mentioned in this paragraph the authors have used our experimental data taken with Lucrecia, the Total Absorption Spectrometer (TAS) at ISOLDE [8-11], for comparison and benchmark.

In the next section, the Total Absorption Spectroscopy technique will be explained and, in the following one, results on ^{64,66}Ga will be shown, along with some preliminary data reduction for the decay of ^{64}Ge.

2 Experimental technique

Even though one might think that the relevant physical quantity, as far as XRB model calculations is concerned, may be the half-life of the nucleus, this gives very limited information on the nuclear structure. In fact, different Gamow-Teller strength distributions (B(GT)) obtained with different models might lead to the same half-life. Therefore, in order to validate a theoretical model, capable of making predictions in a wide region of the nuclear chart and at stellar temperature and densities, one needs accurate experimental B(GT) distributions at terrestrial conditions, rather than β-decay half-lives [13]. Over the years it has been shown that the best tool to perform such an experimental study, in medium mass and heavy nuclei, is the so-called Total Absorption Spectroscopy (TAS) technique. Details on this technique can be found in [16].

The TAS technique is based on the use of gamma detectors of very high efficiency to absorb entire gamma cascades, rather than individual gamma rays, following the beta decay. The analysis of the data, to obtain a reliable B(GT) distribution, is based on the unfolding procedure described in [17]. One such detector is Lucrecia, at ISOLDE, described in [16]. It consists of a large NaI(Tl) cylindrical crystal, 38 cm diameter × 38 cm long, plus some ancillary detectors such as a HPGe telescope for X rays and gamma rays, and a plastic scintillator for betas. In spite of being the most sensitive way of measuring the missing beta strength, unseen in traditional high-resolution beta-decay experiments, the TAS technique provides data that require complex reduction and unfolding methods [17]. In particular, a proper TAS analysis requires an accurate estimation of all contaminants in the spectrum (room background, daughter activity, pileup contribution…).

With the aim of covering the the rp-process path down to its bottleneck ^{64}Ge, the experiment IS570 was carried out at ISOLDE, where we measured the decays of ^{64,65,66}Ge and their daughters ^{64,65,66}Ga.

3 Results

In the process of analysis of ^{64}Ge decay, we need to estimate the ^{64}Ga decay contribution and, with this aim, we measured clean ^{64}Ga samples during the IS570 run. Fortunately, ^{64}Ga, with $T_{1/2}$ = 2.6 min and Q_{EC} = 7.2 MeV, is interesting *per se*, and not only as a contamination in the decay spectrum of ^{64}Ge. In particular, ^{64}Ga decay is very relevant for isospin-mixing consideration since one can measure the beta population of the 0^+ ground state of the daughter nucleus ^{64}Zn that is, in principle, isospin forbidden Fermi $0^+ \rightarrow 0^+$, but populated as a consequence of isospin impurity of ^{64}Ga ground state [18]. The mixing of T=0 and T=1 in the ground state of ^{64}Ga is quantified by an admixture amplitude α that can be derived experimentally from the measured beta strength.

Fig. 2 shows the measured spectrum and resulting beta strength (integral) from our analysis of ^{64}Ga beta decay. In the left panel we can see, in black colour, the TAS spectrum as measured by the Lucrecia spectrometer at ISOLDE. This spectrum is overlaid to the Geant4 simulation of the same decay measured with the same detector, in red colour. As we can see, the fit is virtually perfect until 4.5 MeV excitation energy, where some discrepancies appear. The biggest difference lies at ~6 MeV, where the TAS data show a resonance that is not present in the simulation and therefore has not been reported or included in the Evaluated Nuclear Structure File ENSDF.

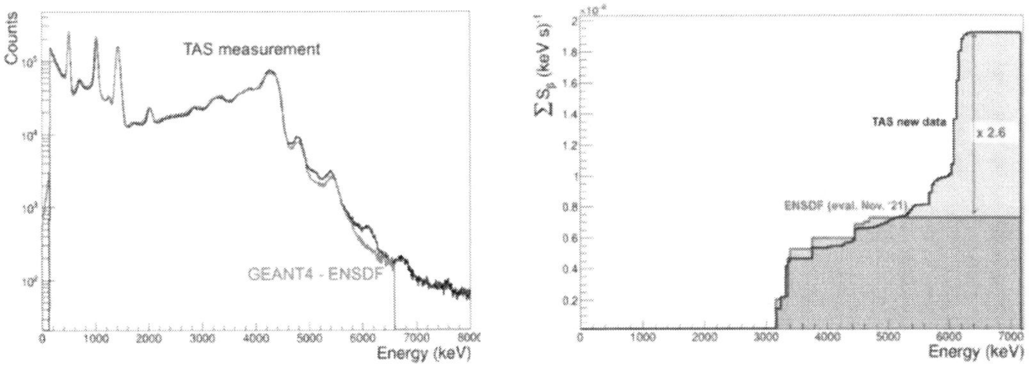

Fig. 2. Left: TAS spectrum of the beta decay of ^{64}Ga (black) compared to the Geant4 simulation of the same decay in the same detector using ENSDF data (red). Right: beta-strength distribution (integral) of the same decay resulting from the analysis of the TAS data (black) compared to the one from ENSDF (red).

The results of the analysis of the TAS spectrum of the left panel of Fig. 2 (black) are plotted in the right panel of the same figure, in black colour, in the form of the sum of the beta strength as a function of excitation energy in the daughter nucleus. In the same graph we see, in red colour, the same beta strength as derived from the evaluated data from previous works [18][19] (ENSDF - Evaluated November 2021). As we can see, the small differences between the ENSDF data and the new TAS data in the spectrum of Fig. 2 left, result in a factor of 2.6 more strength detected with the TAS measurement than in previous measurements. This difference results in a different isospin impurity derived from the beta strength and is, therefore, very relevant for isospin-mixing considerations. In this context, the preliminary admixture amplitude derived from our work $|\alpha| = 19.9\times10^{-3}$ is slightly smaller than the one from previous works $|\alpha| = 21.7\times10^{-3}$ [18].

The beta decay of ^{64}Ge, that is, according to [4], perhaps the most relevant in the region as far as rp-process calculations is concerned, is still under analysis. Fig. 3 shows the direct TAS spectrum of this decay, in black, compared to the Geant4 simulation of the same decay measured with the same detector, in red. Although no further analysis has been performed so far, already from the TAS spectrum one can see striking differences between the two spectra, and obvious new resonances appear from 2 MeV on. This decay is presently under analysis and we expect new results soon that will be published elsewhere.

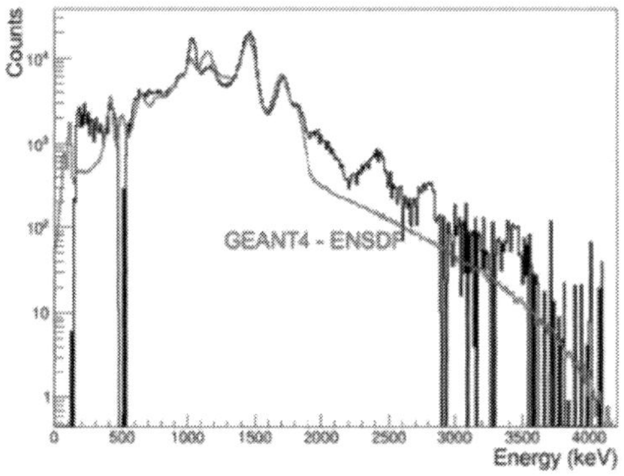

Fig. 3. TAS spectrum of the beta decay of ^{64}Ge (black) compared to the Geant4 simulation of the same decay in the same detector using ENSDF data (red).

This work has been partially supported by the Spanish Ministry of Science MCIN/AEI/10.13039/501100011033 through grant nº PID2019-104714GB-C21 and the Programa de Excelencia Generalitat Valenciana PROMETEO/2019/007 (Spain).

References

1. R.K. Wallace et al., Astrophys. J. Suppl. **45**, 389 (1981)
2. H. Schatz et al., Physics Reports **294**, 167 (1998)
3. H. Schatz et al., Nucl Phys A **777**, 601 (2006)
4. A. Parikh et al., Prog. Part. Nucl. Phys. **69**, 225 (2013)
5. P. Bult et al., Astrophys. J. Lett. **935**, L32 (2022)
6. S. Woosley et al., Astrophys. J. Suppl. **151**, 75 (2004)
7. J. Jose et al., Astrophys. J. Suppl. **189**, 204 (2010)
8. E. Poirier et al., Phys. Rev. C **69**, 034307 (2004)
9. E. Nacher et al., Phys. Rev. Lett. **92**, 232501 (2004)
10. A.B. Pérez et al., Phys. Rev. C **88**, 014324 (2013)
11. J.A. Briz et al., Phys. Rev. C **92**, 054326 (2015)
12. P. Sarriguren, Phys. Lett. B **680**, 438 (2009)
13. P. Sarriguren, Phys. Rev. C **83**, 025801 (2011)
14. Jameel-Un Nabi, Astrophys. Space Sci. **339**, 305 (2012)
15. A. Petrovici et al., Phys. Rev. C **100**, 015810 (2019)
16. B. Rubio et al., J. Phys. G **44**, 084004 (2017)
17. J. L. Taín et al., Nucl. Inst. and Methods A **571**, 728 (2007)
18. S. Raman et al., Nuclear Physics A **254**, 131 (1975)
19. L.G. Mann et al., Nucl. Phys. A **97**, 425 (1967)

EPJ Web of Conferences **279**, 13001 (2023) https://doi.org/10.1051/epjconf/202327913001
NPA-X 2022

New detection systems for an enhanced sensitivity in key stellar (n,γ) measurements

J. Lerendegui-Marco[1]*, V. Babiano-Suárez[1], J. Balibrea-Correa[1], C. Domingo-Pardo[1], I. Ladarescu[1], A. Tarifeño-Saldivia[1], V. Alcayne[2], D. Cano-Ott[2], E. González-Romero[2], T. Martínez[2], E. Mendoza[2], C. Guerrero[3], F. Calviño[4], A. Casanovas[4], U. Köster[5], N. M. Chiera[6], R. Dressler[6], E. A. Maugeri[6], D. Schumann[6], O. Aberle[7], S. Altieri[8,9], S. Amaducci[10], J. Andrzejewski[11], M. Bacak[7], C. Beltrami[8], S. Bennett[12], A. P. Bernardes[7], E. Berthoumieux[13], R. Beyer[14], M. Boromiza[15], D. Bosnar[16], M. Caamaño[17], M. Calviani[7], F. Cerutti[7], G. Cescutti[18,19], E. Chiaveri[7,12], P. Colombetti[20,21], N. Colonna[22], P. Console Camprini[23,24], G. Cortés[4], M. A. Cortés-Giraldo[3], L. Cosentino[10], S. Cristallo[25,26], S. Dellmann[27], M. Di Castro[7], S. Di Maria[28], M. Diakaki[29], M. Dietz[30], E. Dupont[13], I. Durán[17], Z. Eleme[31], S. Fargier[7], B. Fernández[3], B. Fernández-Domínguez[17], P. Finocchiaro[10], S. Fiore[24,32], F. García-Infantes[33], A. Gawlik-Ramięga[11], G. Gervino[20,21], S. Gilardoni[7], F. Gunsing[13], C. Gustavino[32], J. Heyse[34], W. Hillman[12], D. G. Jenkins[35], E. Jericha[36], A. Junghans[14], Y. Kadi[7], K. Kaperoni[29], G. Kaur[13], A. Kimura[37], I. Knapová[38], M. Kokkoris[29], M. Krtička[38], N. Kyritsis[29], C. Lederer-Woods[39], G. Lerner[7], A. Manna[23,40], A. Masi[7], C. Massimi[23,40], P. Mastinu[41], M. Mastromarco[22,42], A. Mazzone[22,43], A. Mengoni[24,23], V. Michalopoulou[29], P. M. Milazzo[18], R. Mucciola[25,44], F. Murtas[†45], E. Musacchio-Gonzalez[41], A. Musumarra[46,47], A. Negret[15], A. Pérez de Rada[2], P. Pérez-Maroto[3], N. Patronis[31], J. A. Pavón-Rodríguez[3], M. G. Pellegriti[46], J. Perkowski[11], C. Petrone[15], E. Pirovano[30], J. Plaza[2], S. Pomp[48], I. Porras[33], J. Praena[33], J. M. Quesada[3], R. Reifarth[27], D. Rochman[6], Y. Romanets[28], C. Rubbia[7], A. Sánchez-Caballero[2], M. Sabaté-Gilarte[7], P. Schillebeeckx[34], A. Sekhar[12], A. G. Smith[12], N. V. Sosnin[39], M. E. Stamati[31], A. Sturniolo[20], G. Tagliente[22], D. Tarrío[48], P. Torres-Sánchez[33], E. Vagena[31], S. Valenta[38], V. Variale[22], P. Vaz[28], G. Vecchio[10], D. Vescovi[27], V. Vlachoudis[7], R. Vlastou[29], A. Wallner[14], P. J. Woods[39], T. Wright[12], R. Zarrella[23,40], and P. Žugec[16]
and the n_TOF Collaboration

[1] *Instituto de Física Corpuscular, CSIC - Universidad de Valencia, Spain*
[2] *Centro de Investigaciones Energéticas Medioambientales y Tecnológicas (CIEMAT), Spain*
[3] *Universidad de Sevilla, Spain*
[4] *Universitat Politècnica de Catalunya, Spain*
[5] *Institut Laue Langevin (ILL), Grenoble, France*
[6] *Paul Scherrer Institut (PSI), Villigen, Switzerland*
[7] *European Organization for Nuclear Research (CERN), Switzerland*
[8] *Istituto Nazionale di Fisica Nucleare, Sezione di Pavia, Italy*
[9] *Department of Physics, University of Pavia, Italy*
[10] *INFN Laboratori Nazionali del Sud, Catania, Italy*
[11] *University of Lodz, Poland*
[12] *University of Manchester, United Kingdom*
[13] *CEA Irfu, Université Paris-Saclay, F-91191 Gif-sur-Yvette, France*

*e-mail: jorge.lerendegui@ific.uv.es
† deceased

[14] *Helmholtz-Zentrum Dresden-Rossendorf, Germany*
[15] *Horia Hulubei National Institute of Physics and Nuclear Engineering, Romania*
[16] *Department of Physics, Faculty of Science, University of Zagreb, Zagreb, Croatia*
[17] *University of Santiago de Compostela, Spain*
[18] *Istituto Nazionale di Fisica Nucleare, Sezione di Trieste, Italy*
[19] *Department of Physics, University of Trieste, Italy*
[20] *Istituto Nazionale di Fisica Nucleare, Sezione di Torino, Italy*
[21] *Department of Physics, University of Torino, Italy*
[22] *Istituto Nazionale di Fisica Nucleare, Sezione di Bari, Italy*
[23] *Istituto Nazionale di Fisica Nucleare, Sezione di Bologna, Italy*
[24] *Agenzia nazionale per le nuove tecnologie (ENEA), Italy*
[25] *Istituto Nazionale di Fisica Nucleare, Sezione di Perugia, Italy*
[26] *Istituto Nazionale di Astrofisica - Osservatorio Astronomico di Teramo, Italy*
[27] *Goethe University Frankfurt, Germany*
[28] *Instituto Superior Técnico, Lisbon, Portugal*
[29] *National Technical University of Athens, Greece*
[30] *Physikalisch-Technische Bundesanstalt (PTB), Bundesallee 100, 38116 Braunschweig, Germany*
[31] *University of Ioannina, Greece*
[32] *Istituto Nazionale di Fisica Nucleare, Sezione di Roma1, Roma, Italy*
[33] *University of Granada, Spain*
[34] *European Commission, Joint Research Centre (JRC), Geel, Belgium*
[35] *University of York, United Kingdom*
[36] *TU Wien, Atominstitut, Stadionallee 2, 1020 Wien, Austria*
[37] *Japan Atomic Energy Agency (JAEA), Tokai-Mura, Japan*
[38] *Charles University, Prague, Czech Republic*
[39] *School of Physics and Astronomy, University of Edinburgh, United Kingdom*
[40] *Dipartimento di Fisica e Astronomia, Università di Bologna, Italy*
[41] *INFN Laboratori Nazionali di Legnaro, Italy*
[42] *Dipartimento Interateneo di Fisica, Università degli Studi di Bari, Italy*
[43] *Consiglio Nazionale delle Ricerche, Bari, Italy*
[44] *Dipartimento di Fisica e Geologia, Università di Perugia, Italy*
[45] *INFN Laboratori Nazionali di Frascati, Italy*
[46] *Istituto Nazionale di Fisica Nucleare, Sezione di Catania, Italy*
[47] *Department of Physics and Astronomy, University of Catania, Italy*
[48] *Uppsala University, Sweden*

Abstract. Neutron capture cross-section measurements are fundamental in the study of astrophysical phenomena, such as the slow neutron capture (s-) process of nucleosynthesis operating in red-giant and massive stars. However, neutron capture measurements via the time-of-flight (TOF) technique on key *s*-process nuclei are often challenging. Difficulties arise from the limited mass (~mg) available and the high sample-related background in the case of the unstable *s*-process branching points. Measurements on neutron magic nuclei, that act as *s*-process bottlenecks, are affected by low (n,γ) cross sections and a dominant neutron scattering background. Overcoming these experimental challenges requires the combination of facilities with high instantaneous flux, such as n_TOF-EAR2, with detection systems with an enhanced detection sensitivity and high counting rate capabilities. This contribution reviews some of the latest detector developments in detection systems for (n,γ) measurements at n_TOF, such as i-TED, an innovative detection system which exploits the Compton imaging technique to reduce the dominant neutron scattering background and s-TED, a highly segmented total energy detector in-

https://doi.org/10.1051/epjconf/202327913001

tended for high flux facilities. The discussion will be illustrated with results of the first measurement of key the *s*-process branching-point reaction ^{79}Se(n,γ).

1 (n,γ) measurements for the *s*-process: challenges and solutions

Neutron capture reactions play a fundamental role in the slow neutron capture (s-) process of nucleosynthesis operating in red-giant and massive stars [1], which is responsible for the formation of about half of the elements heavier than iron. However, neutron capture measurements via the time-of-flight (TOF) technique on key *s*-process nuclei are challenging. As a consequence, for many nuclei, uncertainties in the stellar (n,γ) cross sections are still significantly larger than the very precise abundance observations [1].

Among the relevant isotopes that still present large uncertainties, one can identify neutron-magic nuclei which act as *s*-process bottlenecks [2]. These nuclei feature very low (n,γ) cross sections and the accuracy of the TOF experiments is limited by the dominant neutron scattering background. Sec. 2 describes a new development aimed at suppressing this background and enhancing the detection sensitivity.

Experimental efforts are also focused on the measurement of unstable nuclei which act as branchings of the *s*-process and yield a local isotopic pattern which is very sensitive to the physical conditions of the stellar environment [1]. The limited sample mass available and the high background induced by the sample activity represent the major challenges to experimentally access the (n,γ) cross sections of these isotopes [2]. To overcome these limitations, new facilities with higher neutron fluxes are needed. Indeed, this was the aim of building n_TOF-EAR2 [3], that thanks to its flight path of only 20 m, became a world-leading facility in terms of instantaneous flux. Moreover, after the recent upgrade of the spallation target, an additional increase of 30-50% in the flux is expected [4]. In order to profit from the high flux of EAR2, one requires also a new generation of radiation detectors with higher granularity, as discussed in Sec. 3.

Recently, the first TOF measurement on the unstable ^{79}Se has been proposed and carried out at CERN n_TOF [5]. The branching at ^{79}Se is particularly well suited for determining the thermal conditions of the stellar environment thanks to the strong thermal dependence of its beta decay rate [1]. For this experiment, 2.7 mg of ^{79}Se were produced by means of neutron irradiation of an enriched

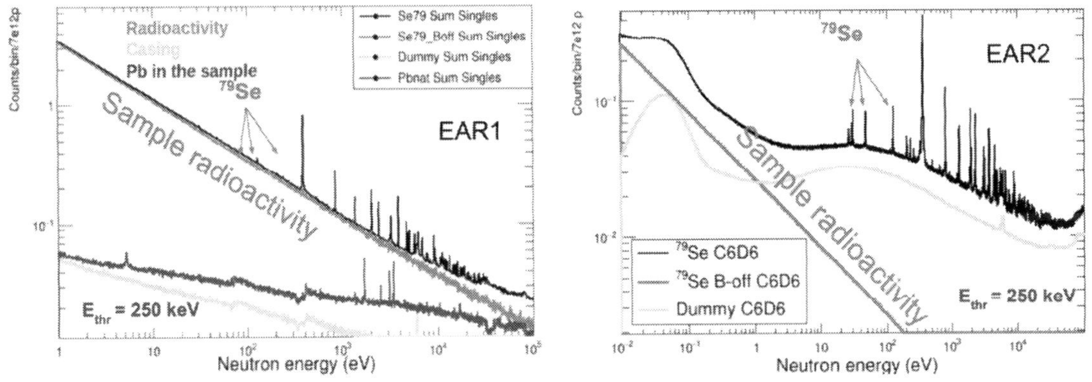

Figure 1. Total counts and background components as a function of the neutron energy measured with the PbSe (^{78}+^{79}Se) sample in EAR1 (left) and EAR2 (right). The first resonances of ^{79}Se are highlighted.

$^{208}Pb^{78}Se$ alloy-sample in the high-flux reactor at ILL [5, 7]. The relevance of having a higher instantaneous flux has been clearly observed in this measurement (see Fig. 1). While in EAR1 [6], the total count rate is dominated by the sample radioactivity, in EAR2 [3], thanks to the enhanced (n,γ)-to-activity ratio in EAR2, ^{79}Se resonances are clearly visible. Some preliminary results of this measurement are discussed in Sec. 4.

2 i-TED: suppressing n-induced background via γ-ray imaging

Neutron-capture time-of-flight (TOF) measurements at CERN n_TOF have been usually carried out with detection systems based on liquid scintillators, such as C_6D_6, which are particularly convenient because of their fast time-response and low intrinsic sensitivity to scattered neutrons [8]. However, they present limited background rejection capabilities. In particular, in TOF capture experiments on nuclei with low capture-to-scattering ratio (see Sec. 1) a large background component arises from scattered neutrons that get subsequently captured in the surroundings of the C_6D_6 detectors [9]. This background has represented the dominant contribution in previous (n,γ) experiments in the energy range of interest for nucleosynthesis studies [10].

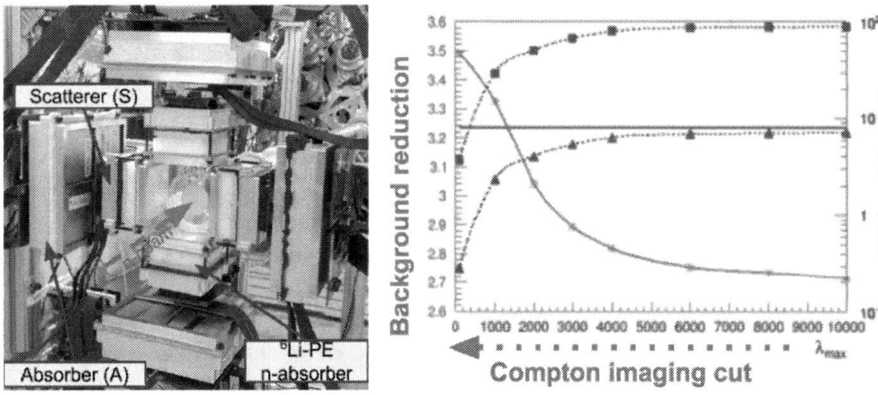

Figure 2. Left: Final i-TED array used in the recent $^{79}Se(n,\gamma)$ experiment at CERN n_TOF EAR1. Right: Background reduction factor at 10 keV and relative (n,γ) efficiency as a function of the Compton imaging cut (see text for details).

To reduce this dominant source of background, a system based on Total-Energy Detection (TED) with γ-ray imaging capability, so-called i-TED, has been recently proposed [11]. i-TED exploits the Compton imaging technique with the aim of determining the direction of the incoming γ-rays. This allows the rejection of events not originating in the sample, thereby enhancing the signal-to-background ration (SBR). This novel detection system has been fully developed and optimized in the recent years [12, 13]. The final i-TED array (see Fig. 2) consists of 4 Compton cameras comprising in total 20 $LaCl_3$ crystals, and it has been used in 2022 at n_TOF EAR1 for the aforementioned $^{79}Se(n,\gamma)$ reaction measurement.

The feasibility of the proposed background rejection method was experimentally demonstrated with an early i-TED prototype by measuring the $^{56}Fe(n,\gamma)$ reaction at CERN n_TOF-EAR1 [14]. A background reduction factor of 3.5 was achieved with respect to state-of-the-art C_6D_6 detectors (see red line in the right panel of Figure 2). The major drawback of the imaging selection is the drop of relative efficiency after applying an imaging cut, shown with blue lines. The line with squares corresponds to the efficiency relative to the situation with no imaging cuts while the line with triangles shows the efficiency in coincidence mode relative to the one of the scatter crystal operated in single mode, which has a maximum value of 8% indicated with the solid blue line. New methods based on Machine Learning (ML) algorithms have been applied to overcome this limitation in efficiency. The reader is referred to Ref. [14] for the details.

3 s-TED: segmented detection volumes for high flux facilities

The high instantaneous flux of n_TOF-EAR2, which has been further enhanced in the latest upgrade of the facility [4], induces counting rates beyond 10 MHz in the existing C_6D_6 detectors (0.6-1 L volume) [8]. Moreover, the intense γ-flash arriving to the experimental hall, leads to severe experimental difficulties [15]. Among others, counting-rate and time-of-flight dependent variations of the photo-multiplier (PMT) gain and large pile-up effects have been observed.

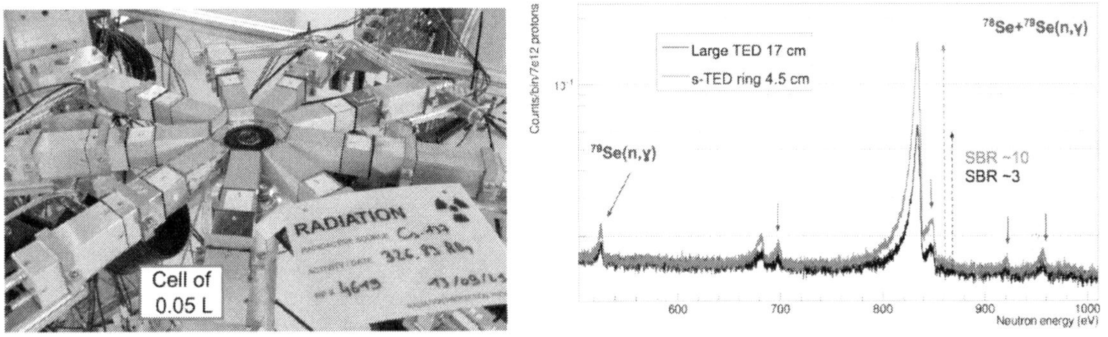

Figure 3. Left: s-TED detector cells in the innovative ring configuration used in EAR2 (right). Right:(n,γ) counting rate measured at EAR2 with the PbSe($^{78+79}$Se) sample in beam using 2 large C_6D_6 detectors (black) and 9 s-TED cells in ring configuration (red).

In order to overcome the aforementioned limitations of conventional TEDs based on relatively large C_6D_6 volumes, a segmented array of small-volume C_6D_6 detectors, so-called s-TED, has been developed [16]. Each detection cell contains only 49 ml scintillation liquid, 12-to-20 times less than previous designs. The original idea of s-TED was to replace each of the conventional C_6D_6 detectors by an array of 3×3 cells [16] in order to preserve a similar overall efficiency. To achieve an optimum efficiency and SBR for the (n,γ) measurements on the unstable ^{94}Nb [17] and ^{79}Se, the s-TED cells have been arranged in a compact-ring configuration around the capture sample shown in the left panel of Fig. 3. This innovative setup minimizes the distance (4.5 cm) to the capture sample under study, and thus enhances the sensitivity in a significant manner compared to the larger conventional C_6D_6 detectors placed at 17 cm [17], as it is shown in the right panel of Fig. 3. The larger SBR achieved with the ring of s-TED cells has been key to identify the weak ^{79}Se(n,γ) resonances measured in a sample containing only 2.7 mg of ^{79}Se embedded in 3.9 g of a eutectic lead-selenide (^{208}Pb^{78}Se) sample.

4 First results of ^{79}Se(n,γ) using new detection systems

The first (n,γ) measurement on ^{79}Se has been just carried out at CERN n_TOF using the two new detection systems described in this work. Choosing the best combination of detection system and experimental area is a key aspect for the success of challenging capture measurements on unstable targets.

Given the small amount of ^{79}Se and the high activity of the sample, n_TOF-EAR2 was the best solution to achieve good statistics and minimize the sample activity background (see Sec. 1). On the other hand, 99.7% of the Se in the sample was ^{78}Se, for which a high resolution measurement was carried out at n_TOF EAR1 [18]. The latter facility was thus better suited for a reliable assessment of the ^{78}Se contribution. Besides the radioactivity, a high background was expected from neutron scattering due to the large ^{208}Pb content (3 g) in the sample. To suppress both sources of background, the usage of the innovative i-TED array was the best solution. This system applies imaging to reduce the n-induced background, as described in Sec. 2, and thanks to the $LaCl_3$ crystals it features a very good energy resolution (5% FWHM at 662 keV), which enables to set precise energy selections to reduce the activity background. However, the maximum acquisition rate of its data acquisition system is of 500 kEvents/s which represents, at this time, a limitation for its use at EAR2. For all of the above, i-TED was chosen for the measurement at EAR1, while the measurement at EAR2 was carried out with the new s-TED detectors to fully exploit the high flux of the facility and optimize the sensitivity, as discussed in Sec. 3.

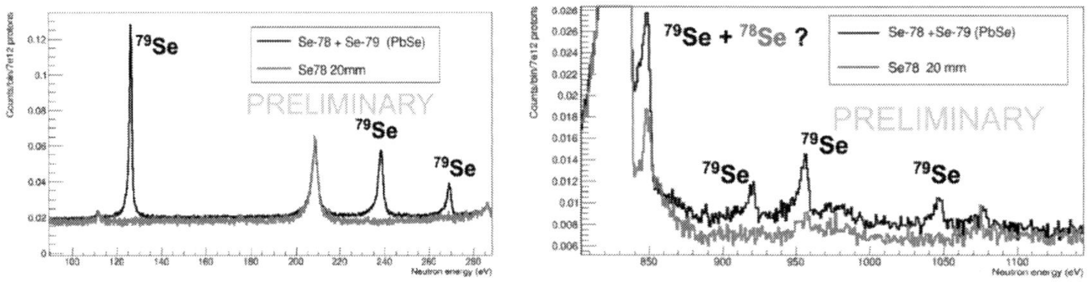

Figure 4. Preliminary (n,γ) counting rates measured at EAR2 with the PbSe($^{78+79}$Se) sample in two neutron energy ranges.

The preliminary analysis and background subtraction of the measurement carried out at EAR2 show promising results. The obtained (n,γ) counting rate on the PbSe($^{78+79}$Se) sample compared to an ancillary measurement of a pure ^{78}Se sample (see Fig. 4) indicates that between 10 and 15 capture resonances in ^{79}Se+n have been measured for the first time in the energy range from 30 eV to 2.5 keV.

5 Summary and outlook

This work has presented recent detector R&D developments carried out at CERN n_TOF, that try to solve existing limitation and shortcomings in previous approaches used for determining (n,γ) cross sections of relevance for the *s*-process. i-TED applies Compton imaging aimed at improving the signal-to-background ratio for measurements affected by large neutron-induced backgrounds, such as the neutron magic nuclei acting as *s*-process bottlenecks. New facilities with higher instantaneous neutron flux, such as n_TOF-EAR2, facilitiate the TOF capture measurements on radioactive *s*-process branching nuclei. However, exploiting the full potential of this facility requires new detectors, such as s-TED, a new array of very small-volume C_6D_6 detectors, capable of dealing with the counting rate conditions and optimizing the signal-to-background ratio.

These two novel detection systems have been used in the challenging (n,γ) measurement on the key *s*-process branching ^{79}Se. The R-Matrix analysis of the final yield, followed by a statistical analysis will allow the calculation of the semi-empirical cross section up to 300 keV, from which the MACS at different $k_B T$ can be determined. This will provide the first experimental constraint to actual spread of theoretical calculations of the ^{79}Se MACS compiled in KaDoNiS.

Acknowledgements

This work has been carried out in the framework of a project funded by the European Research Council (ERC) under the European Union's Horizon 2020 research and innovation programme (ERC Consolidator Grant project HYMNS, with grant agreement No. 681740). J. Lerendegui-Marco and J. Balibrea were supported, respectively, by grants FJC2020-044688-I and ICJ220-045122-I funded by MCIN/AEI/ 10.13039/501100011033 and by European Union NextGenerationEU/PRTR. The authors acknowledge support from the Spanish Ministerio de Ciencia e Innovación under grants PID2019-104714GB-C21, FPA2017-83946-C2-1-P, FIS2015-71688-ERC, CSIC for funding PIE-201750I26. In line with the principles that apply to scientific publishing and the CERN policy in matters of scientific publications, the n_TOF Collaboration recognises the work of V. Furman and Y. Kopatch (JINR, Russia), who have contributed to the experiment used to obtain the results described in this paper.

References

[1] F. Käppeler et al., Reviews of Modern Physics **83**, 157 (2011), 1012.5218
[2] C. Domingo-Pardo, et al., Eur. Phys. Jour. A (submitted) (2022)
[3] C. Weiß et al., Nucl. Inst. Methods A **799**, 90 (2015)
[4] J. Lerendegui-Marco. et al., *New perspectives for neutron capture measurements in the upgraded CERN-n_TOF Facility* (2022), ND2022 Conference, *EPJ Web Conf. (submitted)*
[5] J. Lerendegui-Marco et al., CERN-INTC-2020-065; INTC-P-580; http://cds.cern.ch/record/2731962 (2021)
[6] C. Guerrero, et al., Eur. Phys. Jour. A **49**, 27 (2013)
[7] N.M. Chiera et al., Nucl. Inst. Methods A **1029**, 166443 (2022)
[8] R. Plag et al., Nucl. Inst. Methods A **496**, 425 (2003)
[9] P. Žugec et al., Nucl. Inst. Methods A **760**, 57 (2014)
[10] G. Tagliente et al. , Phys. Rev. C **87**, 014622 (2013)
[11] C. Domingo-Pardo, Nucl. Inst. Methods A **825**, 78 (2016)
[12] V. Babiano et al., Nucl. Inst. Methods A **953**, 163228 (2020), 1908.08533
[13] J. Balibrea-Correa et al., Nucl. Inst. Methods A **1001**, 165249 (2021)
[14] V. Babiano-Suárez, J. Lerendegui-Marco et al., Eur. Phys. Jour. A **57**, 197 (2021), 2012.10374
[15] V. Alcayne et al., *Results of the ^{244}Cm, ^{246}Cm and ^{248}Cm neutron-induced capture cross sections measurements at EAR1 and EAR2 of the n_TOF facility* (2022), ND2022 Conference, *EPJ Web Conf. (submitted)*
[16] V. Alcayne et al., *A segmented total energy detector (sTED) for (n,γ) cross section measurements at n_TOF EAR2* (2022), ND2022 Conference, *EPJ Web Conf. (submitted)*
[17] J. Balibrea-Correa, et al., *Fist measurement of the 94Nb neutron capture cross-section at the CERN n_TOF Facility* (2022), NPA-X Conference, *EPJ Web Conf.*
[18] C. Lederer et al., CERN-INTC-2017-038 ; INTC-P-509; http://cds.cern.ch/record/2265570 (2017)

Development of a jet gas target system for the Felsenkeller underground accelerator

Anup Yadav[1,2,*] *Konrad* Schmidt[1,**] *Daniel* Bemmerer[1,2]

[1]Helmholz-Zentrum Dresden-Rossendorf, Institute of Radiation Physics, 01328 Dresden, Germany
[2]Technische Universität Dresden, 01069 Dresden, Germany

Abstract. For direct cross section measurements in nuclear astrophysics, in addition to suitable ion beams and detectors, also highly pure and stable targets are needed. Here, using a gas jet as a target offers an attractive approach that combines high stability even under significant beam load with excellent purity and high localisation. Such a target is currently under construction at the Felsenkeller underground ion accelerator lab for nuclear astrophysics in Dresden, Germany. The target thickness will be measured by optical interferometry, allowing an in-situ thickness determination including also beam-induced effects. The contribution reports on the status of this new system and outlines possible applications in nuclear astrophysics.

1 Introduction

The Felsenkeller underground laboratory is suitable to study, amongst others, two important astrophysical reactions of interest: ^{3}He$(\alpha,\gamma)^{7}$Be and ^{12}C$(\alpha,\gamma)^{16}$O. The first reaction controls the hydrogen burning in the Sun and also affects lithium production in Big Bang Nucleosynthesis (BBN). In order to measure the astrophysical S-factor and angular distribution of this reaction, we are planning to use a ^{3}He gas jet target for comprehensive data coverage of the entire BBN range.

The ^{12}C$(\alpha,\gamma)^{16}$O reaction affects the abundance ratio of the two elements carbon and oxygen. Due to the low cross-section in the Gamow window, significant statistics for this reaction will be achieved with ^{12}C beam incident on a static windowless ^{4}He gas target, and a 4π detector.

To measure reactions of astrophysical interest targets are needed that are locally well confined, highly dense, thermally stable, and chemically pure. Measurements of the angular distribution require a highly localized target for a high-resolution of the reaction energy and the emission angle. A solid target often contains other elements through contaminations and requires backing materials like tantalum for mechanical stability, which may increase the beam induced background and energy straggling in case of implantation. Solid targets can also degrade over time or exposure to a beam resulting in issues like overheating, and blistering.

A wall-jet gas-target is an alternative solution, which can provide a high gas density, high localization, chemically purity, and a thermal stability [1, 2, 4, 5].

*e-mail: a.yadav@hzdr.de
**e-mail: konrad.schmidt@hzdr.de

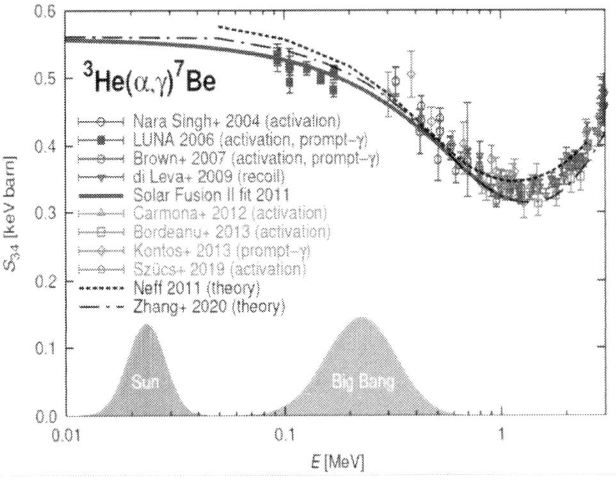

Figure 1. Astrophysical S-factor of the $^3He(\alpha, \gamma)^7Be$ reaction as a function of energy. The Felsenkeller gas-jet will help to perform comprehensive measurements of the reaction covering the entire BBN range.

2 Experimental setup

2.1 Differential pumping system

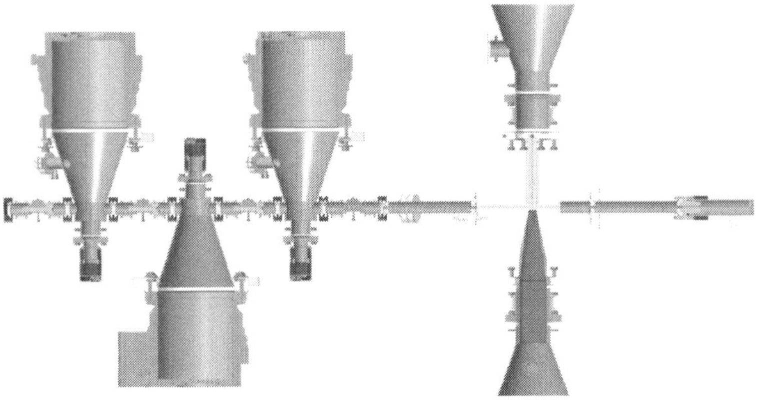

Figure 2. Technical drawing of differential pumping stages. The beam enters at the left and passes through apertures in between the pumping stages, the jet chamber and the static on the most right where it is stopped in a beam calorimeter. The pumping stages are evacuated by three 700 l/s and the jet chamber by a 2300 l/s turbomolecular pump.

A recirculating gas-jet target (Fig. 2) creates a supersonic jet based on the JENSA gas jet target [1]. Here, the target density is 10 times less than at JENSA due to the different physics case. The exhaust of the nozzle has a cross-sectional area at the neck of less than 1 mm. About 12 mm below the nozzle a cone-shaped gas catcher receives most of the gas jet. Catcher is connected to a cone-shaped chamber, and this chamber is evacuated by a 2300 liter turbomolecular pump. The exhaust of all the pumps are connected to a multi-stage root pump and this pump will then compress the gas back to the initial inlet pressure. Before

Table 1. Aimed characteristics of the gas wall jet

Parameter	Aim	Remark
Areal density	10^{18} atoms/cm^2	Single resonance measurement
Thickness	<1 mm	Reaction localization for angular distribution measurements
Width	10 mm	Beam (5 mm diameter) crosses jet within homogeneous center
Inlet pressure	~1 bar	Needed to get a supersonic jet
Mass flow	1.4 bar l/s	Result of bottle neck cross section of ~1 mm^2

return back to the nozzle, the gas is also passing a chemical getter that will clean it from residual contamination that might have enter during the circulation.

Part of the gas jet diffusing in the jet chamber and return to the beam line. This residual gas is evacuated by a 2300 liter turbomolecular pump on the top of the jet chamber and three 700 liter turbomolecular pumps along the beam line, all separated by copper apertures. These collimators reduce the pressure down to 10^{-7} mbar, that is necessary for the ion accelerator. Further, the apertures also reduce the beam size to a diameter of 5 mm or less, in order to provide a well defined beam spot. Downstream of the jet, a extended static gas chamber with a beam calorimeter (see the section 2.3) is installed to determine the ion beam intensity.

In order to monitor the system, all the pumps, pressure gauges, and valves are controlled by a LabVIEW system and parameters of the system are logged. The gas system can be operated either in jet mode or in static mode.

2.2 Wall jet gas target

For measurements of low cross-section reactions, a high density of target nucleui within the confined region is needed to maximize count rate. On the other hand, a limited target thickness reduces reaction product energy loss and straggling. The size of the jet should be comparable to the full beam spot so the beam crosses the jet within the homogeneous center. The inlet pressure should be sufficiently high to get a supersonic jet. The characteristics of the gas wall jet that are aimed for, are summarized in Table 1.

The nozzle is placed vertically with the jet flowing downward with two inlet connections. The shape of the nozzle is a converging-diverging, de Laval type of nozzle [6]. The distance between the nozzle and catcher is around 12 mm as suggested by the literature [4, 8] and by the beam size. This value can be changed for future measurements.

2.3 Static gas target with beam calorimeter

Downstream of the gas wall jet, a windowless, extended gas target of static type (Fig. 3) is installed that contains the beam calorimeter. The concept is based on the LUNA gas target[2]. In static mode, the gas enters the target chamber from the right through the pipes shown in grey color. The gas flow is regulated by two parallel metered valves (EVR 116) to keep a constant pressure in the chamber. The expected target pressure is around 2.0 mbar.

A collimator between the jet target chamber and the entrance of extended target limits the gas flow in the direction of the jet chamber. Any remaining target gas is removed by the jet chamber pumps and three differential pumping stages (see section 2.1).

It is not possible to use a Faraday cup to measure the beam intensity while the target chamber is filled with gas because of secondary electrons, therefore a different technique in form of a power compensation calorimeter is used. The calorimeter is made from copper and consists of hot side with heating resistor (12Ω, 60W) inside the copper structure and cold side.

The hot and cold sides are always kept at a constant temperature with the help of temperature regulated heating resistors at the hot side and circulating low-temperature coolant from a high precision temperature stability chiller (Julabo 600F) at the cold side. When the ion beam hits the calorimeter, the resistor will reduce the heating to maintain a constant temperature measured by a PT100 sensors. The reduction of the power consumption of the resistor, in turn, enables the determination of the beam intensity at known beam energy.

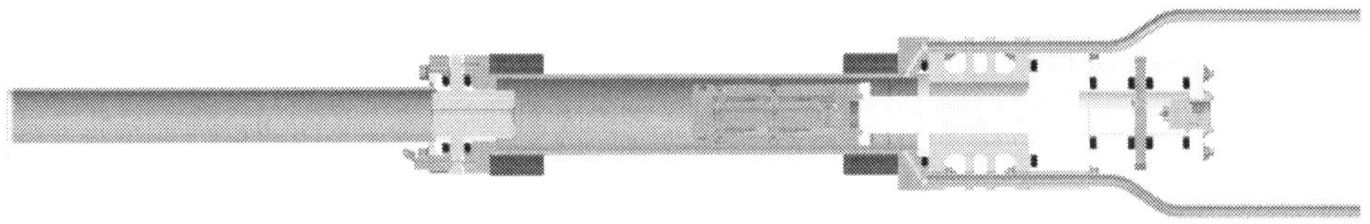

Figure 3. Extended gas target chamber and a top view of beam calorimeter are shown. Two grey pipes at the right side of the figure, to inlet the gas inside the chamber.

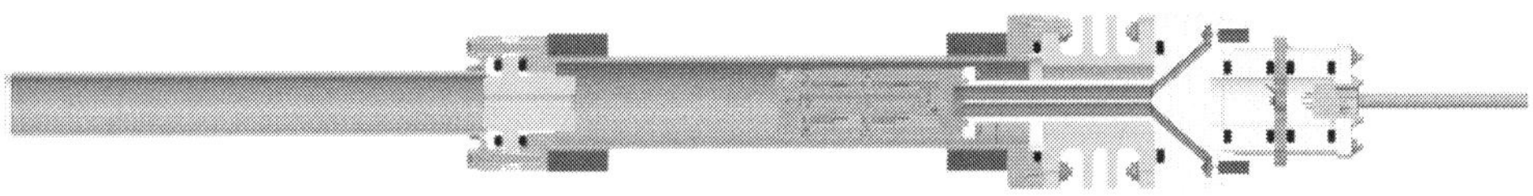

Figure 4. Extended target chamber with the side view of beam calorimeter are shown. Two blue pipes at the right side of the figure are for water cooling of calorimeter.

3 Measurement

3.1 Jet density measurement

For a continuous gas jet density measurement, a laser interferometry technique [3] is used, where the presence of gas induces an optical path length difference between the signal arm (passing through the jet) and the reference arm resulting in a phase shift on the interferogram. The optical path difference depends on the density distribution of the gas jet and its refractive index. The laser beam is expanded to illuminate the entire gas jet with the help of lenses to see the shape of the jet.

With the use of laser interferometry, the relative density and the shape of a jet are measured. The absolute density is determined by the energy loss method, where the energy loss of alpha particles traversing the jet are combined with tabulated experimental stopping powers at different points of the jet. For this absolute thickness measurement, a silicon detector will be mounted at the end of the beam line at one side of the jet and at the other side a triple alpha source (^{239}Pu, ^{241}Am, ^{244}Cm) will be mounted. In this way, the energy of the alpha particles are measured with and without the jet and the difference in energy gives the energy loss of the alpha particles in the jet and hence the absolute density.

3.2 Pressure profile measurement

For first jet tests, a needle valves is placed at the nozzle inlet to vary the gas flow rate. To monitor the pressure at different regions, pressure gauges are placed on all the pumping stages,

Figure 5. Technical drawing of the gas jet interferometry setup. the laser and beam are showing in the red and blue colour respectively. The laser is passing through the jet at 15° with respect to the beam axis inside a jet chamber that is not shown here for the clarity.

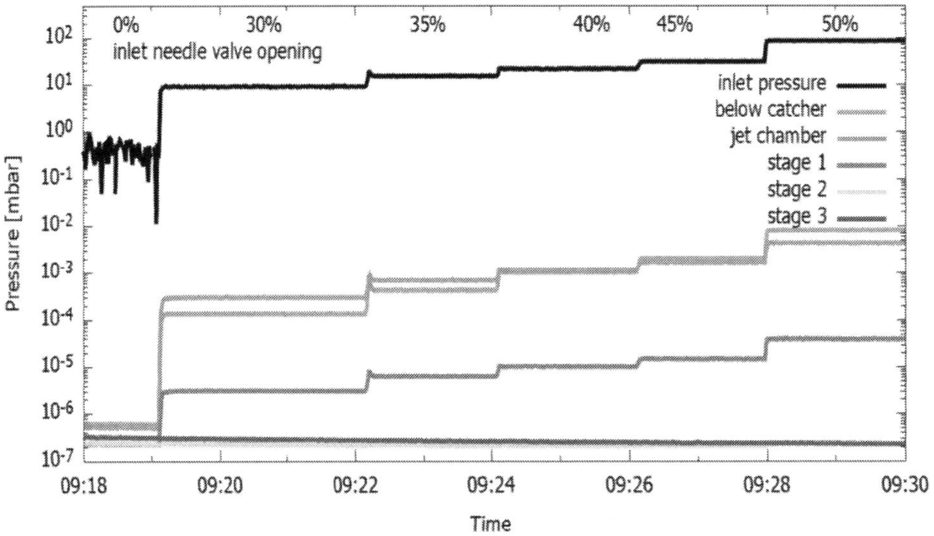

Figure 6. Pressure profile measurements inside the chamber and beam line at different needle valve opening connected to the nozzle.

the lower and upper cones of the expansion chamber, the nozzle inlet, and at the exhaust of all pumps. The pressure profile across the gas target setup at different needle valve opening for a first jet test with a preliminary nozzle are shown in Fig. 5. The pressure of the system without an inlet gas is in the order of 10^{-7} mbar in the beam line and expansion chamber. The pressures at the second and third pumping stages are not changing for different openings of needle valves.

At an opening of 30%, the pressure in the jet chamber is higher than below the catcher, indicating that there is more gas diffusing inside the chamber than going directly in the catcher. This is caused by the too small opening of the needle valve that is reducing the amount of gas to an inlet pressure of ~10 mbar. This low inlet pressure is not able to accelerate the gas to supersonic velocity and hence not creating a jet. This behaviour changes at an opening of 50% when the inlet pressure increases to ~100 mbar and most of the gas streams as a jet inside the gas catcher where about twice as much gas flows compared to the gas chamber. At an opening of 100% (not yet tested) the inlet pressure will reach 1 bar and create a supersonic jet into the catcher, with less then 1% of the gas drifting into the jet chamber.

4 Outlook

The combined gas target has been built at the HZDR Rossendorf campus and will be transferred to the Felsenkeller underground laboratory in 2023. First jet tests have been performed to start a series of jet characterizations regarding pressure, density, and thickness. These are continuously measured by an interferometry setup that is in turn calibrated by an energy loss measurement of alpha particles across the jet.

There is further development of enhanced nozzles. Computational Fluid Dynamic (CFD) simulations in ANSYS Fluent are currently being carried out to optimize the shape of those upcoming nozzles.

Finally, in-beam tests are foreseen, as soon as the target system is attached to the beam line of the 5-MV Pelletron accelerator. This combination of a wall-jet and a static gas target will be a unique in nuclear astrophysics and might become a role model for future gas target developments.

5 Acknowledgement

Financial support by the European Union (ChETEC-INFRA, 101008324) is gratefully acknowledged.

References

[1] K. Schmidt, *et al.*, Nucl. Instrum. Methods Phys. Res. A **911** (2018) 1-9.
[2] F. Ferraro, *et al.*, Eur. Phys. J. A (2018) **54:** 44.
[3] J. P. Couperus, *et al.*, Nucl. Instrum. Methods Phys. Res. A **830** (2016) 504-509.
[4] K. A. Chipps, *et al.*, Nucl. Instrum. Methods Phys. Res. A **763** (2014) 533
[5] F. Favela, *et al.*, Phys. Rev. ST Accel. Beams **18** (2015) 123502.
[6] G.P. de Laval, U.S. Patent 522066, 1894.
[7] A. Kontos, *et al.*, Nucl. Instrum. Methods Phys. Res. A **664** (2012) 272.
[8] D. Rapagnani, *et al.*, Nucl. Instrum. Methods Phys. Res. B **407** (2017) 217.
[9] Z. Meisel, *et al.*, Nucl. Instrum. Methods Phys. Res. A **828** (2016) 8-14.

Background gamma measurement for low energy astrophysical reaction at FRENA

Sukhendu Saha[1,2,*], *Tanmoy* Bar[1,2], and *Chinmay* Basu[1]

[1]Saha Institute of Nuclear Physics, 1/AF Bidhannagar, Kolkata 700064, INDIA
[2]Homi Bhabha National Institute, Anushaktinagar, Mumbai 400094, INDIA

Abstract. The majority of astrophysical reactions have very low cross-sections (nanobarn to picobarn range) and significant error bars [1, 2]. Hence γ-ray background studies up to 3 MeV have been performed inside the Facility for Research in Experimental Nuclear Astrophysics (FRENA), Saha Institute of Nuclear Physics, Kolkata, India. A meticulous background calculation has been performed at different positions inside the accelerator hall and beam hall using two NaI(Tl) detectors [3]. An XPS measurement has been carried out to identify elements present in the FRENA wall concrete. Activity calculations done in [4] shows that neutrons produced by various experiments may combine with any nearby elements and form various radioactive isotopes with long half-lives, which may produce a γ-ray background .

1 Introduction

Gamma-ray measurement especially those dedicated to the study of astrophysical reactions, require a close attention to measuring and minimizing the radiation background. Most astrophysical reactions occur far below the coulomb barrier, which results in very low cross-sections (order of nanobarn to picobarn) with significant error bars [1, 2]. Therefore, the precious cross-section measurements from γ-spectroscopy greatly depend on the γ-ray background. A detailed background measurement will also help to minimise the uncertainty in the observed data.

Background γ-rays having energy less than 3 MeV, comes mainly from naturally present radioactive isotopes, energy higher than 3MeV are mainly due to the cosmic ray interaction at the upper atmosphere. In astrophysical studies low energy γ-rays play major role. An active shielding of higher energy γ-ray spectrum has been done by sophisticated techniques in data acquisition and detector systems. For the study of higher energy γ-ray spectrum and the rejection of the backgrounds, underground labs are preferable [5].

Facility for Research in Experimental Nuclear Astrophysics (FRENA) is an upcoming tandem accelerator located at Saha Institute of Nuclear Physics (SINP), Kolkata, India. FRENA is a high current and low energy accelerator, which is primarily designed for nuclear astrophysics studies. The accelerator can operate with the terminal voltage of 0.2 to 3 MV with the standard voltage resolution of 3×10^{-5} (ΔE/E) and the typical DC current at 3 MV terminal voltage for the H^+ ion 300 μA, for the He^{2+} ion is 100 μA and for the heavier ions between 20-60 μA [6].

*e-mail: sukhendu.saha@saha.ac.in

In this work, background measurements of the γ-ray spectrum below 3 MeV, most relevant to nuclear astrophysics study, have been performed. It primarily originate from radioactive naturally occurring isotopes. A details counts per seconds(CPS) measurement have been performed. Two NaI detectors have been used for the background measurements due to their higher efficiency [3]. An XPS analysis of the indoor wall concrete has been conducted for elemental search.

2 Sources of γ-ray background

The earth's crust is composed of a variety of elements and isotopes with different abundances [7]. These naturally occurring isotopes continuously produce lower energy γ-ray. This represents the background for any laboratory performing lower energy γ-ray spectroscopy. These γ-ray spectrum need to be measured precisely and to reduce background effects, detectors need to be passive shielding with high Z materials (Pb, Z=82).

The elements potassium, lead, thorium, and uranium are abundantly found in nature. Natural potassium has 0.012% of ^{40}K, which emits γ-rays with an energy of 1.460 MeV and a half-life of 1.248×10^{9} years. ^{204}Pb decays through a long chain of gammas and alpha particles, effectively producing γ-ray with energies of 899.15 keV and 911.74 keV with relative intensities of 99.17% and 91.5%. ^{227,232}Th and ^{235,238}U have a long gamma chain with a wide energy range, also Gd, Dy, Ta, and Tl have a large number of isotopes with larger abundances and emit γ-rays.

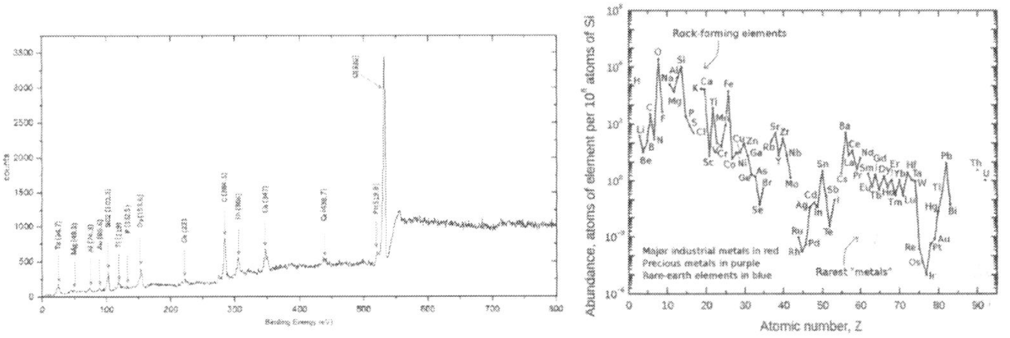

Figure 1. (Left) XPS analysis of indoor concrete powder confirms the presence of C, O, Si, Ca and Mg with the other elements. (Right) Chemical element abundance (fraction of atoms w.r.t 10^{6} Si atoms) in the Earth's upper crust as a function of atomic number [7]

.

The existence of C, O, Si, Ca, Mg, Al, Ta, P, and Ce in the concrete powder of indoor blocks has been verified by comparing experimental X-ray photoelectron spectroscopy(XPS) data with NIST database (Figure 1). The primary components of concrete are C, O, Si, Ca, H, and Mg and K, Ca, Si, Mg and O are the rock forming elements (Figure 2).

At FRENA, the gamma background has been measured under beam-off conditions. While the accelerator is operating, the neutron emitted from an experiment may interact with nearby stable elements (elements of concrete and machine parts) and may produce a long-lived activity. Which may be a source of gamma background for the experiments being performed close-by [4].

3 Experimental details

3.1 Venues of the measurement

The FRENA building has 1.2 m thick walls, which effectively isolate the indoor and outdoor environments from radiations. However, naturally occurring radioactive isotopes (^{40}K, ^{211}Pb, ^{214}Bi and so on) in the wall concrete itself contribute to the gamma-ray background in the experimental measurements performed in the FRENA building. The same concrete wall divides the beam hall from the accelerator hall. The roof is made of 60 cm thick concrete to reduce cosmic background radiation [5].

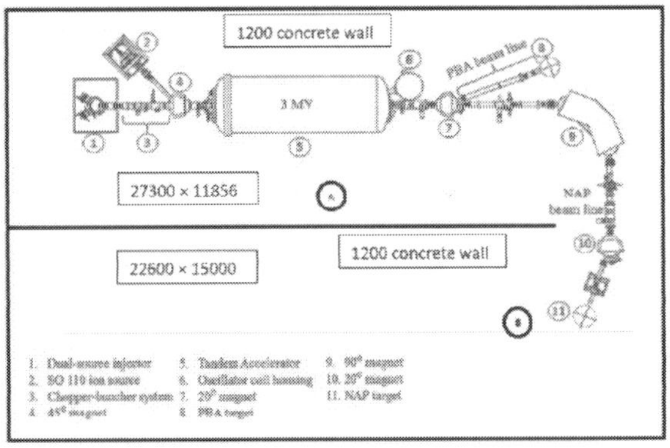

Figure 2. Schematic sketch of FRENA indoor buildings and the machine parts. spot A and spot B are the positions of the detectors at the Accelerator hall and the Beam hall for the background γ-ray measurement. Dimensions are in mm.

The background measurements has been performed at positions A and B, 132 cm and 265 cm from the concrete wall, respectively, in the accelerator hall and the beam hall (Figure 3).

3.2 Detector, Power supplies and data acquisition system

Two 1.5"×1.5" (diameter×length) NaI(Tl) crystal scintillator detectors have been used. Both have glass photomultiplier tubes. Detectors numbered detector 1 and detector 2. A specific power supply has been used for each detector.

For detector 1, the SENS-TECH PM30D power supply has been used. The detector bias was given 2100 V and the leakage current was 60 μA. For detector 2, Canberra 3002D power supply has been used. The bias supplied to the detector was 2000 V and the leakage current was 1.36 mA. Both the detectors have been placed at both positions alternatively. ^{60}Co was used as a known gamma source for the calibration of the detectors.

CAEN-DT5730 is an 8-channel 14-bit @ 500 MS/s digitizer used for the data acquisition system. CoMPASS multiparametric DAQ software was used to run the digitizer with DPP-PHA firmware.

3.3 Experimental setup

At positions A and B, detectors 1 and 2 have been placed (Figure 3). After a long run with bare detectors calibration has been performed by ^{60}Co (γ-ray energies, 1173 keV and 1332 keV). The different channels of the digitizer have concurrently been connected to detectors 1 and detector 2. The data were stored in a computer by CoMPASS multiparametric DAQ software. The background was measured in both spots by switching between the two detectors with their fixed power sources.

The detectors were shielded with lead bricks (3"×2"×1") after a long run with bare detectors to compare counts per second. In addition to these two locations, background measurements have also been made in the control room, detector lab, and close to the entrance gate of the FRENA building.

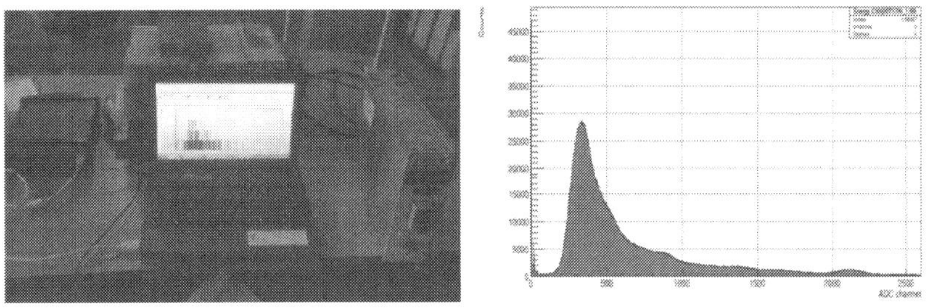

Figure 3. (Left) Experimental setup when Detector was shielded by lead bricks. (Right) CoMPASS DAQ software interface during background measurement at beam hall by detector 2.

4 Results and Discussions

The background γ-ray spectrum at various energy ranges has been identified after a run of more than 7 days for background measurement. The notable peaks at the spectrum were mainly coming from isotopes of thorium, bismuth, lead, and potassium. From the calculation, ^{40}K has a background effect of roughly 72 counts per hour at 1.460 MeV of energy [8].

The possible isotopes from which the contribution to the background is significant are listed in Table 1. The energy range defined in Table 1 is determined by considering peak position and its energy uncertainty from Full-width Half Maximum (FWHM) fitting. The value is calculated for each peak energy range as peak value$\pm$(FWHM/2). The range is given only for data collected from measurements done in the beam hall, and during the study the Compton backscatter peak was ignored. Possible isotopes and their strongest γ-rays with their relative intensities have been compared to the JAERI-Data/Code 98-008 [8]. A count drop by a factor of 100 was observed for detectors shielded with Pb bricks(3"$\times$2"$\times$1") with respect to bare detectors. It confirms that gamma-ray background is coming from external elements, not due to any internal activity of the detectors.

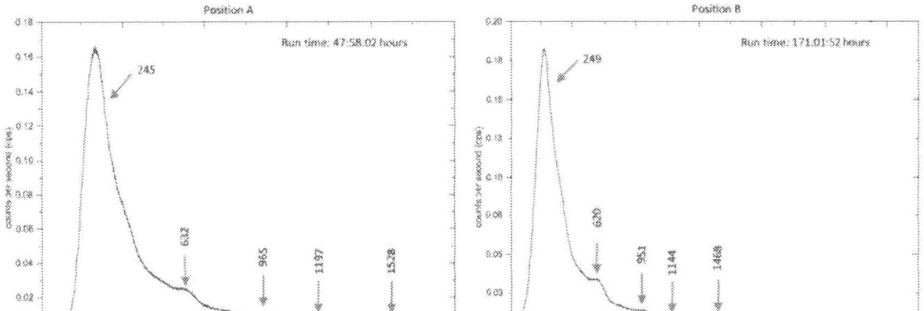

Figure 4. (Left) The background γ-ray spectrum at the position A in the accelerator hall. NaI(Tl) detector has been placed at 132 cm from the FRENA concrete wall. (Right) The background γ-ray spectrum at the position B in the beam hall. NaI(Tl) detector has been placed at 265 cm from the FRENA concrete wall, where the other detectors supposed to be placed during the beam operating experiments.

The background measurement was done while the machine was off. Now, neutrons produced during experiments can interact with ^{181}Ta and form ^{182}Ta, which has a half-life of

Table 1. The contribution of background γ-ray energy spectrum from the possible isotopes of sufficient relative γ intensity(%).

Energy range (keV)	Possible isotopes with γ-ray energy(keV) having large γ intensity(%)
405 - 795	^{208}Tl(510.77, 8.13%), ^{210}Tl(670, 4.2×10^{-4}%), ^{211}Pb(427.09, 1.76%), (704.64, 0.462%), ^{214}Pb(487.09, 0.422%), ^{212}Bi(727.33, 6.58%), ^{214}Bi(609.31, 46.1%), (768.36, 4.94%), ^{227}Th(432.33, 4.6×10^{-3}%), ^{230}Th(620, 8×10^{-7}%), ^{231}Pa(410.3, 3.2×10^{-3}%), ^{234}Pa(416.1, 5.8×10^{-5}%), ^{223}Fr(493.1, 3.7×10^{-5}%), ^{224}Ra(645.5, 5.4×10^{-3}%), ^{235}U(433, 4×10^{-3}%), ^{223}Ra(445.03, 1.27%), ^{228}Ac(463, 4.4%), ^{212}Pb(415.2, 0.143%), ^{215}Po(438.8, 0.04%), ^{222}Rn(510, 0.076%), ^{220}Rn(549.73, 0.114%)
747 - 1154	^{212}Bi(785.37, 1.1%), ^{214}Bi(768.3, 4.94%),(934.06, 3.03%), ^{227}Th(812.2, 2.5×10^{-3}%), ^{211}Pb(832.01, 3.52%), ^{214}Pb(785.96, 1.07%), ^{223}Fr(908.2, 2.2×10^{-4}%), ^{234}Pa(821.18, 3×10^{-3}%), ^{208}Tl(860.56, 4.47%), ^{234}Pa(821.18, 3×10^{-3}%), ^{208}Tl(1165, 0.1%), ^{228}Ac(911.2, 25.8%), (968.97, 15.8%), (794.95, 4.25%), ^{235}U(794.7, 6×10^{-4}%), ^{214}Po(799.7, 0.01%), ^{238}La(788.74, 33.6%),
1094 - 1194	^{214}Bi(1120.6, 15.1%),(1155.19, 1.63%), ^{211}Pb(1109.48, 0.115%), ^{234}Pa(1126.8, 4.8×10^{-4}%), ^{208}Tl(1185.2, 6.1×10^{-3}%), ^{210}Tl(1110, 1.5×10^{-3}%), ^{206}Tl(1165, 0.1%), ^{228}Ac(1110.61, 0.304%)
1452 - 1484	^{228}Ac(1459.14, 0.83%), ^{40}K(1460.83, 10.67%), ^{214}Bi(1479.15, 0.051%)
2000 - 3000	^{214}Bi(2204.21, 5.08%), ^{208}Tl(2614.53, 35.64%)

114.74 days and produces several cascaded gamma rays having energy below 2 MeV with a sufficiently high relative intensity. These rays can serve as a γ-ray background for the next performing experiment [4].

5 Acknowledgements

The authors would like to thank Prof. Chandi Charan Dey and Prof. Mala Das, Applied Nuclear Physics division, SINP for their kind help and cooperation during the measurement. The author would also like to thank Prof. Satyajit Hazra and Mr. Gautam Sarkar of Surface Physics and Material Science division at SINP for providing XPS facility. Sukhendu Saha receives his fellowship grant from Council of Scientific & Industrial Research (CSIR), Government of India as Senior Research Fellow (SRF) with the fellowship Grant No. 09/489(0119)/2019 EMR-I

References

[1] A. Caciolli et al., Eur. Phys. J. A **39** (2009) 179.
[2] Claus E. Rolfs, *Cauldrons in the Cosmos*, University of Chicago Press, Chicago, 1988.
[3] Glenn F. Knoll, *Radiation Detection and Measurement* (Wiley, USA, 2000) 335-341
[4] T. Bar et al., EPJ Web of Conferences **275**, 02015 (2023).
[5] Sharma, S., JINST **15**, Issue 09, pp. T09003 (2020).
[6] http://www.saha.ac.in/web/frena-accelerator
[7] Anderson, Don L, *Theory of the Earth* (Wiley–Blackwell, Oxford, 1989) 147–175.
[8] https://inis.iaea.org/collection/NCLCollectionStore/_Public/29/033/29033666.pdf?r=1
[9] R. Brittonet al., J Radioanal Nucl Chem **298** (2013) 1491.

^{40}Ar proposed as probe of neutron-induced reactions in a high-density stellar-like plasma at the National Ignition Facility

M. Paul[1,*], *R. N.* Sahoo[1], *M.* Tessler[2], *J.* Jeet[3], *C.* Velsko[3], *A.* Zylstra[3], *M.* Avila[4], *C.* Dickerson[4], *C.* Fougères[4], *H.* Jayatissa[4], *R. C.* Pardo[4], *K. E.* Rehm[4], *R.* Scott[4], *I.* Tolstukhin[4], *R.* Vondrasek[4], *T.* Bailey[5], *L.* Callahan[5], *A. M.* Clark[5], *P.* Collon[5], *Y.* Kashiv[5], *A.* Nelson[5], *U.* Köster[6], *H. F. R.* Hoffmann[7], *M.* Pichotta[7], *K.* Zuber[7], *T.* Döring[8], and *R.* Schwengner[8]

[1]Hebrew University of Jerusalem, Israel 91904
[2]Soreq Nuclear Research Center, Yavne Israel
[3]Lawrence Livermore National Laboratory, Livermore CA USA
[4]Argonne National Laboratory, Argonne IL USA
[5]University of Notre Dame, Notre Dame IN USA
[6]Institut Laue-Langevin, Grenoble France
[7]Technical University Dresden, Dresden Germany
[8]Helmholtz Zentrum Dresden Rossendorf, Dresden Germany

Abstract. The thermodynamical conditions and the neutron density produced in a laser-induced implosion of a deuterium-tritium (DT) filled capsule at the National Ignition Facility (NIF) are the closest laboratory analog of stellar conditions. We plan to investigate neutron-induced reactions on ^{40}Ar, namely the ^{40}Ar$(n, 2n)^{39}$Ar($t_{1/2}$ =268 y), the ^{40}Ar$(n, \gamma)^{41}$Ar(110 min) and the potential rapid two-neutron capture reaction ^{40}Ar$(2n, \gamma)^{42}$Ar(33 y) in an Ar-loaded DT capsule. The chemical inertness of noble gas Ar enables reliable collection of the reaction products.

1 Introduction

Reactions of astrophysical interest in the *s*-process regime have been extensively studied in the laboratory [1] but *r*-process rapid neutron captures occurring in explosive nucleosynthesis at very high neutron densities (in the range 10^{22}–10^{25} neutrons/cm^3) [2] have been beyond the realm of terrestrial experiments. Deuterium-tritium (DT) inertial confinement fusion implosions at laser facilities like the National Ignition Facility (NIF) are a unique environment for approaching the conditions of explosive nucleosynthesis in the laboratory. High-power laser beams, directed to a DT-filled capsule at the center of a large chamber, compress and heat the DT fuel to thermodynamical conditions (density, temperature, and pressure) comparable to or exceeding those in the center of stars. Recent experiments at NIF first passed the burning-plasma threshold [3–6], where self-heating exceeds the external heating applied to the fuel, and then produced record fusion yields of > 1 MJ [6], see also [7] for a review article. Neutrons are produced by the DT thermonuclear process in a volume with a radius of

*e-mail: paul@vms.huji.ac.il

$\approx 50\ \mu$m within ≈ 100 ps, representing neutron fluxes of 10^{31} cm^{-2}s^{-1} and a neutron density approaching 10^{22} cm^{-3}, for the first time close to those of the astrophysical r process [2] in a laboratory setting. Although DT neutrons are produced at ≈ 14 MeV (unlike stellar neutrons in the energy range of 10–100 keV), they are partly energy-downgraded by scattering in the high-density environment (> 1000 g/cm^3) of the NIF implosion. The study of neutron capture reactions, which have a strong energy-dependence, is expected to reveal the extent of the neutron thermalization in the medium and provide also information on areal density, a critical implosion parameter.

A laser shot and experiment dedicated to these studies was approved as part of the Discovery Science program at NIF and was performed in October 2022 (not yet analyzed). Our experiment is aimed specifically at the study of neutron-induced reactions on ^{40}Ar seeds incorporated in the DT-filled capsule. Noble gas argon was selected for its chemical inertness ensuring reliable collection of its gaseous isotopic reaction products after dispersion of the capsule debris. The ^{40}Ar$(n, 2n)^{39}$Ar$(t_{1/2} = 268$ y), the ^{40}Ar$(n, \gamma)^{41}$Ar(110 min) and ^{40}Ar$(2n, \gamma)^{42}$Ar(33 y), a potential rapid two-neutron capture reaction, will be investigated. Long-lived ^{39,42}Ar products can be detected at ultra-high sensitivity by the Noble Gas Accelerator Mass Spectrometry (NOGAMS) technique (see [8] for details) and ^{41}Ar by γ spectrometry.

2 Capsule and laser shot design and collection of argon reaction products

For our NIF experiment, the argon dopant is loaded inside the DT-filled capsule, imploding in a so-called indirect drive and shown schematically in Fig. 1. The NIF lasers (1.8 MJ, 400 TW) irradiate the inside of a small high-Z cell, termed ‘hohlraum’, cantilevered at the center of the large NIF vacuum chamber (10 m diameter). The hohlraum has 3.45 mm diameter apertures on either end through which the laser beams enter. The laser energy is converted into a thermal X-ray bath inside the hohlraum, which drives a high-density-carbon (HDC) capsule located at the center. Ablation of the HDC compresses by inertial confinement the DT fuel to fusion conditions. The fuel is equimolar DT gas at 4mg/cc with 0.1% atomic Ar dopant. As the implosion stagnates and fusion conditions are reached, the D and T react producing 14 MeV neutrons, which irradiate the Ar.

After the capsule implosion, the remaining target capsule explodes outward, releasing residual target gases (D, T, H atoms or molecules, He and the gaseous reaction products from the Ar component) into the NIF target chamber. To help the collection of the reaction products, a 15-30 cm^3 STP stable isotope carrier containing enriched ^{38}Ar and ^{nat}Ar (^{38}Ar/Ar$\approx 1/3$) is released into the NIF chamber 10 seconds after the shot time. The ^{38}Ar component is needed for differentiation from Ar present in the residual gas of the NIF chamber. The chemical inertness of noble gas argon enables reliable collection of the Ar isotopic products and the Ar carrier gas. The Radiochemical Analysis of Gaseous Species (RAGS) diagnostic system at NIF [9] collects gas from the target chamber for up to 15 minutes after shot execution. Based on the ^{39}Ar abundance sensitivity of the NOGAMS technique (^{39}Ar/Ar $\lesssim 10^{-15}$) [8, 10], the experiment is able to detect Ar reaction products down to the range of 10^4-10^5 atoms.

3 Detection of ^{39}Ar from the ^{40}Ar$(n, 2n)^{39}$Ar reaction

In order to establish conditions of the NOGAMS analysis of ^{39}Ar produced in the NIF shot, the following samples were prepared in separate *ad hoc* irradiations and analyzed. They will be used as calibration samples in our NIF-shot experiment.

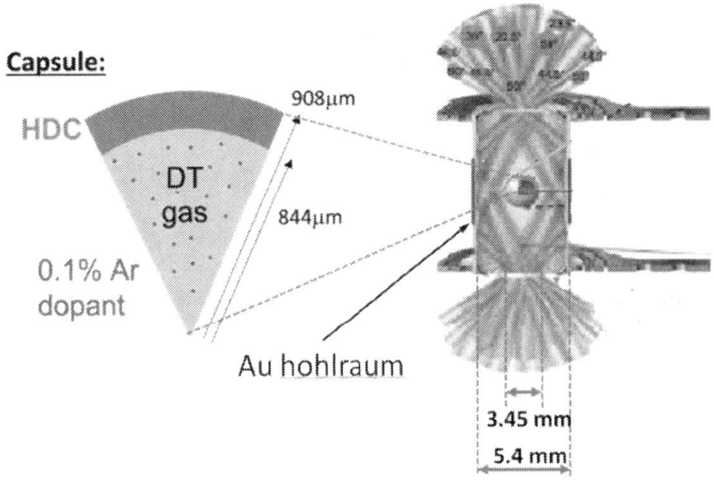

Figure 1: Target used for the NIF experiment: the hohlraum is made of Au, with a HDC capsule at the center. The laser beams are depicted to enter the hohlraum through the top and bottom apertures. The capsule is filled with DT gas that contains a 0.1% Ar dopant. The figure is adapted from [7].

A ^{39}Ar sample was produced at Soreq NRC (Israel) by thermal neutron activation of enriched ^{38}Ar (99.961%) contained in a quartz ampoule. The sample (denoted by SNRC) was diluted with ^{38}Ar and ^{40}Ar (^{38}Ar/^{40}Ar= 9.5%) to produce a sample measurable by AMS with a ratio ^{39}Ar/Ar≈ 10^{-12}.

A 20 mm diameter stainless steel spherical ampoule (denoted by HZDR) made of thin stainless steel (0.5 mm thickness) was cryogenically filled with Ar gas enriched in ^{40}Ar. The sphere was activated during four hours in the 14 MeV neutron flux of the DT generator [11] of Technical University Dresden at Helmholtz Zentrum Dresden Rossendorf, together with appropriate neutron monitors (Al, Zr, Nb) to produce ^{39}Ar by the ^{40}Ar$(n, 2n)$ reaction.

Fig. 2 shows identification spectra for the SNRC, HZDR and the blank (non-irradiated) Ar samples in a NOGAMS experiment performed at Argonne National Laboratory for ^{39}Ar accelerated at charge state 8+ under conditions similar to those described in [10, 12]. The newly commissioned MONICA detector [13] was used in this experiment. The ^{39}Ar group is clearly identified, both in position and in ΔE4 energy loss and separated from beam contaminants (^{34}S accelerated as $^{34}S^{7+}$ nearly identically as $^{39}Ar^{8+}$ $(\delta(m/q)/(m/q) \approx 0.4\%)$ and stable ^{39}Kr isobar partly deflected outside the detector acceptance). The results are currently being analyzed to extract the ^{40}Ar$(n, 2n)^{39}$Ar reaction cross section.

A recent test experiment was conducted at NIF before the dedicated October 2022 laser shot in a configuration similar to that shown in Fig. 1 with an Ar content of 0.05% in the DT capsule. A first signal of ^{39}Ar, produced by the ^{40}Ar$(n, 2n)^{39}$Ar reaction within the inertially-confined fusion environment, was observed and is illustrated in Fig. 3. The signal is being studied quantitatively together with possible background production of ^{39}Ar.

In future experiments, we aim to search for a ^{42}Ar signal produced by the rapid two-neutron capture ^{40}Ar$(2n, \gamma)$ reaction. The detection of ^{42}Ar by the NOGAMS technique was demonstrated for the first time in a calibration sample produced in a high-fluence thermal neutron irradiation of ^{40}Ar. Preliminary results were presented in [13].

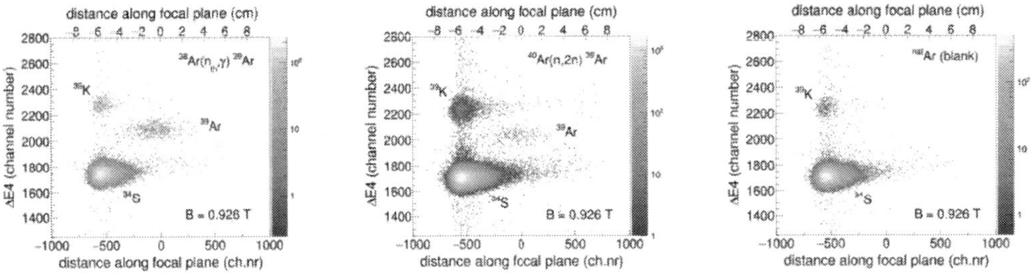

(a) (b) (c)

Figure 2: Identification spectra (energy loss vs. focal-plane position) of ions accelerated to 5.5 MeV/u and detected in the focal plane of the Argonne Enge gas-filled spectrograph (see [8, 12] for details): ^{39}Ar ions detected in (a) sample SNRC thermal-neutron activated ^{38}Ar (^{39}Ar/^{40}Ar= 6.3×10^{-12}); (b) 14-MeV neutron-activated ^{40}Ar (sample HZDR); (c) non-activated (blank) ^{nat}Ar under same conditions in (a), (b).

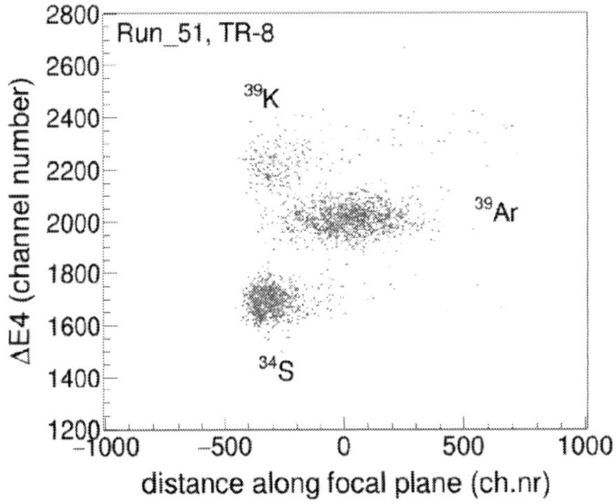

Figure 3: First signal of ^{39}Ar produced by the ^{40}Ar(n, 2n)^{39}Ar reaction in a NIF laser shot on a DT-Ar(0.05 at%) filled capsule. The axes in the figure are the same as in Fig. 2.

4 Summary

We described an experimental platform based on Noble-Gas Accelerator Mass Spectrometry (NOGAMS) for the study of neutron-induced reactions in the high-density plasma of a NIF implosion of a DT-Ar filled capsule. ^{39}Ar produced in calibration samples by the ^{38}Ar(n, γ)^{39}Ar and the ^{40}Ar(n, 2n)^{39}Ar reactions were identified. A first signal of ^{39}Ar produced by the ^{40}Ar(n, 2n)^{39}Ar reaction in a NIF shot on an DT-Ar filled capsule was observed. The results are under quantitative analysis.

5 Acknowledgements

This talk was dedicated to the memory of our long-time collaborator Franz Käppeler whose enthusiasm and approach to laboratory nuclear astrophysics remain for us as a beacon of guidance.

We gratefully acknowledge the Discovery Science program of the NIF facility (LLNL) for the approval of a high-power laser shot for this research and for the laboratory support of the experiment. Support of Pazy Foundation (Israel) and USA-Israel Binational Science Foundation (BSF Grant Nr. 2020136) and support of Israel Science Foundation (ISF Grant Nr.876/19) for the building of the MONICA detector are acknowledged. This work was supported in part by the U.S. Department of Energy, Office of Nuclear Physics, under Contract No. DE-AC02-06CH11357. This research used resources of ANL's ATLAS facility, which is a DOE Office of Science User Facility. A.Z. acknowledges support from U.S. DOE Early Career Research Program (Fusion Energy Sciences) under Grant No. FWP SCW1658. The U. of Notre Dame co-Authors are supported by National Science Foundation, Grant No. NSF PHY-2011890 and the Nuclear Regulatory Commission, Award No. 31310019M0037;

References

[1] F. Käppeler, R. Gallino, S. Bisterzo, W. Aoki, Rev. Mod. Phys. **83**, 157 (2011)
[2] J.J. Cowan, C. Sneden, J.E. Lawler, A. Aprahamian, M. Wiescher, K. Langanke, G. Martínez-Pinedo, F.K. Thielemann, Rev. Mod. Phys. **93**, 015002 (2021)
[3] A.B. Zylstra, O.A. Hurricane, D.A. Callahan, A.L. Kritcher, J.E. Ralph et al., Nature **93**, 542 (2022)
[4] A.L. Kritcher, C.V. Young, H.F. Robey, C.R. Weber, A.B. Zylstra et al., Nature Phys. **18**, 251 (2022)
[5] R.M. Wilson, Physics Today **75**, 16 (2022)
[6] H. Abu-Shawareb et al. (Indirect Drive ICF Collaboration), Phys. Rev. Lett. **129**, 075001 (2022)
[7] C.J. Cerjan et al., J. Phys. G: Nucl. Part. Phys. **45**, 033003 (2018)
[8] M. Paul, R.C. Pardo, P. Collon, W. Kutschera, K.E. Rehm, R. Scott, R.C. Vondrasek, Nuclear Instruments and Methods in Physics Research Section B: Beam Interactions with Materials and Atoms **456**, 222 (2019)
[9] D.A. Shaughnessy, C.A. Velsko, D.R. Jedlovec, C.B. Yeamans, K.J. Moody, E. Tereshatov, W. Stoeffl, A. Riddle, Rev. Sci. Instr. **83**, 10D917 (2012)
[10] P. Collon, M. Bichler, J. Caggiano, L.D. Cecil, Y.E. Masri, R. Golser, C. Jiang, A. Heinz, D. Henderson, W. Kutschera et al., Nucl. Instrum. Methods Phys. Res. B **223-224**, 428 (2004), proceedings of the Ninth International Conference on Accelerator Mass Spectrometry
[11] A. Klix, T. Döring, A. Domula, K. Zuber, EPJ Web of Conferences **170**, 02004 (2018)
[12] M. Tessler, M. Paul, S. Halfon, B.S. Meyer, R. Pardo, R. Purtschert, K.E. Rehm, R. Scott, M. Weigand, L. Weissman et al., Phys. Rev. Lett. **121**, 112701 (2018)
[13] L.K. Callahan, P. Collon, M. Paul, M. Avila, B. Back, T. Bailey, A. Clark, C. Dickerson, J. Greene, H. Jayatissa et al., Nuclear Instruments and Methods in Physics Research Section B: Beam Interactions with Materials and Atoms **532**, 7 (2022)

A technique for studying (n,p) reactions of astrophysical interest using radioactive beams with SECAR

Pelagia Tsintari[1,9,*], *Georg P. A.* Berg[3,10], *Jeff* Blackmon[6], *Kelly* Chipps[5], *Manoel* Couder[3,10], *Catherine* Deibel[6], *Nikolaos* Dimitrakopoulos[1,9], *Ruchi* Garg[2,8,9], *Uwe* Greife[7], *Kirby* Hermansen[2,8,9], *Ashley* Hood[6], *Rahul* Jain[2,8,9], *Cavan* Maher[2,8,9], *Caleb* Marshall[4,9], *Zach* Meisel[4,9], *Sara* Miskovich[2,8,9], *Fernando* Montes[2,9], *Georgios* Perdikakis[1,2,9], *Jorge* Pereira[2,9], *Thomas* Ruland[6], *Hendrik* Schatz[2,8,9], *Kiana* Setoodehnia[2], *Michael* Smith[5], *Louis* Wagner[2,9], and *Remco G. T.* Zegers[2,8,9]

[1]Department of Physics, Central Michigan University, Mt. Pleasant, MI 48859, USA
[2]Facility for Rare Isotope Beams, Michigan State University, East Lansing, MI 48824, USA
[3]Department of Physics and Astronomy, University of Notre Dame, Notre Dame, IN 46556, USA
[4]Department of Physics & Astronomy, Ohio University, Athens, OH 45701, USA
[5]Physics Division, Oak Ridge National Laboratory, Oak Ridge, TN 37831, USA
[6]Department of Physics and Astronomy, Louisiana State University, Baton Rouge, LA 70803, USA
[7]Department of Physics, Colorado School of Mines, Golden, CO 80401, USA
[8]Department of Physics and Astronomy, Michigan State University, East Lansing, MI 48824, USA
[9]The Joint Institute for Nuclear Astrophysics - Center for the Evolution of the Elements, Michigan State University, East Lansing, MI 48824, USA
[10]The Joint Institute for Nuclear Astrophysics - Center for the Evolution of the Elements, University of Notre Dame, Notre Dame, IN 46556, USA

Abstract. The formation of nuclei in slightly proton-rich regions of the neutrino-driven wind of core-collapse supernovae could be attributed to the neutrino-p process (νp-process). As it proceeds via a sequence of (p,γ) and (n,p) reactions, it may produce elements in the range of Ni and Sn, considering adequate conditions. Recent studies identify a number of decisive (n,p) reactions that control the efficiency of the νp-process. The study of one such (n,p) reaction via the measurement of the reverse (p,n) in inverse kinematics was performed with SECAR at NSCL/FRIB. Proton-induced reaction measurements, especially at the mass region of interest, are notably difficult since the recoils have nearly identical masses as the unreacted projectiles. Such measurements are feasible with the adequate separation level achieved with SECAR, and the in-coincidence neutron detection. Adjustments of the SECAR system for the first (p,n) reaction measurement included the development of new ion beam optics, and the installation of the neutron detection system. The aforementioned developments along with a discussion on the preliminary results of the p(^{58}Fe,n)^{58}Co reaction measurement are presented.

1 Introduction

Advances in the the modeling of the core collapse of supernovae have demonstrated that neutrino-driven winds may be proton-rich immediately after the collapse [1]. As such, a

*e-mail: tsint1p@cmich.edu

neutrino-p (νp) process [2] has been inferred as a possible formation mechanism of neutron deficient isotopes close to the valley of stability. The strong neutrino fluxes allow for electron anti-neutrino absorption on protons to yield a small residual neutron density. These neutrons are then used in (n,p) reactions to overcome the typical reaction-flow delay caused by the nuclei with relatively long β-decay half-lives (such as the ^{56}Ni and ^{64}Ge) [3, 4].

Research efforts for the development of a suitable technique to study and experimentally constrain the (n,p) and (p,n) reaction rates of short-lived nuclei [5] are ongoing at FRIB (Facility for Rare Isotope Beams). The measurement of the ^{58}Fe$(p,n)^{58}$Co reaction in inverse kinematics at SECAR (SEparator for CApture Reactions) described in this work is the most recent experimental effort towards this direction.

2 The SECAR system

The SEparator for CApture Reactions at NSCL/FRIB, is a sequence of a total of 8 dipole (B1-B8), 15 quadrupole (Q1-Q15), 3 hexapole (Hex1-Hex3), 1 octupole (Oct1) magnets, and 2 velocity filters (WF1 and WF2). It has been designed with the appropriate sensitivity to study capture reactions in inverse kinematics of astrophysical interest [6]. A first direct measurement of the known ^{58}Fe$(p,n)^{58}$Co reaction [7] in inverse kinematics with SECAR demonstrates the capability of the system to accommodate such cross section measurements.

A polyethylene foil placed at the beginning of the SECAR beam-line was irradiated with a ^{58}Fe^{21+} beam at 3.8 MeV/u energy, and an average intensity of 3.0x10^6 pps, that was delivered by ReA3/NSCL [8]. It also included a 30%, on average, contamination of ^{58}Ni^{21+} ions. In addition, a carbon target foil was used for background characterization.

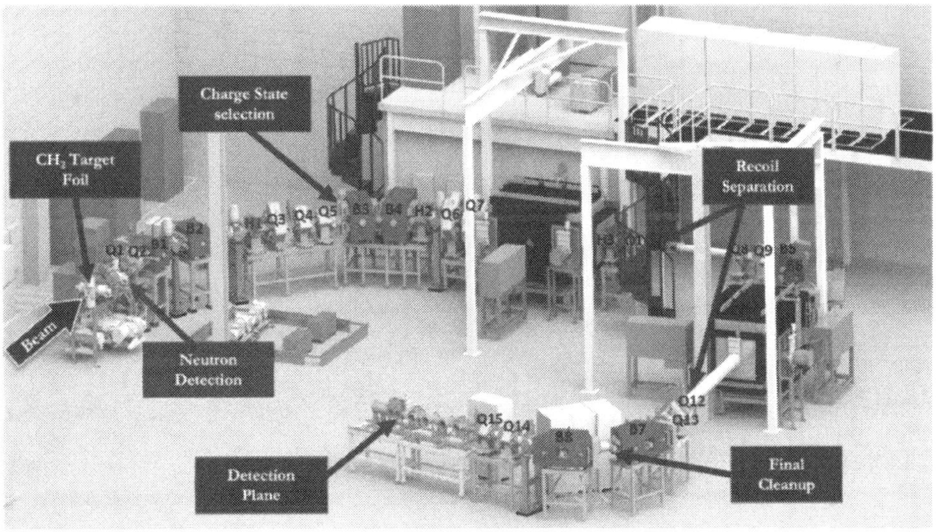

Figure 1. The SECAR layout. Beam enters SECAR at the target location, shown on the left of the picture. A neutron detection system surrounds the target. It is constituted of four liquid scintillators (EJ-301, 2x2 inch) in a ring configuration, and 21 LENDA bars (Low-Energy Neutron Detection Array) [10]. Thereafter, the consecutive sequence of dipole pairs, focusing magnetic elements, and velocity (Wien) filters follows, aiming to separate the recoils from the unreacted projectiles. Note that the Wien filters fields were not used for this measurement as the difference in magnetic rigidity was deemed adequate for separation. Refer to Section 2 for a detailed description of the system.

As the ^{58}Fe^{21+} ions impinge the target foil, both the heavy-ion recoils and the unreacted beam particles enter SECAR shown in Figure 1. Since all the ions going through the target reach charge-state equilibrium, the selection of a single charge state occurs using magnetic analysis in the first couple of dipoles (B1-B2). Thereafter, a consecutive sequence of two dipole pairs and velocity (Wien) filters follow (B3-B4 and WF1, B5-B6 and WF2, respectively), separating the recoils from the unreacted projectiles. An additional pair of dipole magnets (B7-B8) performs a final momentum analysis for the rejection of scattered beam particles ("leaky beam"), near the end of the SECAR beam-line. Although the Wien filters provide the necessary separation for radiative capture reaction studies, the magnetic rigidity difference between the beam and recoils for the (p,n) measurement was sufficient and the Wien filters fields were set to zero.

3 Experimental details

The separation between the recoils and unreacted beam particles from a (p,n) reaction, is limited solely to their energy difference as they have nearly identical masses. More specifically, the energy difference between the ^{58}Fe and ^{58}Co ions (as well as with the ^{58}Ni contaminant), originates from the reaction threshold, kinematics, and energy-loss effects from target.

The standard SECAR ion optics, specifically designed for radiative capture studies [6], was modified to accommodate for the requirements for the (p,n) measurement. The code COSY Infinity [9] with multi-objective optimization routines was used to find the best SECAR tune. In Figure 2 characteristic rays in the horizontal and the vertical plane, are shown in the left and right graph, respectively. The red lines represent the $^{58}Co^{22+}$ recoils, while the unreacted $^{58}Fe^{22+}$ particles are in blue (both ions were assumed to have a charge-state distribution that is in equilibrium and mean calculated energy after the target). The notable difference compared to the standard SECAR beam optics (see Figure 3 in [6]) is the deflection of the beam observed at the four focal planes (FP1-FP4 in Figure 2). This shift provided a sufficient magnetic separation at the original positions of the focal planes (FP1-FP4) and enabled us to use the existing slits at those focal planes to achieve an adequate beam rejection.

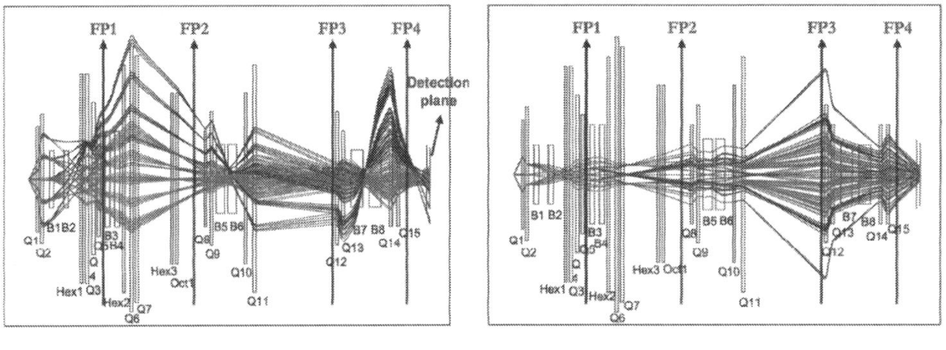

Figure 2. The ion optics of SECAR for the (p,n) measurement. The characteristic rays in the horizontal and vertical planes are shown in the left and right panel, respectively. The red lines represent the $^{58}Co^{22+}$ recoils, and the blue ones the unreacted $^{58}Fe^{22+}$ ions. The arrows show the focal plane positions where the existing mass slits were used to cut down the majority of the unreacted beam.

Although, the unreacted beam-particle distribution (central blue lines Figure 2) is not completely rejected, and the "tail" of the distribution may still reach the detection plane, it could still be eliminated with the use of neutron (at the target) - recoil (at FP4) coincidence. A neutron detection system at the start of the SECAR beam-line is installed. It includes four organic liquid scintillators (EJ-301, 2x2 inch) in a ring configuration around the target, and 21 LENDA bars (Low-Energy Neutron Detection Array) [10]. This system was used in-coincidence with the SECAR detection plane that consists of a gas ionization chamber (IC) and a double-sided silicon strip detector (DSSD) acting as a ΔE-E particle identification. The two detection systems allow for an event-by-event in-coincidence detection between neutrons emitted at the target location and recoils reaching the end of SECAR beam-line.

Lastly, the beam intensity was monitored with two fully depleted planar silicon surface barrier detectors of 75 μm thickness and 300 mm^2 active area located inside the target chamber, via the detection of scattered target particles.

4 Discussion

The analysis of the experimental data is still ongoing however, preliminary results show that the unreacted beam particles and "leaky beam" reaching the SECAR detection plane do not exceed 0.01% of the average beam intensity. While this performance is adequate, the essential separation of the recoil ions will emerge from the neutron detection in coincidence with the focal plane detection system.

As the analysis is still progressing, multiple crucial factors are yet to be determined (e.g. neutron efficiency, transmission, rejection and acceptance of the system) for the reaction cross-section calculation to be completed. The complete cross-section calculation along with a detailed description of the method developed of (p,n) measurements with SECAR will be be presented in a future publication.

5 Acknowledgements

This material is based upon work supported by the U.S. Department of Energy, Office of Science, Office of Nuclear Physics under Award Number DE-SC0014285. This work was supported by the National Science Foundation through grant no. PHY-2209429. The SECAR project is supported by the U.S. Department of Energy, Office of Science, Office of Nuclear Physics, under Award Number DE-SC0014384 and by the National Science Foundation under grant No. PHY-1624942.

References

[1] T. Fischer, et al., Astronomy and Astrophysics, **517**, A80+, (2010)
[2] C. Froehlich, et al., Phys. Rev. Lett., **96**, 142507, (2006)
[3] S. Wanajo, et al., APJ, **729**, 46, (2011)
[4] A. Arcones, et al., APJ, **750**, 18, (2012)
[5] P. Gastis, et al., Nucl. Inst. Meth. Phys. Res. A, **985**, 164603, (2021)
[6] G. P. A. Berg, et al., Nucl. Inst. Meth. Phys. Res. A, **877**, 87–103 (2018)
[7] S.G. Tims, et al., Nucl. Inst. Meth. Phys. Res. A, **563**, 473-493, (1993)
[8] D. Leitner, Nucl. Inst. Meth. Phys. Res. B., **317**, 235-241, (2013)
[9] M. Berz, COSY Infinity, $http://www.bt.pa.msu.edu/index_files/cosy.html$
[10] G. Perdikakis, et al., Nucl. Inst. Meth. Phys. Res. A, **686**, 117-124, (2012)

GEANT4 simulations for γ-ray detector array design for in-plasma β-decays studies of nuclear astrophysics interest

Eugenia Naselli[1,*], *Domenico* Santonocito[1], *Simone* Amaducci[1], *Luigi* Celona[1], *Alessio* Galatà[2], *Giorgio* Mauro[1], *Maria* Mazzaglia[1], *Bharat* Mishra[1], *Angelo* Pidatella[1], *Giuseppe* Torrisi[1], and *David* Mascali[1]

[1]Istituto Nazionale di Fisica Nucleare - Laboratori Nazionali del Sud (INFN-LNS), Catania, Italy
[2]Istituto Nazionale di Fisica Nucleare - Laboratori Nazionali di Legnaro (INFN-LNL), Legnaro, Italy

Abstract. The PANDORA project aims to investigate, by a new experimental approach, the β-decays lifetimes of isotopes of nuclear astrophysics interest as a function of thermodynamic conditions of a laboratory plasma able to mimic some stellar-like conditions. A γ-ray detection system was designed by GEANT4 simulations to tag the in-plasma β-decays via the γ-rays emitted from the excited states of the daughter nuclei. The feasibility of PANDORA, in terms of significance levels, was checked by a "virtual experiment run", also investigating the sensitivity for discriminating among different theoretical predictions.

1 Introduction

Laboratory plasmas can be generated by Electron-Cyclotron-Resonance (ECR) heating - through electromagnetic waves interacting with gases or vapours in presence of a magnetic field - which allows to produce highly charged ions, very relevant for studies of nuclear astrophysical interest. Theoretical predictions have in fact shown that the ionization state of nuclei can dramatically modify, even in orders of magnitudes, the lifetimes [1, 2], due to the bound state β-decay mechanism [3]. Only few experimental evidences have been collected up to now mainly using as setups storage rings [2–4]. If confirmed, these evidences are expected to have a huge impact in the study of nuclear-astrophysics processes and cosmology (BBN, s-processing, Cosmo-chronometers, Early Solar System formation) [5–9].
The PANDORA (Plasma for Astrophysics, Nuclear Decay Observation and Radiation for Archaeometry) project [10], supported by the INFN, proposes a new experimental approach to measure the variation of the half-life of β-decaying nuclei in the magnetically confined ECR plasma, correlating the decaying rate with thermodynamic plasma properties. ECR plasmas can be maintained in a dynamic-equilibrium for weeks with on-average constant local electron density n_e and temperature T_e ($n_e \sim 10^{11}-10^{13} cm^{-3}, T_e$–$0.1-100keV$), which univocally determine the ion charge state distribution (CSD). The β-decaying isotopes can be directly fluxed inside a plasma chamber (where a buffer plasma is preliminary ECR heated) to be turned into plasma-state; while the isotopes decay, the daughter nuclei emit γ-rays of energy of hundreds of keV that can be detected with an array of γ-ray detectors. Thus, the in-plasma radioactivity can be correlated with T_e and n_e, monitored by a multi-diagnostic setup [11].

*e-mail: eugenia.naselli@lns.infn.it

The sketch of the system that will equip the PANDORA trap [12] is shown in Fig. 1(a), with the array of HPGe detectors [13] and several diagnostic tools surrounding the trap. Fig. 1(b) shows the whole superconducting magnetic system (red), the cryostat (blue) and the external iron yoke (grey).

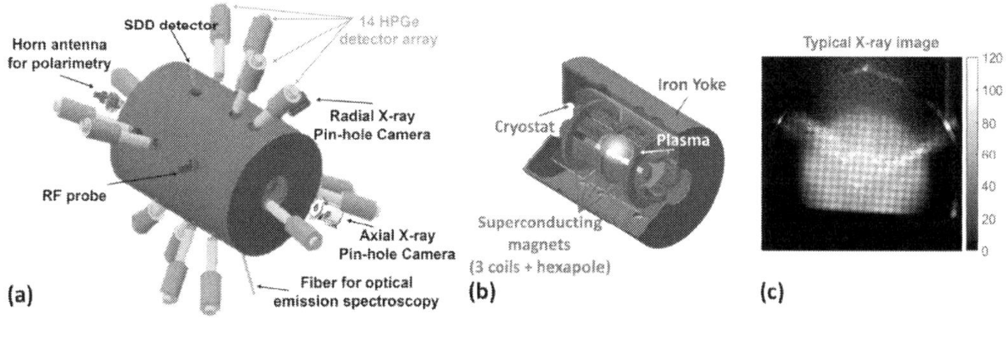

Figure 1. (a) Rendering of the PANDORA trap with the experimental setup for measuring the in-plasma activity (14 HPGe detector array) and on-line monitoring n_e and T_e (i.e., two X-ray CCD cameras, optical spectrometer, interfero-polarimetry, two-pin RF probe, SDD). (b) Sketch of the superconducting magnetic system (red), cryostat (blue) and iron yoke (grey). (c) Typical Argon energy-filtered X-ray image.

The total amount of decays can be determined according to the following formula:

$$N(t_{meas}) = \lambda n_i V_p t_{meas} \tag{1}$$

where λ is the decay constant, n_i is the isotope ion density, V_p is the plasma volume and t_{meas} is the overall measurement time over which the γ-rays produced by the excited states of the decays products are counted by the detector array. Due to dynamical equilibrium, the number of measured decays scales linearly with t_{meas}. Equation 1 implies that the in-plasma ion density and volume must be known all along the measurement to deconvolve λ. The plasma volume can be diagnosed with high spatial resolution ($\sim 400\ \mu m$) by means of the soft X-ray imaging through the pin-hole CCD camera technique [14]. This system allows spectrally-resolved imaging [15]. A typical X-ray image, energy-filtered selecting the fluorescence K_α Argon lines, is shown in Fig. 1(c). The plasma radius thus can be estimated from imaging, with an uncertainty of 5%. A model to link the experimental information to local plasma parameters (n_e, n_i, T_e) is under development [16]. In PANDORA, two CCD pin-hole cameras will be used simultaneously (along the axial and radial lines) to estimate the plasma volume and thermodynamic conditions in each voxel.

2 GEANT4 simulations for the PANDORA detector array design

In order to design the PANDORA γ-ray detector array, GEANT4 [17] simulations were carried out (geant4-10-06-patch-02 version and reference physics list QBBC). The system, as designed in [12], consists of a stainless steel chamber (external radius of 15 cm, length of 70 cm and thickness of 1 cm) surrounded by a magnetic system made of three NbTi superconducting coils and a NbTi hexapole. The cryostat was simulated as an Al cylindrical structure and an external ARMCO iron yoke was implemented. Twelve azimuthal conical holes (lower diameter of 4 cm) were performed along the conductor hexapole interspaces both in the cryostat and in the plasma chamber. The goal is to use the hollowed cryostat as multi-collimator to suppress the photon flux coming from the walls and not directly from the plasma core, improving the signal-to-background ratio. At the end of each collimator a quartz window of thickness of 0.3 cm was placed. An array of 14 γ-ray HPGe detectors [18] (2 axially and 12 radially placed) has been placed collinearly at each collimator.

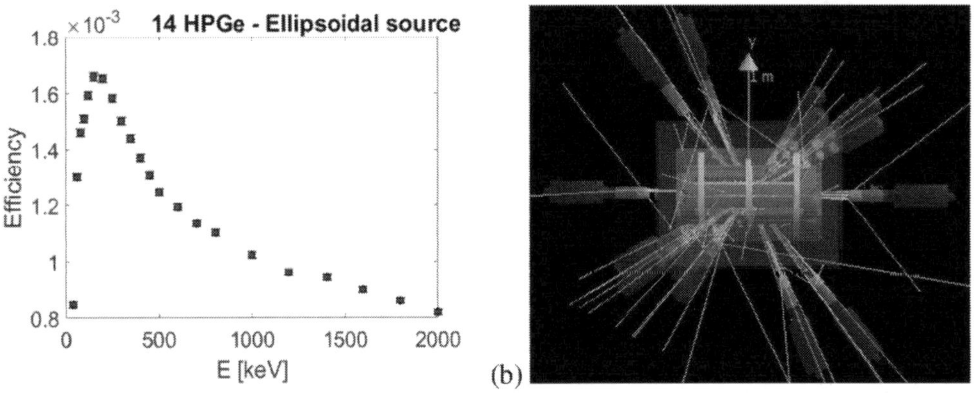

Figure 2. (a) Efficiency vs. γ-ray energy for the HPGe array. (b) Ray-tracing simulations for γ-rays.

Simulations were performed considering an isotropic ellipsoidal source placed around the center of the plasma chamber, having semi-axes lengths of 79 mm, 79 mm and 56 mm, respectively for the x, y and z axis (corresponding to the plasma volume and shape provided by the magnetic field profiles in the PANDORA plasma trap). The γ-ray energy range explored extends from 40 keV to 2 MeV. The trend of the array efficiency as a function of the incident γ-ray energy is shown in Fig. 2(a) assuming 14 HPGe detectors. An example of ray-tracing simulations for γ-rays in the 100 keV – 1 MeV range and trajectories of the γ-rays impinging in the detectors is sketched in Fig. 2(b). For the evaluation of the background spectrum due to plasma self-emission, a density of $n=10^{13}cm^{-3}$ and a volume of 1500 cm^3 were considered. It was evaluated starting from measurements on existing traps and rescaling them to higher densities and volumes according to an emissivity model [19]. Starting from this spectrum it has been possible, once the total efficiency vs. γ-ray energy was determined by GEANT4 simulation, to evaluate the background-rate that would be measured by the PANDORA setup. Such a rate was compared to the rate of the γ-rays emitted from the daughter nuclei after the β-decay and detected in the array in order to determine the measurement time needed to reach a 3σ significance (see section 3.1).

3 Virtual experimental run for the physics case of the ^{134}Cs isotope

A subset of physical cases were chosen to start the study. The selection was based on the scientific relevance, on the expected effects on the lifetime due to CSD and on the maximum ionization state that can be reached by the trap design. Moreover, being the identification of the decay products based on the γ–ray detection, isotopes whose daughter nuclei emit γ–rays were chosen. The selection procedure has given as output three isotopes: the ^{176}Lu (which might play a crucial role as cosmo-clock), the ^{134}Cs (involved in s-processes and relevant for the production of the s-only isotopes ^{134}Ba and ^{136}Ba) and the ^{94}Nb (relevant for the abundance of ^{94}Mo in single or binary systems of stars). In this work we only focus on the ^{134}Cs case.

A scheme of the reaction branching at ^{134}Cs in s-process [5, 7] is shown in Fig.3. The s-process contribution to the Ba starts from neutron captures on the stable isotope ^{133}Cs and

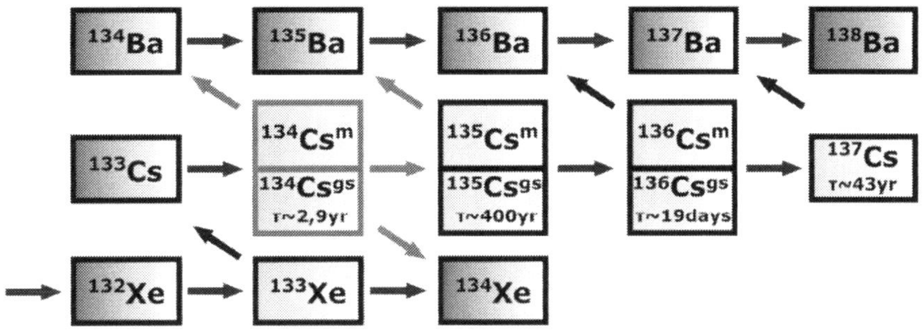

Figure 3. The s-process path trough Cs and Ba isotopes. The stable isotopes are shown in grey boxes, the unstable ones in yellow boxes. The ^{134}Cs and ^{135}Cs branching points are highlighted by red arrows.

proceeds through a branching point (red squared) at the radioactive ^{134}Cs, where n-captures compete with β-decay (laboratory τ ~ 2.9 yr) to excited states of ^{134}Ba and, much less effectively, with electron captures to ^{134}Xe. From ^{134}Cs, neutron captures feed the longer-lived ^{135}Cs and then ^{136}Cs and ^{137}Cs, which are sites of branching points for the s-process path. According to theoretical predictions [1] the ^{134}Cs lifetime is reduced to 48 days at T=3 10^8K, whilst the ^{137}Cs decay rates remain unchanged for varying temperatures. However the rate of ^{134}Cs decay estimated from [1] seems currently unsatisfactory [20], as it inhibits the proper reproduction of the two s-only isotopes ^{134}Ba, ^{136}Ba in suitable proportions. In particular, in [5] one needs to assume that the temperature dependence of the decay rate is less steep by around a factor of 8 (as modelled in [21]) to level the production of ^{134}Ba and ^{136}Ba.
Hence, in-plasma measurements could shed light on the role played by the ionization state affecting the $\beta-$decay and a simulated virtual experiment was carried out to assess the feasibility and sensitivity of the PANDORA setup and experimental approach to be adopted.

3.1 Results of a virtual experimental run for the ^{134}Cs physical case

In order to estimate the measurement time needed to reach 3σ significance taking into account different lifetimes, i.e. different rates in the detector array, a dedicated plot showing the total counts and the expected significance in function of the measurement time was made.

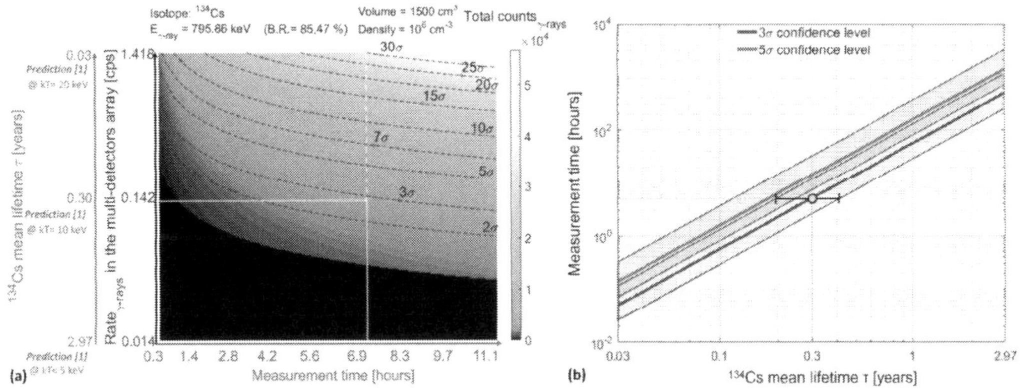

Figure 4. (a) Confidence levels that it is possible to obtain after a given t_{meas}, reported on the x axis, vs the decay-rate (or the lifetime) of ^{134}Cs. (b) - Iso-significance level at 3σ and 5σ reached after a given measurement time, reported on the y axis, versus the decay-rate (or the lifetime) of the ^{134}Cs isotope.

The results for the ^{134}Cs case are shown in Fig. 4(a). The green vertical axis reports on the lifetime expressed in years, starting from the mean lifetime of the neutral isotope (2.97 years) to the values of lifetimes predicted by the theory; the expected collapse of the lifetime [1] is about 2 orders of magnitude at electron plasma temperatures around 20 keV, about 1 order of magnitude at around 10 keV while no variations are expected for values lower than 5 keV. Considering the in-plasma activity (assuming an ellipsoidal plasma of 1500 cm^3 in volume with a relative concentration of 10^{-5}% of ^{134}Cs with respect to the buffer gas plasma density) and taking into account the efficiency estimated by GEANT4 simulation, the counting rate expected in the γ detector array was obtained (black vertical axis). The x-axis indicates the measurement time. Pseudo-colors give the total number of counts at the γ peak of interest (at E_γ=795.86 keV). The error on the background was estimated as the square-root of counts in the same energy window, whilst the counts due to the real decays occurring in the plasma grow up linearly with time (see equation 1). The condition that the signal overcomes in 3σ the background level defines the measurement time needed to have a 3σ significance. Black dashed lines shown in the plot in Fig.4(a) represent iso-significance contours at each given combination of expected activity and t_{meas}. The black region is such that the significance is worse than 1σ. These simulations results indicate that a measurement of several hours is needed. The possibility to discriminate among different models will depend on the uncertainties in the measured λ values. The plot in Fig. 4(b) shows the iso-significance curves at 3σ (blue) and 5σ (red) in log-log scale with the corresponding uncertainties intervals. The in-plasma λ can be in fact estimated by equation 1 with an uncertainty of around 28% (considering errors propagation). The relative uncertainties for the parameters that will be measured are: dN/N~0.3% (3σ significance measurement), dV/V~15% (the error in estimating the radius is around 5% [15]) and dn_i/n_i ~25% (can be measured by the interfero-polarimentric tool [22] or by optical [23] and X-ray spectroscopy). New techniques of analysis are being developed to further reduce the afore mentioned uncertainties, improving the performances of the measurement.

4 Conclusion

PANDORA represents a promising experimental setup to verify the theoretical predictions on the dependence of β-decay lifetimes on plasma temperature. To evaluate the feasibility of the measurements, the run duration to get statistically meaningful results in terms of significance was estimated through GEANT4 simulations for the physical case of ^{134}Cs. It was demonstrated that PANDORA will have the adequate sensitivity to discriminate between different sets of theoretical predictions of the expected variation of the mean lifetime and to eventually better fit the data of nucleosynthesis which disagrees with the current decay rate predictions.

5 Acknowledgements

Authors are grateful to the INFN 3rd Nat. Comm. of INFN for the Financial Support under the Grant PANDORA_Gr3.

References

[1] K. Takahashi et al., Phys Rev C **36**, 1522 (1987)
[2] Y. Litvinov et al., Phys. Rev. Lett. **99**, 262501 (2007)
[3] F. Bosch et al., Phys. Rev. Lett. **77**, 5190 (1996)
[4] M. Jung et al., Phys. Rev. Lett. **69**, 2164 (1992)

[5] S. Palmerini, ApJ **921**, 7 (2021)
[6] M. Busso et al., Front. Astron. Space Sci. **9**, 956633 (2022)
[7] S. Cristallo et al., Astrophys J Suppl Ser **197**, 17 (2011)
[8] N. Klay et al., Phys Rev C **44**, 2839 (1991)
[9] R. Spartà et al., Front. Phys. **10**, 896011 (2022)
[10] D. Mascali et al., Universe **80**, 8 (2022)
[11] E. Naselli et al., J. Instrum. **14**, C10008 (2019)
[12] G. Mauro et al., Front. Phys. Sec. Nucl. Phys. **10**, 931953 (2022)
[13] E. Naselli et al., Front. Phys. Sec. Nucl. Phys. **10**, 935728 (2022)
[14] S. Biri et al., J. Instrum. **16**, P03003 (2021)
[15] E. Naselli et al., J. Instrum. **17**, C01009 (2022)
[16] B. Mishra et al., Phys. Plasmas. **28**, 102509 (2021)
[17] S. Agostinelli et al., Nucl. Instrum. Methods Phys. Res. A **506**, 250 (2003)
[18] A. Goasduff et al., Front. Phys. **10**, 936081 (2022)
[19] A. Gumberidze et al., Rev. Sci. Instrum. **81**, 033303 (2010)
[20] M. Busso et al., ApJ **908**, 55 (2021)
[21] S. Taioli et al., ApJ, **933**, 158 (2022)
[22] D. Mascali et al., Rev. Scient. Instr. **93**, 033302 (2022)
[23] A. Pidatella et al., Front. Astron. Space Sci. **9**, 931744 (2022)

X17 search project with EAR2 neutron beam

Gianpiero Gervino[1,*], *Carlo* Gustavino[2], *Evaristo* Cisbani[3], *Nicola* Colonna[4], *Giorgio* Cotto[1], *Marco* D'Incecco[5], *Salvatore* Fiore[6], *Cristian* Massimi[7], *Pierfrancesco* Mastinu[8], *Annamaria* Mazzone[9], *Elizabeth* Musacchio-Gonzalez[8], *Cristina* Petrone[10], and *Michele* Viviani[11]

[1]Dipartimento di Fisica and INFN Torino (TO), Italy
[2]INFN Roma1 (Roma), Italy
[3]ISS, Istituto Superiore di Sanità and INFN Roma1 (Roma), Italy
[4]INFN Bari (BA), Italy
[5]INFN-LNGS Assergi (AQ), Italy
[6]INFN Roma1 and ENEA Frascati (Roma), Italy
[7]Dipartimento di Fisica e Astronomia and INFN Bologna (BO), Italy
[8]INFN-LNL Legnaro (PD), Italy
[9]CNR and INFN Bari (BA), Italy
[10]IFIN-HH, Magurele, Romania
[11]Dipartimento di Fisica and INFN Pisa (PI), Italy

Abstract. We present the state of the art of the n_TOF Collaboration Working Group activity dedicated to study how to solve the puzzle about the existence of the so called new particle X17, spotted for the first time few years ago by a team at ATOMKI in Hungary and since then never confirmed by other independent experimental collaborations but also never refuted. An "ad hoc" detection set-up is under realization for this goal, in order to reach an angular resolution of the two emerging trajectories from the X17 decay and an energy resolution for the invariant mass reconstruction enough to cast light in a definitive way about this puzzle. To design the present detection setup we work in close contact with the Pisa Nuclear Theory team, that has deeply studied the implication of X17 existence and extracted by the ATOMKI results its eventual nature, kinematics and general properties.

1 Introduction

Two significant anomalies have been recently observed in the emission of electron-positron pairs in the ^{7}Li(p,e^{-} e^{+})^{8}Be and ^{3}H(p,e^{-} e^{+})^{4}He reactions[1],[2]. These anomalies have been deciphered as the signature of the existence of a new boson called X17 of mass $M_{X17} \simeq 16.8$ MeV that could be a mediator, if it exists, of a fifth force, characterized by a strong coupling suppression of protons compared to neutrons (protophobic force)[3]. The evidence for such a particle was first reported in 2012 at a workshop in Italy by the group led by Attila Krasznahorkay of ATOMKI Institute for Nuclear Research at Debrecen (Hungary): they reported about unexpected results observed in their nuclear physics experiments. In April 2015, the same group of ATOMKI confirmed the existence of the new light boson, giving an evaluation

*e-mail: gianpiero.gervino@to.infn.it

of its mass (nearly 17 MeV): this time the sigma levels were 6.8 excess. According to the team, this was the simplest way to explain the experimental results of anomalies in Beryllium transitions and such excess of sigma levels cannot be related to any statistical fluctuations[1]. Recently the ATOMKI group published new results about $^{11}B(p,\gamma)^{12}C$ nuclear reaction ($J^{\pi}=1^{-}$, $\Gamma=1.15$ MeV, ^{12}C excited state) at $E_p=1.388$ MeV incident proton energy, that also suggest the existence of X17 boson[4]. In the first ATOMKI measurements the target was irradiated by protons of energy around 1.03 MeV, thus exciting lithium nuclei up to the 18.15 MeV beryllium state. The excited beryllium nucleus is expected to reach its ground state throughout the transition $0^{+} \rightarrow 0^{-}$ by internal pair conversion (IPC), the only possible way according to nuclear spectroscopy rules (angular momentum and parity conservation). The angular distribution between two emerging leptons from IPC has an apparent peak around 10° - 15°, after this angular interval is going fast to zero, i.e. the probability to measure a relative angle for the e^{-} e^{+} pairs over 30° is negligible. What it has been observed instead is a clear peak (up to 6.8 excess from the $e^{+}e^{-}$ background) in the angular range of 100° - 130°, with a maximum around 120° (see Fig. 1). The importance of such a discovery - if confirmed – is beyond nuclear physics. Every new particles, specially bosons, could be associated with a new force or at least with a new, unknown and unexpected aspect of one of the known forces. X17 could be related to the role of quarks, trapped in the hadronic bag and to the nucleon binding energy inside nucleus. Preliminary theoretical studies [3] has pointed out the hypothesis of different behaviour for up and down quarks unveiled by a much stronger cross section when neutron transitions are observed with respect to the ones of protons (protophobic scenario). This innovative scenario could provide new ideas to explain, at least partially, the long-standing anomaly on the muon magnetic moment. More in general, the possible existence of a new particle is of paramount importance in particle physics and in cosmology (dark matter). Therefore, the ATOMKI claim [1], [2], [4], [5] clearly calls for new experimental studies.

2 Physics case

Presently we are engaged to carry on a first series of measures at n_TOF, where the excited levels of ^{4}He, ^{8}Be can be populated via the conjugated $^{3}He(n,e^{-}e^{+})^{4}He$ and $^{7}Be(n,e^{-}e^{+})^{8}Be$ reactions using the spallation neutron beam EAR2 at CERN. This approach has two relevant advantages: (i) for the first time X17 existence is investigated through neutron induced reactions exploiting the world unique properties of EAR2 beam [6] and (ii) the experimental setup is completely different with respect to the one used by the ATOMKI group.

Our efforts aim to realize a suited detection setup for the determination of particle kinematics with a strong capability to discriminate the reaction ejectiles (assumed to be e^{-} e^{+} pairs) from background in a wide energy range. If the existence of X17 is confirmed, we would be able to establish quantum numbers and mass of the X17 boson, and to shed light on the so-called protophobic nature of a fifth force. In fact, state-of-the-art "ab-initio" calculations are in good agreement with present literature data (in particular for the "few body" ^{4}He nucleus) and would provide quantitative predictions to establish the X17 nature, e.g. if it is a scalar, pseudoscalar, vector or axial boson [8] and to get information on the interaction of the X17 boson with quarks and gluons [3]. EAR2 station of the n_TOF facility at CERN provides a pulsed neutron beam in a wide energy range, which broadly covers the region of interest for this experiment, i.e. 10^{3} eV $< E_n <$ 10^{7} eV [6], [7]. Counting rate estimations show that the neutron intensity at EAR2 is high enough to carry on a conclusive experiment within approximately one measurement month. Working at a spallation neutron beam line means

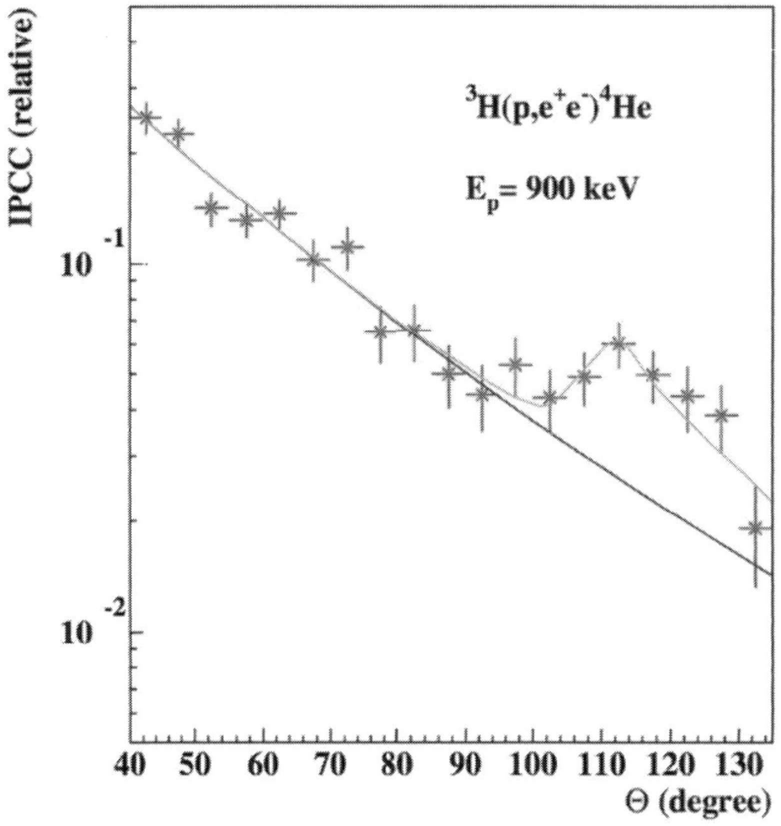

Figure 1. Angular correlations for the e^-e^+ pairs measured in the ^{3}H(p,e$^-$ e$^+$)^{4}He reaction at E_p=900 keV. The red points are the experimental data showing the e^- e^+ pairs excess around 115°. The blue is the expected angular distribution of e^-e^+ pairs based on standard IPC physics, while the green line is the best data fit assuming M_{X17}=16.87 MeV

to have to tackle the problem of beam induced background. It is well-known that the dominant outgoing channel in the neutron-induced reaction is the proton final state, in the case of ^{3}He(n,p)^{3}H with a reaction Q-value = 764 keV. The sub-dominant process ^{3}He(n,γ)^{4}He produces single photons with E_γ > 20 MeV. Finally, virtual photons can produce e^- e^+ pairs through the internal pair conversion (IPC) channel ^{3}He(n,e$^-$ e$^+$)^{4}He. For this reaction, with incident neutrons of $E_n \geq$ 21 MeV, it would be possible to excite the $0^- \rightarrow 0^+$ level of ^{4}He with $\Gamma \sim$ 0.84 MeV, the $2^- \rightarrow 0^+$ ^{4}He excited level with $\Gamma \sim$ 2.01 MeV can be reached at E_n $\geq$ 21.8 MeV. The IPC pairs and the e^- e^+ pairs generated by the interaction of real photons with the material surrounding the ^{3}He target represent a relevant background for the process of interest. However, the amount of e^- e^+ pairs produced by (virtual or real) photons rapidly decreases while increasing their emission relative angle. Instead, the ^{4}He* $\rightarrow$ ^{4}He + X17 process and successive X17 $\rightarrow$ e^- e^+ decay produces pairs with a large relative angle, determined by the X17 mass and energy. As a consequence, the excess of pairs at large relative angles (see Fig. 1) represents a clear signature of new physics. It is worth pointing out that a possible existence of a vector X17 boson would alleviate the present tension between data and calculations concerning the muon magnetic moment (see for instance the recent [9]). The prediction for the value of the (electron) muon anomalous magnetic moment includes three parts:

$$a_\mu^{SM} = a_\mu^{QED} + a_\mu^{EW} + a_\mu^{hadron} = 0.001\ 165\ 918\ 04(51)$$

The first two components that represent the photon and lepton loops (a_μ^{QED}) and the W boson, Higgs boson and Z boson loops (a_μ^{EW}) can be both calculated precisely. The third term (a_μ^{hadron}) represents hadron loops and it cannot be calculated accurately from theory alone, it is estimated from experimental measurements of the ratio of hadronic to muonic cross sections in electron-positron ($e^+ e^-$) collisions [10]. As of July 2020, the a_μ^{SM} measurements disagree with the Standard Model by about 3.5 standard deviations, suggesting physics beyond the Standard Model may be having an effect (or that the theoretical/experimental uncertainties are not completely under control). This is one of the long-standing discrepancies between the Standard Model and experiments. If X17 exists and has a hadron nature, the hadronic term of the anomalous magnetic moment must be increased taking into account this new contribution that it could be evaluated roughly speaking from 10^{-6} up to 10^{-5}.

3 Experimental

To measure the particle kinematics observables, to reach a high level of particle identification in a wide energy range and to optimize the signal-to-noise ratio, thus disentangling the different contributions eventually leading to e^+e^- pairs, we have studied a detection setup based upon two trackers of rectangular shape placed at the side of the target at around 5 cm far, one in front of the other. Back to each tracker it is placed an outer electromagnetic calorimeter made of two planes of EJ-200 slabs, 50 cm long, 1 cm wide and 0.6 cm thick. The two EJ-200 slab planes are mounted in a perpendicular way, one along x-axis and the other along y in order to strength the incident particle trajectory reconstruction that starts into the tracker. Back to the EJ-200 slabs the volume is fulfilled by 16 cubes of EJ-200 scientillators, 10 cm side, able to stop and to absorbe all the incident leptons energy. The detection setup is the same on both side of the target. In the following all the items that jointly cooperate to build up the experimental apparatus are discussed and analyzed.

3.1 Targets

The first data collection campaign will be dedicated to measure $^3He(n, e^- e^+)^4He$ nuclear reaction. The ^3He target would be a cylinder 3 cm diameter, 3.5 cm long of carbon fiber 0.6 mm thick. According to preliminary tests this target could hold up to 350 atm of ^3He, we plan to fill the target up to 300 atm to keep a safety margin. The carbon fiber cylinder is placed at close contact inside another cylinder of about the same dimensions made of Al 1.2 μm thick, to avoid or at least to minimize any ^3He leaks throughout the very thin carbon fiber. As far as $^7Be(n, e^+ e^-)^8Be$ reaction is concerned, we plan to have an enriched ^7Be solid target 4-5 mg/cm^2 thick evaporated onto a 4-5 μm thick Ta backing. The experimental apparatus is designed to minimize the dead layers that the emerging leptons have to pass through with the aim to reduce multiple scattering effect. Multiple scattering is a source of systematic error in the mutual $e^+ e^-$ angle measures, it could affect the trajectory reconstruction and dead layers could also spoil the overall energy resolution.

3.2 Tracker

In order to dissolve all the doubts about the existence or not of X17, the most of kinematic observables of the induced reactions must be measured at low uncertainties. Then the core of the detection set-up must be a tracker, in Fig. 2 (see caption) a block diagram is shown. The outer blue sector represents an electromagnetic calorimeter dedicated to the total energy of the emerging leptons and then allowing a precise measure of X17 mass.

Each tracker is realized by TPC that could deliver direct 3D track information for pattern recognition even in high multiplicity events and particle identification over a wide momentum range. It is even suited to work, if needed, inside a magnetic field that deflects the tracks enabling the measure of the particle momentum. The drift time of each track segment could be measured to provide the vertical coordinate Z. The position in the XY projection is obtained by recording the induced charge profiles on the segmented readout plane after amplification, the recorded charge provides dE/dx useful for particle identification (PID). The TPCs have rectangular shape, 50 cm long 30 cm wide, with a radial electric field, read out by Micro Pattern Gaseous Detectors (MPGD) like μRwell. The very low material budget of a TPC with a helium-based gas mixture and an anode of 3 mm Cu minimizes the multiple scattering experienced by the ejectiles, while the radial field, compared to a more traditional longitudinal configuration, reduce the diffusion experienced by the ionization electrons drifting toward the anode, so improving the space resolution of the detector. These features are expected to produce a substantial improvement in the measurement of the e^+e^- relative angle. The active path available for tracking is of 3 cm. For the segmented readout the use of bulk-μRwell technology (a well-known technology with limited ion feedback) shows the following advantages: i) all-in-one detector: blind areas minimized (including edges and corners), ii) simple design, cheap & robust, iii) good uniformity of performances.

The performances of the prototype have been investigated with and without the target on the beam. The results of the test provide us with information on the response to the γ-flash and helps to evaluate the highest neutron energy that can be reached (around few MeV), together with a first measure of the background rate induced by the neutron beam in the target.

3.3 Electromagnetic calorimeter

The use of a dedicated segmented calorimeter composed by two arrays of scintillating slabs placed perpendicular to each other and back to the two slab planes, in the outer position, 16 cubic tiles 10x10x10 cm^3 of EJ-200 (equal situation on both target sides), together with the action of the tracker, allows a trajectories reconstruction of $\Delta\Theta = \Delta\phi = 5$ deg. or better. The cubic tiles are thick enough to fully absorb all the electron energy and stop them. According to theoretical model [8] the leptons emerging from X17 share an energy fraction between 2 (minimum) and 15 (maximum) MeV. By GEANT4 simulation, electrons of up of 20 MeV energy, that could be considered a safe upper limit for our measures, will have a range slightly over 11 cm in EJ-200 scintillator and each electron has to get across two slabs of 0.6 cm thick and a cube of 10 cm side, in the minimum path when they are emitted perpendicular to the electromagnetic calorimeter (without considering the tracker). The slab readout is provided by SiPM array placed at each ends and related electronics. The EJ-200 cubes will be read by fast PMTs with active divider in order to ensure constant gain up to the maximum allowable current: this is necessary because of the impact of γ flash on the detection system, the main source of beam induced background. All the electron energy would be fully contained inside the calorimeter and the energy loss inside the slabs and the cubes will be summed up: we expect to reach an energy resolution for X17 around 10%. The electromagnetic calorimeter provides complementary measurements with respect to the tracker detector. It is worth mentioning that the electromagnetic calorimeter would make possible to study in detail the two nuclear reactions ^{3}He(n,γ)^{4}He and ^{3}He(n,e^+e^-)^{4}He differential cross section: these measurements are very important to firmly establish the structure of excited levels of ^{4}He, providing a robust experimental footing for the ab-initio calculations. Moreover, in case it is confirmed the scenario in which X17 decays into two unknown particles (and with unknown charge), it is

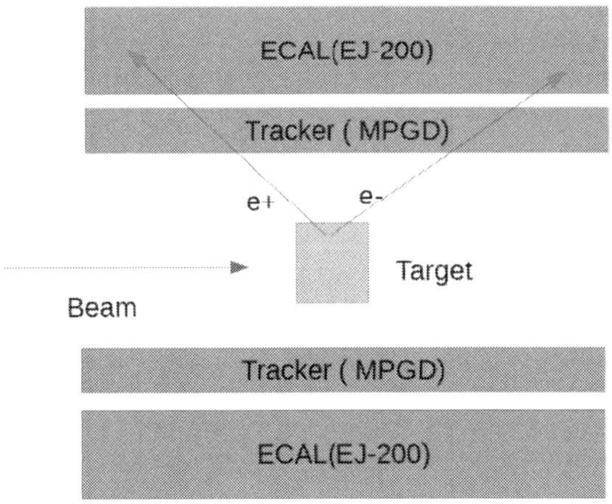

Figure 2. Conceptual design of the $^3He(n,e\text{-}e\text{+})^4He$ at EAR2. The neutron beam strikes on a ^3He target, done in special carbon fiber, 0.6 mm thick, 300 - 350 atm. The tracks direction are measured for the first 3 cm with the inner detector (TPC) then by the EJ-200 slab planes, the energy of each particle is measured with an e.m. calorimeter with adequate granularity (EJ-200).

possible to observe the different response of the calorimeter with respect to pairs at small relative angle, uniquely due to e^+e^- pairs from IPC, providing an independent check concerning particle identification. Of course, the scintillator is also sensitive to the protons produced by the $^3He(n,p)^3H$ channel, that have approximately the same energy of the interacting neutrons (Q=764 keV), but we are confident to identify the protons with the joint action of tracker, electromagnetic calorimeter and TOF measure of the pulsed neutron beam. In fact, neutrons with $E_n > 15$ MeV reach the EAR2 experimental area not later than 350 ns after the spill time, while events generated by neutrons with $E_n < 10$ MeV are detected not before 400 ns in the EAR2 experimental hall. Therefore, events due to the residual protons with $E_p > 15$ MeV can be easily rejected using the TOF technique.

4 Conclusion

X17 is offering an exciting challenge for experimental nuclear physicists, with the unique opportunity of a real fundamental discovery or at least to contribute to cast light in a promising new physics field. Even if the existence of X17 would be not confirmed by the new experimental efforts, the data taken in the efforts to solve this puzzle would be useful to investigate effects that for too long have been neglected, like the IPC phenomenon on light nuclei and testing experimentally the validity of the ab-initio nuclear theories, thanks to a new enthusiastic collaboration from nuclear theorists and experimentalists. Our experimental approach to X17 hunting is focussed on leptons trajectory reconstruction, high precision $e^+ e^-$ angle (we plan to reach an overall uncertainty ≤ 3 deg.) and X17 mass measurements. From the phenomenological side we would investigate the possible protophobic effect that could be seen comparing the X17 cross section production (if any) by proton and neutron induced reactions (neutron cross section is expected to be around 30% higher than proton one [3], [8]): this is the main motivation in our choice to start the measurements at n_TOF neutron spallation beam (at the moment there are no data on neutron induced reactions). In the optimistic side we could claim that these efforts triggered by X17 hunting, independent from the success or not of the chase, would start a new era for the study of strong interaction and nuclear structure.

References

[1] A.J. Krasznahorkay et al., Phys. Rev. Lett. **116(4)**, 042501, (2016)
[2] A.J. Krasznahorkay et al., Phys. Rev. C. **104(4)**, 044003, (2019)
[3] J. L. Feng et al., Phys. Rev. Lett. **117**, 071803 (2016)
[4] A.J. Krasznahorkay et al., Phys. Rev. C. **106**, L061601, (2022)
[5] A.J. Krasznahorkay et al., Journal of Physics G: Conf. Series **1643(1)**, 012001, (2020)
[6] M. Sabaté-Gilarte et al., Eur. Phys. J. A **53**, 210, (2017)
[7] R. Esposito et al., Phys. Rev. Accel. Beam **24**, 093001, (2021)
[8] M. Viviani et al. Phys. Rev. C **105(1)**, 014001, (2022)
[9] B. Abi et al., Phys. Rev. Lett. **126(14)**, 141801, (2021)
[10] Sz. Borsanyi et al., Nature **593**, pag. 51-54, (2021)

AUTHOR INDEX

AUTHOR INDEX

AUTHOR INDEX

AUTHOR INDEX

AUTHOR INDEX

AUTHOR INDEX

AUTHOR INDEX

AUTHOR INDEX

AUTHOR INDEX